NO LONGER
PROPERTY OF
OLIN LIBRARY
WASHINGTON UNIVERSITY

TRENDS IN
PHOTOBIOLOGY

TRENDS IN PHOTOBIOLOGY

Edited by
C. Hélène
Laboratoire de Biophysique, INSERM U201
Muséum National d'Histoire Naturelle
Paris, France
and
Centre de Biophysique Moléculaire
CNRS, Orléans, France

M. Charlier
Centre de Biophysique Moléculaire
CNRS, Orléans, France

Th. Montenay-Garestier
Laboratoire de Biophysique, INSERM U201
Muséum National d'Histoire Naturelle
Paris, France

and
G. Laustriat
Laboratoire de Physique, Faculté de Pharmacie
Université Louis Pasteur
Strasbourg, France

PLENUM PRESS • NEW YORK AND LONDON

Library of Congress Cataloging in Publication Data

International Congress on Photobiology, 8th, Strasbourg, 1980.
  Trends in photobiology.

  "Proceedings of the Eighth International Congress on Photobiology [and] the Colloque International du CNRS on 'Les Effets Biologiques et la Bioconversion du Rayonnement Solaire', held July 20-25, 1980, in Strasbourg, France."
  Includes index.
  1. Photobiology—Congresses. I. Hélène, C. II. France. Centre national de la recherche scientifique. III. Title.
  QH515.I57 1980   574.19'153   80-29512
  ISBN 0-306-40644-6

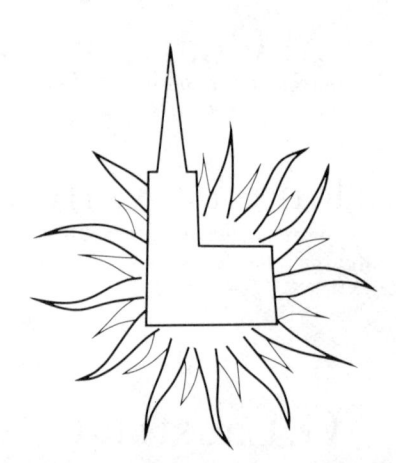

Proceedings of the Eighth International Congress on Photobiology and of the
Colloque International du CNRS on 'Les Effets Biologiques et la Bioconversion
du Rayonnement Solaire," held July 20–25, 1980, in Strasbourg, France

© 1982 Plenum Press, New York
A Division of Plenum Publishing Corporation
233 Spring Street, New York, N.Y. 10013

All rights reserved

No part of this book may be reproduced, stored in a retrieval system, or transmitted,
in any form or by any means, electronic, mechanical, photocopying, microfilming,
recording, or otherwise, without written permission from the publisher

Printed in the United States of America

PREFACE

The Eight International Congress on Photobiology and the *Colloque International du Centre National de la Recherche Scientifique (C.N.R.S.)* entitled *"Effets biologiques et bioconversion du rayonnement solaire"* have been held in Strasbourg (France) on July 20-25, 1980. "TRENDS IN PHOTOBIOLOGY" is a collection of the lectures which were presented during these two scientific manifestations. This book also contains a summary of several round table discussions, together with "summing-up" reports by experts in the field.

Photobiology is a very active, multidisciplinary field of research which plays a growing role in modern science. More attention is being given to the beneficial as well as the detrimental effects of sunlight on living organisms, and especially on mankind.

Light is a natural and essential "cofactor" of a large number of biological processes. Vision, photosynthesis, photomorphogenesis, photoregulation... are among the most documented areas of research. On one hand, new technologies are developed which are aimed at understanding the succession of events which take place on a shorter and shorter time scale (down to picoseconds). On the other hand, the relationships between photochemical events and physiological responses are attracting more and more interest.

Solar energy conversion has received fresh impetus during the past several years due to an increasing demand for renewable energy sources. The science of Photobiology should provide the basic knowledge which is needed to devise efficient molecular machines which could convert light energy into chemical, mechanical or electrical energy. Obtaining energy from biomass has also received considerable interest in the last few years.

Ultraviolet and visible radiations may have detrimental effects on man and other living species. Photoinduced skin cancer has been a main concern of many photobiological studies. Attention has also been paid to the photosensitizing effects of chemicals in the environment and of pharmaceutical drugs. Damages may be created in the genetic material of living cells. The knowledge that is accumulating on the different mechanisms that cells have evolved to repair these lesions is based largely on the results of photobiological investigations.

Natural and artificial sunlight have beneficial effects. The astute combination of chemicals and light is receiving increasing attention from a medical point of view. Therapeutic applications in the treatment of several diseases, i.e., photochemotherapy, now appear as an important part of the science of Photobiology.

Light can also be used as a probe to investigate structural properties of biological systems. The dynamics of biological macromolecules and membranes can be studied by using the known excited-state characteristics of intrinsic and/or extrinsic labels. Photochemical reactions, especially photoinduced crosslinking, are also becoming a tool for the determination of the three-dimensional structure of complex multimolecular systems.

Although it was not possible to cover equally well all domains of Photobiology during the International Congress and the Associated C.N.R.S. colloquium, it is hoped that TRENDS IN PHOTOBIOLOGY will provide an up-to-date view of this field of science and highlight areas of investigations which are in the beginning stages of development.

C. Hélène
M. Charlier
T. Montenay-Garestier
G. Laustriat

ACKNOWLEDGEMENTS

The Eighth International Congress on Photobiology and the associated *Colloque International* have been sponsored by the *Association Internationale de Photobiologie (A.I.P.)* and the *Centre National de la Recherche Scientifique (C.N.R.S.)*, respectively. The organization of these two scientific manifestations has been made possible thanks to the generous support from the following organizations : *Ministère des Universités, Ligue Nationale Française contre le Cancer, Muséum National d'Histoire Naturelle, Institut National de la Recherche Agronomique, Commissariat à L'Energie Atomique, Commissariat à l'Energie Solaire*, European Research Office (USARSG)[*], *Sociétés Greiter A.G., Roussel-Uclaf* and *L'Oréal*.

The Board of Directors of the *Association Internationale de Photobiologie* is very much obliged to Professor Dr. Franz Greiter and Greiter Company of Vienna, Austria, for their generous donation to the *Association Internationale de Photobiologie*. The support of Professor Greiter has made the publication of this volume of the Congress possible. Professor Greiter has a long and close interest, and professional as well as scientific relationship with photobiology and photobiologists, and his generous contribution to the *Association Internationale de Photobiologie* will significantly help the Association sponsor international photobiologic activities.

We wish to express our sincere thanks to Mrs J. Florian and Mrs C. David for retyping some of the manuscripts.

---

[*]*"The views, opinions, and/or findings contained in this report are those of the authors and should not be construed as an official Department of the Army/Department of the Air Force position, policy, or decision, unless so designated by other documentation".*

CONTENTS

I - PHOTOPHYSICS AND PHOTOCHEMISTRY OF BIOLOGICAL MOLECULES

Time-Resolved Chromophore Resonance Raman and Protein
    Fluorescence of Intermediates in some Photobio-
    logical Changes 1
  M.A. El Sayed

Synchrotron Radiation Sources for Photobiology and
    Ultraviolet, Visible and Infrared Spectroscopy 11
  J.C. Sutherland

Photoacoustic Methods Applied to Biological Systems 21
  D. Cahen, G. Bultz, S.R. Caplan, H. Garty and
    S. Malkin

Dynamics of Heme Proteins 33
  H. Frauenfelder

Measurement of Rotational Diffusion of Membrane Pro-
    teins Using Optical Probes 43
  R. J. Cherry

Flow Sorting on the Basis of Morphology and Topology 51
  T. M. Jovin and D.J. Arndt-Jovin

Primary Processes in the Photochemistry of Proteins 67
  L.I. Grossweiner, A. Blum and A.M. Brendzel

Models of Photoregulation 81
  B.F. Erlanger and N.H. Wassermann

U.V. Induced Formation of Polynucleotide-Protein Cross-
    linkages as a Tool for Investigation of the
    Nucleoprotein Structure and Function 93
  E.I. Budowsky

Round Table Summary : Endogeneous and Exogeneous Inhi-
    bitors and Sensitizers. Fundamental Aspects 109
  J.D. Spikes

Round Table Summary : Prebiotic Photochemistry and
    Photochemical Reactions in Space 123
  F. Raulin

Bioluminescence and its Applications 133
  A.M. Michelson

The Trends and Future of Photobiology : Physical and
    Biophysical Aspects 147
  J.W. Longworth

II - MUTAGENESIS, CARCINOGENESIS AND DNA REPAIR

Cell Inactivation and Mutagenesis by Solar Ultraviolet
    Radiation 155
  R.M. Tyrrell

Photoreactivation of Pyrimidine Dimers Generated by a
    Photosensitized Reaction in RNA of Insect Embryos
    (*Smittia Spec.*) 173
  K. Kalthoff and H. Jäckle

Molecular Aspects of Error Prone Repair in *Escherichia
    coli* 189
  K. McEntee

Round Table Summary : Genetic Engineering and DNA Repair . . . 205
  W. Dean Rupp

Aspects of Radiation-Induced Mutagenesis and Malignant
    Transformation 217
  E. Moustacchi

Genetic Aspects of Repair Deficiency and Skin Cancer 229
  H. Takebe

Immunologic Aspects of U.V. Carcinogenesis 235
  M.L. Kripke

The Trends and Future of Photobiology : Biochemical
    and Genetic Aspects 243
  K.C. Smith

CONTENTS

III - PHOTOMEDICINE

Skin : Structure, Natural and Therapeutical Targets of Ultra-Violet Radiation 253
    K. Wolff

Photochemical Reactions of Furocoumarins 267
    F. Dall'Acqua

A Photochemical Characterization of Reactions of Psoralen Derivatives with DNA 279
    S.T. Isaacs, C. Chun, J.E. Hyde, H. Rapoport and J.E. Hearst

Photobiology of Furocoumarins 295
    D. Averbeck

Photochemotherapy with Furocoumarins (Psoralens) 309
    H. Hönigsmann

Advances in Phototherapy of Skin Diseases 321
    J.A. Parrish

Advances in Phototherapy of Neonatal Hyperbilirubinemia 339
    T.R.C. Sisson

Photodynamic Therapy of Infections 349
    L.R. Caldas, S. Menezes and R.M. Tyrrell

L'Utilisation du Laser en Ophtalmologie 367
    G. Coscas, G. Quentel et M. Binaghi

Ultraviolet Prophylaxis of Adverse Effects of Environmental Chemicals on Organisms 375
    J.I. Prokopenko

Round Table Summary : Usage and Testing of Sunscreens ... 385
    F. Greiter

The Trends and Future of Photobiology : Medical Aspects .. 389
    T.B. Fitzpatrick

IV - PHOTOPHYSIOLOGY

Visual Rhodopsin and Phototransduction in the Vertebrate Retina 399
    M. Chabre

Photoregulation of *E.coli* Growth 413
   A. Favre

Photomovements of Microorganisms 421
   F. Lenci

The Mechanism of the Circadian Rhythm of Photosynthesis . . 437
   H.G. Schweiger

Photoregulation of Neuroendocrine Rhythms 451
   I. Assenmacher, A. Szafarczyk, J. Boissin and
    M. Jallageas

The Effects of Artificial and Natural Sunlight upon
   some Psychosomatic Parameters of the Human
   Organism 465
   F. Greiter, P. Bilek, N. Bachl, L. Prokop,
   R. Maderthaner, H. Bauer, G. Kroyer, I. Steiner,
   P. Riederer, J. Washüttl and G. Guttmann

Intracellular Location of Phytochrome 485
   P.H. Quail

The Role of Phytochrome in the Natural Environment 501
   H. Smith

Phytochrome and Gene Expression 515
   H. Mohr

The Trends and Future of Photobiology : Physiological
   Aspects 531
   D. Vince Prue

V - PHOTOSYNTHESIS AND BIOCONVERSION OF SOLAR ENERGY

Organization of the Photosynthetic Pigments and
   Transfer of the Excitation Energy 539
   G. Paillotin, A. Vermeglio and J. Breton

Laser Studies of Primary Processes in Photosynthesis 549
   N.E. Geacintov and J. Breton

Systematic Modification of Electron Transfer Kinetics
   in a Biological Protein : Replacement of the
   Primary Ubiquinone of Photochemical Reaction
   Centers with other Quinones 561
   P.L. Dutton, M.R. Gunner and R.C. Prince

CONTENTS

Direct Measurement of Light Induced Currents and
    Potentials Generated by Bacterial Reaction
    Centers 571
  N.K. Packham, P.L. Dutton and P. Mueller

Primary and Associated Reactions in Photosystem II 579
  H.J. van Gorkom and A.P.G.M. Thielen

Biophotolysis of Water for H_2 Production using
    Immobilized and Synthetic Catalysts 587
  D.O. Hall, P.E. Gisby and K.K. Rao

Solar Energy Bioconversion at the Ecosystem Level 597
  P. Duvigneaud

Utilization of Solar Radiation by Phytoplankton 619
  J.F. Talling

Limiting Factors in Photosynthesis. From the
    Chloroplast to the Plant Canopy 633
  J.L. Prioul

Bioconversion of Solar Energy 645
  M. Calvin

List of Contributors 661

Index 665

TIME-RESOLVED CHROMOPHORE RESONANCE RAMAN AND PROTEIN FLUORESCENCE

OF INTERMEDIATES IN SOME PHOTOBIOLOGICAL CHANGES[†]

M. A. El-Sayed

Department of Chemistry
University of California
Los Angeles, California 90024 USA

This paper summarizes the work of my group over the past four years. Different resonance Raman techniques are described which are useful in studying intermediates of photolabile systems in the millisecond, microsecond, nanosecond, and picosecond time domains. These techniques are used to study two important photobiological systems: bacteriorhodopsin (bR) and carbonmonoxyhemoglobin (HbCO). The summary of the results of applying these techniques to study the retinal system in bR is given and discussed in terms of what is known about its photochemical proton pump cycle. The main results are: (1) the largest retinal configurational changes occur in the first step (the absorption step) and (2) the Schiff base proton in

$$-C=\overset{+}{N}H-$$

ionizes in 40 μsec (in the $bL_{550} \rightarrow bM_{412}$ step).

The picosecond Raman spectra of the porphyrin system of the intermediate produced in the photolysis of HbCO in the picosecond time domain suggest that photodissociation of CO might proceed

---

[†]Modified versions of this paper are being submitted to three different invited publications: (a) The Proceedings of the Sergio Porto Laser Memorial Meeting, Rio de Janeiro, Brazil, June 29-July 3, 1980; (b) The Proceedings of the VIIIth International Congress on Photobiology, Strasbourg, France, July 21-25, 1980; (c) a volume edited by Lester Packard on Visual Pigments and Purple Membranes. It is also being submitted as an abstract to the International Congress on Raman Spectroscopy, Ottawa, Canada, August 4-9, 1980.

from a quintet state of HbCO to give a high spin Fe(II) with the Fe(II) being coplanar with the porphyrin plane. The iron moves out of the plane in the μsec time scale, suggesting that its motion is probably controlled by the protein motion.

In the last section of the paper, two pulsed laser systems, fired at different delay times from one another, are used to follow the protein fluorescence of the different intermediates in the bR photocycle. The results suggest that no fluorescence intensity changes take place during the first step (when the retinal undergoes the largest configuration changes). A discussion of the implication of the observed protein fluorescence changes in time is given in terms of the proton pumping process.

INTRODUCTION

While picosecond lasers have been used previously to obtain the resonance Raman spectra of stable molecules[1] (in particular to reduce interference from fluorescence radiation), the first resonance Raman spectra of transients formed in the picosecond time domain have just been reported this year by our group[2] and others.[3] Of course optical absorption and emission spectroscopy of picosecond transients has been useful during the past decade in determining the number, rise time, and decay time of picosecond intermediates in photolabile systems. However, due to the broad nature of the optical absorption of the system studied, these spectra did not yield the kind of structural information one would like to have in order to identify the exact structural changes taking place in these processes. Vibration spectra, as obtained from the Raman scattering process, are expected to give more structural information. For this reason, as well as the development of cavity dumped picosecond lasers and the imaging detection systems, time-resolved Raman spectroscopy is expected to be an active field of research over the coming decade or two.

I. Time-Resolved Resonance Raman (TRRR) Techniques[4]

A. <u>The millisecond, microsecond, and nanosecond transients</u> (with J. Terner, A. Campion, L. Hsieh, and A. Burns): Different time-resolved resonance Raman techniques have been developed[4] for determining the resonance Raman spectra of the intermediates formed from photolabile systems, e.g., the proton pump system of bacteriorhodopsin[5-8] and the CO hemoglobin.[2] The method used varied and depends on the time scale in which the photointermediate builds up in concentration. All the methods that we have used have the following features in common:

1. One laser is used which acts both as the photolytic as well as the Raman probe light source. This is especially applicable to systems of broad adsorption bands that one expects to have on overlap of the absorption band of the photolabile parent compound and that for the intermediate whose resonance Raman is being examined.

2. The experiment is carried out in a time scale appropriate for the rise and decay times of the intermediates being studied.

3. The laser used is adjusted at a wavelength which gives high photolytic probability and large R. Raman enhancement for the intermediate examined. Furthermore, the scattered Raman radiation should have minimum overlap with any fluorescence present.

4. For obtaining the spectrum of a certain intermediate, chemical or physical perturbations are used, if possible, to maximize its concentration.

5. Satisfying the above conditions, two Raman spectra are then recorded using the optical multichannel analyzer for detection: one spectrum is obtained at very low powers (to obtain the spectrum of the unphotolyzed parent compound at minimum photolysis), and the second spectrum is obtained at high powers (to maximize the concentration of the photoproduct under examination).

6. Computer subtraction techniques are then used to subtract out the low power spectrum from the high power spectrum to obtain the Raman spectrum of the intermediate, having maximum concentration in the time scale of the experiment used and which has the maximum enhancement at the wavelength of the laser used.

In order to satisfy condition (2) above, i.e., adjust the time scale of the experiment to maximize the concentration of a certain intermediate, one of the following techniques is used:

 i. <u>Pulsed lasers</u>:[5,9,10] Only intermediates appearing in a time equal to or shorter than the pulse width could be detected by Raman spectroscopy if the wavelength is adjusted for maximum enhancement. Intermediates with rise times shorter, and decay times longer, than a few nanoseconds can be studied[5,9,10] by using the N_2-pumped (e.g., Molectron) or Nd YAG-pumped (e.g., Quanta-Ray) dye lasers which have few nanoseconds pulse width. Chromatics could be used for intermediates with decay times in the microsecond time scale. In principle, picosecond intermediates could be detected by using the high power pulsed picosecond lasers. However, the low duty cycle of these lasers and sample destruction from multiphoton processes could hamper the observation of good signal-to-noise signals in these experiments.

ii. <u>Modulation of c.w. lasers</u>:[6,7] Electric or mechanical modulation of c.w. lasers can produce pulses with different pulse width and at a given modulation frequency. Electric modulation could give short pulses with high duty cycles. Mechanical modulation could give longer pulses with lower duty cycles. We have used mechanical choppers (rotating disks) fixed with variable size slits. A c.w. laser could function as a pulsed laser with variable pulse width. The fact that the laser can be brought up to a small focus in the micron range makes slit width of a few microns usable in these experiments. With the available practical motors usable in this experiment, intermediates in the 50-100 nanosecond time scale could be detected.[6] Of course with slow motors and large slit width, millisecond intermediates could be easily observed. Two slits, one for the photolysis and one for the probe laser could also be used[7] with variable time delays (i.e., separation between the slits). The duty cycle in these experiments is determined by the number of slits in the rotating disk as well as the motor speed. In any case, they could be much better than in the pulsed laser experiment.

iii. <u>Flow techniques</u>:[4,8] Instead of pulsing the laser, the sample could be "pulsed" by flowing it across a focused c.w. laser beam. Actually the flow technique was first used by Mathies, et al.,[11] to determine the Raman spectrum of the unphotolyzed rhodopsin. Marcus, et al.,[12] were the first to use it for kinetic studies in bacteriorhodopsin by varying the flow rate of the sample (which could be changed by a factor of ten) to obtain different time scales for different intermediates. We have extended the time scale of this method by realizing[4,8] that, for the same flow rate, the time scale of the experiment could be varied by varying the laser focus itself. By using a microscope objective and a flow rate of 10-40 m/sec, the experimental time resolution (determined by the time it takes the flowing sample to cross the focused laser beam) could be in the 50-nanosecond time scale! More importantly, the scattered Raman relation is being collected continuously in this experiment.

B. <u>Resonance Raman of Picosecond Intermediates</u>[2] (with J. Terner, T. G. Spiro, M. Nagumo, and M. F. Nicol): By replacing the c.w. laser in experiment (A-iii) above with a mode-locked-cavity dumped Argon or Krypton ion pumped dye laser, the resonance Raman spectrum of intermediates in the picosecond time scale could be recorded. In this case, the time resolution of the experiment is no longer determined by the sample residence time in the laser beam as in (A-iii), but rather by the pulse width of the picosecond laser used. This is true only if the time between the laser pulses (∼1 μsec) is longer than the residence time of the sample in the beam (∼0.1 μsec).

II. Resonance Raman Results on the Proton Pump System of Bacteriorhodopsin[6-8,13] (with J. Terner, C.-L. Hsieh, and A. Burns)

Bacteriorhodopsin absorbs visible light (λ_{max} = 570 nm) and passes through a number of intermediates (at least four) before the bR returns to its initial form:[14]

$$bR_{570} \xrightarrow{h\nu(psec)} bK_{590(610)} \xrightarrow{2\ \mu sec} bL_{550} \xrightarrow{40\ \mu sec} bM_{412} \xrightarrow{} bO_{640} \xrightarrow{msec} bR_{570}.$$

As a result of this cycle, (one or two) protons are pumped out of the cell, thus creating proton gradients across the bR cell membrane.[15] It is this electric free energy that is believed to be used in the synthesis of the high energy molecules (ATP).

We have attempted to use the TRRR technique in order to examine two problems concerning the above photocycle. The first one is concerned with the retinal conformation changes. Based on chemical reconstitution studies,[16] it is believed that the retinal in bR_{570} is all-trans retinal. In rhodopsin, resonance Raman studies showed that[11] the retinal is in the 11-cis configuration, and it has been the common belief that the absorption process leads to isomerization to the all-trans form in the first step. This change in the retinal configuration leads to changes in the retinal-protein interaction energy as well as entropy. It is the change in the free energy upon the absorption that leads to the storage of the free energy necessary to drive the system through the latter process. The question then arises: If the retinal in the bR_{570} already contains an all-trans retinal, what is the mechanism of converting solar energy into free energy in the absorption act of the bR system? Could it be that even if retinal in bR_{570} is in the all-trans configuration, further configuration changes take place in the first step of the cycle? (It should be pointed out that the all-trans retinal inside the protein might not necessarily have the lowest value of the free energy of the system.) Another possibility is that when the all-trans retinal combines with the bacteriorhodopsin to form the retinal-protein complex it does not retain its all-trans configuration. In support of this is the difficulty in comparing the fingerprint region of the retinal in bR_{570} with that of all-trans model compounds,[13] unlike the rhodopsin system when good agreement between the 11-cis and the rhodopsin system is obtained.[11] In any case, one would like to investigate whether or not the first step in the cycle involves a change in the retinal configuration. By using TRRR one can examine the fingerprint region (1000-1400 cm^{-1}) (sensitive to retinal configuration changes) for the $bK_{590(610)}$ intermediate and compare it to that for bR_{570}. The $bK_{590(610)}$ intermediate is formed in the picosecond time domain; however, it lasts a few microseconds. Thus, the TRRR experiment can be carried out in the 0.1-1 μsec time scale.

The second problem concerns the origin of the protons pumped during the photochemical cycle. The retinal in bR_{570} is bonded to the lysine in the protein via a Schiff base linkage:

$$\begin{array}{c} | + \\ -C=N- \\ | \\ H \end{array}$$

While not definitely proven, it has been assumed[15] that one of the protons pumped out in the $bR_{570} \rightarrow bM_{412}$ transformation comes from this group. We have followed by TRRR techniques the

$$\begin{array}{c} | + \\ -C=N- \\ | \\ H \end{array}$$

vibration at ~1640 cm^{-1} to determine the step at which this band disappears (due to the formation of the unprotonated Schiff base):

$$\begin{array}{c} | \\ -C=N- \end{array},$$

which has a frequency below 1620 cm^{-1}.

The important results in the fingerprint region (which is sensitive to the retinal configuration) and the C=N stretching region (to follow the deprotonation of the Schiff base) can be summarized as follows:

1. Similar to rhodopsin,[17] large changes in the fingerprint region are observed during the first

$$bR_{570} \xrightarrow{h\nu} bK_{590}$$

transformation.

2. Like in rhodopsin,[17] large enhancement is observed for the low frequency vibrations in the ~980 cm^{-1} region for bK_{590}. This might suggest twisting of the polyene system.

3. Unlike in rhodopsin,[11] a complete identification of the isomeric form of the different intermediates is difficult to achieve by comparing our spectra with those of model compounds in solution. This could result from one or more of the following reasons: (a) Some distortion of the spectra obtained by subtraction techniques. (b) Larger perturbation of the spectra of retinal by the retinal-opsin interaction in bacteriorhodopsin than that present in rhodopsin. (c) The small difference between the all-trans and 13-cis fingerprint spectra in solution.

4. In the $bK_{590} \rightarrow bL_{550}$ transformation, the 980 cm^{-1} region becomes normal. This might suggest a relaxation of the twisted polyene structure.

5. Small changes in the fingerprint region take place in the $bL_{550} \rightarrow bM_{412}$ process, the process which is accompanied by the largest change in the position of the optical absorption maximum in the cycle. It also involves the deprotonation of the Schiff base. This might suggest[13] that the theories based on ionic interactions[18] rather than retinal configurational changes[19] might be the correct ones in explaining the origin of the red shift in the retinal absorption upon combining with the opsin.

6. The C=N stretching vibration is greatly reduced in frequency during the first step ($bR_{570} \rightarrow bK_{590}$). However, it is found that it is reduced further in D_2O solvent. This might suggest that in the first step of the cycle, the interaction between the Schiff base nitrogen and the proton has been reduced. This is unlike the results in the rhodopsin[17] system, in which the C=N frequency in the batho form has remained similar to that in the parent compound.

7. The C=N frequency in the bL_{550} form is found to be similar to the parent compound (and is thus protonated). This suggests that deprotonation of the nitrogen Schiff base takes place during the $bL_{550} \rightarrow bM_{412}$ process, i.e., in ~40 μsec.

III. The Carbonmonoxyhemoglobin Results[2]

Using optical picosecond spectroscopy,[20,21] it is found that HbCO dissociates in the picosecond time scale and the intermediate that appears lasts for a longer time than[22] 680 picoseconds. What we[2] wanted to investigate was to determine the vibration spectrum of this intermediate. The Fe(II) in HbCO is in the low spin state and is coplanar with the porphyrin plane. In deoxyhemoglobin (Hb), however, the Fe is off the porphyrin plane and is in the high spin state. In the high spin state, an electron occupies the $d_{x^2-y^2}$ orbital which forms σ^* antibonding molecular orbitals with the pyroll nonbonding atomic orbitals. This leads to expansion of the porphyrin ring and to a large reduction in the frequency of some of the Raman active C=C stretching vibrations in the Hb molecule as compared to the HbCO molecule. Thus, these bands could be used to label the spin state of the Fe in any Fe-porphyrin system.[23] Of course the more in-plane the Fe becomes, the stronger the σ^* repulsion becomes and the further is the reduction in the C=C stretching frequency.

Carrying out the picosecond experiment described in Section I-B, it is found that when the laser is focused (high intensity condition), three new bands appear in the C=C stretching region at 1603, 1552, and 1540 cm^{-1} for the picosecond intermediate; with the 1552 cm^{-1} band being anomalously polarized. These bands are red shifted greatly from the corresponding ones in HbCO molecules and are further reduced by only 4-6 cm^{-1} from those for the Hb molecule itself. This suggests that in this intermediate, Fe(II) is in the high spin state (to explain the fact that these bands are closer in frequency to Hb than to HbCO) with the Fe being more in the porphyrin plane in this intermediate than in Hb (to account for the observed small frequency reduction from those for Hb).

Low spin to high spin conversion in the ground state of Fe porphyrin is known to take place in the microsecond[24] time domain. The fact that the low spin (in HbCO) to high spin (in this intermediate) conversion takes place in the 25 picosecond time domain (the pulse width of our laser) suggests that the conversion must have taken place in the excited states. This then suggests that in HbCO, $S_0 \to S_1$ absorption is followed by $S_1 \leadsto Q$ (quintet state) intersystem crossing process. Dissociation might then take place from the quintet state itself. Some quintet states are theoretically predicted to fall below S_1 in iron porphyrins.[22]

IV. Time-Resolved Protein Fluorescence Results[25] (with J. Fukumoto, W. Hopewell, and B. Karvaly)

One interesting question that immediately arises concerning the cycle in bR is: When does the protein learn about the absorption act by the retinal in the bacteriorhodopsin photocycle? By using two pulsed lasers, the first to initiate the bR cycle and the second (delayed by different times) to excite the protein fluorescence, the intensity of the fluorescence emission of the protein in the different intermediates has been monitored. The important results can be summarized as follows:

1. While the largest configuration is observed in the first step, no change in the protein fluorescence is observed in this step.

2. The fluorescence intensity decreases during the $bK_{590} \to bL_{550} \to bM_{412}$ transformation, with the bM_{412} having the lowest fluorescence intensity.

3. The quenching of the bM_{412} fluorescence could be explained by deprotonation of 1-2 tyrosine molecules or charge perturbation of 1-2 tryptophane molecules, as was recently suggested to explain the changes in the protein absorption by Hess, et al.[26]

The Raman and fluorescence results on the bR system suggest that the absorption act

$$(bR_{570} \xrightarrow{h\nu} bK_{590})$$

causes a change in the configuration of the retinal with some change in the nitrogen-proton binding of the Schiff base. These changes could be accompanied by changes in the energetics of the retinal-protein pocket. The protein then responds to this change in the $bK_{590} \rightsquigarrow bL_{550} \rightsquigarrow bM_{412}$ transformation. During this transformation, two protons are released, the Schiff base proton is ionized and there is a possibility that 1-2 tyrosine protons might also be ionized during this half of the photocycle.

ACKNOWLEDGEMENT

The author wishes to thank Drs. Stoeckenius and Bogomolni for supplying some of the bacteriorhodopsin samples. The financial support of the U.S. Department of Energy (Office of Basic Energy Sciences) is gratefully acknowledged.

REFERENCES

1. M. Bridoux, C. R. Acad. Sc. Fr. 258, 620 (1964); M. Bridoux, A. Chapput, M. Crunelle and M. Delhaye, Adv. in Raman Spectr., ed. by J. P. Mathieu (1973) pp. 65-69; M. Delhaye, Proceedings of the Fifth International Conference on Raman Spectroscopy, ed. by Schmid et al. (1976) pp. 747-752; M. Bridoux, A. Deffontaine and C. Reiss, C.R. Acad. Sci. 282, 771 (1976); M. Bridoux and M. Delhaye, in "Advances in Infrared and Raman Spectroscopy," Vol.2, eds. R. J. H. Clark and R. E. Hester (Heyden, 1976) p. 140; P. P. Yaney, J. Opt. Soc. Am. 62, 1297 (1972); R. P. Van Duyne, D. L. Jeanmaire and D. F. Shriver, Anal. Chem. 46, 213 (1974); F. E. Lyttle and M. S. Kelsey, Anal. Chem. 46, 855 (1974); M. Nicol, J. Wiget and C. K. Wu, Proceedings of the Fifth International Conference on Raman Spectroscopy, ed. by Schmid et al. (1976) pp. 504-505.
2. J. Terner, T. G. Spiro, M. Nagumo, M. F. Nicol and M. A. El-Sayed, J. Am. Chem. Soc. 102, 3238 (1980).
3. M. Coppey, H. Tourbez, Pierre Valat and B. Alpert, Nature 284, 568 (1980).
4. For a previous review see: Time-Resolved Resonance Raman Spectroscopy in Photochemistry and Photobiology. In Multichannel Image Detectors in Chemistry, ACS SYMPOSIUM SERIES Bk 102, Chpt. 10 (1979) pp. 215-227.
5. A. Campion, J. Terner and M. A. El-Sayed, Nature 265, 659 (1977).
6. A. Campion, M. A. El-Sayed and J. Terner, Biophys. J. 20, 369 (1977).
7. J. Terner, A. Campion and M. A. El-Sayed, Proc. Natl. Acad. Sci. USA 74, 5212 (1977).
8. J. Terner, C.-L. Hsieh, A. R. Burns and M. A. El-Sayed, Proc. Natl. Acad. Sci. USA 76, 3046 (1979).
9. W. H. Woodruff and S. Farquharson, Science 201, 831 (1978).

10. (a) K. B. Lyons, J. M. Friedman and P. A. Fleury, Nature 275, 565 (1978).
    (b) R. F. Dallinger, J. R. Nestor and T. G. Spiro, J. Am. Chem. Soc. 100, 6251 (1978).
11. R. Mathies, T. B. Freedman and L. Stryer, J. Mol. Biol. 109, 367 (1977).
12. M. A. Marcus and A. Lewis, Science 195, 1328 (1977).
13. M. A. El-Sayed and J. Terner, J. Photochem. Photobiol. 30, 125 (1979).
14. R. H. Lozier, R. A. Bogomolni and W. Stoeckenius, Biophys. J. 15, 955 (1975).
15. D. Oesterhelt, Angew Chem. (Int. ed.) 15, 17 (1976).
16. M. J. Pettei, A. P. Yudd, K. Nakanishi, R. Henselman and W. Stoeckenius, Biochemistry 16, 1955 (1977).
17. G. Eyring and R. Mathies, Proc. Natl. Acad. Sci. USA 75, 4642 (1979).
18. B. Honig, A. D. Greenburg, V. Dinur and T. Ebrey, Biochemistry 15, 4593 (1976).
19. R. Kornstein, K. Muszkat and S. Sharafy-Ozeri, J. Am. Chem. Soc. 95, 6177 (1973).
20. L. J. Noe, W. G. Eisert and P. M. Rentzepis, Proc. Natl. Acad. Sci. USA 75, 573 (1978).
21. C. V. Shank, E. P. Ippen and R. Bersohn, Science 193, 50 (1976).
22. B. J. Greene, R. M. Hochstrasser, R. B. Weisman and W. A. Eaton, Proc. Natl. Acad. Sci. USA 75, 5255 (1978).
23. T. G. Spiro and J. M. Burke, J. Am. Chem. Soc. 98, 5482 (1976).
24. (a) J. K. Beattie, N. Sutin, D. H. Turner and G. W. Flynn, J. Am. Chem. Soc. 95, 2052 (1973).
    (b) J. K. Beattie, R. A. Binstead and R. J. West, ibid. 100, 3044 (1978).
25. J. M. Fukumoto, W. D. Hopewell, B. Karvaly and M. A. El-Sayed, "Time-Resolved Protein Fluorescence Studies of the Intermediates in the Photochemical Cycle of Bacteriorhodopsin." Proc. Natl. Acad. Sci. USA, accepted for publication.
26. B. Hess and D. Kuschmitz, FEBS Letters 100, 334 (1979).

SYNCHROTRON RADIATION SOURCES FOR PHOTOBIOLOGY AND ULTRAVIOLET,

VISIBLE AND INFRARED SPECTROSCOPY

J. C. Sutherland

Biology Department
Brookhaven National Laboratory
Upton, New York 11973, USA

WHAT IS SYNCHROTRON RADIATION?:

Maxwell's equations show that an accelerating electrical charge emits electromagnetic radiation. The emission of radio waves by electrons oscillating within an antenna is a familiar example of this effect. Electrons moving through a vacuum can be accelerated radially by a magnetic field oriented perpendicularly to their direction of motion. The high energy electrons circulating within a synchrotron experience a centripetal acceleration each time they pass through a bending magnet. Electrons whose path is bent by the magnetic field emit photons along the direction tangential to their path. Thus at each bending magnet around a synchrotron ring, radiation is emitted in a fan-shaped distribution; the limits of the "fan" are the directions of travel of the electrons before entering and after leaving the bending magnet. The radiation covers a very broad spectral range as shown in Fig. 1; this is the most important feature of synchrotron radiation.

The shorter the wavelength, the greater the fraction of the radiation with linear polarization of the electric vector in the plane containing the curved orbit of the electrons, and the greater the extent to which the photons are confined in or close to this plane. For x-rays, synchrotron radiation is effectively plane polarized while for ultraviolet radiation, a significant fraction of the photons radiated above or below the orbital plane will be polarized with the electric vector perpendicular to the orbital plane.

The short wavelength limit of the synchrotron emission

spectrum is proportional to cube of the kinetic energy of the electrons and inversely proportional to the radius of curvature of their trajectory through the magnetic field. Thus high-energy photons are produced by high-energy electrons and intense magnetic fields. For a given wavelength, the photon flux is proportional to the number of electrons passing through the magnet per unit time. Thus, the number of ultraviolet, visible and infrared photons is strongly dependent on the number of electrons in the ring - i.e., the circulating current - and correspondingly independent of the energy of the electrons (provided their energy is greater than about 10^8 electron volts, which is the case for all modern synchrotron storage rings).

The "time structure" of synchrotron radiation is also important in uv, visible and infrared applications. Photons emitted when electrons pass through bending magnets carry with them energy which must be replaced if the electrons are going to travel around and around the storage ring for long times - typically several hours. The electrons are accelerated by passing through a hollow electrode to which a high voltage oscillating electric field is applied. The accelerating field must oscillate so that the electrons are attracted as they approach the electrode and repulsed as they leave. Typically an oscillating frequency in the range used for radio transmissions is used and the accelerating device is termed an "rf" cavity.

For the electrons to receive the correct "boost" each time they pass through the rf cavity and thus maintain a stable orbit, they must arrive at just the correct time relative to the sinusoidally changing rf field. Thus electrons travel around the ring in a number of "bunches"; the maximum number of possible bunches is given by pf/c where p is the distance the electrons travel in going all the way around the ring, f is the frequency of the rf field and c is the velocity of the electrons - which is only very, very slightly less than the velocity of light in vacuum. Light is thus generated in pulses. The duration of a pulse is the "length" of the bunch of electrons divided by c. The distribution of the electrons within the bunch and hence the temporal profile of the light pulse is approximately gaussian. For most storage rings, the full-width-at-half-maximum of the light pulse is one nanosecond or less.

The minimum time between flashes is achieved by populating all of the potentially stable bunches with electrons; the pulse-to-pulse separation is $1/f$; values of 5 to 30 nanoseconds are typical. The maximum flash-to-flash separation is realized by loading all of the electrons into a single bunch resulting in a dark interval given by p/c. Typical values for this mode range from 10 nanoseconds to a few microseconds depending on the size of the synchrotron.

For some experiments it is desirable to have still longer interpulse periods. Besides using very fast shutters–such devices exist for use in the uv and visible but not for x-rays, it may be possible to perturb the path of the electrons so that light from only every n^{th} flash traverses the experimental optical system and reaches the sample (Blumberg, 1979). Since n could be adjusted experimentally, a broad range of interpulse periods could be achieved. This scheme has the advantage of working equally well for all wavelengths.

SYNCHROTRON RADIATION AS A PROBE OF BIOLOGICAL STRUCTURE AND DYNAMICS:

Biological applications of synchrotron radiation involve ultraviolet, visible and infrared radiation as well as x-rays. This paper is mainly concerned with the former, but I will also mention the different ways in which x-rays from a synchrotron may be used to probe the structure and dynamics of biological materials. All of these applications are discussed in the volume edited by Castellani and Quercia (1979).

Biological applications of synchrotron radiation in the ultraviolet, visible and infrared (UVISIR) regions of the spectrum can be divided into two classes based on the use which is made of the time structure of the radiation. Experiments such as circular dichroism, magnetic circular dichroism, photoacoustic spectroscopy, and action spectroscopy make use of the broad spectrum and high intensity of synchrotron radiation. For these experiments, time structure is not important. In contrast, fluorescence lifetime and anisotropy experiments require a pulsing source. Of course, the broad spectrum and high intensity will also be valuable for the time-structure-dependent class of experiments.

Circular dichroism is one of the most important spectroscopic experiments which will benefit from the broad tunability of synchrotron radiation. Commercial dichromators with xenon arc sources operate marginally below 200 nm and not at all below 180 nm. CD experiments using hydrogen discharges have shown the spectral region below 200 nm to be extremely important in the study of proteins, nucleic acids, and sugars. The greater intensity available from synchrotron radiation will improve the precision and sensitivity of far ultraviolet CD experiments and probably permit kinetic measurements of changes of CD in this region. Synchrotron radiation may also be the best source for extending CD measurements into the infrared.

Circular dichroism is sensitive to the conformation of molecules. Thus the far ultraviolet CD spectrum of proteins

(165-240 nm) is sensitive to the secondary structure of peptide bonds (Johnson and Tinoco, 1972; Brahms and Brahms, 1980). In particular, the region from 165-200 is important for quantitating β sheet and reverse turns in proteins (Brahms and Brahms, 1980). In the case of nucleic acids, the strong band observed between 180 and 190 nm appears more sensitive to helical structure and base-base interactions than the more extensively studied 260 nm band (Sprecher et al., 1979). Far ultraviolet CD has also been important in studies of polysaccharides since all of the electronic absorption bands of these molecules lie below 200 nm. Johnson (1978) and Pysh (1976) have reviewed the early work on vacuum ultraviolet circular dichroism.

Magnetic circular dichroism (MCD) is sensitive to the configuration, i.e. the covalent and electronic structure, of a molecule; generally MCD, in contrast to CD, is insensitive to conformation. Far ultraviolet MCD will thus be useful in probing the higher excited states of biological molecules and assigning the origin of bands observed in absorption and CD. One important application will be in the nucleic acids where theoretical predictions have yet to be tested.

To date, CD experiments using synchrotron radiation have been performed at the Aladdin storage ring (Stoughton, WI, USA) by Patricia Snyder and her colleagues (Snyder and Rowe, 1980) and at the SURF II ring (Gaithersburg, MD, USA) by my group. The latter experiment will be moved to the National Synchrotron Light Source (Brookhaven, NY, USA) sometime in 1981. I expect that CD and MCD experiments will be built at several of the other synchrotron facilities.

Synchrotron radiation is a useful source for probing the biological effects of far ultraviolet radiation. In contrast to longer wavelength ultraviolet where the most important biological target is DNA, wavelengths below 185 nm are absorbed very near the surface of a cell. Experiments by Ito and his colleagues using radiation from the Tokyo synchrotron as well as conventional sources indicate that these short wavelengths can kill cells but do not induce mutations (Ito et al., 1980). Perhaps focused beams of synchrotron radiation can be used to probe surface structure. It will also be interesting to see if some organisms have evolved repair mechanisms for this type of damage.

Photoacoustic spectroscopy is another measurement which can be extended to shorter wavelengths by the use of synchrotron radiation. Dehydrated samples and materials adsorbed to surfaces can be analyzed.

Fluorescence spectroscopy uses to advantage both the broad

spectrum and the pulsing nature of synchrotron radiation. Excitation spectra can be extended to at least 185 nm (where water starts to absorb). At about 220 nm, for example, phenylalanine absorbs as strongly as tryptophan and tyrosine, in contrast to longer wavelengths where absorption by phenylalanine is usually unimportant in photochemical and photophysical processes if tryptophan or tyrosine is present. The ability to extend excitation spectra to shorter wavelengths will also be valuable for measurements of phosphorescence and delayed fluorescence.

The time structure inherent in synchrotron radiation can be used to measure fluorescent lifetimes (Alpert and Lopez-Delgado, 1976), time resolved polarization anisotropies (Munro et al., 1979) and time resolved emission spectra. These measurements reveal molecular dynamics with sub-nanosecond resolution. The advantages of using synchrotron radiation compared to flash discharge or pulsed laser radiation are the ease with which any excitation wavelength can be chosen, the excellent temporal stability of the pulse and the high repetition rate.

The most dramatic impact to date of synchrotron radiation on structural biology has involved Extended X-ray Absorption Fine Structure (EXAFS) spectroscopy. This technique permits determination of the radial distances from a unique atom to its neighbors - e.g., the distances from the iron in an iron-sulfur protein to the sulfur ions surrounding it. Samples can be either dehydrated or in solution; crystals are not required. Since interatomic distances can be determined with subangstrom precision, EXAFS spectroscopy is an important complement to diffraction methods. For a review see Shulman et al. (1978).

Synchrotron radiation may become an important tool in x-ray crystallographic studies of biological structure by virtue of the changes in a diffraction pattern which occur for x-ray wavelengths above and below the absorption edge of a unique component - again consider the iron ion in an iron-sulfur protein. Changing the wavelength of the x-rays may provide the same information which presently must be obtained from tedious isomorphic substitutions. As for small angle x-ray scattering, the higher fluxes of collimated radiation obtainable with synchrotron radiation have made possible the measurement of changes in scattering associated with time dependent events - e.g., the contraction of a muscle - with a resolution of milliseconds.

X-ray microscopy may become a practical reality as a consequence of the degree of collimation inherent in synchrotron radiation. For biological materials the principal advantage of x-ray microscopy is the relative opacity of organic materials and transparency of water between the carbon absorption edge (4.36 nm) and the oxygen absorption edge (2.33 nm). Thus biological materials could be studied with higher resolution than is achievable with visible radiation, but without the dehydration and staining required by electron microscopy. The theoretical resolution of an x-ray microscope is below that achievable with an electron microscope by about an order of magnitude. The experimental challenge facing the designers of these instruments is finding some means of focusing soft x-rays; at present the use of Fresnel zone-plates seems to be the best approach. For recent reviews, see the volume edited by Wright (1980).

SYNCHROTRON LIGHT SOURCES

The generation of electron synchrotron storage rings which presently are being used as light sources was originally built to perform experiments in nuclear and particle physics. Several of the older, lower energy storage rings were converted to "dedicated" radiation sources when they became obsolete for their original mission. These facilities are useful sources of short wavelength ultraviolet and soft x-rays. Higher energy rings, the ones at Hamburg and Stanford for example, are still used for elementary particle research. Simultaneously, they serve as sources of synchrotron radiation in what is called a "symbiotic" mode of operation. The higher energy of the electrons in these rings means that they are good sources of hard x-rays as well as longer wavelength radiation.

A new generation of storage rings designed specifically to serve as sources of synchrotron radiation is nearing completion in several European countries, Japan and the United States. As an example of the facilities which will be available at these new "dedicated" synchrotron radiation centers, I will describe the National Synchrotron Light Source (NSLS) which is being built at Brookhaven National Laboratory in the United States; it is located on Long Island about 100 km east of New York City. The NSLS will have a total of 44 ports. Of these, 28 will receive radiation from a high energy storage ring, the electrons in which will have an energy of up to 2.5×10^9 electron volts. These ports will be used primarily for experiments which require hard x-rays. A smaller, lower energy storage ring (the electrons will have energy of 7×10^8 electron volts) will supply soft x-rays, ultraviolet, visible and infrared radiation to 16 additional ports. The radiation from many of these 44 ports will be divided between two, three and sometimes even four experiments. Thus, in a few years, there may be 100 or more different experiments in operation. We expect five experiments to be designed primarily for biophysical research, specifically 1) UVISIR spectroscopy, 2) x-ray spectroscopy, 3) x-ray crystallography,

4) small angle x-ray scattering and 5) x-ray microscopy. Roughly an equal number of experiments designed primarily for some other purpose will occasionally be used for biophysical research. To put these data in perspective, I must also point out that the NSLS represents about 50% of the presently planned synchrotron radiation capabilities of the United States, indeed of the western hemisphere. In Europe, more ports will be available at more research centers in a smaller geographical area. Even here, and certainly in the rest of the world, many scientists who use synchrotron radiation will have to travel some distance from their home institution. If we are to realize the fullest potential of synchrotron radiation, experimental apparatus and support facilities and personnel must be available at the synchrotron radiation facility. The need for support facilities and personnel is even more important for biological research than in the physical sciences. The inconvenience of performing one's experiments at a central facility is, however, a small price to pay for the advantages inherent in the use of synchrotron radiation.

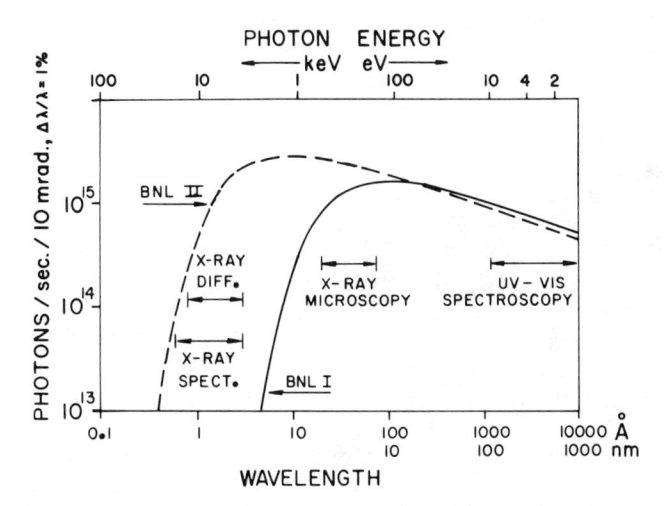

Fig. 1 : The spectra anticipated to be emitted by the two synchrotron storage rings of the National Synchrotron Light Source. The smaller ring (BNL I) will be used for ultraviolet, visible and infrared spectroscopy while the larger ring (BNL II), which will circulate higher energy electrons, will mainly be used for x-Ray research. The spectra were calculated assuming a horizontal acceptance of 10 mrad, although some experiments will use three to five times this amount.

SUMMARY

The advantages of synchrotron radiation in several types of spectroscopy, microscopy and diffraction studies are clear. The availability of synchrotron radiation will expand rapidly in the early 1980's as experimental programs start at the new generation of dedicated storage rings.

ACKNOWLEDGMENTS

Preparation of the article was supported by the United States Department of Energy and a Research Career Development Award from the National Cancer Institute, United States Department of Health and Human Resources (CA-00465). Its presentation at the Eight International Congress on Photobiology was made possible by a travel grant from the Centre National de la Recherche Scientifique, Ministère des Universités, République Française.

REFERENCES

Alpert, B., and Lopez-Delgado, R., 1976, Fluorescence lifetimes of haem proteins excited into the tryptophan absorption band with synchrotron radiation, Nature, 263:445-446.

Blumberg, L. N., 1979, Vertical kicker for fluorescence decay experiments in the NSLS VUV ring, "BNL Report - 26856".

Brahms, S., and Brahms, J., 1980, Determination of protein secondary structure in solution by vacuum ultraviolet circular dichroism, J. Mol. Biol., 138:149-178.

Castellani, A., and Quercia, I. F., 1979, "Synchrotron radiation applied to biophysical and biochemical research", Plenum Press, New York.

Ito, T., Kobayashi, K., and Ito, A., 1980, Effects of broad-band vacuum-UV synchrotron radiation on wet yeast cells, Radiation Research, 82:364-373.

Johnson, W. C. and Tinoco, I., 1972, Circular dichroism of polypeptide solutions in the vacuum ultraviolet, J. Am. Chem. Soc., 94:4389-4390.

Johnson, W. C., Jr., 1978, Circular dichroism spectroscopy and the vacuum ultraviolet region, Ann. Rev. Phys. Chem. in the press.

Munro, I., Pecht, I., and Stryer, L., 1979, Subnanosecond motions of tryptophan residues in proteins, Proc. Natl. Acad. Sci. U. S., 76:56-60.

Pysh, E. S., 1976, Optical activity in the vacuum ultraviolet, Ann. Rev. Biophys. Bioengr., 5:63-75.

Shulman, R. G., Eisenberger, P. and Kincaid, B. M., 1978, X-ray absorption spectroscopy of biological molecules, Ann. Rev. Biophys. Bioeng., 7:559-578.

Snyder, P. A. and Rowe, E. M., 1980, The first use of synchrotron radiation for vacuum ultraviolet circular dichroism measurements, Nuclear Instrumentation and Methods, in press.

Sprecher, C. A., Baase, W. A. and Johnson, W. C., Jr., 1979, Conformation and circular dichroism of DNA, Biopolymers, 18:1009-1019.

Wright, F., 1980, "Ultrasoft x-ray microscopy," Annl. N. Y. Acad. Sci., 342.

PHOTOACOUSTIC METHODS APPLIED TO BIOLOGICAL SYSTEMS

David Cahen, Gerard Bults, S. Roy Caplan, Haim Garty

and Shmuel Malkin

The Weizmann Institute of Science, Rehovot, Israel

When intensity-modulated light is absorbed by a sample contained in a constant volume cell filled with gas or liquid, the resulting periodic heating and cooling of the sample will cause, after transfer to the surrounding medium, periodic pressure changes in the cell. These changes can be detected by microphones or any pressure transducer and will, after frequency specific ("lock-in") amplification, yield a signal proportional to the heat produced by the light absorption. Fig. 1 illustrates this effect for chromophores contained in a liquid or solid phase. If the modulating frequency is an acoustic one, the pressure waves, having the same frequency, will thus be acoustic waves. This explains the name of the effect, first described a century ago by A.G. Bell, "photoacoustic", or "optoacoustic". Since the revival of this method, mainly by workers from Bell Labs. (Robin, and Rosencwaig, in ref. 1), great emphasis has been put on photoacoustic spectroscopy (PAS), for the optical study of (semi) solids and liquids. Because only radiation that, in fact, is absorbed by the sample contributes to the PA signal, the technique can be used to study highly scattering samples, often as an alternative to diffuse reflection. Results on biological suspensions, tissues and leaves have been reported. It should be borne in mind, though, that quite often, if only information on optical properties is desired, conventional reflectance measurements can be used as well.

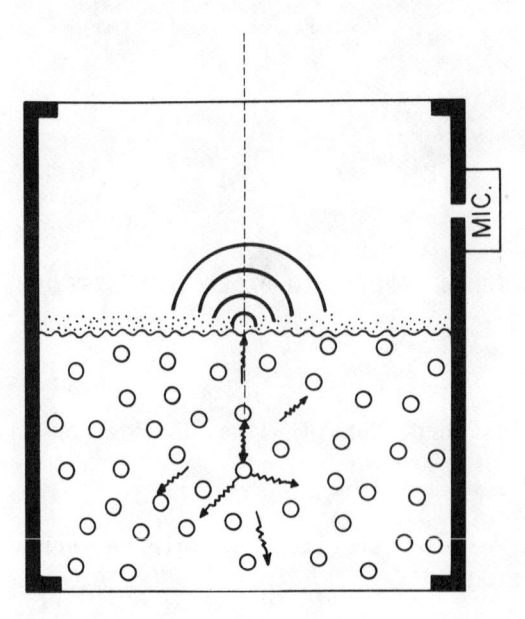

Fig. 1

Schematic description of the PA effect. Circles indicate absorbing material (here shown as e.g. a suspension). Arrows show the directions of heat diffusion. Part of the heat reaches the interface of the sample with the acoustic coupling medium (e.g. air) and causes, through microthermal changes in the boundary layer (points) pressure changes in the cell, which are sensed by a transducer (MIC : microphone).

Because the PA signal depends a) on that amount of radiation that is converted into heat within the time scale of the experiment and b) on the thermal properties of the sample, it is possible to extract calorimetric information on a photoprocess involving the sample, such as quantum yield of product formation and enthalpy changes between reactants and products. Also information on the sample thermal properties can be obtained, which in turn can yield data of direct biological relevance, such as hydration parameters.

After a short description of the experimental set-up and theoretical background, we will give examples of some of these uses of the PA technique.

Fig. 2 shows a schematic illustration of a PA set-up, single beam in terms of detection, with possibility for double beam sample excitation. With such a set-up both wavelength and modulation frequency spectra can be measured. Rather intense light sources are needed to obtain acceptable PA S/N values (e.g. 450 W Xe - arc).

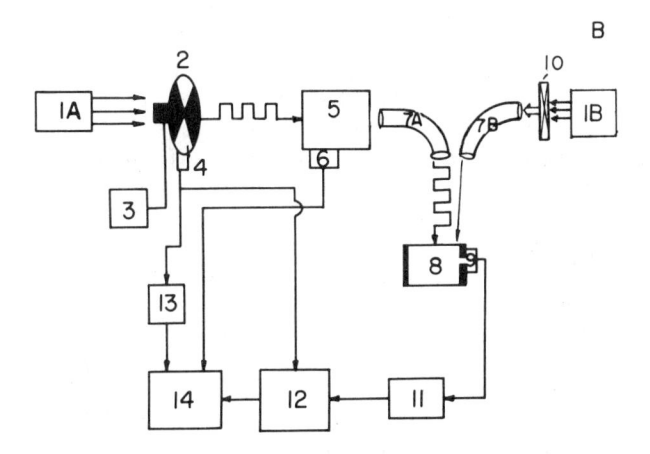

Fig. 2

PA set-up. 1A, 1B : light sources ; 2,3 : mechanical chopper + driving mechanism ; 4 : source for reference signal to lock-in amplifier (12), and frequency to voltage converter (13) ; 5,6 : monochromator + driving mechanism ; 7A, 7B : light-guides for modulated and continuous illumination ; 8,9 : PA cell + transducer ; 10 : shutter and/or filters ; 11 : preamplifier ; 14 : recorder.

The mechanical chopper can be replaced by direct electronic modulation of the lamp power supply to minimize vibrations (though, audio frequency electrical noise may be introduced then). The double illumination is useful for e.g. excited state spectra and to saturate photochemical activity (vide infra). A variety of PA cells have been described in the literature. If non-resonant types are used, the ratio of the illuminated area to volume of the acoustic coupling gas should be minimized (2). All the experiments described here have been carried out using an air-filled cell with 50% duty cycle modulated excitation. Systems using completely liquid-filled cells, or samples attached directly to a piezoelectric transducer, have been described (3). Also short light pulses have been used to improve S/N and to measure fast processes (3, 4).

An optically excited sample may undergo any of several deexcitation processes, e.g. luminescence, photochemistry, and radiationless decay. The PA signal originates from this last process only, and we will ignore, for the moment, the other decay routes. Then the PA signal will be some function of the amount of absorbed light energy. The final PA signal will depend, in general, on the depth within the sample where heat generation occurs, the sample bulk thermal properties and the efficiency of heat transfer from the sample to the surrounding medium. Only heat generation occuring within roughly one thermal diffusion length (µs) will yield a PA signal detected by the phase sensitive, "lock-in" detector. The optical absorption length (reciprocal of the optical absorption coefficient, β), μ_β will, in general limit the thickness of the sample layer participating in the PA signal generation. μ_s depends on the modulation frequency (as $1/\sqrt{\omega}$) and on the bulk thermal properties of the sample. Depending on the relative values of μ_β and μ_s, and on the sample thickness, Rosencwaig and Gersho (5) distinguished several special cases. To be able to obtain PA spectra that represent the wave-length dependence of β, $\mu_\beta > \mu_s$ (the case for fig. 3). However when μ_β becomes comparable to μ_s, and especially once $\mu_\beta < \mu_s$, the PA signal wavelength dependence does not reflect the optical absorption spectrum accurately, and finally, becomes wavelength independent. We now have the case of PA saturation, i.e. all of the light is absorbed within μ_s. Because of the modulation frequency dependence of μ_s, increasing ω may avoid saturation ; however experimental difficulties, especially a decrease in S/N with increasing ω, limit this possibility.

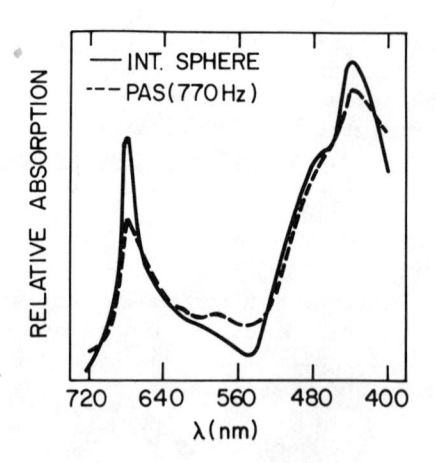

Fig. 3

Comparison between photoacoustic and reflectance spectra of broken lettuce chloroplasts. The reflectance spectrum was obtained with an integrating sphere. The spectra have been scaled arbitrarily. The large difference at ∼680 nm is due to "photochemical loss" in the PA signal (see text).

OPTICAL SPECTRA

In fig. 3 we show the PA spectrum of chloroplasts and compare it with the absorption spectrum, obtained from reflectance measurements. Because of the occurence of photochemistry, part of the absorbed radiation is not converted into heat and the PA signal is reduced accordingly. This effect is felt especially around the ~680 nm peak where the quantum efficiency for photosynthesis is optimal (6).

Fig. 4 shows how PA spectra can yield information on steady-state populations of excited states, to which the chromophore is excited by illumination with an additional, non-modulated beam of light.

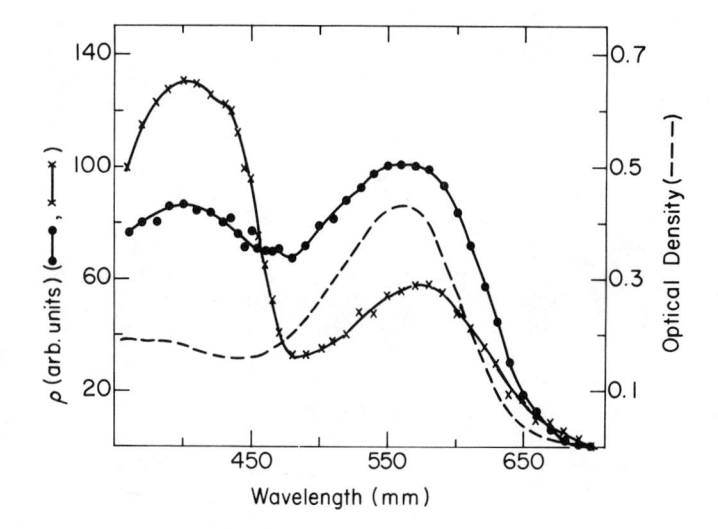

Fig. 4

PA spectra of films of purple membrane fragments of H. halobium dried at 63% relative humidity. Dots represent PA signals (ρ) obtained with modulated light only ; crosses show PA signals obtained in the presence of strong, >460 nm, background light. The absorption spectrum of the film (dashed line) is shown for comparison.

Such additional illumination will not interfere with the PA measurement (only modulated signals are sensed) and no detector saturation occurs, enabling side-illumination also at the same wavelength as that of the modulated beam. Because the photocycle in H. halobium is slowed down upon drying, dried samples show very drastic spectral changes during strong side-illumination.

Actually, fig. 4 shows that, even with modulated excitation only, a measurable excited state population is obtained (M_{412}). But in the presence of continuous illumination, the M_{412} species becomes the main absorbing one, while the 570 nm species is strongly depleted. Such spectra are similar to those obtained by more conventional techniques at low temperatures, as long as care is taken to avoid photoacoustic saturation. Also the decay time of the excited state can be measured photoacoustically, by observing the PA signal time dependence upon interrupting the side-illumination (7).

PHOTOCALORIMETRY

As noted above, the PA signal will be reduced if not all the absorbed radiation is converted into heat, but, for example, drives a photochemical reaction. We called this reduction "photochemical loss" (8). Because such a process of photoactivity will be wavelength dependent, the PA wavelength spectrum will be affected only in those regions where the quantum yield for photoactivity is appreciable. If we now take the ratio of the PA spectrum to that of the absorbed radiation (obtainable from normal absorption spectra) we get the wavelength dependence of heat-dissipation and such data can provide information on the amount of energy stored in photochemistry, or on the quantum yields. Fig. 5 shows such spectra for H. halobium preparations.

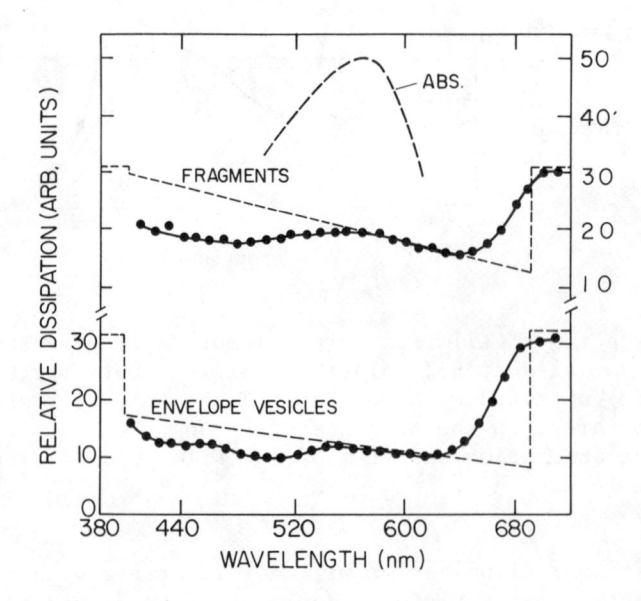

Fig. 5

Heat dissipation wavelength dependence of H. halobium preparations (i.e. PA signal/absorbed light intensity). Straight dashed lines show theoretical heat dissipation spectra if $\Sigma\phi \cdot \Delta E$ is constant between 420 and 690 nm and zero at other wavelengths.

We see that between 540 and 640 nm the relative dissipation is a
linear function of wavelength, indicating a constant amount of
energy stored in the photoprocess. The increase in dissipation at
higher wavelengths indicates a decreased efficiency for the
photoprocess. The low dissipation at shorter wavelengths is not
well understood at the moment, but could be caused by direct
absorption by photointermediates. From extrapolation of such
spectra, values for stored energy can be obtained (e.g. 5% energy
storage of 600 nm photons at 10 Hz modulation frequency).

An alternative way for obtaining such data is provided by
modulation frequency spectra. This is illustrated in fig. 6.

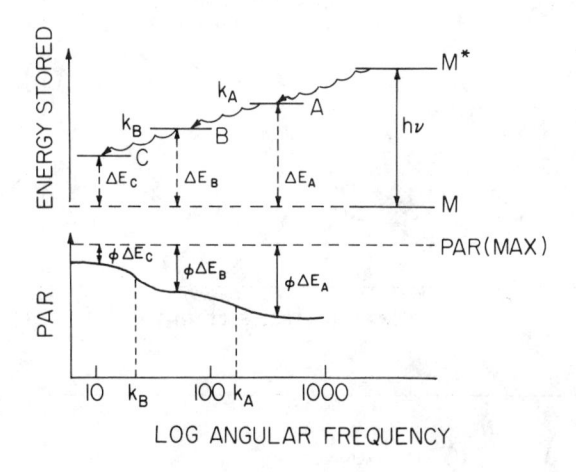

Fig. 6

Scheme showing the PA signal amplitude as function of modulation
frequency (ω) (bottom), for simple photochemical reaction involving
intermediates A, B and C. For formation of A, a quantum yield ϕ is
assumed while all other steps occur with a quantum yield of unity.
Differences between the maximal PA signal, PAS (MAX), and those
shown by the curve, represent "photochemical losses".

Here we show how intermediates are "sensed" photoacoustically, if
they have a life time longer than, and a buil-up time shorter than
the reciprocal of the angular modulation frequency. Thus earlier,
shorter lived, intermediates are sensed at higher frequencies, and
later, longer lived ones at lower frequencies, while stable end
products will be sensed at frequencies approaching zero. Then the
steps in the frequency spectrum provide kinetic information, while
the "photochemical loss" gives information on the quantum yield
and enthalpy of the intermediates. Fig. 7 illustrates this for
chloroplasts. With $Fe(CN)_6^{3-}$ as electron acceptor one single step,

corresponding to an intermediate with ~0.5 msec decay time, is observed. Energy storage of the 686 nm light is 12% for this intermediate and 6.5% for the one into which it decays. With Methylviologen we observe, in addition, an increasing step at low frequencies which may be related to an endothermic step in the photosynthetic reaction chain (9).

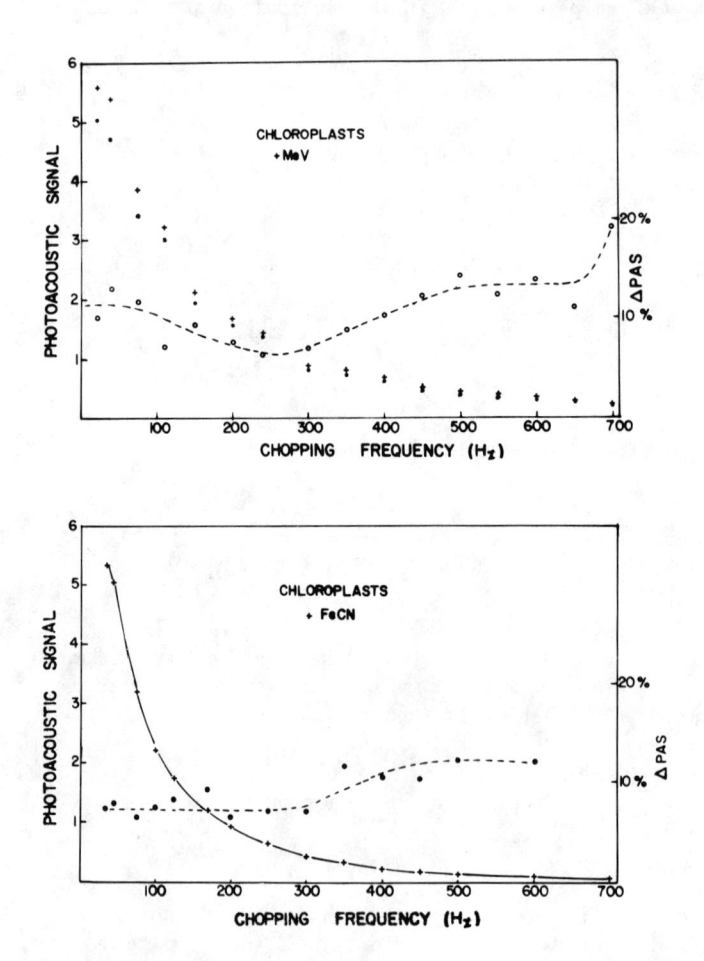

Fig. 7

Photoacoustic response (PAR) (+), and the relative change of PAR upon continuous side illumination (·), as function of modulation frequency for broken lettuce chloroplasts ; λ = 686 nm. Δ(PAR) signals corrected for cell and microphone frequency response and for cell background signal. Fe CN = K_3 Fe $(CN)_6$; MeV = methylviologen. The higher Δ(PAR), the more energy is stored in the system.

PHOTOACOUSTIC METHODS APPLIED TO BIOLOGICAL SYSTEMS   29

Similar experiments on whole R. rubrum cells, which were carried out to higher frequencies, show well over 30% energy storage for the first intermediates sensed, which is over 3 times more than what is stored at low frequencies, i.e. in later intermediates (fig. 8).

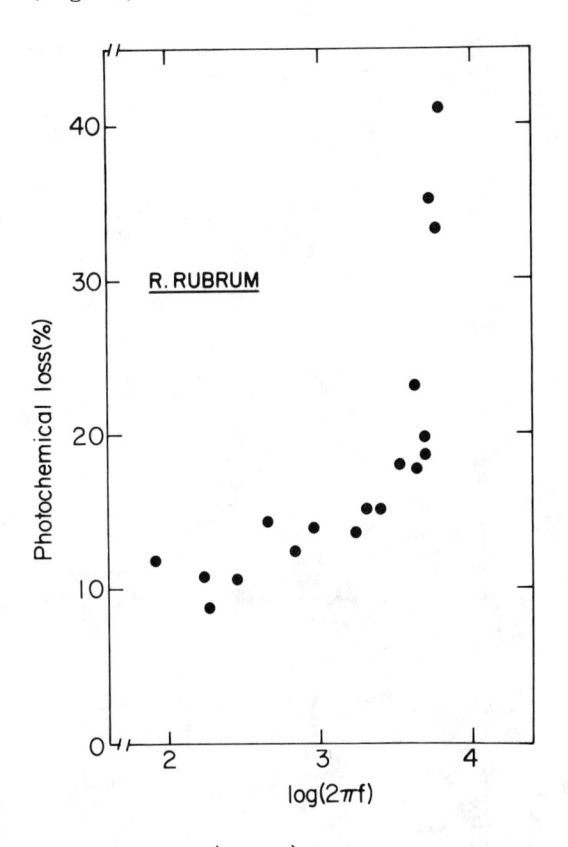

Fig. 8

Photochemical loss ($\sim \Delta$(PAR)) as function of modulation frequency for whole cells of R. rubrum at λ = 800 nm. Background light : λ > 700 nm at saturating intensity.

Because of the rather low light levels at which R. rubrum starts to become saturated photochemically, these data show considerable scatter. More refined experiments are presently in progress. For the data shown in figs. 7 and 8 the maximal PA signal, i.e. PAS(MAX) from fig. 6, was obtained by using a strong continuous actinic light beam to saturate the system photochemically. This is a convenient way of referencing (8,9), because the reference signal originates from the very sample under investigation, which assures similar thermal and passive optical properties (see however ref. 10).

Modulation frequency spectra of H. halobium preparations are shown in fig. 9A. Here bleached purple membranes, to which black ink was added to give them the same O.D. at 570 nm as the sample, were used as reference. The increase in signal down to 30 Hz reflects mainly the increase in μ_s, while the decrease below 30 Hz is due to decreasing microphone sensitivity. Fig. 9B shows that even at 400 Hz most of the absorbed light is dissipated as heat, something which correlates well with quantum yield data for this system. Furthermore we see at least 4 enthalpy changes, three of which are exothermic, and one of which, the last one, is distinctly endothermic. Data such as those shown in fig. 9B can be used to construct a scheme for the enthalpy changes that occur during the photocycle, and this is shown in fig. 10, where we also identify, tentatively, some of the intermediates (11).

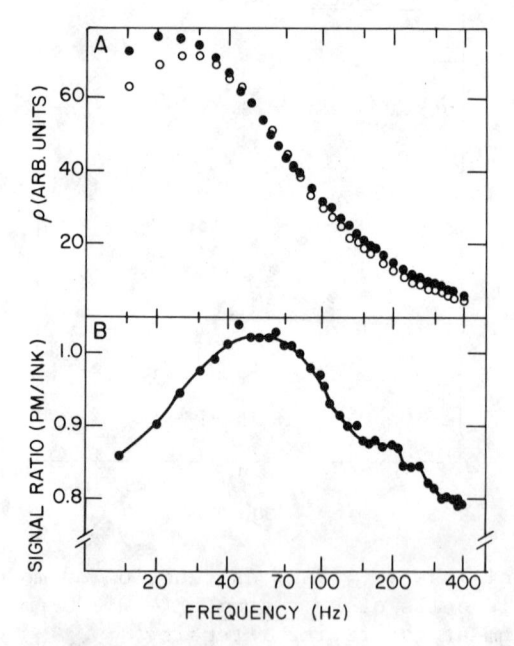

Fig. 9

(A) Modulation frequency dependence of PA signal (ρ) of H. halobium purple membrane fragments (o—o) and of bleached, ink-blackened membranes of equal density (•—•) at 570 nm.

(B) Ratio of signals from A. Up to ~5% error at high frequencies. Steps indicate decay of one intermediate and growth of subsequent one. Data at pH = 7. Change in pH, and especially addition of e.g. KCl, will greatly alter the frequency spectrum.

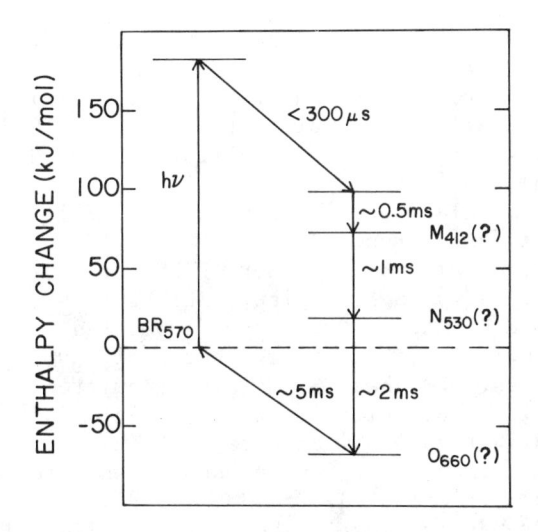

Fig. 10
Scheme of approximate changes in enthalpy during the photocycle of H. halobium purple membrane fragments, based on several experiments, such as that shown in fig. 9. The intermediate at ∿ +100kJ/mole is not detected by normal optical spectroscopy.

Although we have stressed here calorimetric applications of photoacoustics, the method can be and has been used for quite different purposes as well. These include tissue studies, depth profiles to determine the relative positions of chromophores, and the use of the PA signal dependence on the sample thermal parameters to determine e.g. hydration conditions of tissues (12).

In summary we should stress that, although, possibly, initial expectations of the application of photoacoustics to biological problems may have been too high, the method can offer additional unique insights into biological systems and problems, especially when used in conjunction with complimentary techniques, if its limitations are borne in mind.

Acknowledgement : Part of this research is supported by the U.S.-Israël Binational Science Foundation, Jerusalem, Israël.

References

1. Y.H. Pao, ed. "Optoacoustic spectroscopy and detection", Acad. Press, N.Y. (1977).
2. D. Cahen, H. Garty, Anal. Chem. 51, 1865 (1979).
3. C.K.N. Patel, A.C. Tam, Appl. Opt. 18, 3348 (1979) ; Appl. Phys. Lett. 34, 467 (1979) ; J.B. Callis, J. Res. N.B.S. A, 80A, 413 (1976).
4. J. Wrobel, M. Vala, Chem. Phys. 33, 93 (1978).
5. A. Rosencwaig, A. Gersho, J. Appl. Phys. 47, 64 (1976).
6. D. Cahen, S. Malkin, E.I. Lerner, FEBS Lett. 91, 339 (1978).
7. D. Cahen, H. Garty, S.R. Caplan, FEBS Lett. 91, 131 (1978) ; H. Garty, S.R. Caplan, D. Cahen, submitted.
8. S. Malkin, D. Cahen, Photochem. Photobiol. 29, 803 (1979).
9. N. Lasser-Ross, D. Cahen, S. Malkin, submitted ; see also J.B. Callis et al. Biochim. Biophys. Acta 267, 348 (1972).
10. Y. Inoue, A. Watanabe, K. Shibata, FEBS Lett. 101, 321 (1979).
11. H. Garty, D. Cahen, S.R. Caplan, submitted ; see also D. Ort, W.W. Parson, Biophys. J. 25, 355 (1979).
12. D. Cahen, Trends in Biochem. Sciences, 4, N240 (1979) ; D. Cahen, G. Bults, H. Garty, S. Malkin, J. Biochem. Biophys. Meth., in press.

DYNAMICS OF HEME PROTEINS

Hans Frauenfelder

Department of Physics
University of Illinois at Urbana-Champaign
Urbana, Illinois 61801

The interior of heme proteins is an excellent and nearly unique laboratory for the investigation of biomolecular dynamics. The investigation of ligand binding to heme proteins by flash photolysis over a wide range of time, temperature and pressure has yielded information on many features of proteins dynamics and has led to a deeper understanding of the relation between structure and function. The thrust of the studies is not a specific description of the binding process, but elucidation of the general features that underly protein reactions. In the present review I will try to briefly describe some of the main ideas. Details, specific results, and proofs can be found in the references.

1. INTRODUCTION AND APPROACH

Myoglobin (Mb) is a globular protein consisting of 153 amino acids, with molecular weight of about 17 200 daltons, that stores dioxygen (O_2) reversibly and also binds carbon monoxide (CO). The active center is a heme group and the binding occurs at its central iron atom. The covalent bond between the iron atom and the ligand can be broken by light. Since the optical spectrum of the free and the bound state differ, rebinding can be followed optically. The basic approach to the study of ligand binding thus is very simple, but the richness and variety of phenomena that can be investigated is unexpected. One reason is the wide range over which crucial parameters can be varied : rebinding after photodissociation can be studied from ps to ks, from 4 K to 300 K, and up to pressures of over 200 MPa. Moreover, ligand binding is sensitive to many features of the protein structure and can thus be used as a probe to explore the dynamic behavior.

The pioneering kinetic work has been performed by Gibson (1) ; his and later experiments are reviewed in the monograph by Antonini and Brunori (2). In the present survey I restrict the discussion to recent work by the Illinois group and to aspects that are of general interest. The flash photolysis set-up is described in ref. 3. In most experiments, Mb with bound ligand L (MbL) is placed into a cryostat at the chosen temperature T and photolyzed by a laser flash. The subsequent rebinding, Mb + L → MbL, is followed optically. A unique feature of the system is the transient analyzer with a logarithmic time base which records from 2 μs to 1 ks in a single sweep(4).

At temperatures below about 180 K, the protein is rigid and the flashed-off ligand remains within the pocket that surrounds the heme group. The studies below 180 K thus explore the immediate surrounding of the active center and provide information about processes that can only be studied with difficulties at higher temperatures. Above about 180 K, the ligand moves farther and farther away from the heme group and the protein begins to relax. Thus pathways to the protein interior, conformational relaxation, and the influence of the external solvent on protein motion can be investigated.

2. MOLECULAR TUNNELING

As stated above, the ligand does not leave the heme pocket below about 180 K. The situation thus is as sketched in Fig. 1 (5).

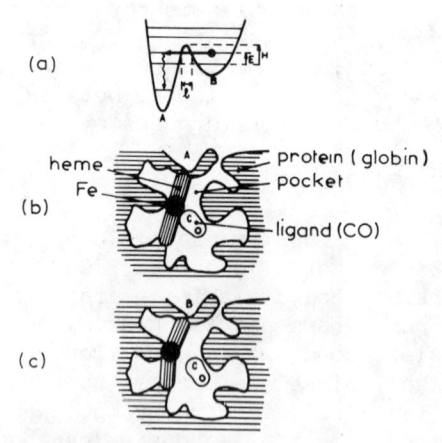

Fig. 1. The states involved in the tunneling of CO within myoglobin. (a) Tunneling occurs from the "free" state B to the bound state A. (b) In state A, the CO is bound to the heme iron, the heme group is planar, and the iron lies in the heme plane. (c) After photodissociation, the Fe-CO bond is broken, the CO molecule has moved away from the iron into the pocket, the iron has moved out of the heme plane, and the heme is domed. (From) Ref. 5)

Table 1. CO stretching frequencies in myoglobin

State	Binding	$\nu(cm^{-1})$ ^{13}CO	Binding enthalpy
Myoglobin A	covalent to Fe	1901	90 kJ mol^{-1}
Myoglobin B$_2$	dipole-dipole (?)	2071	1.5
Myoglobin B$_1$	dipole-dipole	2082	1
Myoglobin B$_0$	"free" (VdWaals)	2096	0
CO gas	--	free	2096	-

4. CONFORMATIONAL SUBSTATES

The rebinding B → A in the Arrhenius region above about 80 K yields clear evidence for the complexity of protein structure. If the barrier between wells B and A in Fig. 1a had a unique height, H, rebinding would be described by a single exponential in time, $N(t) = N(0) \exp(-kt)$, with the rate coefficient k satisfying the Arrhenius equation $k = A \exp(-H/RT)$. N(t) is the fraction of Mb molecules that have not rebound CO at the time t after the photo-dissociation, and $R = 8.3$ J mol^{-1}K^{-1} is the gas constant. When we first measured N(t) carefully, we were surprised to find that it deviates very much from an exponential and, in fact, can be approximated by a powerlaw, $N(t) = N(0) (1 + t/t_0)^{-n}$ where n and t_0 are two temperature-dependent parameters (14). The observed time dependence can be explained by assuming that the activation enthalpy H of the barrier between B and A is not sharp, but given by a distribution g(H), where g(H)dH denotes the probability of finding a Mb molecule with barrier height between H and H + dH. From the observed N(t), the distribution function g(H) can be found and it turns out to be characteristic of a particular heme protein (10).

The existence of an activation enthalpy spectrum at low temperatures agrees with what is known about protein structure (15,16) : a protein consists typically of 100-200 amino acids. Each residues can, on the average, assume a few (say 2) states of equal energy. The entire protein thus is expected to possess of the order of 2^{100} to 2^{200} conformational substates, states of about equal energy and approximately equal overall structure. At low temperatures, each protein molecule will be frozen into a particular substate. At sufficiently high temperatures, conformational relaxation will occur and each protein molecule will rapidly move from one substate to another. Rebinding then becomes exponential (3).

The indirect observation of conformational substates in low-temperature ligand binding raises the question if the spatial distribution of these states can also be observed. X-ray diffraction

In state A, the ligand (CO) is bound to the heme iron, the heme group is planar, and the iron lies in the heme plane. The laser flash breaks the Fe-CO bond, the CO moves into the heme pocket, the iron moves out of the heme plane, and the heme becomes domed (state B). The situation is idealized in Fig. 1a. Rebinding corresponds to the transition B → A. Between 80 and 160 K, the rebinding process satisfies an Arrhenius relation and consequently is interpreted as an over-the-barrier transition (3). Below 80 K, rebinding is faster than the extrapolation of the data above 80 K predicts and we interpret the deviation from the classically expected behavior as quantum-mechanical tunneling through the barrier (6-11). In the limit T → 0, the transition rate should become temperature independent and the data obtained with some heme proteins indeed show that this criterion is satisfied (9). At first it appears impossible that a heavy system such as CO should tunnel. However, tunneling occurs because the barrier height H and the tunneling distance l (Fig. 1a) are sufficiently small. Moreover, with time-resolved Fourier transform infrared spectroscopy we have studied the isotope effect by comparing the tunneling of $^{12}C^{16}O$ with both $^{13}C^{16}O$ and $^{12}C^{18}O$ (5). Two main results emerge : (i) An isotope effect of about the expected magnitude exists and the heavier isotope rebinds slower than the light one. (ii) The structure of the system is important for tunneling ; $^{12}C^{18}O$ rebinds faster than $^{13}C^{16}O$ even though it is heavier. These results imply that detailed studies of molecular tunneling in different proteins may yield new information about tunneling and also give considerable insight into the last step of ligand binding.

3. STATES IN THE POCKET

The stretching frequency of the CO molecule is a sensitive probe that indicates how CO is bound (12). Fig. 1 consequently suggests a new approach to the study of the interior of a heme protein : by observing the stretching frequency of the CO in the photodissociated state B we can find out if and how CO is bound to the walls of the pocket. Indeed, such experiments are feasible. We have investigated the infrared spectrum of the bound and the photodissociated states of Mb-^{12}CO and Mb-^{13}CO from 12 to 100 K between 1800 and 2200 cm^{-1} (13). The photodissociated state shows three forms (B_0, B_1, B_2) with relative populations that depend on temperature. The temperature dependence yields binding enthalpies. Preliminary parameters for the B states are listed in Table I, together with corresponding values for the bound state A. For comparison, the free (gas) value is also listed. The data in Table I suggest the following interpretation : B_0 corresponds to CO moving essentially freely (but rotationally constrained) within the pocket. In B_1 and B_2, CO may be bound to the pocket walls by dipole-dipole forces. This preliminary experiment indicates that infrared studies of photodissociated states at low temperatures may be a rich source of information on states within proteins.

indeed permits the detailed exploration of spatial fluctuations in proteins (17-19). The basic idea is straightforward. If each atom in a protein crystal occupied its "ideal" position and did not move, the diffraction spots would show their "ideal" intensities. In reality, each atom i is spread over a distance approximately characterized by $<x_i^2>^{1/2}$ and this spread leads to decreased spot intensities (Debye-Waller factor). From the observed intensities, the mean-square displacements for all non-hydrogen atoms of myoglobin (17,19) and lysozyme have been determined. In many parts of the proteins, the observed displacements are much larger than expected if the atoms vibrated around a unique position. The discussion of the displacements yield surprisingly detailed information and provides very strong support for the concept of conformational substates.

5. MULTIPLE BARRIERS AND STOCHASTIC THEORY

The topics treated so far have been related to the low-temperature experiments and consequently yielded information about the protein interior. At higher temperatures, rebinding after flash photolysis involves the entire reaction path, from solvent to the final binding site at the heme iron (3). Fig. 2 shows as example N(t), the fraction of Mb molecules that have not rebound O_2 at the time t after photodissociation with a 30 ns laser at 280 K in three different solvents (20). Three different processes can be discerned ; the two faster ones (I and II) are independent of the O_2 concentration in the solvent, III is proportional to it. Experiments at lower temperatures prove that process I persists to well below 180 K and becomes identical to the transition B → A shown in Fig. 1a.

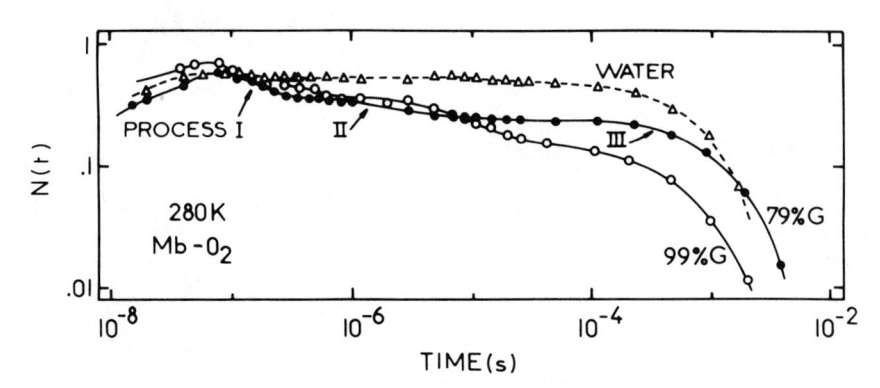

Fig. 2. Binding of dioxygen to myoglobin at 280 K after flash photolysis. N(t) is the fraction of Mb molecules that have not rebound O_2 at time t after photodissociation. G denotes glycerol-water solvents (% glycerol by weight). The rise at the beginning shows the photodissociation during the 30 ns laser pulse. (After ref. 20).

We interpret the data by postulating that O_2, on its way from the solvent to the binding site, must surmount three barriers in succession. The outermost barrier could be between solvent and protein, the second between the outer protein parts and the pocket, and the third at the heme as discussed in Section 2.

The conventional treatment of the migration of a ligand molecule over the multiple barriers is based on two assumptions (3) : (i) At any given temperature, the migration can be described by a set of deterministic linear rate equations, (ii) The enthalpy barriers are temperature independent and the rate coefficients for surmounting the barriers satisfy Arrhenius relations. With these two assumptions, the multiple-barrier model describes a wide range of processes satisfactorily ; it accounts in particular also for the occurence of a fast "geminate" recombination at around 300 K (21,22). This fast component is denoted by I in Fig. 2.

Despite the success of the multiple-barrier model with assumptions (i) and (ii), both assumptions must be generalized. Consider first (i). If the ligand concentration in the solvent is so large that the probability of finding two or more ligand molecules within one protein becomes appreciable, linear rate equation are no longer adequate. The stochastic approach provides the desired generalization (23). Assumption (ii) is inconsistent with the dynamic protein structure sketched in Section 4. If the protein is indeed a dynamic, breathing system, then it is unlikely that the potential barriers are temperature independent. In the following section, this inconsistency will be resolved.

6. VISCOSITY AND PROTEIN DYNAMICS

If proteins are dynamic systems that continuously fluctuate among very many conformational substates, then the solvent surrounding the protein should influence these fluctuations. The solvent viscosity, in particular, should affect protein reactions. Fig. 2 indeed shows that the binding of O_2 to myoglobin depends on solvent. How can this dependence be described and understood ? Two problems have to be faced : What is the proper equation for the rate parameter k in a medium with viscosity η ? How can the change of solvent viscosity with temperature be taken into account in the evaluation of protein reactions ? Both of these problems have clear answers which will now be sketched.

The temperature dependence of reactions is usually evaluated with the help of transition state theory. In this theory, the influence of the surrounding is not taken into account explicitly and it therefore does not solve the first problem. A solution was given in 1940 by Kramers, who considered the escape of a particle from a potential well as a function of the coupling to a heat bath (24,25). In the strong-coupling (high-damping) limit, which applies to nearly all solvents, he obtained for the rate coefficient for escape over a barrier with height H

$$k = A\eta^{-1} e^{-H/RT}.$$

The solvent viscosity η plays the role of the coupling coefficient to the heat bath and appears explicitly. For a protein the situation is complicated by the fact that the internal viscosity is not known, may vary from position to position, and also differ from the solvent viscosity. We assume that the internal viscosity is related to the solvent viscosity by $\eta_{internal} = \eta^x$ and then write for the rate coefficient for a raction in a protein

$$k = A\eta^{-x} e^{-H/RT}.$$

To answer the second question we measure a particular protein reaction in a wide variety of solvents with widely differing viscosities, but keep all other parameters (pH, permittivity...) constant. The reaction is then characterized by a rate coefficient $k(T,\eta)$ that is a function of the two variables T and η. If the dominant solvent parameter is the viscosity, we can find the activation enthalpy characteristic of the protein alone by plotting log $k(T,\eta)$ versus 1/T for constant viscosity. The parameter x is found from a graph of log k versus log η at constant temperature.

Flash photolysis experiments with Mb-O_2, Mb-CO, and bacteriorhodopsin show that the approach just described can characterize data in a wide range of solvents in a unified way, that the generalized Kramers equation fits the data better than transition state theory, and that reactions inside proteins are to a large extent controlled by the solvent viscosity (20,26,27).

The examples treated in the present survey -molecular tunneling, states within the heme pocket, conformational substates, conformational relaxation, multiple barriers, influence of solvent viscosity- demonstrate that flash photolysis and related experiments on a simple problem, namely ligand binding, can provide deep insight into protein dynamics.

ACKNOWLEDGEMENTS

This work was supported in part by the U.S. Department of Health, Education and Welfare under Grant GM 18051 and the National Science Foundation under Grant PCM 79-05072. I should like to thank my collaborators, who are listed in the papers from the Illinois group, for their stimulation and their dedicated efforts.

REFERENCES

1. Q.H. Gibson, J. Physiol. 134:112 (1956).
2. E. Antonini and M. Brunori,"Hemoglobin and Myoglobin in their Reactions with Ligands", North-Holland Publ. Co., Amsterdam (1971).
3. R.H. Austin, K.W. Beeson, L. Eisenstein, H. Frauenfelder and I.C. Gunsalus, Biochemistry 14:5355 (1975).
4. R.H. Austin, K.W. Beeson, S.S. Chan, P.G. Debrunner, R. Downing, L. Eisenstein, H. Frauenfelder and T.M. Nordlund, Rev. Sci. Instrum. 47:445 (1976).
5. J.O. Alben, D. Beece, S.F. Bowne, L. Eisenstein, H. Frauenfelder, D. Good, M.C. Marden, P.P. Moh, L. Reinisch, A.H. Reynolds and K.T. Yue, Phys. Rev. Letters 44:1157 (1980).
6. F. Hund, Z. Phys. 43:805 (1927).
7. B. Chance et al., "Tunneling in Biological Systems", Academic, New-York (1979).
8. V.I. Goldanskii, Chem. Scr. 13: 1 (1978-79) and Nature (London) 279:109 (1979).
9. N. Alberding, R.H. Austin, K.W. Beeson, S.S. Chan, L. Eisenstein, H. Frauenfelder and T.M. Nordlung, Science 192:1002 (1976).
10. N. Alberding, S.S. Chan, L. Eisenstein, H. Frauenfelder, D. Good, I.C. Gunsalus, T.M. Nordlund, M.F. Perutz, A.H. Reynolds and L.B. Sorensen, Biochemistry 17:43 (1978).
11. H. Frauenfelder, in : "Tunneling in Biological Systems", Chance et al. ed., Academic, New York, pp. 627-649 (1979).
12. J.O. Alben and W.S. Caughey, Biochemistry 7:175 (1968). Note, however, that the structure Fe-O-C favored in this paper is not correct.
13. J.O. Alben et al., to be published.
14. R.H. Austin, J. Beeson, L. Eisenstein, H. Frauenfelder, I.C. Gunsalus and V.P. Marshall, Phys. Rev. Letters 32:403 (1974).
15. G. Careri, P. Fasella and E. Gratton, Crit. Rev. Biochem. 3:141 (1975).
16. A. Cooper, Proc. Nat. Acad. Sci. USA 73:2740 (1976).
17. H. Frauenfelder, G.A. Petsko and D. Tsernoglou, Nature 280:558 (1979).
18. P.J. Artymiuk, C.C.F. Blake, D.E.P. Grace, S.J. Oatley, D.C. Phillips and M.J.E. Sternberg, Nature 280:563 (1979).
19. H. Frauenfelder and G.A. Petsko, Biophys. J., in press.
20. D. Beece, L. Eisenstein, H. Frauenfelder, D. Good, M.C. Marden, L. Reinisch, A.H. Reynolds, L.B. Sorensen and K.T. Yue, in print.
21. B. Alpert, S. Mohsni, L. Lindqvist and F. Tfibel, Chem. Phys. Letters 64:11 (1979).
22. D.A. Duddel, R.J. Morris and J.T. Richards, Biochim. Biophys. Acta 621:1 (1980).
23. N. Alberding, H. Frauenfelder and P. Hänggi, Proc. Nat. Acad. Sci. USA 75:26 (1978).
24. H.A. Kramers, Physica 7:284 (1940).

25. J.L. Skinner and P. Wolynes, J. Chem. Phys. 69:2143 (1978).
26. D. Beece, L. Eisenstein, H. Frauenfelder, D. Good, M.C. Marden, L. Reinisch, A.H. Reynolds, L.B. Sorensen and K.T. Yue, Biochemistry, in press.
27. D. Beece, S.F. Bowne, J. Czégé, L. Eisenstein, H. Frauenfelder, D. Good, M.C. Marden, J. Marque, P. Ormos, L. Reinisch and K.T. Yue, Photochem. Photobiol., in press.

MEASUREMENT OF ROTATIONAL DIFFUSION OF MEMBRANE PROTEINS USING OPTICAL PROBES

Richard J. Cherry

Laboratorium für Biochemie
ETH-Zentrum
CH-8092 Zürich
Switzerland

INTRODUCTION

Optical spectroscopy has been used for many years for investigating the structure and function of biological systems. In particular, motional properties have frequently been studied by observing fluorescence depolarization of probe molecules. However, because of the short lifetime of fluorescence, such measurements are confined to the nanosecond time range. More recently, it has become possible to measure much slower motion through the introduction of "triplet" probes. Rotation is measured by observing either transient dichroism (Cherry and Schneider, 1976; Cherry et al., 1976) or phosphorescence depolarization (Austin et al., 1979; Moore et al., 1979) following excitation of the triplet state by a linearly polarized light pulse. Because of the long lifetime of the triplet state, rotational relaxation times in the microsecond-millisecond time range can be determined. These probes are currently being used to investigate rotational motion of proteins embedded in biological membranes (for recent review, see Cherry, 1979). In addition to triplet probes, a few membrane proteins possess intrinsic chromophores where excitation produces a long-lived spectroscopic species. Rotation of these proteins can also be investigated by transient dichroism. This paper summarizes our recent studies of protein rotation in membranes, using both triplet probes and intrinsic chromophores.

METHODS

The flash photolysis apparatus used for measuring transient dichroism is described in detail elsewhere (Cherry, 1978). Excitation is by a vertically polarized flash of duration 1-2 us from a flashlamp-pumped dye laser. Absorption changes $A_\parallel(t)$, $A_\perp(t)$ for light polarized parallel and perpendicular to the polarization of the exciting flash are recorded simultaneously. The absorption anisotropy at time t after the flash is calculated from the expression:

$$r(t) = \frac{A_\parallel(t) - A_\perp(t)}{A_\parallel(t) + 2A_\perp(t)} \quad (1)$$

Normally signal averaging is used to improve the signal to noise ratio.

BACTERIORHODOPSIN

Bacteriorhodopsin (BR) contains an intrinsic chromophore, retinal, which in the light adapted state is in the all-*trans* form. Following excitation, a decrease in absorption is observed at 570 nm due to BR entering the photochemical cycle. Transient dichroism of these depletion signals may be used to measure rotational motion of the protein.

We have measured the rotation of BR in vesicles reconstituted from detergent solubilized purple membrane and added phosphatidylcholine (Heyn et al., 1977; Cherry et al., 1978). Typical r(t) curves obtained with such vesicles are shown in Fig. 1. At temperatures below the T_c (temperature of lipid gel to liquid-crystalline phase transition), only a very slow decay of r(t) attributable to vesicle tumbling is observed. Immobilization of BR occurs as a result of recrystallization into a hexagonal lattice indistinguishable from that exisiting in the native purple membrane (Cherry et al., 1978). Above the T_c, the decay of r(t) indicates that BR is mobile and hence that the proteins are disaggregated. Provided the lipid: protein mole ratio is greater than about 80, the r(t) curve above the T_c is in good agreement with the following

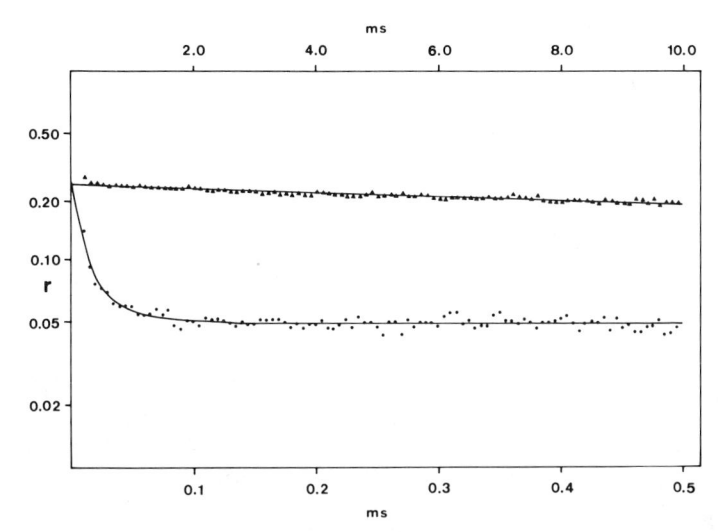

Fig. 1. r(t) curves calculated from 570 nm depletion signals from BR-dimyristoylphosphatidylcholine vesicles above and below lipid phase transition. ▲ 9°C (upper time scale); ● 25°C (lower time scale) The slight decay at 9°C is due to vesicle rotation.

theoretical equation derived by assuming that rotation of BR only occurs about an axis normal to the plane of the membrane:

$$r(t) = \frac{r_o}{A_1 + A_2 + A_3} [A_1 \exp(-t/\phi_\parallel) + A_2 \exp(-4t/\phi_\parallel) + A_3] \quad (2)$$

where $A_1 = [6/5][\sin^2\theta\cos^2\theta]$, $A_2 = [3/10][\sin^4\theta]$, $A_3 = [1/10][3\cos^2\theta - 1]^2$, r_o is the experimental anisotropy at t = 0, θ is the angle between the transition dipole moment and the membrane normal and $\phi_\parallel$ is the relaxation time for rotation around the membrane normal. $\phi_\parallel$ is in the range 10- 200 µs, the precise value depending on composition and temperature. θ is found to be $78 \pm 3°$ in reasonable agreement with the value $71 \pm 4°$ determined from polarized absorption spectra of oriented purple membranes (Heyn et al., 1977).

CYTOCHROME OXIDASE

Cytochrome oxidase is the terminal enzyme of the respiratory chain of the inner mitochondrial membrane. Rotational motion may in this case be measured by detecting transient dichroism after photolysis of the heme a_3-CO complex. Previously Junge and Devault (1975) observed no decay of flash-induced transient dichroism for cytochrome oxidase in mitochondrial membranes. Taken together with ESR and optical studies of Erecińska et al (1977) which indicate that the heme a_3 plane is perpendicular to the plane of the membrane, these experiments appear to show that cytochrome oxidase is immobilized. In contrast, Höchli and Hackenbrock (1979) have observed lateral diffusion of cytochrome oxidase and other integral membrane proteins using freeze-fracture electron microscopy.

In order to resolve the above conflicting observations, we have made transient dichroism measurements with improved signal to noise ratio (Kawato et al., 1980). We find that a decay of the absorption anisotropy does occur, the shape of the $r(t)$ curve being qualitatively similar to that seen with BR above the T_c. The relaxation time is in the order of 300-400 µs. However, the value of the residual anisotropy at long time is considerably higher than predicted for orientation of the heme plane perpendicular to the plane of the membrane. On this basis, it is deduced that about 40% of cytochrome oxidase is immobile ($\phi_\parallel > 30$ ms).

ERYTHROCYTE MEMBRANES

Band 3, the anion transport system of the human erythrocyte membrane, may be selectively labeled with the triplet probe eosin maleimide (Nigg and Cherry (1979). In Fig. 2, curve I shows $r(t)$ measured at 37°C from transient dichroism of ground state depletion signals. In contrast to BR above the T_c, the $r(t)$ curve for band 3 does not decay to a constant residual anisotropy as predicted by equ. (2). Curve fitting analysis indicates the presence of a fairly rapidly decaying component (decay time ~200 µs) and a slowly decaying component (decay time ~4 ms).

Recently we have obtained evidence that at 37°C, the slowly decaying component is largely due to a population of band 3 whose rotational motion is restricted by interactions with cytoskeletal proteins (Nigg and Cherry, 1980). Curve III in Fig. 2 shows $r(t)$ measured after treatment

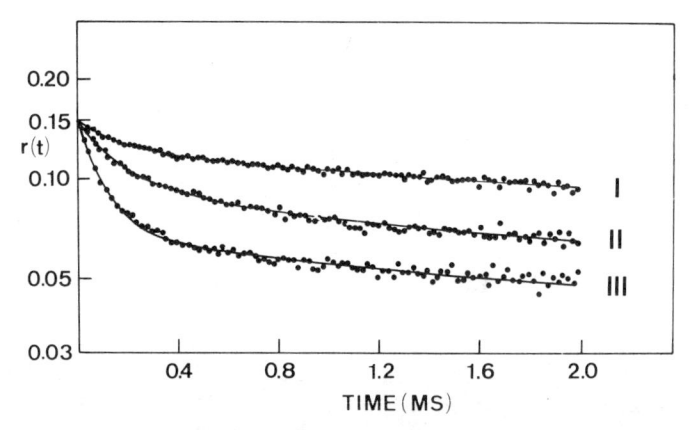

Fig. 2 r(t) curves for eosin maleimide-labeled band 3 in erythrocyte membranes at 37°C. Membranes were in 5 mM phosphate buffer pH 7.4 containing 10% Kollidon 90 to eliminate effects due to tumbling of small vesicles produced by salt and trypsin treatments. Curve I: control. Curve II: ghosts subjected to low salt/high salt extraction procedure similar to that described by Tyler et al (1979). Curve III: ghosts (2 mg membrane protein /ml) incubated with trypsin (0.5 μg/ml) for 1 hr at 22°C in 5 mM phosphate buffer pH 7.4.

of the membranes with trypsin at low concentrations. This procedure selectively removes the 40,000 dalton segment of band 3 which protrudes from the cytoplasmic side of the membrane. Curve-fitting analysis shows that after trypsin treatment, a further 40% of band 3 contributes to the rapidly decaying component of r(t) with relatively little change in the decay time. A qualitatively similar enhancement of band 3 rotational mobility is seen after a sequential low salt/high salt extraction which removes most of the cytoskeletal proteins (Curve II). From these results we conclude that up to 40% of band 3 proteins have restricted rotational mobility due to interaction of their cytoplasmic segment with the erythrocyte cytoskeleton. These experiments provide a good illustration of the usefulness of protein rotation measurements for studying protein-protein associations in membranes.

Ca^{2+}-ATPase IN SARCOPLASMIC RETICULUM

We have labeled the Ca^{2+}-ATPase in sarcoplasmic reticulum vesicles with the probe iodoacetamidoeosin. As with band 3, rotation is measured by observing transient dichroism of eosin ground state depletion signals. In this case the absorption anisotropy measured at 20 µs after the flash is small at 37°C but increases considerably with decreasing temperature and upon fixation with glutaraldehyde (Bürkli and Cherry, 1980). The low value of the anisotropy at 37°C is due to the existence of a fast motion, which in part may be assigned to segmental motion of the protein. This internal flexibility of the ATPase may have considerable significance for its functional properties. At times longer than 20 µs, the r(t) curve decays to a constant level with a decay time which varies from about 90 µs at 0°C to about 40 µs at 37°C. This decay is assigned to rotation of the protein about the membrane normal.

ACKNOWLEDGEMENTS

I wish to thank A. Bürkli, Dr. R. Godfrey, Dr. S. Kawato, Dr. U. Müller, Dr, E. Nigg, E. Sigel and C. Zugliani for their contributions to the work described here. I am also grateful to the Swiss National Science Foundation for financial support.

REFERENCES

Austin, R.H., Chan, S.S. and Jovin, T., 1979, Rotational diffusion of cell surface components by time resolved phosphorescence anisotropy, Proc. Nat. Acad. Sci. USA, 76:5650.
Bürkli, A. and Cherry, R.J., 1980, Rotational motion and flexibility of Ca^{2+},Mg^{2+}-dependent ATPase in sarcoplasmic reticulum membranes (submitted).
Cherry, R.J. and Schneider, G., 1976, A spectroscopic technique for measuring slow rotational diffusion of macromolecules. 2: Determination of rotational correlation times of proteins in solution, Biochemistry, 15:3657.
Cherry, R.J., Bürkli, A., Busslinger, M., Schneider,

G. and Parish, G.R., 1976, Rotational diffusion of band 3 proteins in the human erythrocyte membrane, Nature, 263:389.

Cherry, R.J., 1978, Measurement of protein rotational diffusion in membranes by flash photolysis, Meth. Enzymol., 54:47.

Cherry, R.J., Müller, U., Henderson, R. and Heyn, M.P., 1978, Temperature-dependent aggregation of bacteriorhodopsin in dipalmitoyl- and dimyristoylphosphatidylcholine vesicles, J. Mol. Biol., 121:283.

Cherry, R.J., 1979, Rotational and lateral diffusion of membrane proteins, Biochim. Biophys. Acta, 559:289.

Ericińska, M., Blasie, J.K. and Wilson, D.F., 1977, Orientation of the hemes of cytochrome c oxidase and cytochrome c in mitochondria, FEBS Lett. 76:235.

Heyn, M.P., Cherry, R.J. and Müller, U., 1977, Transient and linear dichroism studies on bacteriorhodopsin:determination of the orientation of the 568 nm all-trans retinal chromophore, J. Mol. Biol., 117:607.

Höchli, M. and Hackenbrock, C.R., 1979, Lateral translational diffusion of cytochrome c oxidase in the mitochondrial energy-transducing membrane, Proc. Nat. Acad. Sci. USA, 76:1236.

Junge, W. and Devault, D., 1975, Symmetry, orientation and rotational mobility in the a_3 heme of cytochrome c oxidase in the inner membrane of mitochondria, Biochim. Biophys. Acta, 408:200.

Kawato, S., Sigel, E., Carafoli, E. and Cherry, R.J., 1980, Cytochrome oxidase rotates in the inner membrane of mitochondria and submitochondrial particles, J. Biol. Chem., 255:5508.

Moore, C., Boxer, D. and Garland, P., 1979, Phosphorescence depolarization and the measurement of rotational motion of proteins in membranes, FEBS Lett., 108:161.

Nigg, E.A. and Cherry, R.J., 1979, Influence of temperature and cholesterol on the rotational diffusion of band 3 in the human erythrocyte membrane, Biochemistry, 18:3457.

Nigg, E.A. and Cherry, R.J., 1980, Anchorage of a band 3 population at the erythrocyte cytoplasmic surface:protein rotational diffusion measurements, Proc. Nat. Acad. Sci. USA, in press.

Tyler, J.M., Hargreaves, W.R. and Branton, D., 1979, Purification of two spectrin-binding proteins: biochemical and electron microscopic evidence for site-specific reassociation between spectrin and bands 2.1 and 4.1, *Proc. Nat. Acad. Sci. USA*, 76:5192.

FLOW SORTING ON THE BASIS OF MORPHOLOGY AND TOPOLOGY

Thomas M. Jovin and Donna J. Arndt-Jovin

Abteilung Molekulare Biologie
Max-Planck-Institüt für biophysikalische Chemie
3400 Göttingen, Federal Republic of Germany

SUMMARY

We describe the application to flow systems of one-dimensional image analysis by slit-scanning and dual laser excitation for increased spectral resolution. The former allows morphological identification of large metaphase chromosomes, increased resolution in flow karyotypes, and detection of chromosome aberrations. The dual laser system and appropriate detection optics facilitates the use of resonance energy transfer to probe the topological properties of cell surface components. As an example, we discuss the receptors for the lectin concanavalin A.

INTRODUCTION

The analysis and fractionation of complex mixtures are essential to the understanding of biological systems. In the case of cells and their organelles, many methods based on different biological and physical principles are available (1). However, the search for increased sensitivity and specificity continues unabated. This circumstance follows directly from the enormous dynamic range required for single cell measurements: a given constituent may be represented by as little as one or as many as 10^{12} molecules. The major success in the case of flow methods for analysis and sorting (1,2) has been in the quantitation of the amount of macromolecules such as DNA and enzymes per cell. However, the functions of a cell are mediated by molecular assemblies and structures so that the distribution of matter is equal in importance to its total quantity. Traditionally structure and location have been investigated by light and electron microscopy, techniques based upon observations of a (generally) static field. The exploitation of of electrooptical and computer technology in flow systems is modifying this situation so that machines are able to observe, measure, classify <u>and</u> <u>sort</u> at speeds far in excess of those attainable by a human observer. The importance of the sorting capability cannot be overstressed for it provides highly purified material for more precise identification and further experimentation.

We report here on recent efforts to extend flow system technology to include <u>morphological</u> information of biological relevance. In the first instance, we demonstrate that normal and aberrant condensed metaphase chromosomes can be identified and sorted on the basis of shape as well as DNA content. Secondly, we present the use of fluorescence resonance energy transfer in the determination of ligand-induced redistribution of cell surface components. This information reveals details of morphology and organization (topology) at the molecular level.

FLOW SYSTEM

The principles of and instrumentation for flow analysis and sorting have been documented extensively (1,2-4). The spectroscopic signals that can be generated and detected simultaneously upon interaction with the focused exciting laser beam(s) include absorption, emission, and light scattering in several modes (intensity, spectra, polarization). It can be anticipated (1) that the triplet state (phosphorescence, delayed fluorescence) will be exploited also, as well as resonance Raman and birefringence phenomena correlated with biological structures and properties of interest.

Flow system technology offers significant advantages over static microphotometric methods:
1) detection sensitivities corresponding to 10^2-10^4 chromophores have been reported. It can be anticipated that single macromolecules will be resolved, given the existence of suitable reagents.
2) an instrumental precision better than 1% (coefficient of variation) is routine.
3) population statistics can be generated at high speed (10^2-10^5 Hz).
4) correlations between numerous measured parameters and their functions can be made in real time or a posteriori.
5) bleaching is minimized due to the short exposure time (1-10 µs).
6) heterogeneity and time-dependent phenomena are assessed quantitatively.
7) light scattering artifacts (e.g. depolarization) inherent in turbid suspensions are avoided.
8) contributions from background luminescence (suspending fluid or unbound ligands) are greatly suppressed due to the small illuminated volume (10-100 picoliters) and the pulse nature (a.c. coupling) of the signals.

CYTOFLUORIMETRY

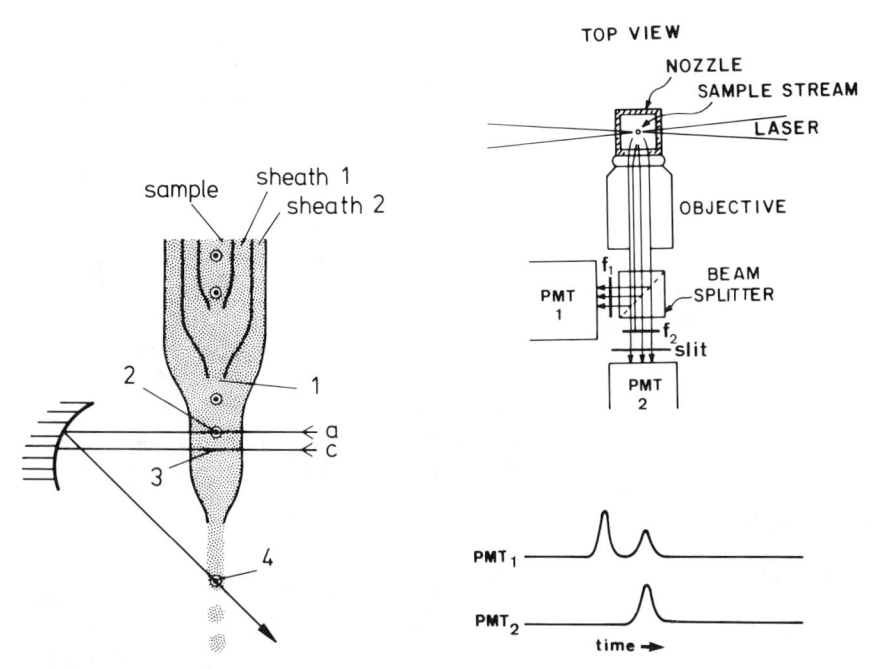

Fig. 1. Flow analyzer/sorter with dual-laser excitation and slit-scanning. The nozzle consists of 3 concentric glass channels which hydrodynamically focus the sample particles/cells and stably position them within the square cross-section (5). There are 3 measuring stations: conductrimetric volume determinations are performed at 1; stations 2 and 3 generate optical signals (emission, absorption, scattering) as the cell/particle traverses the two laser beams, a,c. The latter originate from a dye laser and an argon ion laser respectively (6). After exiting from the glass nozzle (50 µ diameter), the liquid column in air is subject to perturbation by an ultrasonic generator attached to the nozzle (not shown) so that drops form uniformly at point 4. The reflected laser beam is focused at the same place and produces an optical signal used by the system (computer-controlled) to charge the drops containing cells/particles meeting the

9) enrichment or cloning by physical sorting can be coupled to the analytical features of the system.

We have recently modified the apparatus developed in our laboratory so as to provide for a) slit scanning of the flowing particle or cell (5), and b) dual laser excitation (6). The essential features of the device are shown in Figure 1. The new nozzle assembly increases the spectral resolution of the flow analyzer/sorter and adds a one-dimensional image analysis capability.

criteria for separation. These are deflected by high voltage plates (not shown). The cross-sectional (top) view in the plane of excitation shows the optical path for the sequential measurement of emission and scattering at stations 2 and 3. The light collected by the high-aperture microscope objective is passed through a beam splitter (which can be selected so as to have dichroic or polarizing properties) and directed into 2 photomultipliers (PMT 1,2). Filters f_1, f_2 further define the spectral bandwidth. The adjustable (1µ-2 mm) slit (long axis as shown) is positioned at the back focal plane of the objective and defines a scanning aperture at the sample stream equal to the slit size divided by the objective magnification (usually 40x). The two signals depicted for PMT_1 correspond to the measuring stations 2 and 3. Depending upon the laser wavelengths and filters, they can be any desired combination of light scattering and fluorescence. In the example shown for PMT_2 a detectable signal is generated only at station 3.

FLOW CYTOGENETICS

The chromosomes of an eucaryotic cell are the fundamental units responsible for regulated genetic expression and cellular division. The condensed morphology achieved during mitosis can be exploited in the research and clinical laboratory using a number of spectroscopic techniques (7). However, the classical methods for chromosome examination based on light microscopy are tedious and of limited statistical significance. Flow systems offer dramatic increases in speed (8,9) and precision, both developments with profound implications for cytogenetic studies (10).

In the case of karyotypes with well-separated chromosome size classes (e.g. Chinese hamster, Muntjac;10) a moderate resolution is adequate for flow analysis or sorting. More complex karyotypes such as the human or mouse place greater demands on instrumental as well as cytochemical resolution and stability. A determination of DNA content alone is not sufficient. This circumstance has spurred the development of improved nozzle designs with image scanning capability (11,5; Figure 1) and the use of multiple dyes and laser sources (12) so as to emphasize differences in base composition (and possibly structure) of otherwise overlapping chromosome classes.

The information derived from flow systems equipped with the slit-scan configuration are shown in Figure 2. If the illuminating beam and emission aperture are appreciably larger than the object being examined, essentially Gaussian profiles are obtained which can be characterized by a pulse height PH, pulse area (integral) PA, and pulse width PW (time-of-flight, see e.g. 13-16).

CYTOFLUORIMETRY					55

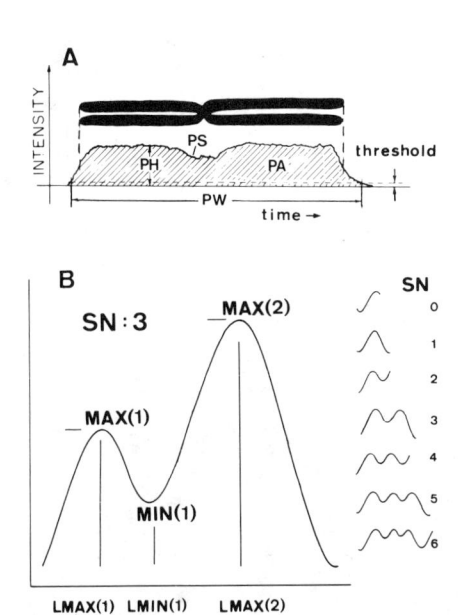

Fig. 2. Information from analog intensity signals (pulses) arising from individual stained chromosomes passing through a focused laser beam at 365 nm and scanned by a 1 μ image slit. A. A typical metacentric chromosome is shown above the abscissa and the pulse generated by it is below. The chromosome is assumed to be oriented along the flow axis. The flow system acquires and stores the pulse height (PH), pulse area (PA), pulse width (PW), and up to 512 points describing the pulse shape (PS) for each object. B. Pulse shape descriptors generated by programs operating in real time on the PS data and used for recognition and sorting. The values and locations of the maxima and minima are determined and a corresponding shape (SN , shape number) assigned.

    The introduction of a limiting aperture (slit) on the excitation (11) or emission (5) paths and sufficiently rapid analog signal detection and digitization (up to 100 MHz) produces a pulse which represents the apparent shape of the object scanned by the slit. For an oriented metacentric chromosome, a bimodal signal results with a dip corresponding to the position of the centromeric constriction

(Fig. 2A). The PH, PA and PW parameters, necessarily modified, are generated by the conventional electronics. However, the pulse profile (PS) can be analyzed more extensively in our system by computer algorithms operating on the digitized data in real time (Fig. 2B). Pulse descriptors are calculated (maxima and minima and their positions, for example) from which one or higher dimensional frequency distributions and sorting decisions can be derived. The SN (shape number) denotes the indicated shapes and is also accessible as a parameter.

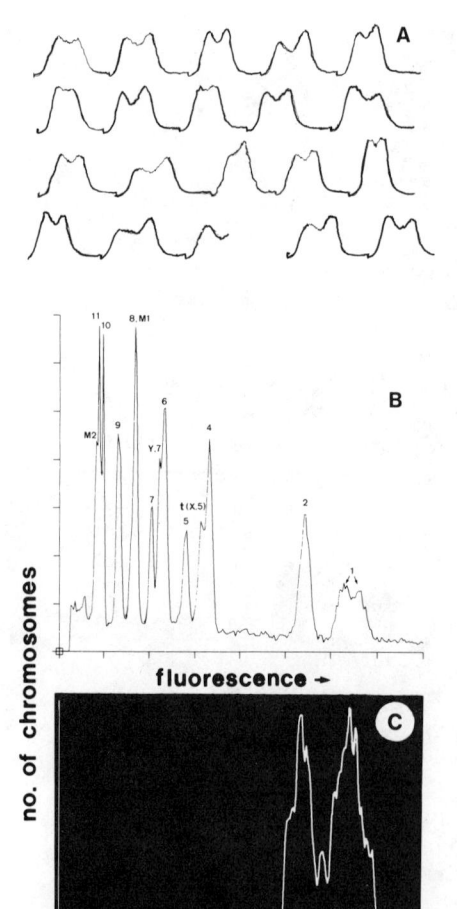

Fig. 3.
Pulse shape analysis of M3-1 Chinese hamster metaphase chromosomes by slit scanning in a flow system.
A) Typical pulse shapes of chromosomes with a pulse area corresponding the number 1 chromosomes, (largest) metacentric scanned with an image slit of 1 μ.
B) Pulse area histogram of the M3-1 chromosomes (see refs. 5, 10-12).
C) The pulse area of all objects scanned with a 1 μ slit and showing a pulse shape descriptor SN = 3, i.e. having two maxima and an intervening minimum. Chromosomes were stained with 10 μM H-33342 (16).

CYTOFLUORIMETRY

A series of typical slit scans of the number 1 chromosomes from the Chinese hamster M3-1 line are shown in Figure 3A. The entire M3-1 karyotype derived from PA signals is shown in Figure 3B and the PA corresponding to objects with a bimodal shape (SN = 3) in Figure 3 C. It can be seen that the metacentric chromosomes numbers 1, 2 and some 4's are within the resolving power of the system.

In another experiment (data not shown) chromosomes were examined from the same cell line after a heavy dose of radiation. In this instance, the flow system was programmed to sort chromosomes with a PA greater than the number 1 peak and a SN of 5, i.e. a shape with two minima, such as might be expected for a dicentric aberrant chromosome. A significant number of end-to-end chromosome fusion products were obtained (Arndt-Jovin et al, to be published). It is anticipated that this approach may offer the sensitivity required for the evaluation of low radiation doses and environmental clastogens.

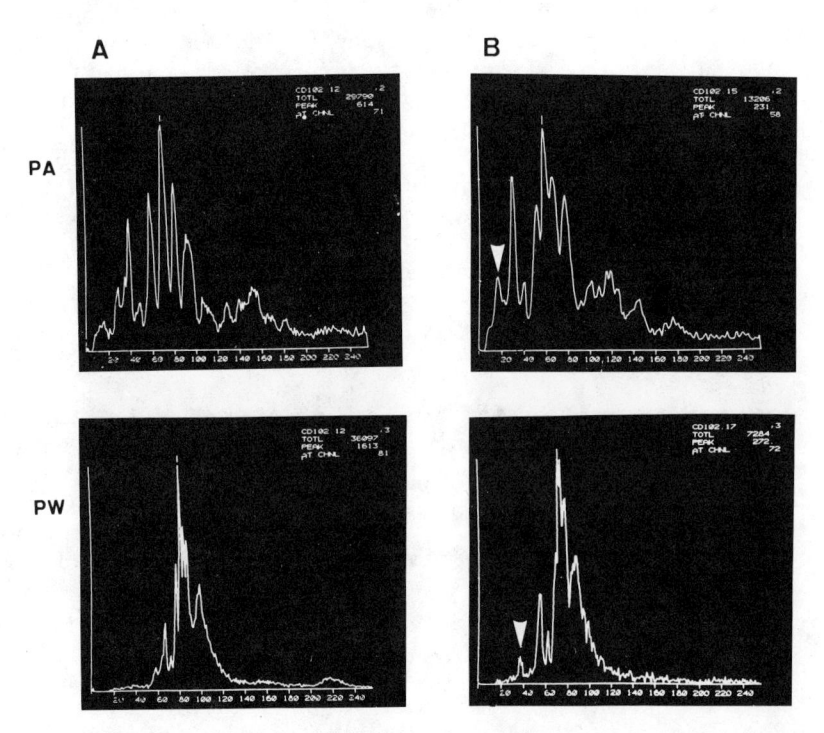

Fig. 4. Flow karyotypes for mouse fibroblast chromosomes. A. Pulse area and pulse width frequency histograms for chromosomes from the Ltk^{-d} line lacking the tk gene. B. Pulse area and pulse width frequency histograms for chromosomes from the 2TG4Tb$_i$ line. The positions in the karyotype of the fragment of the human 17 chromosome responsible for the transformation of the Ltk^{-d} cells to this line are indicated by the arrows.

Slit scanning can also be exploited using the PH, PA, and PW parameters without the more elaborate pulse shape analysis. A flow karyotype of the Ltk$_i^{-d}$ mouse line (17) obtained with a 2 μ slit is shown in Figure 4A. These cells are deficient in the thymidine kinase gene (tk-). The same line after transformation with a fragment of the human 17 chromosome carrying the human thymidine kinase gene (2TGT4b$_i$.) has the karyotype shown in Figure 4B. It is apparent that the fragment produces a distinct peak on the PW but not the PA distribution which (allowing for differences in resolution) is essentially equivalent to that for the parent Ltk^{-d} line. The small fragment has been sorted and its biological/genetic properties are being assessed (Arndt-Jovin et al., to be published).

LIGAND-INDUCED REDISTRIBUTION OF CELL SURFACE RECEPTORS

Many regulatory processes of eucaryotic cells are mediated by agents acting upon the cell surface and inducing changes in ion or metabolite permeation, enzymatic activity, and cytoskeletal organization. In the case of certain polypeptide hormones and growth factors (insulin, epidermal growth factor EGF, nerve growth factor NGF; 18) as well as cell surface immunoglobulins (IgE on mast cells; 19), the biological effect(s) appear to be initiated by aggregation (microclustering?) of the specific ligand-receptor complexes. The mitogenic stimulation of resting cells by lectins is another process involving the redistribution of cell surface components and, in common with many other systems (18,19), the potentiation of transmembranar Ca^{++} transport (20). The dynamic properties (e.g. lateral and rotational diffusion) of the membrane components involved in these processes (primarily glycoproteins and glycolipids) are undoubtedly of great importance and interest (21).

Resonance energy transfer by the Förster mechanism (22) can supply valuable additional information about interactions on and within the membrane due to its a) high spatial resolution; b) temporal sensitivity; c) chemical sensitivity; and d) biochemical selectivity. The physical principle is the exchange of energy between an excited singlet state and a nearby ground state chromophore, a process competing with other radiative and non-radiative deactivation processes (Figure 5).

Resonance energy transfer methods have been applied extensively in studies of lipid bilayers and membranes (for references and theory see 23, 24). We have previously reported flow system determinations of the proximity of lectin (concanavalin A, Con A) receptors on living erythroleukemia cells using lectins modified with chromophores constituting an efficient donor/acceptor pair (fluorescein /rhodamine, respectively) (25). The existence of significant energy transfer was established both by emission intensity and anisotropy measurements, indicating that the Con A molecules liganded to the cell (and by implication their receptors) were tightly aggregated.

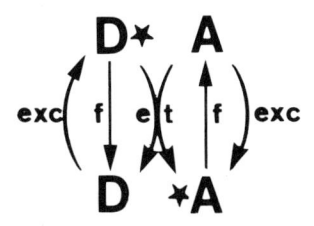

Fig. 5. Resonance energy transfer between donor D and acceptor A. Direct excitation and emission (f) pathways are shown for both species. They interconnect through the nonradiative resonance energy transfer (et) reaction. The energy transfer efficiency T for an isolated donor/acceptor pair is given by

$T = 1 - q_D^r = [(R/R_o)^6 + 1]^{-1}$ where q_D^r is the donor quantum yield normalized to the value in the absence of transfer. R and R_o are the separation distance and critical transfer distance, respectively (22).

Recently, we have extended these studies to lower levels of receptor saturation and more accuracy in the estimation of energy transfer at the single cell level by exploiting the dual-laser excitation system (6). The latter permits the correction of the observed acceptor emission for contributions from direct donor and acceptor excitation so as to obtain the purely sensitized component, i.e. that arising from the energy transfer pathway alone (Figure 6).

The donor and acceptor moieties were labeled with the isothiocyanate derivatives of the fluorophores to a mean stoichiometry per tetrameric protein of 2.2 for fluorescein Con A (F-Con A) and 3.6 for tetramethyl rhodamine Con A (R-Con A) respectively. Cells exposed to saturating levels of a 1:3 molar ratio of F-Con A and R-Con A showed a very significant degree of energy transfer when measured at 4° (Fig. 7A). The intensity of sensitized emission did not increase after capping at 37° for 30 minutes although about 30% of the bound lectin apparently was shed. Cells labeled with lectin to a mean of 30-40% of maximal occupancy showed very similar behavior. However, the energy transfer signal was absent when a 2.3-fold excess of unlabeled native Con A was added to the R-Con A and F-Con A mix-

Donor Alone   Acceptor Alone

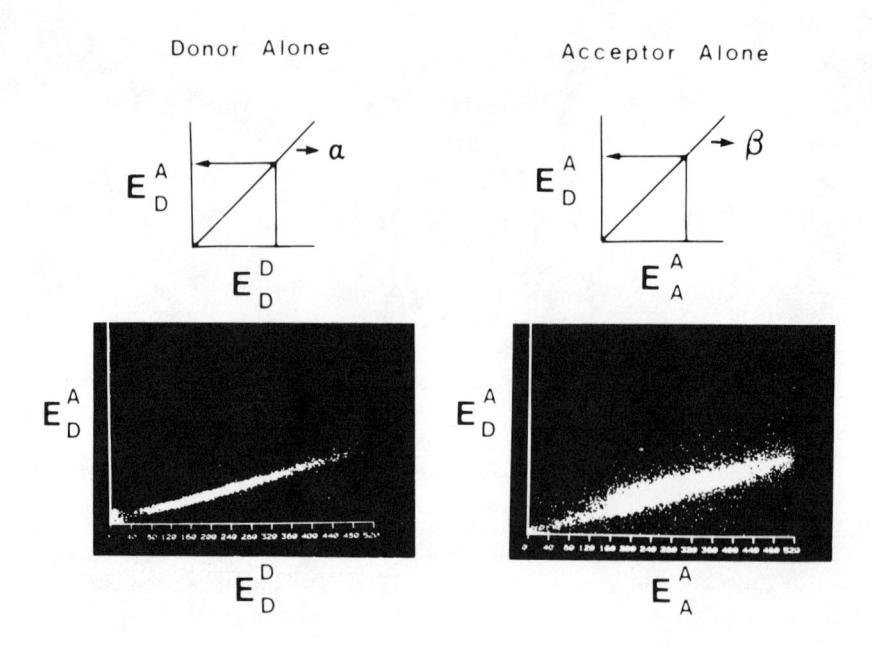

Donor + Acceptor (double labelled)

$$ET = E_D^A - \alpha E_D^D - \beta E_A^A$$

Fig. 6. Scheme for the calculation of resonance energy transfer emission (ET) from flow sorting data. For the case of fluorescein as donor fluorophore and rhodamine as acceptor fluorophore the optical components shown in Figure 1 are as follows: beam splitter = dichroic mirror passing emission above 580 nm. Laser beam c is an argon ion 488 nm line (1 watt), beam a is the dye laser output at 555 nm (470 mW). PMT 1 measures light scattering at 90° from objects intersecting the 555 nm beam and is used to gate the measurements of the fluorescence emissions. E_D^D, measured by PMT 1, 515-545 nm emission from 488 nm excitation (donor emission). E_A^A, measured by PMT 2, >590 nm emission from 555 nm excitation (acceptor emission by direct excitation). E_D^A, measured by PMT 2 in gated mode, >590 emission from 488 nm excitation. ET is calculated from the E_D^A signal after correction for spectral spillage from direct fluorescein emission and direct rhodamine excitation by the 488 nm laser beam (see equation). The correction factors α and β are obtained as shown by the diagrams from data measured on cells labelled with a single species of fluorescent Con A molecules.

ture prior to binding (Figure 7B). The latter control establishes that energy transfer requires the proximity of the donor and acceptor molecules on the cell surface.

In order to calculate the transfer efficiency from data such as in Figure 7A an additional piece of information is required on a cell-to-cell basis, i.e. the donor concentration. The observed donor emission intensity E_D necessarily reflects the quenching due to energy transfer and quantitation of donor amount by absorption is not feasible in the flow system. However, the population distributions of E_D for the double labeled cells (F-Con A and R-Con A) and single labeled cells (F-Con A and equivalent dark Con A to match rhodamine in the previous mixture) indicate that the mean donor quenching and thus transfer efficiency was 40% (see below).

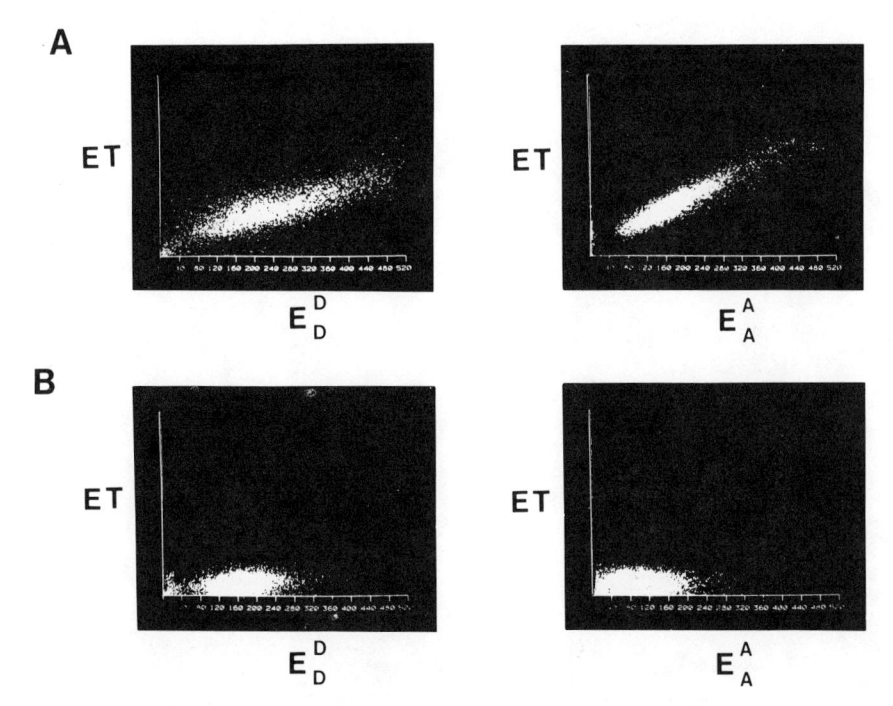

Fig. 7. Energy transfer from cells labeled with R-Con A and F-Con A at saturating concentrations. A. Cells were labeled with 110 μg/ml of Con A in a molar ratio of 3:1, R-Con A to F-Con A. ET on a cell by cell basis was calculated as in Fig. 6 and is plotted as a function of the donor emission on the left and as a function of the acceptor emission on the right. B. The cells were labeled with 2.3-fold excess native Con A in addition to R-Con A and F-Con A. The ET signals are clustered about the zero axis with equal numbers below the abscissa and not seen on the plot.

A detailed discussion of the above results and related data will be presented elsewhere (Arndt-Jovin et al., in preparation). We restrict ourselves here to some general observations. The problem of energy transfer between moieties distributed in a two-dimensional planar matrix has been treated analytically (23,24,26) under the set of assumptions:

1) the distribution of (point-like) donors and acceptors in the planar surface is random and mutually independent.

2) translational motion during the excited state lifetime is negligible.

3) there is no donor-donor transfer.

Under these conditions

$$q_D^r = \int_0^\infty \exp(-\lambda - \varepsilon c R_o^2 \sqrt[3]{\lambda}) d\lambda = \sum_{j=0}^\infty \{(-\varepsilon C)^j \Gamma(j/3+1)/j!\} \quad (1)$$

where c is the average number of acceptors per unit area, C is the normalized acceptor concentration, i.e., the number of acceptors per area element R_o^2, ε is a constant ($\pi\Gamma(2/3) = 4.25...$), and λ is time normalized to the excited state lifetime ($\lambda = t/\tau$). Equation 1 is plotted in Figure 8.

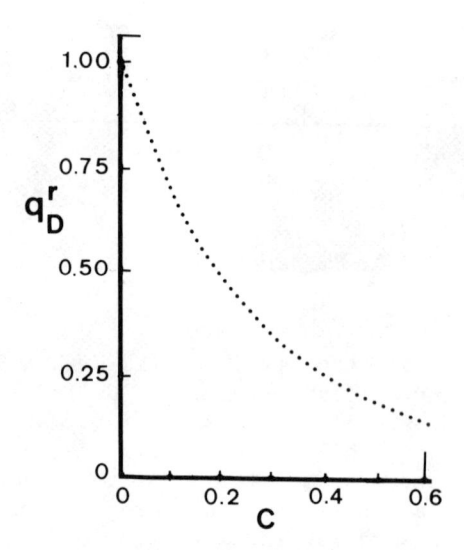

Fig. 8. Quenching of randomly distributed donors by randomly distributed acceptors (Equation 1). See Eq. 1 and Figure 5 for definitions of q_D^r and C.

Deviations from the stated assumptions of Equation 1 lead to the following effects (24):

1) finite donor and acceptor size such that the distance of closest approach is comparable to R_o. This results in a time-dependent scaling of ε so that the transfer efficiency decreases. Thus, the curve in Figure 7 shifts to the right.

2) acceptor binding to the donor. The resultant quenching reflects the weighted contributions from the bound and unbound acceptor. The curve (Figure 7) shifts to the left.

3) non-statistical dipole-dipole orientation factor. The two limit cases are dynamic averaging (reorientation in a time short compared to the excited state lifetime) and static averaging (stationary orientation). A scaling of the acceptor concentration variable C results which can shift the curve (Figure 7) in either direction.

4) donor and acceptor are at different levels in the plane of the membrane. This case is equivalent to (a) above.

We will now attempt to interpret our results in terms of the above formalism. Friend cells have a diameter of about 8 μ and 2×10^7 Con A binding sites (see 25). R_o for the fluorescein-rhodamine donor/acceptor pair is 5.6 nm (25). Due to the mean stoichiometry of 3.6 rhodamine molecules/lectin tetramer, this value has to be increased by the factor $\sqrt[6]{3.6}$ so that R_o = 7.0 nm. (For a more statistical treatment, see 25). From the above quantities, C in Equation 1 equals about 5. The ratio of the maximum dimension of the Con A molecule (25) to R_o is about 1.3. Thus, at saturation, the lectin molecules must be very close to touching each other. The previously measured transfer efficiency of 50-60% was considered to be in accordance with this conclusion (25).

There are numerous factors that affect the above calculation of C and, therefore, our interpretation of Equation 1 (Figure 8). The surface area of the cell may have been underestimated by as much as one order of magnitude. C would then be correspondingly smaller. The transfer efficiency would also be lower due to acceptor exclusion as a consequence of the size factor discussed above. It would increase, however, if the lectin molecules tend to cluster. In fact, the last point constitutes the central question of this study: do the lectin-receptor complexes microaggregate in the absence of metabolic energy?

The key experimental results are a) the transfer efficiency is quite high and independent of saturation (in the range 0.3-1) and microscopically observable state of aggregation (diffuse, patchy, capped); and b) the addition of a moderate excess (2.3-fold) of unlabeled Con A virtually abolishes the transfer. We conclude tentatively that clustering does occur under all the conditions we have examined so far, probably due to the multivalent nature of both the ligand (lectin) and receptor (glycoproteins, glycolipids).

A further consideration is the detailed architecture of the regions of aggregation. Ultrastructural studies indicate that the complexes may not be constrained to a plane but may be piled up to a considerable degree (27,28). If so, the three-dimensional formalism for energy transfer may then apply. This point has been discussed by Dale et al. in other studies of lectin proximity on cell surfaces using energy transfer (29)

CONCLUSIONS

We have not attempted to review the entire scope of morphological and cell surface studies carried out in our laboratory or elsewhere using flow systems. However, the potential in these areas are clearly established. We would anticipate that technical developments in the near future will permit the acquisition of true two-dimensional images of flowing objects and the use of this information for sorting purposes. With respect to energy transfer and related measurements, the introduction and perfection of time-resolved emission measurements in flow will be of great value.

ACKNOWLEDGEMENTS

We are greatly indebted to our colleagues George Striker, Brin Grimwade and Dr. L. Scott Cram for collaborative efforts during many stages of this work.

REFERENCES

1. T. M. Jovin, and D. J. Arndt-Jovin, Cell separation, TIBS 5:214 (1980).
2. M. R. Melamed, P. F. Mullaney, and M. L. Mendelsohn, eds., "Flow Cytometry and Sorting," Wiley, New York (1979).
3. P. K. Horan, and L. L. Wheeless, Jr., Quantitative single cell analysis and sorting, Science 198:149 (1977).
4. D. J. Arndt-Jovin, and T. M. Jovin, Automated cell sorting with flow systems, Ann. Rev. Biophys. Bioeng. 7:527 (1978).
5. L. S. Cram, D. J. Arndt-Jovin, B. G. Grimwade, and T. M. Jovin, One dimensional image analysis of micro-spheres and Chinese hamster cells in a flow cytomer/flow sorter, Acta Path. Microbio. Scand. , in press (1980).
6. D. J. Arndt-Jovin, B. G. Grimwade, and T. M. Jovin, A dual laser flow sorter utilizing a cw pumped dye laser, Cytometry, in press (1980).
7. S. A. Latt, Optical studies of metaphase chromosomes, Ann. Rev. Biophys. Bioeng. 5:1 (1976).
8. J. W. Gray, A. V. Carrano, L. L. Steinmetz, M. A. Van Dilla, D. H. Moore II, B. H. Mayall, and M. L. Mendelsohn, Chromosome measurement and sorting by flow systems, Proc. Nat. Acad. Sci. USA 72:1231 (1975).

9. E. Stubblefield, L. Deavan, and L. S. Cram, Flow microfluorometric analysis of isolated Chinese hamster chromosomes, Exp. Cell. Res. 94:464 (1975).
10. A. V. Carrano, M. A. Van Dilla, and J. W. Gray, Flow cytogenetics: a new approach to chromosome analysis, in Ref. 2.
11. J. W. Gray, D. Peters, J. T. Merill, R. Martin, and M. A. Van Dilla, Slit scan flow cytometry of mammalian chromosomes, J. Histochem. Cytochem. 27:441 (1979).
12. J. W. Gray, R. G. Langlois, A. V. Corrano, K. Burkhart-Schulte, and M. A. Van Dilla, High resolution chromosome analysis: one and two parameter cytometry, Chromosoma 73:9 (1979).
13. T. K. Sharpless, and M. R. Melamed, Estimation of cell size from pulse shape in flow cytofluorometry, J. Histochem. Cytochem. 24:257 (1976).
14. L. L. Wheeless, Jr., Slit-scanning and pulse width analysis, in Ref. 2.
15. J. F. Leary, P. Todd, J. C. S. Wood, and J. H. Jett, Laser flow cytometric light scatter and fluorescence pulse width and pulse rise-time sizing of mammalian cells, J. Histochem. Cytochem. 27:315 (1979).
16. L. S. Cram, D. J. Arndt-Jovin, B. G. Grimwade, and T. M. Jovin, Fluorescence polarization and pulse width analysis of chromosomes by a flow system, J. Histochem. Cytochem. 27:445 (1979).
17. L. A. Klobutcher, and F. H. Ruddle, Phenotype stabilisation and integration of transferred material in chromosome-mediated gene transfer, Nature 280:657 (1979).
18. J. Schlessinger, The mechanism and role of hormone-induced clustering of membrane receptors, TIBS 5:210 (1980).
19. H. Metzger, The IgE-mast cell system as a paradigm for the study of antibody mechanisms, Immunol. Rev. 41:186 (1978).
20. R. B. Mikkelsen, and R. Schmidt-Ullrich, Concanavalin A induces the release of intracellular Ca^{2+} in intact rabbit thymocytes, J. Biol. Chem. 255:5177 (1980).
21. R. J. Cherry, Rotational and lateral diffusion of membrane proteins, Biochim. Biophys. Acta. 559:289 (1979).
22. L. Stryer, Fluorescence energy transfer as a spectroscopic ruler, Ann. Rev. Biochem. 47:819 (1978).
23. T. N. Estep, and T. E. Thompson, Energy transfer in lipid bilayers, Biophys. J. 26:195 (1979).
24. P. K. Wolber, and B. S. Hudson, An analytic solution to the Förster energy transfer problem in two dimensions, Biophys. J. 28:197 (1979).
25. S. S. Chan, D. J. Arndt-Jovin, and T. M. Jovin, Proximity of lectin receptors on the cell surface measured by fluorescence energy transfer in a flow system, J. Histochem. Cytochem. 27:56 (1979).
26. A. G. Tweet, W. D. Bellamy, and G. L. Gaines, Jr., Fluorescence quenching and energy transfer in monomolecular films containing chlorophyll, J. Chem. Phys. 41:2068 (1964).

27. S. de Petris, M. C. Raff, and L. Malluci, Ligand-induced redistribution of concanavalin A receptors on normal, trypsinized and transformed fibroblasts, Nature New. Biol. 244:275 (1973).
28. S. de Petris, Immunoelectron microscopy and immunfluorescence in membrane biology, in:"Methods in Membrane Biology, Vol. 9," E. D. Korn, ed., Plenum Press, New York (1978).
29. R. E. Dale, J. Novros, S. Roth, M. Edidin, and L. Brand, Application of Förster long-range excitation energy transfer to the determination of distributions at the surfaces of cells: distribution of fluorescently-labeled concanavalin A-receptor complexes at the surfaces of yeast and of normal and malignant fibroblasts, in press (1980).

PRIMARY PROCESSES IN THE PHOTOCHEMISTRY OF PROTEINS

L.I. Grossweiner, A. Blum and A.M. Brendzel

Biophysics Laboratory, Physics Department, Illinois
Institute of Technology, Chicago, Illinois
60616, U.S.A.

INTRODUCTION

The effect of ultraviolet irradiation on proteins offers one of the most interesting challenges in the field of photochemistry. The overall objective is to follow the pathway of energy deposition from the initial absorption act through the sequence of physical and chemical alterations leading to permanent changes in structure and biological function. Most of the recent emphasis has been given to enzymes, globular proteins whose catalytic activity depends on the specific sequence of amino acid residues leading to an active conformation in the appropriate ranges of temperature and pH. Proteins provide excellent systems for studying intramolecular energy and electron transfer under conditions where diffusional reactions of the amino acids are inhibited. Similarly, the photolysis reactions of amino acid residues in proteins may differ from the free amino acids in solution because of fixation in the polypeptide chain. The paper first reviews the pre-1970 theories of enzyme inactivation by ultraviolet (UV) and proposes a modified form based on the early work of McLaren[1,2,3] and Dose[4,5] in good agreement with quantum yield measurements. Next, flash photolysis studies on enzymes carried out in this laboratory from 1971 to 1978 are reviewed and related to the inactivation model. Finally, current laser flash photolysis research on aromatic amino acids and enzymes is described, that provides new evidence for multiple pathways of aromatic residue photolysis, depending on the exciting wavelengths and microenvironment of the residue.

RESULTS

1. <u>Ultraviolet Inactivation of Enzymes</u>

The earlier photochemical studies on enzymes involved measuring quantum yields of inactivation by UV and permanent residue destruction, usually at 254 nm. A correlation between the cystine (Cys) content and the inactivation quantum yield (Φ_{in}) was identified by Setlow[6] leading to the conclusion that disruption of the disulfide bridges is the key inactivating act. This approach was extended by McLaren and his co-workers[1,2,3] who proposed that each absorbing residue and the peptide bonds contributes to Φ_{in} by an extent depending on its extinction coefficient and the quantum yield for destruction of the free chromophore in aqueous solution. Although "McLaren's Rule" gave reasonably good values of Φ_{in} for a number of enzymes[3], Dose[5] emphasized that this approach ignores the fact that only some residues of a given type are essential for the activity. He proposed an alternative theory in which essential Cys residues are photolyzed by direct absorption and also by intramolecular processes initiated by light absorption in the aromatic residues. However, the values of Φ_{in} calculated by Dose[5] are not very different from those calculated by McLaren and Hidalgo-Salvatierra[3].

A simple modification of the early theories leads to predicted quantum yields in quite good agreement with measurements at 254 nm and 280 nm. It is assumed that Φ_{in} depends on the photolysis of only essential Cys and tryptophan (Trp) residues as determined from the photolysis of these amino acids in proteins:

$$\Phi_{in} = \Gamma_{trp} f_{trp} \eta_{trp} + \Gamma_{cys} f_{cys} \eta_{cys} \qquad (1)$$

where Γ is the fraction of essential residues of the given type in the enzyme, f is the fractional absorption by all residues of that type, and η is the quantum yield for destruction of the residue in the enzyme. The results in Table I were obtained with η_{cys} = 0.20 at 254 nm and η_{cys} = 0.13 at 280 nm, based on the experimental quantum yields for disulfide destruction in glutathione at these wavelengths[6]. A constant value η_{trp} = 0.05 was used at 254 nm and 280 nm, based on Trp destruction in proteins (e.g., the review of Grossweiner[8]) and the initial flash photolysis yields of Trp photoionization in aqueous enzymes (Table II). The values of f were calculated from the extinction coefficients of the chromophoric amino acids at each wavelength, as described by McLaren and Luse[1]. The values of Γ were deduced from conventional biochemical studies on these enzymes and also from residue sensitivities identified with the radical-anion probe technique (e.g., the review of Grossweiner[8]). The calculated quantum yields are in quite good agreement with experimental data, except for subtilisin Carlsberg that contains no Cys and a single, non-essential Trp residue.

Equation (1) implies that photolysis of essential Trp and Cys residues accounts for inactivation of the enzymes considered (excepting subtilisin Carlsberg) and that energy transfer from aromatic residues to disulfide bridges is not an important factor. However, it is not required that the essential Trp residues be involved with substrate binding or the catalytic reactions per se. As discussed by Grossweiner et al.[9], the photolysis of a Trp residue adjacent to a catalytic residue can lead to inactivation, e.g. Trp 199 in trypsin is adjacent to serine 198, the side chain of Trp 177 in papain is in contact with histidine 159 of the active site, and Trp 73 in carboxypeptidase A is adjacent to glutamic acid 72 which is a ligand of essential zinc. In lysozyme, Trp residues 62, 63 and 108 are part of the active crevice, whereas in ribonuclease A none of the tyrosine residues is essential and Cys photolysis accounts for Φ_{in}.

2. Flash Photolysis Investigation of Enzymes

The key objective in flash photolysis studies on enzymes is to identify the initial photochemical reactions and deduce their relationship to the changes in catalytic activity and permanent residue destruction. The 1971 investigation of Grossweiner and Usui[13] on hen lysozyme with a xenon flash lamp apparatus (5 μs response, λ>250 nm) led to the observation of three transient products: the neutral radical formed by one-electron oxidation of Trp (Trp·, λ_{max} 510 nm), the hydrated electron (e^-_{aq} λ_{max} 720 nm) and the one-electron adduct to a cystyl residue (RSSR·⁻, λ_{max} 420 nm). The decay lifetime of e^-_{aq} was less than 5 μs, Trp· decayed via a fast stage (<500 μs) and a slower process (~0.01 s) and RSSR·⁻ was very long-lived and could be detected after 10 s. The formation of RSSR·⁻ within the time resolution of the system

TABLE I

Calculation of Enzyme Inactivation Quantum Yields Based on Essential Cystyl and Tryptophyl Residues

Enzyme	Γ_{trp}*	Γ_{cys}*	254 nm# Φ_{in}(calc)	Φ_{in}(exptl)	280 nm** Φ_{in}(calc)	Φ_{in}(exptl)
lysozyme	3/6	2/4	0.028	0.024[10]	0.023	0.023[13]
trypsin	1/4	2/6	0.015	0.020[4]	0.009	0.010[14]
papain	1/5	0/3	0.006	0.006[11]	0.006	0.003[11]
carboxypeptidase A	1/6	0/1	0.005	0.005[12]		
subtilisin Carlsberg	0/1	-	0.000	0.007[3]		
ribonuclease	-	2/4	0.029	0.030[7]	0.004	0.007[7]

* essential residues/total residues
# Φ_{in}(calc) = f_{trp}(254 nm)Γ_{trp}0.05 + f_{cys}(254 nm)Γ_{cys}0.20
** Φ_{in}(calc) = f_{trp}(280 nm)Γ_{trp}0.05 + f_{cys}(280 nm)Γ_{cys}0.13

was explained by postulating that the electron was transferred to Cys by a fast, intramolecular process and not by the reactions of e^-_{aq} with lysozyme molecules. The radical yields were estimated by comparison with equivalent mixtures of the chromophoric amino acids, leading to the conclusion that 1-2 Trp residues were photolyzed in the initial act, comparable to the permanent destruction of Trp at the inactivating dose. It was concluded that Trp photolysis is a major inactivating process in lysozyme and that the e^-_{aq} produced by photoionization of Trp does not contribute to inactivation irrespective of its subsequent reactions.

The general approach was extended to investigations on ribonuclease A[15], trypsin[16] and papain [17]. These measurements were repeated in 1976[9] with refined techniques and the advantages of new information about the photolysis of aqueous Trp and tyrosine (Tyr), including the radical extinction coefficients and photoionization quantum yields. In 1972 Santus and Grossweiner[18] identified the triplet state of aqueous Trp (^{3}Trp, λ_{max} 460 nm), which has been observed in many proteins subsequently, and the radical-cation precursor of Trp˙ (Trp˙, λ_{max} 580 nm), which we have not seen in any enzyme at ~50 ns resolution. The amino acid comparison technique was quantified by assuming that aromatic residues in enzymes are photoionized at the same quantum efficiency as the aqueous amino acids or not at all, making it possible to estimate the number of "photolabile" aromatic residues (Δn) by comparing the transient spectra for the enzymes with the corresponding amino acid mixtures. The results in Table II show that Δn is approximately equal to the number of exposed residues as measured with conventional techniques. Furthermore, multiplying the transient radical absorptions in the enzymes relative to the amino acid mixture by the quantum yield for photolysis of the aqueous amino acid gives the quantum yield for residue photolysis in the enzyme as calculated for absorption by all residues of that type(Φ'). The results in Table II were obtained with Φ(Trp˙) = 0.10 from Baugher and Grossweiner[20] and Φ(Tyr˙) = 0.095 from Bent and Hayon[21], where Tyr˙ is the phenoxyl radical formed by photoionization of aqueous tyrosine[22]. (These initial photolysis quantum yields were obtained by laser flash photolysis at 265 nm.) The values of Φ'_{Trp} are approximately equal for these enzymes except subtilisin Carlsberg, and comparable to the quantum yields for permanent Trp residue destruction[7], in support of the value η_{trp} = 0.05 used in Eq.(1)

3. Current Work - Laser Flash Photolysis

Flash photolysis with pulsed 265 nm lasers made it possible to measure the initial transient product quantum yields with higher accuracy and faster time response than with xenon flash lamps. A summary of results in Table III shows that the initial quantum yields of Trp˙ and/or Tyr˙ approximately equals the quantum

PRIMARY PROCESSES IN THE PHOTOCHEMISTRY OF PROTEINS

TABLE II

Flash Photolysis of Aromatic Residues in Enzymes

Enzyme	Φ'_{Trp} *	tryptophan $\Delta n/n$ #	n_{exp} **	Φ'_{Tyr} *	tyrosine $\Delta n/n$ #	n_{exp} **
lysozyme	0.042	2.5/6	~2			
trypsin	0.055	2.2/4	2.4			
papain	0.060	3.0/5	2.2			
carboxypeptidase A	0.076	5.3/6	5			
subtilisin Carlsberg	0.10	1/1	0.8	0.058	8/19	8.8
ribonuclease				0.038	2.4/6	~2

* initial photoionization quantum yield based on absorption by trytophyl or tyrosyl residues only
\# photolabile residues/total residues
** exposed residues[9,19]

TABLE III

Initial Product Yields From 265 nm Laser Flash Photolysis

Enzyme	$\Phi(Trp\cdot)$ #	$\Phi(Tyr\cdot)$ #	$\Phi(RSSR\cdot^-)$ #	$\Phi(e^-_{aq})$ #	$\Sigma(\Phi_e)$ **
lysozyme[23]	0.031	0.000	0.007	0.019	-0.005
trypsin*	0.026	0.000	0.006	0.021	+0.001
carboxypeptidase A[14]	0.057	0.000	0.013	0.042	-0.002
subtilisin* Novo	0.054	0.018	-	0.059	-0.013
subtilisin* BPN	0.031	0.018	-	0.045	-0.004
ribonuclease[24]	-	0.016	0.000	0.016	0.000

* unreported results
\# transient product quantum yield based on absorbtion by enzyme
** electron balance

yields of e^-_{aq} plus $RSSR^{\cdot-}$, based on light absorption by the entire enzyme. The results support the earlier conclusion that photoionization is the principal initial photolysis process of aromatic residues. However, the question still remains as to whether photoionization is significant in proteins at ordinary low light intensities. New results obtained in this laboratory indicate that aqueous Trp can be photoionized by a monophotonic and a biphotonic process, under conditions where the monophotonic pathway should be applicable to ordinary light intensities.

A recent study on aqueous Trp[25] showed that the photoionization yield at 265 nm (UV) was enhanced by simultaneous excitation at 265 nm and 530 nm (UV+G). The UV process is thermally activated, whereas the initial e^-_{aq} yield for UV+G excitation is constant from 20°C to 80°C. These results are shown in Fig.1 plus new data for aqueous Tyr. In the case of Trp, the UV yield at high temperatures approaches the constant UV+G yield, indicating a competition between biphotonic photoionization and monophotonic photoionization with thermal assistance. This is not the case for Tyr, however, suggesting a biphotonic contribution for UV excitation. These conclusions are supported by measurements of the initial e^-_{aq} yields from aqueous Trp and Tyr compared with the e^-_{aq} yield from ferrocyanide ion which was shown to be monophotonic at 265 nm. The results for Trp in Fig.2 show a linear dependence for UV excitation and definite curvature for UV+G. However, for Tyr there is evidence of curvature for UV excitation as well as UV+G.

The Trp results are explained by postulating that UV excitation generates a short-lived intermediate state (X) that is ionized by absorption of a photon and by thermal activation according to the scheme:

$$Trp \xrightarrow{UV, \Phi_x} X \xrightarrow{\Phi_{th}, \Phi_G} Trp^+ + e^-_{aq}$$
$$X \xrightarrow{\tau_x} Trp$$

The straight-forward kinetics analysis leads to the initial e^-_{aq} yield:

$$(e^-_{aq}) = [\Phi_{th}/\tau_x + I_G \alpha_G \Phi_G] f_{uv} \Phi_x \tau_x D_{uv} \qquad (3)$$

where Φ_{th} and Φ_x are the quantum yields for thermal and optical ionization of X, respectively, τ_x is the thermal decay lifetime of

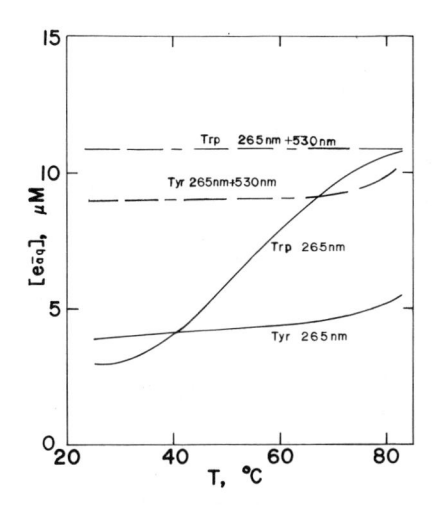

Figure 1. Dependence of initial hydrated electron yield on temperature for 0.4 mM tryptophan (aq) and 1.6 mM tyrosine (aq) irradiated simultaneously at 265 nm and 530 nm and at 265 nm only. The irradiation source was a Nd:glass laser providing 17 ns pulses; the pulse energy was 20 mJ at 265 nm and 60 mJ at 530 nm.

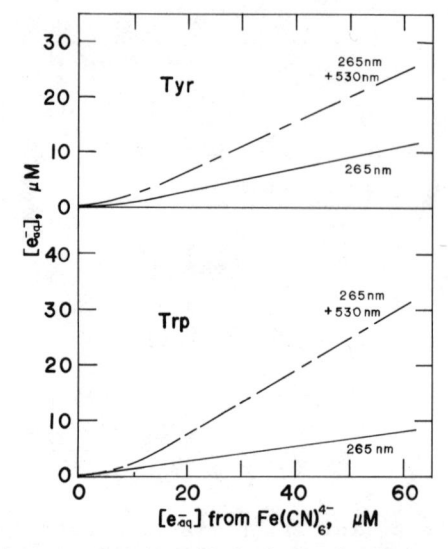

Figure 2. Dependence of initial hydrated electron yield on pulse energy for 0.4 mM tryptophan (aq) and 1.6 mM tyrosine (aq). The electron yield is plotted relative to the electron yield from 0.66 mM ferrocyanide ion irradiated at the same intensity.

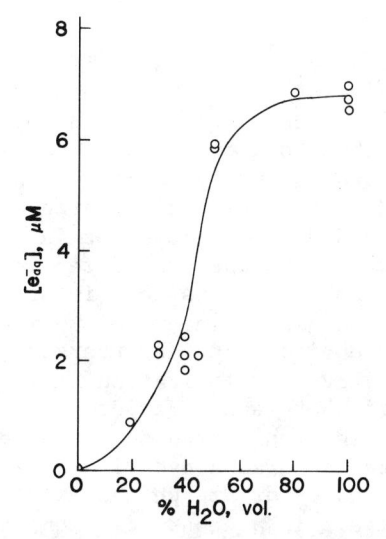

Figure 3. Initial solvated electron yield from 0.4 m$\underline{M}$ tryptophan irradiated at 265 nm plus 530 nm in water-ethanol solutions.

X, D_{uv} is the UV pulse dose, I_G is the mean G intensity in the pulse, f_{uv} is the fractional UV absorption and α_G is absorption coefficient for G excitation. (It is assumed that $\tau_x I_G \alpha_G \Phi << 1$). According to Eq.(3), a plot of $(e_{aq})_0$ for Trp vs $(e_{aq}^-)_0$ for $Fe(CN)_6^{4-}$ should have a constant slope equal to $\Phi_{th}\Phi_x/\Phi'$ for UV excitation, where Φ' is the monophotonic photoionization quantum yield of $Fe(CN)_6^{4-}$ at 265 nm, equal to 0.52[26]. However, for UV+G excitation on the slope should increase with I_G as observed for both Trp and Tyr. The initial e_{aq}^- quantum yields calculated from the slope of the UV excitation lines are 0.073 for Trp and 0.095 for Tyr, although the latter may include a biphotonic contribution.

The species X was postulated as a CTTS state populated via the second excited singlet state, in competition with internal conversion to the fluorescent state[25]. This mechanism is consistent with the wavelength threshold of 275 nm reported for low intensity photoionization of aqueous Trp [27] and the diminished Trp fluorescence below 240 nm in high dielectric constant media[28]. The importance of the dielectric constant in UV+G photoionization is shown in Fig.3, where Trp was photolyzed in water-ethanol solutions. The threshold suggests that internal Trp residues in proteins are not photoionized because of an inadequate dielectric constant in the microenvironment. Furthermore, the necessity of populating the CTTS state for ionization explains the correlation between the number of photolabile residues and exposure (Table II), whether the electron is transferred to the aqueous medium as e_{aq}^- or to an internal trap such as a cystyl bridge. The occurrence of a wavelength threshold for Trp photoionization implies that permanent Trp photolysis in aqueous solutions at wavelengths above about 290 nm[29,30,31] must be caused by very inefficient photoionization or another process leading directly to Trp· radicals[32]. However, this mechanism would not be significant at 254 nm where most data for residue destruction in proteins has been obtained.

Preliminary results on enhancement of photoionization in peptides and enzymes are given in Table IV. The occurrence of higher e_{aq}^- yields for UV+G excitation provides direct evidence for the biphotonic pathway in these compounds. Furthermore in the case of trypsin and papain, where only Trp photolysis has been observed[16,17] the UV excitation process should be monophotonic, implying that photoionization of Trp residues takes place at low light intensities as well. The situation may be more complicated in enzymes where Tyr residues are photoionized because of the possibility that 265 nm excitation induces a biphotonic photoionization process. However, the constant value of $\Phi(e_{aq}^-)$ = 0.095(±10%) obtained in two separate 256 nm laser flash photolysis investigations, where the intensity in one system[21] was about 7 times higher than the other[24] indicated that there must be a

TABLE IV

Photoionization of Aromatic Amino Acids at 265 nm and Simultaneous 265 nm and 530 nm Laser Pulses

Photolyte	$\Phi(e^-_{aq})^*_{265}$	$\Phi(e^-_{aq})_{265+530}/\Phi(e^-_{aq})_{265}$
tryptophan (aq)	1	3.7
tryptophyl-glycine	0.3	6
tyrosine (aq)	1.3	2.3
glycyl-tyrosine	0.9	4
trypsin	0.03	12
papain	0.08	10
carbonic anhydrase	0.03	5
subtilisin Novo	0.1	5
subtilisin BPN	1.4	3

* relative to aqueous trytophan as unity

significant monophotonic component which could be controlling at low intensities. It is interesting to note in Table IV that the enhancement by UV+G tends to be higher for lower UV yields, suggesting that monophotomic photoionization is limited by the thermal ionization of the CTTS state (Φ_{th}) and not the quantum yield for populating the CTTS state (Φ_x).

CONCLUSIONS

The photoionization of aromatic residues is the major initial process identified by UV flash photolysis of proteins. The ejected electrons are stabilized in the aqueous medium as hydrated electrons and also may be trapped at disulfide bridges. The correlation between photoionized tryptophy and tyrosyl residues and exposure to the aqueous medium is explained by new results indicating that the electrons are released from CTTS states formed only in high dielectric constant environments. The tryptophan CTTS state can be ionized by green light (biphotonic pathway) and thermal excitation (monophotonic pathway), where the latter should be applicable at ordinary low light intensities. Tyrosine photoinonization may have a biphotonic contribution at flash photolysis intensities but should be monophotonic at ordinary intensities. The quantum yields for low intensity inactivation of five enzymes are predicted by assuming that direct and irreversible photolysis of essential tryptophyl and cystly residues are the most important reactions. Structural considerations indicate that destruction of cystyl bridges alters the active conformation, whereas tryptophyl residue destruction is inactivating if the residue is part of the active center or is located adjacent to a key catalytic residue. The success of this model argues against the importance of intramolecular energy and electron transfer process in UV inactivation of enzymes.

ACKNOWLEDGMENTS

We are pleased to acknowledte the research support of the U.S. Department of Energy on Contract N° DE-ASOE-76EV02217 and the Department of Health, Education and Welfare on Grant N° 20117. This is COO-2217-35.

REFERENCES

1. A. D. Mclaren and R. A. Luse, Mechanisms of Inactivation of Enzyme Proteins by Ultraviolet Light, Science 134:836 (1961).
2. R. A. Luse and A. D. McLaren, Mechanism of Enzyme Inactivation by Ultraviolet Light and the Photochemistry of Amino Acids (At 2537 A), Photochem Photobiol. 2:343 (1963).
3. A. D. McLaren and O. Hildalgo-Salvatierra, Quantum Yields for Enzyme Inactivation and the Amino Acid Composition of Proteins, Photochem.Photobiol. 3:349 (1964).
4. S. Risi, K. Dose, T. K. Rathinasamy and L. Augenstein, The Effect of Environment on Cystine Disruption by Ultraviolet Light, Photochem.Photobiol. 6:423 (1967).
5. K. Dose, Theoretical Aspects of the U.V. Inactivation of Proteins Containing Disulfide Bonds, Photochem.Photobiol. 6:437 (1967)
6. R. Setlow, A Relationship Between Cystine Content and Ultraviolet Sensitivity of Proteins, Biochem.Biophys. Acta 16:444 (1955).
7. T. K. Rathinasamy and L. G. Augenstein, Photochemical Yields in Ribonuclease and Oxidized Glutathione Irradiated at Different Wavelengths in the Ultraviolet, Biophys.J. 8:1275 (1968).
8. L. I. Grossweiner, Photochemical Inactivation of Enzymes, Curr.Topics Radiat.Res. Quart. 11:141 (1976).
9. L. I. Grossweiner, A. G. Kaluskar and J. F. Baugher, Flash Photolysis of Enzymes, Int.J.Radiat.Biol. 29:1 (1976).
10. D. Shugar, The Measurement of Lysozyme Activity and the Ultraviolet Inactivation of Lysozyme, Biochem.Biophys. Acta 8:302 (1952).

11. K. Dose and S. Risi, The Action of U.V. Light of Various Wavelengths on Papain, Photochem.Photobiol. 15:43 (1972).
12. R. Piras and B. L. Vallee, Carboxypeptidase A. Quantum Yields on Ultraviolet Irradiation, Biochemistry 6:2269 (1967).
13. L. I. Grossweiner and Y. Usui, Flash Photolysis and Inactivation of Aqueous Lysozyme, Photochem.Photobiol. 13:195 (1971).
14. W. A. Volkert and C. A. Ghiron, The Destruction of Tryptophanyl Residues in Trypsin by 280-nm Radiation, Photochem.Photobiol. 17:9 (1973).
15. W. A. Volkert and L. I. Grossweiner, Flash Photolysis of Ribonuclease A, Photochem.Photobiol. 17:81 (1973).
16. A. G. Kaluskar and L. I. Grossweiner, Photochemical Inactivation of Trypsin, Photochem.Photobiol. 20:329 (1974).
17. J. F. Baugher and L. I. Grossweiner, Ultraviolet Inactivation of Papain, Photochem.Photobiol. 22:163 (1975).
18. R. Santus and L. I. Grossweiner, Primary Products in the Flash Photolysis of Tryptophan. Photochem.Photobiol. 15:101 (1972).
19. J. Y. Lee and L. I. Grossweiner, Laser Flash Photolysis Laser Flash Photolysis and Inactivation of Carboxypeptidase A, Photochem.Photobiol. 27:635 (1978).
20. J. F. Baugher and L.I. Grossweiner, Photolysis Mechanism of Aqueous Tryptophan, J.Phys.Chem. 81:1349 (1977).
21. D. V. Bent and E. Hayon, Excited State Chemistry of Aromatic Amino Acids and Related Peptides.I. Tyrosine, J.Am.Chem.Soc. 97:2599 (1975).
22. L. I. Grossweiner, G. W. Swenson and E. F. Zwicker, Photochemical Generation of the Hydrated Electron, Science 141:805 (1963).
23. J. F. Baugher, L. I. Grossweiner and J. Y. Lee, Laser Flash Photolysis of Lysozyme, Photochem.Photobiol. 25:305 (1977).
24. J. F. Baugher and L. I. Grossweiner, Photolysis Mechanism of Aqueous Tyrosine and Tyrosyl Peptides, Photochem.Photobiol. 28:175 (1978).
25. L. I. Grossweiner, A. M. Brendzel and A. Blum, Multiple Pathways of Aqueous Tryptophan Photolysis, Chemical Physics (submitted).
26. M. Shirom and G. Stein, Excited State Chemistry of the Ferrocyanide Ion in Aqueous Solution.I.Formation of the Hydrated Electron, J.Chem.Phys. 55:3372 (1971).
27. E. Amouyal, A. Bernas and D. Grand, On the Photoionization Energy Threshold of Tryptophan in Aqueous Solutions, Photochem.Photobiol. 29:1071 (1979).
28. I. Tatischeff and R. Klein, Influence of the Environment on the Excitation Wavelength Dependence of the Fluorescence Quantum Yield of Indole, Photochem.Photobiol. 22:221 (1975).
29. M. T. Pailthorpe and C. H. Nicholls, Indole N-H Bond Fission During the Photolysis of Tryptophan, Photochem.Photobiol. 14: 135 (1971).

30. R. F. Borkman, Ultraviolet Action Spectrum for Tryptophan Destruction in Aqueous Solution, Photochem.Photobiol. 26:163 (1977).
31. K. P. Ghiggino, G. R. Mant, D. Phillips and A. J. Roberts, Photodegradation and Fluorescence of Aqueous Tryptophan, J.Photochem. 11:297 (1979).
32. R. F. Evans, C. A. Ghiron, W. A. Volkert, R. R. Kuntz, R. Santus and M. Bazin, Flash Photolysis of N-Acetyl-L-Tryptophanamide; Evidence for Radical Production Without Hydrated Electron Formation, Chem.Phys.Lett. 42:39 (1976).

MODELS OF PHOTOREGULATION

B. F. Erlanger and N. H. Wassermann

Dept. of Microbiology, Cancer Research/Cancer Center
Columbia University
New York, N. Y. 10032

INTRODUCTION

Physiological processes that are influenced by light abound in nature. Organisms as complex as the human being and as "simple" as bacteria show evidence of having developed photoresponsive systems. Examples are the phototactic behavior of bacteria, plants and animal cells, diurnal variations in social and sexual behavior of animals (including man), and season-related migration patterns in birds. The process of vision is a highly sophisticated photoresponsive process as are the processes in plants that are controlled by the phytochrome systems. (Two reviews that can prove useful are Erlanger, 1976 and Briggs and Rice, 1972).

Some photoresponsive systems occur in highly differentiated organs or organelles and have their primary response to light converted into a physiological response by complex pathways. Others occur in simple structures and in some cases, appear to be directly linked to the transducing mechanisms. A good example of the latter is the photoresponsive proton pump of the "purple membrane" of Halobacterium halobium (cf. Oesterhelt and Stoeckenius, 1971).

The ubiquity of photoresponsive systems suggests, first of all, that it must be "easy" for nature to design them. We will have more to say about that later. Another implication is that many of the systems, complex and simple, might share common biochemical aspects. With respect to the latter, all of them that have been studied in some detail contain a low molecular weight photoresponsive prosthetic group, structural changes of which influence the properties of associated macromolecules or membranes. Examples are retinal of rhodopsin of the eye and of bacteriorhodopsin of H. halobium, and the tetrapyrrole moiety of the phytochrome systems of plants (Shropshire, 1972). These prosthetic groups invariably exhibit the property of photochromism, either when pure (retinal) or when associated with the macromolecule (the tetrapyrrole of phytochrome).

PHOTOCHROMISM

Photochromism can be defined as a light-stimulated reversible change of a chemical structure between two states having different absorption spectra. The driving force in at least one direction should be light; the reverse change can be caused by thermal energy or by light (Brown, 1971).

If we deal with the system

$$A \underset{\text{dark}}{\overset{\text{light}}{\rightleftarrows}} B$$

the process can be described as in Figure 1.

Other systems are driven in both directions by light of appropriate wavelengths and most of our work is with this kind of sys-

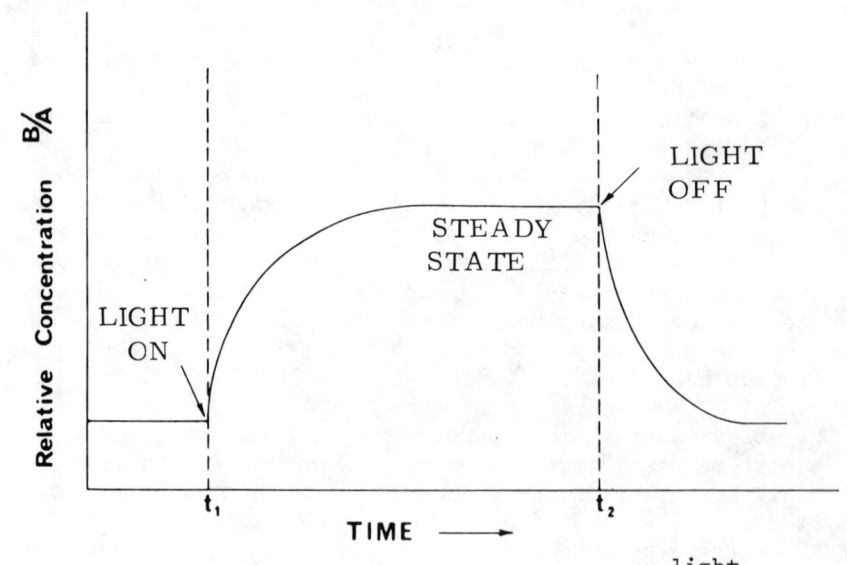

Fig. 1. Behavior of a system in which $A \underset{\text{dark}}{\overset{\text{light}}{\rightleftarrows}} B$. (From Brown, 1971, with permission).

tem. Physicochemically, the difference between these two types of behavior is primarily the potential energy barrier between the two chemical states. If the barrier is high with respect to the reaction in either direction, photons are needed for both conversions. If it is low in one direction, thermal energy is sufficient (Margerum and Miller in Brown, 1971).

Photochromism is a property of a large number of organic (and inorganic) compounds including spyropyrans, anils, hydrazones, stilbenes, and azobenzene derivatives. Our studies have delt solely with azobenzene derivatives but our observations of the action of azobenzene compounds in biochemical systems (below) can be applied to other photochromic compounds as well.

Azobenzene derivatives exhibit photochromism by shifting reversibly between trans and cis isomers under the influence of light as follows:

As with other photochromic compounds, one or both of the isomeric conversions will require light. There are examples of seemingly non-photochromic azobenzenes. However, upon examination of them by fast reaction techniques they were found to undergo cis ⇌ trans conversions at very rapid rates (Wettermark et al., 1965).

It is crucial to understand that the cis and trans isomers of azobenzene derivatives are entirely different compounds, having marked differences in their physical and frequently in their chemical properties, which allow them to be separated from one another when present in a mixture. This becomes readily apparent if one makes models of cis and trans-azobenzene (Figure 2).

It is clear that the structural differences between the two isomers should be easily distinguishable by specific macromolecules, such as enzymes or receptors, either free or when incorporated into a membrane.

PHOTOREGULATION OF ENZYME ACTIVITY

Our interest in model systems of photoregulation began accidentally in 1967 when Dr. H. Kaufman, who was a graduate student in our laboratory at that time, was studying the inactivation of chymotrypsin by a potent active site reagent p-phenylazodiphenylcarbamyl chloride (PADPC) (Figure 3).

For a period of time, circumstances made it necessary for Kaufman to carry out much of his experimental studies at night, in particular, experiments in which the kinetics of the inactivation of chymotrypsin by PADPC were being determined. A later attempt to reproduce these studies (as it turned out, in the daytime) gave different kinetic constants. It took us quite a few months to realize that the problems were caused by the photochromic behavior of PADPC as shown in Figure 4.

PADPC isolated by crystallization is 100% trans isomer, and remains so in the dark (at night). On exposure to 320 nm UV light present in daylight, about 15% of the cis isomer is formed. We were able to separate the isomers from one another by thin-layer chromatography. The cis isomer was about 5× more potent an inactivator of chymotrypsin than the trans isomer (5300 M^{-1} sec^{-1} vs. 1150 M^{-1} sec^{-1}). Their UV spectra also differed (Figure 5).

We were able to establish conditions of pH, concentration and temperature of the inactivation process such that inactivation by the trans isomer occurred at a very low rate. Then, stimulation by long UV light initiated the inactivation process as a result of the

Fig. 2. Models of cis (left) and trans (right) azobenzene.

Fig. 3. Structure of p-phenylazodiphenylcarbamyl chloride (or fluoride).

Fig. 4. Photochromic behavior of azobenzene derivative of the type used in our studies.

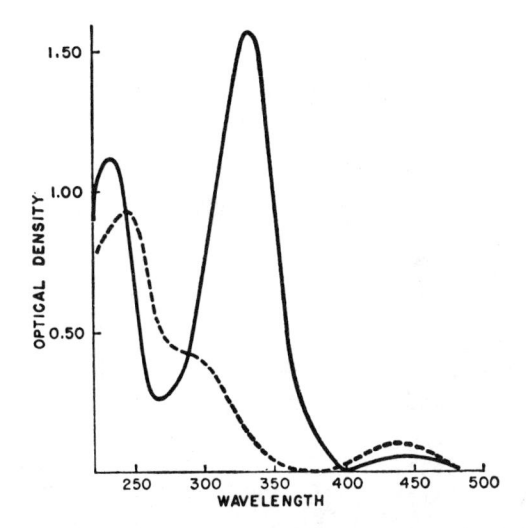

Fig. 5. UV spectra of cis (dashes) and trans (solid line) PADPC (from Kaufman et al., 1968, with permission).

formation of cis isomer. Or if the reaction were begun with pure cis-PADPC, it could be terminated by a flash of blue light (Figure 6).

Thus we were able to show (Kaufman et al., 1968) that an enzymic process, itself insensitive to light, could be made photoresponsive through the use of a light-sensitive effector molecule. We pointed out the analogies between our model system and some photosensitive systems found in plants and animals in which enzyme activities were shown to be responsive to light, and the process of vision, which is also potentiated by a photochromic cis to trans isomerization.

Subsequent to the above findings a number of laboratories have studied model systems in which enzymes could be photoregulated.

These include the laboratories of Suzuki, of Berezin and Martinek, and of Montagnoli (Aizawa et al., 1977; Berezin et al., 1978; Montagnoli et al., 1978 and references cited in these papers).

For our part we extended these studies as follows:

(a) Photochromic inactivators related to PADPC were developed for acetylcholinesterase and were used as photoswitching devices for this enzyme (Bieth et al., 1969, 1973).

(b) Photochromic reversible inhibitors could be used to photoregulate levels of acetylcholinesterase activity (Bieth et al., 1969, 1970). Included in these studies was a compound whose conversion from the trans isomer to a 50-50 cis-trans mixture could be effected by ordinary sunlight (rather than requiring an intense artificial source of light of 320 nm). Thus the diurnal variation of acetylcholinesterase levels found in some insects (Venkatacheri et al., 1968) could be mimicked.

(c) Photochromic substrates of trypsin were synthesized. Their rates of hydrolysis could be controlled by light (Wainberg and Erlanger, 1971).

PHOTOREGULATION IN MEMBRANES

Because a number of naturally occurring photoresponsive processes occur in membranous systems (e.g. rhodopsin in the outer rod segments of the eye, and the phytochrome system in plants) we sought to extend our studies by preparing model membrane-associated photoresponsive systems. In collaboration with Prof. D. Nachmansohn and his colleagues we carried out experiments using electroplax preparations of the electric fish, Electrophorus electricus. Im-

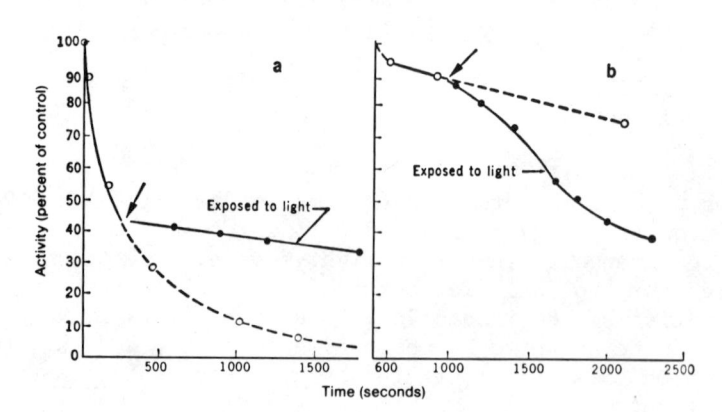

Fig. 6. (a) Effect of light on reaction of chymotrypsin starting with cis PADPC; (b) Starting with trans PADPC (from Kaufman et al., 1968, with permission).

MODELS OF PHOTOREGULATION 87

bedded in the electroplax membrane, which is rich in synaptic junctions, is an acetylcholine receptor (AChR) that controls ion flux and, hence, the electrical potential across the membrane. Reaction of AChR with acetylcholine triggers depolarization of the membrane. We designed photochromic competitive inhibitors of acetylcholine, the cis and trans isomers of which had different binding constants for AChR. We could then photoregulate the potential across the membrane (Figure 7). In these experiments carbamyl choline (CarCh) was used in place of acetylcholine because of the latter's susceptibility to acetylcholinesterase present in the preparation. The photochromic competitive inhibitor was N-p-phenylazophenyl-N-phenylcarbamyl choline (azo-CarCh).

We had mimicked the process of vision in that photons of light produced a neural impulse (Deal et al., 1969). In later experiments (Bartels et al., 1971), photochromic agonists were developed which could react directly with AChR in the membrane to produce depolarization. The structures of two of them are in Figure 8.

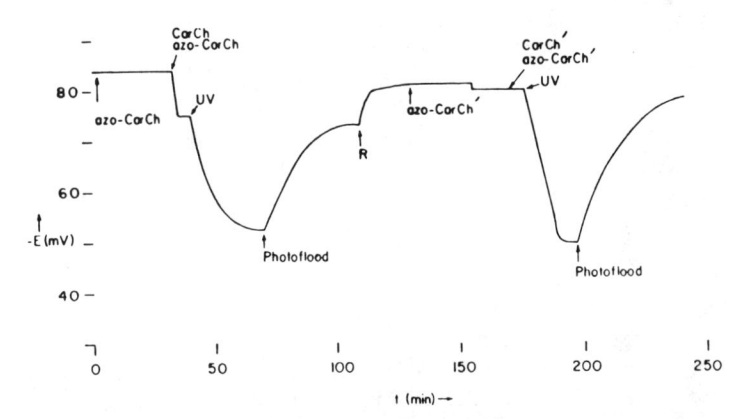

Fig. 7. Photoregulation of the potential across the innervated membrane. Ultraviolet (UV) and photoflood lamps were turned on where indicated. CarCh, 20 µM; azo-CarCh, 1 µM; azo-CarCh', 3 µM; CarCh', 50 µM. R, Ringer solution. (Deal et al., 1969 with permission.)

Fig. 8. Agonists of the electroplax membrane of Electrophorus electricus.

Bis Q, in its trans conformation is the most potent cholinomimetic compound known in the Electrophorus electricus system; the cis isomer is inactive. QBr can be covalently linked to AChR, converting the electrogenic system into a photoelectrogenic system, requiring no additional reagents to control the potential other than light (also see Lester et al., 1980a).

PHOTOCONTROL OF ION BINDING

Most recently we have begun to design compounds whose ability to bind ions, including hydrogen ions, can be regulated by light. The first compound being studied is 4,4'-bis(α-iminodiacetic acid) azotoluene (Figure 9). It is photochromic.

If one examines molecular models of the two isomers (as in Fig. 3) it can be seen in the cis isomer that the two iminodiacetic acids are in close proximity; this is not so for the trans isomer. Dr. M. Blank of the Department of Physiology at Columbia University has found by polarographic techniques that the cis isomer binds one Zn^{++} with $K_a = 1.1 \times 10^5 M^{-1}$. The trans isomer shows no measurable binding of Zn^{++}. Thus, we can photoregulate ion binding. Although we have not yet carried out hydrogen binding studies in detail, preliminary experiments show that the dissociation constants of the carboxyl groups in the cis isomer are different from those in the trans isomer. In collaboration with Dr. Blank, we plan to design a model membrane-bound photoresponsive proton pump based on these principles.

MODEL SYSTEMS, NATURAL SYSTEMS AND EVOLUTION

The model systems described above mimic photoregulated phenomena found in nature. A flash of light (red) can induce seed germination (Borthwick et al., 1952); brief exposure to light can cause a sharp increase of nitrate reductase activity in Sinapsis albus (Karube et al., 1976); a light-induced effector molecule (cinnamic acid) controls enzyme activity in the potato (Lamb, 1977); cGMP levels in outer rod segments of the eye are controlled by a light-sensitive phosphodiesterase (Bitensky et al., 1975); ion flux is

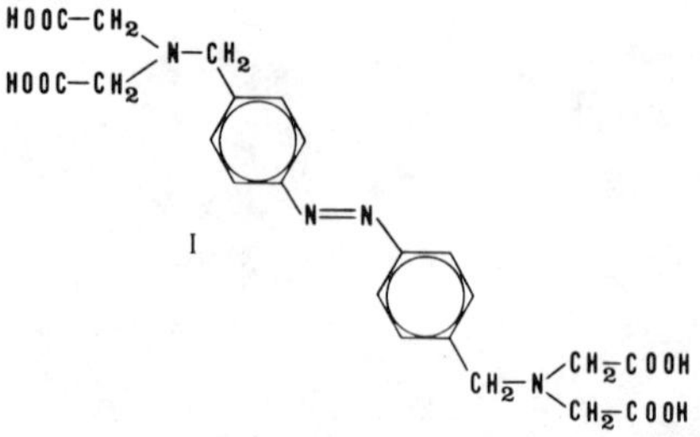

Fig. 9. Structure of 4,4'-bis(α-iminodiacetic acid) azotoluene.

photoregulated by a rhodopsin-like protein in the membrane of halophilic bacteria (Lozier et al., 1978); similarly, ion flux in plant cells is controlled by light-sensitive phytochrome imbedded in a membrane (Hendricks and Borthwick, 1967), and so on.

In the model systems, and in many naturally-occurring systems, low-molecular-weight effector molecules regulate biologically active macromolecules. This kind of regulation is common in biochemical systems and is, for example, the basis of allostery (Koshland and Neet, 1968; Monod et al., 1963). The additional dimension introduced in photosensitive systems is the control of the stereochemistry of the effector molecule by light.

With respect to regulation of enzyme activity in free solution, it is necessary to postulate specific binding of the photochromic effector molecule to an active or controlling site of the macromolecule. Although specificity undoubtedly is important in membrane-bound systems, also, some interesting experiments of Balasubramanian et al. (1975) show that it is not always necessary. They found that the activity of chymotrypsin imbedded in a membrane could be photoregulated by the co-presence of an azobenzene compound that did not bind specifically to the enzyme. Photo-isomerization of the azobenzene derivative apparently affected the organization of the membrane and, hence, the microenvironment of the enzyme, thereby altering its activity.

The experiments of Balasubramanian et al. indicate how photoregulated processes might have evolved in primitive systems. Photochromic lipophilic compounds are distributed widely in nature, the most prominent example being the carotenoids. Their incorporation into membranes containing active macromolecules or their interaction with apolar regions of proteins could have set the stage for the earliest photoregulated systems. Subsequent perfection of the photoregulation mechanisms by the development of specificity could have occurred by natural selection, resulting in the highly sensitive, specialized systems found in organisms today.

EPILOGUE: A BONUS

The development of photochromic reagents whose isomers differ in their ability to influence membrane excitability in the electroplax system has led to very elegant studies carried out in the laboratories of Dr. H. Lester of the California Institute of Technology. Using flash lamps and pulsed lasers, Lester and his colleagues have been able to produce cis-trans and trans-cis isomerizations in high quantum yields in less than a microsecond. The reagents therefore have been utilized as rapid photoswitches to study early events in the synaptic delay of nicotinic transmission (Lester et al., 1980a, 1980b), and also to investigate the mechanism of ion channel blockade of AChR channels by a photochromic compound that resembles a local anaesthetic in mechanism (Lester et al., 1979).

It is not unlikely that appropriately designed photochromic reagents will find this kind of applicability in other systems. The unique value of these reagents lies in their ability to be generated rapidly into a biologically active form, in situ. Ideally one of the photochromic isomers should be biologically inactive. However, in most cases all that is necessary is difference of activity between the two isomers. Rapid generation of the more active isomer is then a phenomenon analogous to that which occurs in temperature-jump experiments.

ACKNOWLEDGEMENTS

We acknowledge the National Science Foundation, the National Institutes of Health, the American Heart Association, the Muscular Dystrophy Foundation and the Irene Heinz Given and John La Porte Given Foundation, Inc., for support of the research carried out in this laboratory.

REFERENCES

Aizawa, M., Namba, K., and Suzuki, S., 1977, Photocontrol of enzyme activity of α-amylase, Arch. Biochem. Biophys., 180:41.
Balasubramanian, D., Subramani, S., and Kumar, C., 1975, Modification of a model membrane structure by embedded photochrome, Nature, 254:252.
Bartels, E., Wassermann, N. H., and Erlanger, B. F., 1971, Photochromic activators of the acetylcholine receptor, Proc. Natl. Acad. Sci. U.S.A., 68:1820.
Berezin, I. V., Varfolomeyev, S. D., Klibanov, A. M., and Martinek, K., 1974, Light and ultrasonic regulation of α-chymotrypsin catalytic activity, FEBS Lett., 30:329.
Bieth, J., Vratsanos, S. M., Wassermann, N. H., and Erlanger, B. F., 1969, Photoregulation of biological activity by photochromic reagents, II. Inhibition of acetylcholinesterase, Proc. Natl. Acad. Sci. U.S.A., 64:1103.
Bieth, J., Vratsanos, S. M., Wassermann, N. H., and Erlanger, B. F., 1970, Photoregulation of biological activity by photochromic reagents, IV. A model for diurnal variation of enzymic activity, Proc. Natl. Acad. Sci. U.S.A., 66:850.
Beith, J., Vratsanos, S.M., Wassermann, N.H., Cooper, A.G., and Erlanger, B.F., 1973, Photoregulation of biological activity by photochromic reagents. Inactivation of acetylcholinesterase, Biochemistry, 12:3023.
Bitensky, M., Miki, N., Keirns, J.J., Keirns, M., Baraban, J.M., Freeman, J., Wheeler, M.A., Lacy J., and Marcus, F.R., 1975, Activation of photoreceptor disk membrane phosphodiesterase by light and ATP, Adv. Cyclic Nucleotide Res., 5:213.

Borthwick, H. A., Hendricks, S. B., Parker, M. W., Toole, E. A., and Toole, V. K., 1952, A reversible photoreaction controlling seed germination, Proc. Natl. Acad. Sci. U.S.A., 38:662.

Briggs, W. R., and Rice, H. V., 1972, Phytochrome: Chemical and physical properties and mechanism of action, Annu. Rev. Pl. Physiol., 23:293.

Brown, G. H., 1971, in: Photochromism, Vol. III, Techniques in Chemistry (Weissberger, A., ed.), Wiley-Interscience, New York, 853 pp.

Deal, W. J., Erlanger, B. F., and Nachmansohn, D., 1969, Photoregulation of biological activity by photochromic reagents, III. Photoregulation of bioelectricity by acetylcholine receptor inhibitors, Proc. Natl. Acad. Sci., U.S.A., 64:1230. Erlanger, B.F., 1976, Photoregulation of Biologically Active Macromolecules, Annu., Rev. Biochem., 45:267.

Hendricks, S. B., and Borthwick, H. A., 1967, The function of phytochrome in regulation of plant growth, Proc. Natl. Acad. Sci. U.S.A., 58:2125.

Karube, I., Nakamoto, Y., Namba, K., and Suzuki, S., 1976, Photocontrol of urease-collagen membrane activity, Biochim. Biophys. Acta, 429:975.

Kaufman, H., Vratsanos, S. M., and Erlanger, B. F., 1968, Photoregulation of an enzymic process by means of a light sensitive ligand, Science, 162:1487.

Koshland, D. E., and Neet, K. E., 1968, The catalytic and regulatory properties of enzymes, Annu. Rev. Biochem., 37:359.

Lamb, C. J., 1977, trans Cinnamic acid as a mediator of the light-stimulated increase in hydroxycinnamoyl-CoA: Quinate hydroxycinnamoyl transferase, FEBS Lett., 75:37.

Lester, H. A., Krouse, M. E., Nass, M. M., Wassermann, N. H., and Erlanger, B. F., 1979, Light-activated drug confirms a mechanism of ion channel blockade, Nature, 280:509.

Lester, H. A., Krouse, M. E., Nass, M. M., Wassermann, N. H., and Erlanger, B. F., 1980a, A covalently bound photoisomerizable agonist: Comparison with reversibly bound agonists at Electrophorus electroplagues, J. Gen. Physiol., 75:207.

Lester, H. A., Nass, M. M., Krouse, M. E., Nerbonne, J. M., Wassermann, N. H., and Erlanger, B. F., 1980b, Electro-physiological experiments with photoisomerizable cholinergic compounds. Review and progress report. Ann. N. Y. Acad. Sci., 346:475.

Lozier, R., Niederberger, W., Ottolenghi, M., Sivorinovsky, G., and Stoeckenius, W., 1968, in: Energetics and Structure of Halophilic Microorganisms (Caplan, R., and Ginzburg, M., eds.), pp. 123-142, Elsevier, N. Y.

Monod, J., Changeux, J. P., and Jacob, F., 1963, Allosteric proteins and cellular control systems, J. Mol. Biol., 6:306.

Montagnoli, G., Monti, S., Nannicini, L., Giovannitti, M. P., and Ristori, M. G., 1978, Photomodulation of azoaldolase activity, Photochem. Photobiol., 27:43.

Oesterhelt, D., and Stoeckenius, W., 1971, Rhodopsin-like protein from the purple membrane of Halobacterium halobium, Nature New Biol., 233:149.

Shropshire, W. Jr., 1972, Phytochrome, a photochromic sensor, Photophysiology, 7:33.

Venkatacheri, S. A. T., and Muralikrishna Dass, P., 1968, Choline esterase activity rhythm in the ventral nerve cord of scorpion, Life Sci., 7:617.

Wainberg, M. A., and Erlanger, B. F., 1971, Investigation of the active center of trypsin using photochromic substrates, Biochemistry, 10:3816.

Wettermark, G., Langmuir, M. E., and Anderson, D. G., 1965, Catalysis of the cis-trans isomerization of 2-hydroxy-5-methylazobenzene, J. Amer. Chem. Soc., 87:476.

UV-INDUCED FORMATION OF POLYNUCLEOTIDE-PROTEIN CROSS-LINKAGES AS A TOOL FOR INVESTIGATION OF THE NUCLEOPROTEIN STRUCTURE AND FUNCTIONS

E.I. Budowsky

Shemyakin Institute of Bioorganic Chemistry

USSR Academy of Sciences, Moscow, USSR

The functions of nucleic acids, their storage and transfer are realized at the level of nucleoproteins. Therefore one of the fundamental problems of molecular biology and bioorganic chemistry is elucidation of the structure and principles of formation and functioning of nucleoproteins.

The storage and transfer nucleoproteins (chromosomes, informosomes, extracellular viruses, etc.) are relatively stable and rigid. Their structure and assembly are essentially independent of the nucleotide sequence and are determined by interactions of proteins or their associates with nucleic acids as such. For instance, DNA and RNA of any sequence could be incorporated into chromosome and informosome respectively, different polynucleotides can form virus-like complexes with TMV protein, DNA with large deletions and insertions could be incorporated into λ or f_d virions, etc...

Formation of functional nucleoproteins is determined by non-covalent short-distance interactions of different types between properly located residues of polynucleotides and proteins, mainly between nucleic bases and amino acid residues. Alterations either of the nature of the interacting residues (e.g. due to point mutations) or of the spatial distribution of the respective residues (e.g. due to some mutations or variation of the conditions) could lead to change (as a rule to decrease) of functional nucleoprotein stability or even prevent their formation at all. Functioning of such nucleoproteins involves interconversion of several distinct states. Any transition from one state into another is caused and/or accompanied by alterations of intermolecular interactions. Therefore functional nucleoproteins represent some kind of machine. Obviously, to know how such machines are designed and how they work one should know not only the structure of the details (primary and higher structures of the respective polynucleotides and proteins) but also their spatial arrangement as well as the presence and the nature of their interactions inside the complex at every stage of self-assembly and functioning. Therefore the study of these interactions could shed light on some principles of formation, as well as functioning of nucleoproteins.

Cross-linking by bifunctional agents and study of energy transfer between some groups located on different macromolecules permit to determine only distances between some fragments of macromolecules giving only limited information pertaining to the spatial arrangement of macromolecules inside a nucleoprotein complex. Since nucleoproteins, especially multicomponent ones, are stabilized due to cooperativity of numerous weak interactions, chemical modification of nucleoprotein components as well as removal of some macromolecules or their parts could result in alteration of nucleoprotein structure. In this connection such attempts for determination of contacts between particular macromolecules or even for the identification of directly interacting fragments and components of polynucleotides and proteins usually fail or lead to ambiguous conclusions. For instance, based on the complex formation between ribosomal protein S1 and isolated 49-nucleotide long 3'-terminal fragment of 16 S RNA it was suggested that this interaction exists also inside 30 S subunit and plays an important role in initiation of translation (1). But recently it has been shown that even removal of this terminal fragment of 16 S RNA does not affect binding of S1 protein to 30 S subunit (2,3). Moreover, S1 protein does not interact at all with 16 S RNA inside 30 S subunit of $E.\ coli$ ribosome (4) and its association with 30 S subunit is determined by protein-protein interactions (5).

More detailed information on the macromolecular interactions and their spatial arrangement into nucleoproteins is given by analysis of the cross-links formed between closely located, and hence directly interacting residues of macromolecules. This approach permits removal of non interacting parts of nucleoprotein and unambiguous determination of interacting macromolecules, even in multicomponent functional complexes, isolation and analysis of interacting fragments of polynucleotides and proteins up to interacting monomers.

The formation of a covalent bond between closely located monomer residues inside macromolecules and their complexes could be provoked by chemical or photochemical activation of one of them.

Chemical induction of polynucleotide-protein cross-links by the action of O-methylhydroxylamine on the phage S_d was discovered in 1970 (6,7). Cytosine nuclei in isolated "native" double stranded DNA are unreactive towards nucleophilic agents due to the shielding of the base planes by neighbouring bases (8). But due to some peculiarities of the intraphage DNA secondary structure (9) part of cytosine nuclei (up to 20 % in the case of S_d phage) in virion are as reactive towards O-methylhydroxylamine as free monomers (6,10). Addition of O-methylhydroxylamine molecule to C5-C6 double bond of cytosine nucleus strongly enhanced reactivity of C4 allowing direct substitution of exocyclic amino group by even weak nucleophiles, such as amino group of amino acid residues of phage proteins closely located to DNA inside the virion. In the presence of an excess of O-methylhydroxylamine these polynucleotide-protein cross-linkages are cleaved (17,10). This allows to determine the transiently cross-linked proteins by conventional methods.

Bisulfite is a stronger nucleophile in comparison with O-methylhydroxylamine and also readily forms an adduct with cytosine (11). In this case even in the presence of large excess of the reagent it is possible to prevent cleavage of the formed polynucleotide-protein cross-links (12). Treatment by bisulfite was shown to induce stable covalent polynucleotide-protein cross-linkages in phages MS2 (13,14), PM2 (15) and S_d (16,17). The proposed minimal cross-linked fragments, particularly, cytosyl-lysin were isolated after bisulfite treatment (14,16). It should be mentioned that the rate of cytosyl-lysine formation inside phage nucleoprotein is more than two orders of magnitude higher than that for the model monomer mixture, containing 1 M of lysine (12). These data evidence that due to the rigidity of the phage nucleoprotein, fixed by numerous cooperative non-covalent interactions, the apparent local concentration of the phage protein lysine residues near reactive RNA cytosines is extremely high.

In favour of such rigidity of storage nucleoproteins are results of the investigation of histones distribution in chromatin. It is known that after alkylation of purines at N_7 the N-glycosidic bond can be cleaved under very mild conditions (18) and the appeared free carbonyl on ribose or deoxyribose residue readily forms Shiff base particularly with ε-amino group of lysine residue. Although this set of reactions leads to removal of several purine bases and formation of DNA-protein cross-linkages at these sites, the distribution of histones was found by this method (19). Hence in this case even the removal of several purine bases does not result in alterations of the nucleosome structure and of histone distribution.

But due to some circumstances the method of chemical activation has a limited application. First, while aforementioned modifications enhanced the reactivity of some residues, they change also the structure of polynucleotides, which even being local could affect the saptial arrangement of the components to be cross-linked. Therefore the validity of the results obtained by these methods must be proved by some independent approach. Second, chemical methods require definite conditions (pH, temperature, ionic strength, etc.) which usually are practicable only for limited number of nucleoproteins, mostly for storage ones.

The photochemical method of activation seems to be much more attractive. The formation of polynucleotide-protein cross-linkages was shown almost 20 years ago (20,21). But much later it was understood that this approach could be used for investigation of the nucleoprotein structure (22,23).

Let us consider now the use of UV radiation for investigation of nucleoproteins.

Electronic excitation due to absorption of a UV photon transforms the respective residues into strong electrophilic, nucleophilic or radical reacting species. Reactions of the excited residue could result in formation of a covalent bond, particularly cross-link between residues of two macromolecules.

The difference of the spectral properties of nucleic acids and protein components permits to accomplish the selective excitation of nucleic bases in nucleoproteins. Hence, at least one of the residues cross-linked by UV irradiation of nucleoprotein must be the nucleic base. The other component could be either an amino acid residue or nucleic base of the same polynucleotide chain (formation of pyrimidine dimers (24,25)) or of the other chain. But the formation of cross-linkages between residues of different polynucleotide chains was observed only in some special cases. For instance UV irradiation of ribosomal complexes containing some N-acetylaminoacyl-tRNA's in the P-site results in cross-linking of tRNA to 16 S RNA (26,27). Such cross-linkages are due most probably to formation of a cyclobutane-type dimer between 5-methoxy- or 5-(carboxymethoxy)-uracil residue of tRNA and some pyrimidine in 16 S RNA, since they appear at long-wave (310-335 nm) irradiation and are cleaved at short-wave (254 nm) irradiation (26) (cf. 24). Under UV irradiation the template poly(U) is cross-linked to tRNA molecules containing wY base (27), and, probably, to 18 S RNA (29). Hence, the main part of intermolecular cross-linkages induced by UV irradiation of nucleoproteins are of polynucleotide-protein type.

The quantum yield of the particular cross-linkage formation depends on several factors - on reactivity and life-time of the respective nucleic base excited state, on the reactivity of the respective amino acid residue and, since the formation of cross-link is a bimolecular reaction, on the distance between an amino acid residue and excited base. In the case of ribosomal complexes the quantum yield for some RNA-protein cross-linkages is as high as 10^{-2} (30), which is comparable with that for dimerisation of pyrimidines in polynucleotides (24), thus evidencing high reactivity and close location inside nucleoprotein of the interacting groups of RNA and protein. In the case of dissociable nucleoproteins quantum yield depends also on the fraction of the respective components in the complex. It was shown that changes of the conditions shifting the equilibrium towards the dissociation of the complex are accompanied by decrease of cross-link formation quantum yield (31,32). Hence, the formation of polynucleotide-protein cross-linkages takes place only between closely located excited nucleic base and amino acid residue.

The life-time of the excited singlet state in liquid aqueous solutions does not exceed several picoseconds, and of the excited triplet, several milliseconds. Therefore the conformational changes in the vicinity of the residue caused by its excitation seems to be insignificant even in flexible functional nucleoproteins and cross-links are formed mainly between residues closely located inside the native starting nucleoprotein. Only in the case of cytosine the cross-links formation could be the secondary dark process, after the photoinduced addition of the water molecule to the C5-C6 double bond (23, Cf. 33). In this case all the above-mentioned disadvantages for chemical activation are also valid.

UV irradiation brings about some damages of polynucleotide and protein residues (24, 25, 34-36) which could influence conformation of macromolecules as well as of complex as a whole. In this connection reliable results on the polynucleotide-protein contacts in nucleoproteins could be obtained only at doses which do not cause changes of the spatial arrangement of not only macromolecules, but even of the interacting residues. The most direct proof for the absence of such changes is an exponential character of the dose-response curve. As it was shown for one of the functional nucleoproteins - complex of 70 S ribosome with mRNA and tRNA, deviation of the dose-response curve for cross-links formation from exponential occurs after absorption of more than 40 quanta per nucleotide (37). Storage nucleoproteins are much more stable and deviation of these curves for phage MS2 (14) and Sendai virus nucleocapsid (38) become observable above 200 quanta per nucleotide. It should be underlined that such absorption must be uniform for all the complexes in solution, which could be achieved in optically thin or sufficiently stirred solution. Hence, polynucleotide-protein cross-linkages are formed inside essentially native functional and storage nucleoproteins at doses below 40 and 200 quanta per nucleotide, respectively. It should be mentioned that even at much greater doses neither cleavage of polynucleotide and polypeptide chain nor formation of polypeptide-polypeptide cross-linkages are detected.

Because of dose limitations one could definitely determine cross-linkages formed at a quantum yield not less than 10^{-4}-10^{-5}. Therefore it is not possible to distinguish between absence and low-yield formation of cross-linkages. Thus definite detection of cross-link undoubtedly evidences the presence of polynucleotide-protein interaction in the starting complex, but negative results do not allow to distinguish between low reactivity of the interacting groups and the absence of this interaction at all. It should be mentioned that there is no correlation between quantum yield for different cross-links and the strength of the respective interactions and therefore absolute values of quantum yields have no quantitative significance. But the quantum yields are very sensitive to the spatial arrangement of the groups to be cross-linked. Therefore the nucleoprotein state alterations, which are caused and/or accompanied by alterations of intermolecular interactions, should strongly affect the ratios of the quantum yields for different cross-links, thus providing straight and very precise control of nucleoprotein state alterations. Sensitivity of the cross-link formation to the environment of the excited residue permits the determination of even minute alterations of this environment and, consequently, of conformational changes accompanying alterations of not only the nucleoprotein functional state, but also of nucleoprotein component conformation, caused by changes of pH, temperature, ionic strength and composition of the solution, etc...

Photochemical reactions only weakly depend directly on the temperature, pH and ionic strength of the solution, thus providing an opportunity to carry out the experiment under the conditions most favourable for stability of the functional state of the nucleoprotein.

In general the ratio of different photoreactions induced by UV irradiation of common nucleoproteins depends significantly on the wavelength of the incident light, on the presence of quenchers and sensitizers. The presence of some base analogs in polynucleotides with significant absorption above 300 nm (4-thiouracil, 5-bromouracil, N^4-hydroxycytosine, etc..) permits also selective excitation of these particular bases in nucleoproteins. Hence, variation of the wavelength, of the concentration and of the nature of quenchers and sensitizers as well as incorporation of some base analogs into polynucleotide provide a degree of control in the specificity of photochemical conversions of nucleoproteins, particularly in specificity of the cross-link formation.

Only a limited number of photochemical reactions of nucleic acid components are studied in detail - photoinduced addition of different molecules to C5-C6 double bond of pyrimidines, including dimerisation (24,25), photoconversions of 5-halogenopyrimidines (25), substitution at C8 of purines (38), degradation of guanine residue (39) etc... But even these reactions were investigated mainly on monomer or polynucleotide level. Much less is known about photoreactions, which could result in the formation of covalent bonds between nucleic acid components and amino acid residues. According to the literature data different nucleic bases - purines (28), particularly adenine (40), uracil (41,42), 5-bromo and 5-iodouracils (43-45), thymine (46-48) and cytosine (14)-could be involved in UV-induced formation of covalent bonds with amino acids. But the mechanisms of such reactions and the structure and properties of the reaction products are studied only in a few cases and as a rule for model mixture of monomers (42,46,47). Only in one case the respective product was isolated also from the irradiated nucleoprotein (14). In spite of the very limited and incomplete data on the mechanisms of UV-induced cross-links formation and on the stability of the cross-linkages, which restrict application of this method, direct UV irradiation of natural nucleoproteins is widely used for the determination of nucleotide- and polynucleotide-protein interaction and of interacting fragments of macromolecules.

The application of UV irradiation for study of different nucleoproteins, including complexes of tRNA and ATP with aminoacyl-tRNA-synthetase, DNA with RNA polymerase, etc.. was reviewed some years ago (49,50). Therefore the following part will be concerned mainly with the use of UV irradiation for study of one of the most complicated and multicomponent functional nucleoproteins : ribosome.

Ribosome of *E. coli* is formed by self-assembly of three types of RNA and more than 50 proteins. During translation the ribosome forms transient complexes with additional proteins (initiation, elongation and releasing factors), with mRNA and tRNA's. Besides selection of the proper mRNA initiation site and the cognate tRNA molecules, ribosomes also translocate mRNA and tRNA molecules. Obviously, these steps are conjugated with alterations of polynucleotide-protein contacts. Variation of the conditions and addition of some antibiotics, which affect not only the rate of translation but also its fidelity, could be also conjugated with the alterations of these contacts. As it was mentioned above the most direct approach for investigation of RNA-protein contacts is UV-induced formation of cross-links between nucleic bases and amino acid residues being in contact. I do not discuss here the investigation of the structure of interacting fragments (51,52), but consider only the use of this method for determination of the direct contacts between RNA's and proteins inside the ribosome and the respective complexes and changes of these contacts along with the change of the nucleoprotein state.

The analysis of polynucleotide-protein contacts, fixed by cross-linkage formation inside multicomponent nucleoproteins, could be performed by different methods, depending on the aim and on the composition of the complex. Analysis of the free components could be performed by the conventional methods (53), but sensitivity and accuracy of this approach is usually low. Much more definite and precise results give the analysis of the cross-linked fraction. If only the presence of particular cross-linkages and kinetic of their formation are to be determined, the most useful approach involves utilisation of nucleoproteins, containing single prelabelled component - polynucleotide or protein. If all the proteins cross-

linked to particular polynucleotide should be determined, it is
necessary either to have nucleoprotein in which this particular
polynucleotide is prelabelled, or to separate this polynucleotide
with cross-linked proteins from free proteins and cross-linked
nucleoproteins containing other polynucleotides. In multicomponent
nucleoproteins, e.g. ribosomes, protein could interact and be cross-
linked with more than one polynucleotide (37,54,55), thus leading
to some complications of the analysis. The next step involves
cleavage of polynucleotides by endonuclease(s) and identification
of proteins, usually by electrophoresis (32,37,55,56). The presence
of the particular protein is as a rule determined by radioactivity
of the respective spot on electrophorogram. The radioactivity appears
due to use of complexes containing prelabelled proteins or polynucleotides (the cross-linked protein contains fragment(s) of polynucleotide), or due to labelling of cross-linked proteins, e.g.
iodination by ^{125}I, before electrophoresis. It should be mentioned
that the separation of cross-linked fraction and subsequent electrophoresis are accompanied by some losses of material, different for
different compounds. On the other hand radioactivity of each spot
in the case of prelabelled polynucleotide depends on the length of
the remaining oligonucleotide, and, in the case of iodination - on
the nature of the particular protein. Therefore only comparison of
the relative spot radioactivit obtained under the similar analysis
conditions, could be taken into account in discussion of the polynucleotide-protein contact alterations.

Functional nucleoproteins are very flexible and photochemical
modification of macromolecules can result in significant changes
of their interactions. On the other hand, to determine the interacting macromolecules and moreover the interacting fragments it is
essential to obtain sufficient amount of cross-linked substance.
Therefore the first aim is the choice of the proper doses of UV
irradiation for getting the reliable results. It was shown (37) that
absorption of up to 40 quanta (254 nm) per nucleotide does not significantly affect the physical parameters and functional properties of the ribosome (sedimentation profile of nucleoprotein, the
ability of subunits to associate, formation of ternary complexes
with mRNA and tRNA etc..). The dose-response curve deviates from
exponential also after absorption of more than 40 quanta per nucleotide (37). From these considerations all the experiments discussed below are performed at doses not exceeding 40 quanta per
nucleotide. It should be mentioned that photosensitivity of functional properties of aminoacyl-tRNA-synthetase (31) and 50 S ribosomal subunit (57) are significantly higher.

Formation of ribosomal nucleoproteins from the respective
RNA's and proteins (self-assembly) in the cell proceeds spontaneously. But in vitro only part of 30 S subunit proteins bind to 16 S
RNA at low temperature, forming ribosomal intermediate (RI) particles (58). Subsequent heating of the RI particles leads to their

Table 1. The relative radioactivity in proteins of 30 S ribosomal subunit cross-linked to RNA by ultraviolet irradiation (λ = 254 nm) of the ribosomal complexes.

Irradiated complex	Type of RNA	Proteins cross-linked	References
RI	16 S RNA	bars at 4, 7	/60/
RI*	16 S RNA	bars at 4, 7, (9/11), 12, (15-17)	/60/
free 30 S subunit	16 S RNA	bars at 4, 7, (9/11), (15-17), (19-21)	/65/
IF3·30 S	16 S RNA	bars at 2, 3, 4, 6, 7, 8/9/11, 10, 12, (15-17), (18-21), JF3	/65/
Q$_\beta$RNA·IF3·30 S	Q$_\beta$RNA	bars at 4, 5, 7, 8/9/11, 12, (15-17), 18, 21	/76/

activation — conversion into RI*, which is accompanied by some changes of 16 S RNA conformation (59) and by appearance of the ability to bind all other proteins, leading to formation of complete active 30 S subunit (58). Alterations of the UV-induced RNA-protein cross-linkages show that conversion of isolated RI particles into RI* is accompanied by strong changes of RNA-protein contacts, and association of the rest of 30 S subunit proteins with RI* particles does not change significantly these contacts (60). It should be mentioned that even split proteins which associated at last with and dissociated at first from the ribosome have direct contacts with 16 S RNA inside the native 30 S subunit (61) (except S1 protein (4), see above).

One of the steps in the formation of the translation complex is association of 30 S subunit with the elongation factor 3 (IF3) (62,63). This step is essential for binding of the natural mRNA to the 30 S subunit (64). In the IF3·30 S complex the initiation factor directly interacts with 16 S RNA and in spite of the small size (in comparison with 30 S subunit) binding of this factor strongly affects RNA-protein contacts in 30 S subunit (65).

The formation of 70 S ribosome by association of 30 S and 50 S subunits also caused pronounced alterations of RNA-protein contacts in 30 S as well as in 50 S subunit (55).

Hence, all the above data evidenced applicability of the method - analysis of the UV-induced cross-links - for the determination of the RNA-protein contacts in ribosomal complexes, and high sensitivity of this method to their alterations in the course of nucleoprotein transition from one state into another.

Investigation of the ribosomal complexes is sometimes performed under different conditions, assuming that such variation does not result in alterations of the nucleoprotein structure. But change of Mg^{+2} concentration from 10 to 20 mM results in strong alterations of polynucleotide-protein contacts in 30 S subunit (to be published elsewhere), evidencing alterations of the nucleoprotein higher structure.

Another problem which could be solved by means of the UV-induced RNA-protein cross-linkages consists in elucidation of the ribosomal proteins directly interacting with RNA of the other subunit and with transient components of translational complexes - mRNA and tRNA's.

The determination of 30 S proteins cross-linked to 23 S RNA and 50 S proteins cross-linked to 16 S inside 70 S ribosome allows elucidation of the proteins taking part in the intersubunit RNA-protein interactions (55). Obviously, these proteins must be located on the contacting surfaces of the subunits, at least in the empty ribosome, containing no mRNA and tRNA. The data obtained coincide with the results of some indirect experiments (66,67). These proteins should be also very significant from the functional point of view, because, as it will be shown below, most of them are also involved into interactions with mRNA and tRNA.

In principle, translation could start from each initiator triplet, but in the native mRNA's it starts preferentially from a few specific sites (68). The efficiency of initiation by the same ribosome is different for different initiation sites in polycystronic mRNA (68-70), and relative efficiencies for these sites are different for different types of ribosomes, e.g. *E. coli* and *B. stearothermophilus* ribosomes (71-73). A long stretch of mRNA, including among other initiator codon and Shine-Dalgarno sequences (74), are involved in formation of the initiator complex (75). All these data evidenced that selection of the proper mRNA initiator site by ribosome involves multiple specific interactions, not only between 16 S RNA and mRNA (74) but also between nucleotide residues of mRNA initiator sites and ribosomal proteins (cf. 70). The first step in this selection is association of mRNA with 30 S ribosomal subunit, which is strongly enhanced by the initiation factor 3 (IF3) (64). UV irradiation of either Q_β RNA.30 S or Q_β RNA.IF3.30 S complexes results in cross-linking to mRNA of the same ribosomal proteins, although in accordance with above-mentioned, the efficiency of cross-linkage formation is about ten times higher in the

Table 2. The relative radioactivity in proteins cross-linked to RNA by ultraviolet (λ = 254 nm) irradiation of the ribosomal complexes.

presence of IF3 (76).

In the course of the translation the ribosome could bind two tRNA molecules - aminoacy-tRNA before transpeptidation and peptidyl-tRNA before translocation are located in the so called A-site, and peptidyl-tRNA after translocation is located in the so called P-site (77). By means of different methods there were determined ribosomal proteins located in the 70 S ribosome nearby tRNA's in the A- and P-sites or essentiel for binding of aminoacyl-tRNA or peptidyl-tRNA (78-80). But UV-induced cross-linking permits to determine the ribosomal proteins directly interacting with a single tRNA molecules in both sites (37). Although the sets of proteins contacting with tRNA in the A and P-sites are different, they include components of both ribosomal subunits and some proteins are in common for both sites. It should be emphasized that the sets of proteins are essentially the same for individual (tRNA^{Phe}) and total tRNA (37).

The amino acid starvation caused reduction of ribosomal components biosynthesis - so called stringent control (81). It was concluded that such control is induced by template-dependent binding of deacylated tRNA to 70 S ribosome containing peptidyl-tRNA in the P-site, in other words by the binding of deacylated tRNA to so called A-like site (82). But after addition of deacylated tRNA to the N-AcPhe-tRNA.poly(U).70 S complex irradiation cross-links to deacylated tRNA three proteins : S7, L5 and L11 (to be published elsewhere). Only S7 is involved in interactions with aminoacyl- and peptidyl-tRNA's, located in A- and P-sites respectively (37). Hence, if the above experiment imitates well the real situation, the stringent control is switched on due to interaction of the respective deacylated tRNA with a new site, which include the above-mentioned proteins and is different from the A- as well as from the P-site, and could be called S-site. It is worth to note that some mutations of protein L11 lead to transition from the stringent to the relaxed control of ribosomal components biosynthesis (83,84).

As it was mentioned above, a number of ribosomal proteins are located on the intersubunit surfaces, as evidenced by their participation in the intersubunit RNA-protein contacts. By the same method -UV-induced formation of RNA-protein cross-linkages- it was shown that most of these proteins interact also with mRNA in the mRNA.IF3.30 S complex, with tRNA's located in the A- and P-sites, with deacylated tRNA bound to post-translocated translational complex. These data are the first evidences confirming significance of the intersubunit surfaces for the principal functions of ribosome. It should be also underlined that a lot of these proteins interact with different RNA's and in this respect are multifunctional.

All the above-mentioned data show not only rigorous substantiation and applicability of the UV-induced formation of polynucleotide-protein cross-linkages for investigation of the structure and functions of a wide range of nucleoproteins, but also high sensitivity of the cross-link formation to functional state of nucleoproteins and to even small local changes of component conformation and their arrangement inside nucleoproteins. Obviously, this approach will be very fruitful for elucidation of functional problems - functional topography of tRNA molecules, of mRNA initiator sites, of ribosome as a whole, as well as for investigation of the polynucleotide-protein interactions in general.

REFERENCES

1. A.E. Dahlberg, J.E. Dahlberg, Proc. Natl Acad. Sci. USA 72:2940 (1975).
2. M. Laughrea, P.B. Moore, J. Mol. Biol. 121:411 (1978).
3. M. Laughrea, P.B. Moore, J. Mol. Biol. 122:109 (1978).
4. I.V. Boni, I.V. Zlatkin, E.I. Budowsky, Bioorg. Khimia (in Russian) 5:1633 (1979).
5. I.V. Zlatkin, I.V. Boni, E.I. Budowsky, Bioorg. Khimia (in Russian) 6:786 (1980).
6. T.I. Tikhonenko, N.P. Kisseleva, B.P. Ivanov, M.L. Andronikova, G.A. Velikodvorskaya, E.I. Budowsky, Voprosy virusologii (in Russian) 6:622 (1970).
7. T.I. Tikhonenko, E.I. Budowsky, V.B. Sklyadneva, I.S. Khromov J. Mol. Biol. 55:535 (1971).
8. E.I. Budowsky, in : "Progress in Nucleic Acid Research and Molecular Biology", W.E. Cohn, ed., Acad. Press, N.Y.-San Francisco-London, 16:126 (1976).
9. T.I. Tikhonenko, Comprehensive Virology 5:1 (1975).
10. V.B. Sklyadneva, N.P. Kisseleva, E.I. Budowsky, T.I. Tikhonenko, Molekularnaya Biologia (in Russian) 4:110 (1970).
11. H. Hayatsu, in : Progress in Nucleic Acid Research and Molecular Biology", W.E. Cohn, ed., Acad. Press, N.Y.-San Francisco-London, 16:75 (1976).
12. I.V. Boni, E.I. Budowsky, J. Biochem. (Tokyo) 73:821 (1973).
13. M.F. Turchinsky, K.S. Kusova, E.I. Budowsky, FEBS Letters 38: 304 (1974).
14. E.I. Budowsky, N.A. Simukova, M.F. Turchinsky, I.V. Boni, Yu.M. Skoblov, Nucleic Acids Res. 3:261 (1975).
15. R. Hinnen, R. Schäfer, R.M. Franklin, Eur. J. Biochem. 50:1 (1974).
16. V.B. Sklyadneva, L.A. Chekanovskaya, T.I. Tikhonenko, Voprosy Med. Khimii (in Russian) 4:436 (1979).
17. V.B. Sklyadneva, L.A. Chekanovskaya, I.A. Nikolaeva, T.I. Tikhonenko, Biochim. Biophys. Acta 565:51 (1979).
18. N.K. Kochetkov, E.I. Budowsky, E.D. Sverdlov, N.A. Simukova, M.F. Turchinsky, V.N. Shibaev, Organic Chemistry of Nucleic Acids, Plenum Press, London-New York, part B, Chapter 8 (1972).

19. A.D. Mirzabekov, V.V. Shick, A.V. Belyavsky, S.G. Bavykin, Proc. Natl Acad. Sci. USA 75:4184 (1978).
20. P. Alexander, H. Moroson, Nature 194:882 (1962).
21. K.C. Smith, Biochem. Biophys. Res. Commun. 8:157 (1962).
22. A. Markovitz, Biochim. Biophys. Acta 281:522 (1972).
23. N.A. Simukova, E.I. Budowsky, FEBS Letters 38:299 (1974).
24. G.J. Fisher, H.E. Johns, in : Photochemistry and Photobiology of Nucleic Acids, S.Y. Wang, ed., Acad. Press, N.Y.-London, 1:169 (1976).
25. S.Y. Wang, in : Photochemistry and Photobiology of Nucleic Acids, S.Y. Wang, ed., Acad. Press, N.Y.-London, 1:296 (1976).
26. J. Offengand, R. Liou, J. Kohut, I. Schwartz, R. Zimmermann, Biochemistry 18:4322 (1979).
27. R. Zimmermann, S.M. Gates, I. Schwartz, J. Offengand, Biochemistry 18:4333 (1979).
28. E. Kuechler, J. Offengand, in : Transfer RNA, part I. Structure, Properties and Recognition, P. Schimmel, D. Söll, J. Abelson, eds, Cold Spring Harbor Laboratories, Cold Spring Harbor, New York (1979).
29. A.M. Reboud, S. Dubost, M. Buisson, J.-P. Reboud, Biochem. Biophys. Res. Commun. 93:974 (1980).
30. E.I. Budowsky, M.F. Turchinsky, N.E. Broude, I.V. Boni, I.V. Zlatkin, K.S. Kussova, N.I. Medvedeva, G.G. Abdurashidova, Kh.A. Aslanov, T.A. Salikhov, in : Frontiers of Bioorganic Chemistry and Molecular Biology, S.H. Ananchenko, ed., Pergamon Press, Oxford-N.Y., pp 389-395 (1980).
31. G.P. Budzik, S.S.M. Lam, H.J.P. Schoemaker, P.R. Schimmel, J. Biol. Chem. 250:4433 (1975).
32. M.F. Turchinsky, N.E. Broude, K.S. Kussova, G.G. Abdurashidova, E.J. Budowsky, Bioorg. Khimia (in Russian) 3:1013 (1977).
33. C. Janion, D. Shugar, Acta Biochim. Polonica 14:293 (1967).
34. A.D. McLaren, D. Shugar, Photochemistry of Proteins and Nucleic Acids, Pergamon Press, Oxford (1964).
35. G.J. Fisher, H.E. Johns, in : Photochemistry and Photobiology of Nucleic Acids, S.Y. Wang, ed., Acad. Press, N.Y.-London, 1:226 (1976).
36. D. Elad, in : Photochemistry and Photobiology of Nucleic Acids S.Y. Wang, ed., Acad. Press, N.Y.-London, 1:357 (1976).
37. G.G. Abdurashidova, M.F. Turchinsky, Kh.A. Aslanov, E.I. Budowsky, Nucleic Acids Res. 6:3891 (1979).
38. R. Raghow, D.W. Kingsbury, Virology 98:267 (1979).
39. D. Elad, in : Ageing, Carcinogenesis and Radiation Biology, K.C. Smith, ed., Plenum Press, New York-London 243 (1975).
40. V.A. Ivanchenko, E.I. Budowsky, N.A. Simukova, N.S. Vulfson, A.I. Tishchenko, D.B. Askerov, Nucleic Acids Res. 4:955 (1977).
41. V.T. Yue, P.R. Schimmel, Biochemistry 16:4678 (1977).
42. J.C. Smith, Biochem. Biophys. Res. Commun. 34:354 (1970).
43. K.C. Smith, R.T. Aplin, Biochemistry 5:2125 (1966).
44. S.-Ya Lin, A.D. Riggs, Proc. Natl Acad. Sci. USA 71:947 (1974).
45. H. Weintraub, Cold Spring Harbor Symp. Quant. Biol. 38:247 (1973).

46. R. Cysyk, W.H. Prusoff, J. Biol. Chem. 247:2522 (1972).
47. T. Jellinek, R.B. Johns, Photochem. Photobiol. 11:349 (1970).
48. A.J. Varghese, in : Agening, Carcinogenesis and Radiation Biology, K.C. Smith, ed., Plenum Press, New York-London, 207 (1975).
49. K.C. Smith, in : Photochemistry and Photobiology of Nucleic Acids, S.Y. Wang, ed., Acad. Press, N.Y.-London, 2:187 (1976).
50. K.C. Smith (ed.), in : Agening, Carcinogenesis and Radiation Biology, Plenum Press, New York-London (1976).
51. P.R. Schimmel, V.T. Yue, Research in Photobiology, A. Castellani, ed., 41 (1977).
52. Ch. Zwieb, R. Brimmacombe, Nucleic Acids Res. 6:1775 (1979).
53. B. Ehresmann, C. Backendorf, Ch. Ehresmann, R. Millan, J.-P. Ebel, Eur. J. Biochem. 104:255 (1980).
54. L. Gorelic, Biochemistry 15:3579 (1976).
55. G.G. Abdurashidova, M.F. Turchinsky, T.A. Salikhov, Kh. Aslanov, E.I. Budowsky, Bioorg. Khimia (in Russian) 4:982 (1978).
56. G.G. Abdurashidova, A.F. Pivarzyan, M.F. Turchinsky, E.I. Budowsky, Bioorg. Khimia (in Russian) 6:626 (1980).
57. O.G. Baca, J.W. Bodley, Biochem. Biophys. Res. Commun. 70:1091 (1976).
58. A.D. Pivazyan, E. Yu. Chirkova, M.F. Turchinsky, E.I. Budowsky, Bioorg. Khimia (in Russian) 5:1642 (1979).
59. M. Nomura, W.A. Held, in : Ribosomes, M. Nomura, A. Tissiers, P. Lengyel, eds, Cold Spring Harbor Lab., Cold Spring Harbor, N.Y., pp. 193 (1974).
60. A.M. Kopylov, E.S. Shalaeva, A.A. Bogdanov, Dokl. Akad. Nauk SSSR (in Russian) 216:1178 (1974).
61. A.D. Pivazyan, G.G. Abdurashidova, M.F. Turchinsky, E.I. Budowsky, Bioorg. Khimia (in Russian) 6:461 (1980).
62. M.F. Turchinsky, N.E. Broude, K.S. Kussova, G.G. Abdurashidova, E.V. Muchamedganova, I.N. Schatsky, T.F. Bystrova, E.I. Budowsky, Eur. J. Biochem. 90:83 (1978).
63. S. Sabol, S. Ochoa, Nature New Biol. 234:233 (1971).
64. M. Revel, Molecular Mechanisms of Protein Biosynthesis, H. Weissbach and S. Pestka, eds, Acad. Press, New York, pp. 245 (1977).
65. C. Vermeer, W. van Alphen, P. van Knippenberg, L. Bosch, Eur. J. Biochem. 40:295 (1973).
66. N.E. Broude, K.S. Kussova, N.I. Medvedeva, E.I. Budowsky, Bioorg. Khimia (in Russian) 5:1352 (1979).
67. G. Stoffler, H.-G. Wittman, in : Molecular Mechanisms of Protein Biosynthesis, H. Wessibach, S. Pestka, eds, Acad. Press, New-York-San Francisco-London, pp. 117 (1977).
68. J.W. Kenny, T.G. Fanning, J.M. Lambert, R.R. Traut, J. Mol. Biol. 135:151 (1979).
69. H. Lodish, J. Mol. Biol. 50:689 (1970).
70. J.A. Steitz, J. Mol. Biol. 73:1 (1973).
71. J.A. Steitz, A.J. Wahba, M. Laughrea, P. Moore, Nucleic Acids Res. 4:1 (1977).

72. M.L. Goldberg, J.A. Steitz, Biochemistry 13:2123 (1974).
73. S. Isono, K. Isono, Eur. J. Biochem. 56:15 (1975).
74. H.F. Lodish, Nature (London) 226:705 (1970).
75. J. Shine, L. Dalgarno, Proc. Natl Acad. Sci. USA 71:1342 (1974).
76. J.A. Steitz, K. Jakes, Proc. Natl Acad. Sci. USA 72:4734 (1975).
77. N.E. Broude, K.S. Kussova, N.I. Medvedeva, E.I. Budowsky, Bioorg. Khimia (in Russian) 6:1303 (1980).
78. R.J. Harris, S. Pestka, in : Molecular Mechanisms of Protein Biosynthesis, H. Weissbach, S. Pestka, eds, Acad. Press, New York-San Francisco-London, pp. 413 (1977).
79. J. Offengand, in : Ribosomes, Structure, Function and Genetics G. Chambliss, Ph. Dat, et al. eds, University Park Press, Baltimore, pp. 497 (1979).
80. A.A. Krayevsky, M.K. Kukhanova, in : Progressin Nucleic Acid Research and Molecular Biology, W.E. Cohn, ed., Acad. Press, New York, London, Toronto, Sydney, San Francisco, pp. 2 (1979).
81. A.E. Johnson, R.H. Fairclough, C.R. Cantor, in : Nuclear Acid-Protein Recognition, H.J. Vogel, ed., Pergamon Press, New York-London, pp. 469 (1977).
82. M. Cashel, I. Gallant, in : Ribosome, M. Nomura, A. Tissiers, P. Lengyel, eds, Cold Spring Harbor Lab., Cold Spring Harbor, New York, pp. 733 (1974).
83. M. Cashel, Annual Rev. Microbiol. 29:301 (1975).
84. I.D. Freisen, N.P. Fül, I.M. Parker, W.A. Haseltine, Proc. Natl Acad.Sci. USA 71:3465 (1974).
85. I. Parker, R.J. Watson, I.D. Freisen, N.P. Fül, Mol. Gen. Genetik 144:111 (1976).

ROUND TABLE SUMMARY : ENDOGENOUS AND EXOGENOUS INHIBITORS AND SENSITIZERS. FUNDAMENTAL ASPECTS

John D. Spikes

Department of Biology
University of Utah
Salt Lake City, Utah 84112 U.S.A.

INTRODUCTION

This Round Table, which served as an introduction to the more clinically oriented Round Table on photosensitization held later in the Congress (Therapeutic and Pathophysiologic Implications) was concerned with the basic aspects of photosensitization and protection against photosensitized damage at the molecular and cellular levels. It was first shown by Raab in 1900 that low concentrations of acridines and certain other dyes, which had no deleterious effects in the dark, give a rapid killing of paramecia on illumination; a few years later von Tappeiner and coworkers demonstrated that dyes such as eosin sensitize the photoinactivation of enzymes (see references in 1). Photosensitized phenomena of this type in biological systems are often termed "photodynamic" reactions, especially in those cases where oxygen is involved.

SENSITIZED PHOTOREACTIONS AT THE CELLULAR AND MOLECULAR LEVELS

Some types of biochemical lesions produced in cells by illumination in the presence of endogenous and exogenous sensitizers are fairly well understood, but little is known in general of the details of the photodynamic killing of cells. Much of the speculation on the molecular aspects of photosensitized damage in cells is based on the results of studies with both substrate (the molecule undergoing photosensitized degradation) and the sensitizer in homogeneous aqueous solution. Biologically important molecules susceptible to such photodegradation include alcohols, aldehydes, amines, some amino acids (cysteine, histidine, methionine, tryptophan and tyrosine), bilirubin, carbohydrates, many drugs, phospholipids, nitrogen heterocyclics, nucleic acids and their components (especially guanine), organic acids, phenols, most proteins, steroids, thiols, unsaturated fats, a number of vitamins, etc. (see references in 1 and 2). Also, a large number of endogenous and exogenous, naturally occurring and synthetic, molecules and ions are photosensitizers for biological systems. These include such organic compounds as acridines, flavins, haloaromatics, phenothiazines (such as methylene blue, some tranquilizers, etc.), polycyclic hydrocarbons, polyenes, iron-free porphyrins, many furocoumarins (psoralens), quinones, xanthene dyes (eosin, rose bengal, etc.); organometallic compounds such as ruthenium bipyridyl; and inorganics such as cadmium sulfide and copper (2+) and iron (3+) ions (see references in 1 and 2).

In most cases, photosensitized damage to biological systems appears to be mediated by the triplet state of the sensitizer, primarily because of its relatively long lifetime. The great majority of sensitized photoreactions in biology involve the consumption of molecular oxygen and are, in fact, sensitized photooxidation processes. There are some important exceptions to this, however. For example, some furocoumarins bind covalently to DNA on illumination in the absence of oxygen,[3] while chlorpromazine sensitizes the photokilling of bacteria[4] and the photohemolysis of erythrocytes[5] under anaerobic conditions.

Two major types of pathways can occur in photosensitized reactions involving oxygen.[1,2] In one (Type I or free radical processes) triplet sensitizer (^{3}SENS) reacts with the substrate (SUBH) by an electron/hydrogen transfer, typically giving semireduced sensitizer (SENSH·) and semioxidized substrate (SUB·) as follows:

$$^3\text{SENS} + \text{SUBH} \rightarrow \text{SENSH}\cdot + \text{SUB}\cdot$$

In many cases, the semioxidized substrate ultimately reacts with oxygen to give a fully oxidized species. Some semireduced sensitizers react efficiently with ground state oxygen ($^3O_2$) to give the superoxide radical (O_2^-) or its conjugate acid ($HO_2^\cdot$) and ground state sensitizer (^{0}SENS), as follows:

$$\text{SENSH}\cdot + {^3O_2} \rightarrow {^0\text{SENS}} + H^+ + O_2^-$$

Superoxide, in turn, can react with certain substrates. In some cases, fully reduced sensitizer (SENSH$_2$) is formed which then reacts with oxygen to give hydrogen peroxide; this species can also result from the dismutation of superoxide. Hydrogen peroxide, of course, reacts with many kinds of biomolecules. Finally, it has been suggested that hydroxyl radical, which reacts with almost any kind of organic molecule, can be produced by illuminated photosensitizers; to what extent this species might be involved in photodynamic reactions is not clear as yet. Under some conditions, reactive intermediates (as formed in Type I reactions, for example) can interact to give various types of covalent adducts (see references in 6). For example, two semioxidized substrate radicals could interact to give a substrate-substrate photoadduct. Reactions of this type might be involved in the sensitized formation of crosslinked proteins in solution.[7] Free radical processes may also participate in the photosensitized generation of protein-nucleic acid adducts.[8] Sensitizer-substrate photoadducts are also frequently observed.[9]

In the other type of reaction pathway (Type II or energy transfer processes), triplet sensitizer reacts with ground state oxygen to give singlet oxygen ($^1O_2$) and ground state sensitizer as follows:

$$^3SENS + {}^3O_2 \rightarrow {}^0SENS + {}^1O_2$$

Most types of biomolecules do not react very rapidly with ground state oxygen, but many react efficiently with singlet oxygen, giving fully oxidized products.[1,2,10]

It is clear from the above that the sensitized photoalteration of even simple biomolecules in homogeneous solution can involve a complex series of competing pathways. A number of experimental approaches have been developed to probe the relative participation of Type I and Type II processes in a given photosensitized reaction. Unfortunately, although often useful in preliminary studies, none of these mechanistic tests appear to give unambiguous results. One commonly used technique is to measure the effects on photooxidation rates of inhibitors such as azide and beta-carotene which are efficient quenchers of singlet oxygen but which do not participate actively in free radical processes. Another approach, although valid only under certain experimental conditions, is to measure the effect of deuterium oxide as solvent on photoreaction rates; rates of singlet oxygen-mediated processes can be markedly higher because of the significantly longer lifetime of this species in deuterium oxide as compared to water. The participation of superoxide and hydrogen peroxide in photoreactions can sometimes be estimated using the enzymes superoxide dismutase and catalase. Perhaps the most promising approach is to compare the pattern of photoreaction product structures with those found to be produced by reaction of the same substrate with chemically/physically generated singlet oxygen or free radical intermediates. In some systems one of the more powerful techniques is that of flash photolysis,[11] which permits quantitative measurements of the rate constants for the interactions of oxygen, substrates and inhibitors with the various excited states of the sensitizer; the technique of pulse radiolysis often serves as a useful adjunct to flash measurements. In general, then, the mechanism(s) of a given sensitized photoprocess can be elucidated only by applying a battery of techniques. Such studies are further complicated by the observation that mechanisms are a function of the sensitizer (chemical structure, physical state), the substrate, the solvent (chemical nature, dielectric constant), and the reaction conditions (reactant concentrations, pH, temperature, etc.). Psoralens, for example, can sensitize in two different ways depending on the reaction system and conditions. Illumination of psoralens complexed with DNA results in the formation of covalent photoadducts between the psoralen and the pyrimidine moieties of the DNA; oxygen is not required for this process and may even be inhibitory. In contrast, in dilute solution, illumination gives triplet psoralen which can react with oxygen giving singlet oxygen, which can then oxidize proteins, for example.[12]

Photochemistry in homogeneous systems is not simple, but sensitized photoreactions in cells must be exceedingly complex. Cells are intricately non-homogeneous with abrupt phase changes and large variations in the chemical composition and polarity of different regions. This results in a large number of membrane-compartmentalized microenvironments with different physical/chemical properties for substrates and for both endogenous and exogenous sensitizers and inhibitors. The photochemical complexity is multiplied by the fact that many sensitizers localize in particular regions of cells or bind selectively to certain subcellular components; subsequent illumination can thus result in selective, localized damage to cell structure and function. As an intermediate step in approaching the complexity of cell photosensitization, studies have been carried out with several model systems. These are, of course, very simple in comparison with living cells, but if properly used, they can hopefully provide some information as to how selected factors might influence the kinetics and mechanisms of sensitized photoreactions in vivo. Since many sensitizers form non-covalent complexes with various cell components, a number of studies have been made on the photochemical effects of such binding (see 6 and references therein). In a number of cases the photochemical efficiency of the sensitizer is increased, apparently as a result of the reduced probability of non-specific collisional deactivation of the light-excited bound sensitizer. Increased efficiency is especially observed with nucleic acid-bound sensitizers,[13] but it also occurs on binding to some proteins.[6] In addition, binding may also alter the photochemical mechanism from that predominating in homogeneous solution; in particular, an enhancement of the direct interaction between excited sensitizer and substrate via Type I processes is observed. Such processes probably enhance the formation of dye-substrate and substrate-substrate covalent photoadducts.[6]

Surfactants in aqueous systems above their critical micellar concentrations have long been used as models of biological membranes. Such micellar structures have both hydrophobic and hydrophilic regions; thus water-insoluble materials can be "solubilized" in aqueous micellar systems by being incorporated into the micellar structure. Recently there has been a burst of research activity in the use of micellar systems as models in photosensitization studies.[6,14-16] Substrates and sensitizers incorporated into micelles and dissolved in the aqueous medium in various combinations have been used in this work. Although the results are somewhat complex, they show that singlet oxygen generated in the solution can penetrate into micelles with good efficiency and oxidize incorporated substrates. Further, both singlet oxygen and electrons generated photochemically in a micelle can diffuse through the aqueous medium and react with substrates incorporated in other micelles. All of these results suggest the importance of using micellar systems as models for cell-level photosensitization mechanisms.[6]

Liposomes (microscopic bilayered phospholipid vesicles) have also been used extensively as models for certain aspects of cell membrane behavior.[17] Non-polar molecules can be incorporated into the liposomal membrane during preparation and water-soluble materials can be entrapped in the interior of the liposome. Singlet oxygen generated photodynamically in the aqueous medium or by microwave discharge can oxidize membrane lipid resulting in liposomal lysis. Incorporation of the protective compounds beta-carotene or alpha-tocopherol into the membrane prevents both lipid peroxidation and membrane lysis; free radical quenchers also inhibit photodynamic attack.[18] A number of the usual photodynamic sensitizers as well as phototoxic phenothiazines[19] and all-trans retinal[20] sensitize liposome damage. The presence of hematoporphyrin in cholesterol containing liposome membranes leads to the formation of the singlet oxygen oxidation product of cholesterol on illumination; if the hematoporphyrin is in the aqueous medium surrounding the liposomes rather than in the membrane, the rate of photooxidation of membrane cholesterol is markedly reduced.[21] These and other studies indicate the utility of using liposomes as models for photosensitized effects on cell membranes.

A very large amount of work has been carried out on photosensitized phenomena in cells. Many kinds of cells have been used as experimental material including procaryotic microorganisms (primarily bacteria), eucaryotic microorganisms (paramecium, yeast, etc.) and various types of cells from multicellular organisms (sperm, eggs, erythrocytes, mammalian cells in tissue culture, etc.). A wide variety of effects has been documented depending on the cell type; the sensitizer; the time of contact of the sensitizer with the cells prior to illumination; the reaction conditions before, during and after illumination; the physiological status of the cells; etc. Effects observed include "killing", delay of cell division, loss of motility, plasma membrane damage, increase in osmotic fragility leading to cell lysis, interference with metabolic processes (decreased active transport, protein and nucleic acid synthesis, respiration, glycolysis, etc.), production of chromosomal aberrations and mutations, etc. (see 1, 6 and 22 for references). Most cell level studies have been carried out with exogenous sensitizers; however some types of cells contain endogenous sensitizers such as chlorophyll and other non-iron porphyrins, etc. In many cases, exogenous protectors (sodium azide, etc.) can decrease photosensitized damage. Also, cells can contain endogenous molecules with a protective function such as beta-carotene and alpha-tocopherol. Replacement of the environmental water with deuterium oxide increases the photosensitized damage to cells in some cases, suggesting the involvement of singlet oxygen.[22]

Many attempts have been made to determine the site(s) of sensitized photodamage in cells as well as the nature of the lesions produced. As indicated above, sensitizers often localize in cells and thus produce damage to limited regions of cells or to particular structures or organelles on subsequent illumination. Unfortunately, in most cases, the site of localization is not well established.[23] Ito[22] distinguishes three major situations with respect to sensitizer localization. In the first, sensitizer remains outside of the cell or penetrates just into the cell membrane and thus produces photodamage only in the membrane. In the second, sensitizer penetrates into the cytoplasm and thus sensitizes photodamage to cytoplasmic proteins and nucleic acids, mitochondria, lysosomes, ribosomes, etc. In the last case, sensitizer penetrates into the nucleus and thus sensitizes chromosomal and DNA damage. Although considerable work has been done on all three of these phenomena, we probably know the most about sensitized reactions in cell membranes; also model studies as discussed above are most applicable to the membrane system.[6,23]

PROGRAM OF THE ROUND TABLE

In the formal part of the Round Table, the first speaker (René V. Bensasson, Paris, France) discussed the present difficulties of attempting to correlate the photochemical and photophysical properties of furocoumarins of different structures with their photobiological activity at the cellular level (see references 24-26). Since furocoumarins are used in the photochemotherapy of several skin disorders in man (psoriasis, vitiligo, etc.) this presentation served as an introduction to the subsequent Round Table on the clinical aspects of photosensitization as well as to the symposium on furocoumarins. Data were presented for a number of furocoumarins with respect to the efficiencies of triplet state formation, triplet state properties, and rate constants for the quenching of triplets by thymine, uracil and tryptophan; the complexing constants of a number of furocoumarins with DNA were also discussed. Furocoumarins intercalate between base pairs in the double helix of DNA and, on excitation with UV-A radiation, form cyclobutane type monoadducts with pyrimidine bases or interstrand crosslinks between two pyrimidine bases. The photochemical properties of the linear, monofunctional furocoumarin, 3-carbethoxypsoralen, were discussed in particular. This sensitizer shows no erythemogenic reaction and no photocarcinogenic action in mice. Since it appears to be effective in the phototherapy of psoriasis, it may be a useful alternative to the usual furocoumarins such as 8-methoxypsoralen which appear to have photocarcinogenic effects after long term use in therapy.

The next speaker, Hermann Berg (Jena, German Democratic Republic), discussed the mechanisms involved in the photodynamic damage of nucleic acids and their components. A number of types of photosensitized lesions can be produced in nucleic acids including non-alkalai labile sites, alkalai labile sites, single strand breaks and double strand breaks. Lesions are produced by both Type I and Type II mechanisms. Guanine is the most sensitive base in nucleic acids and the end products of guanine photodegradation are similar with both Type I and Type II mechanisms. With Type I reaction pathways, it appears that sugar radicals are formed leading to strand breaks in the nucleic acid backbone. In the Type I pathways, sensitizer radicals are produced from sensitizer triplet states. It was suggested that these then react with guanine to give guanine radicals which can break down or which can react with sugar moieties to produce sugar radicals which in turn give strand breaks. A number of the factors involved in the sensitized photodegradation of nucleic acids were discussed, including the effects of different binding states of the sensitizer (intercalation, stacking, very low binding) on the generation of sensitizer radicals and singlet oxygen with different sensitizers, the effect of the electron donor oxidation potential on reaction rates with sensitizer radicals, possible mechanisms of double strand breaks, etc. (see references 27, 28).

René Santus (Paris, France) discussed some aspects of the photochemistry of amino acids as well as the problems associated with attempting to extend the results of photochemical studies with amino acids, simple peptides, etc., especially if carried out under anaerobic conditions, to proteins and to in vivo systems with oxygen present. Free tryptophan can be excited directly by the shorter wavelengths of the UV-B region with the formation of N-formylkynurenine as the major photoproduct. Tryptophan is converted to the same product using exogenous photodynamic sensitizers. N-Formylkynurenine, which absorbs at longer wavelengths than tryptophan, in turn is an efficient photodynamic sensitizer which mediates both Type I and Type II reactions with the usual biological substrates (amino acids, DNA, vitamins, biomembranes, etc.). Tryptophan occurs in blood and interstitial fluid and is possibly subject to photooxidation by endogenous and exogenous sensitizers as well as to direct UV-B photolysis in body areas exposed to solar radiation. The resulting N-formylkynurenine could, in turn, amplify photodynamic damage at the cellular level. Tryptophan incorporated into proteins, including lens protein, is also converted to N-formylkynurenine by UV-B radiation or by sensitized photooxidation (see references in 29). In the latter case, the N-formylkynurenine is suspected of being responsible for protein-protein crosslinking and aging of the human lens and for cataract formation. The involvement of structural effects in the photooxidation of proteins, which makes it often impossible to generalize from results obtained with model systems, is well exemplified by the photosensitized oxidation of carotenoproteins such as crustacyanins. These are blue carotenoproteins extracted from marine invertebrates in which the prosthetic group of the protein is a beta-carotene derivative (astaxanthin), tightly bound to hydrophobic sites of the apoprotein. The strains introduced by the binding, as well as by the interactions with the protein, shift the carotenoid absorption to the red and lead to new chemical properties. Thus, while free beta-carotene derivatives are well-recognized physical quenchers of singlet oxygen in solution, the blue carotenoprotein is irreversibly bleached by either singlet oxygen or superoxide at rates approaching the diffusion controlled limit.

The next speaker (Giulio Jori, Padova, Italy) discussed some aspects of photosensitized processes in cell membranes and in membrane model systems,[6,15] pointing out that the cell membrane is the subcellular organelle whose photosensitized degradation has been studied in the greatest detail.[6,23] In studies of photodynamic effects on membranes of nerve cells with fluorescein sensitizers[31] and on mammalian cells with porphyrin sensitizers,[32] it is found that the most effective sensitizers of membrane damage are rather hydrophobic and thus penetrate into the membrane, localizing near its inner surface. Highly water soluble porphyrins, which remain outside of the membrane, are much less effective than those which bind in the membrane; this is surprising, since externally generated singlet oxygen has ample time to diffuse through the cell membrane during its lifetime.[6] Further, experiments with aqueous dispersions of micelles show clearly that singlet oxygen generated in a micelle can diffuse through the aqueous medium, penetrate other micelles and attack substrates there.[15] Illumination of photosensitized cell membranes leads to photodegradation of unsaturated lipids, cholesterol, certain amino acids in membrane proteins, etc. Some of the radical type reaction pathways probably lead to the formation of adducts between adjacent protein molecules and between protein and lipid molecules.[6,32-34] As a result, membrane function is severely altered, e.g., certain enzymes are inactivated, and sugar and amino acid transport systems and ionic pumps are inhibited. Membrane fragility is increased and cells often lyse (especially erythrocytes); the mechanisms involved here are not completely resolved but probably involve steroid photooxidation products and cross-linked membrane proteins.[32-34]

Takashi Ito (Tokyo, Japan) then discussed cell level photodynamic effects with emphasis on the relation of damaged sites to the location of the sensitizer. Comparative studies with dye photosensitizers have led to a scheme in which cell level photodynamic effects are classified in three types: (a) attack from outside of the cell with the sensitizer remaining outside, (b) attack from inside the cell with the sensitizer localized mainly in the cytoplasm, and (c) attack from inside the cell with the sensitizer being preferentially bound to DNA.[22] The exogeneous photosensitizers toluidine blue, thiopyronine and acridine orange belong to the first, second and third categories, respectively. Acridine orange penetrates rapidly into cells; recent experiments detect acridine orange reaching DNA within 20 seconds after the dye is added to yeast cells. The acridine orange in the nucleus mediates photochemical genetic changes and cell killing; that part of the dye remaining outside of the cell has a negligible sensitizing effect. In contrast, thiopyronine accumulates in membranous structures in the cytoplasm; on illumination, the fermentation and respiratory systems are damaged. Mutants resistant and sensitive to thiopyronine-sensitized photokilling have been isolated. With toluidine blue, a combined method of fast mixing of dye and cells (< 20 milliseconds) and a short period of illumination (2-30 seconds) shows that the sensitizer has two modes of action with yeast cells. One is mediated by toluidine blue in a free form outside the cell and the other by dye bound or adsorbed to the membrane; the latter form sensitizes more effectively. Interestingly, in similar experiments, hematoporphyrin shows only one mode of action, i.e., free in solution; there is no evidence of adsorption or binding to the cells. Neither toluidine blue nor hematoporphyrin sensitize any genetic effects.

Norman I. Krinsky (Boston, U.S.A.), the last Round Table speaker, reminded the audience that little information on the molecular mechanisms of photosensitized cell damage can be obtained by counting colonies or surviving cell numbers. A better approach is to use model systems in which one or only a few parameters are investigated. Such model systems have included individual proteins or enzymes, nucleic acids and their components, and model cell systems such as liposomes or mammalian red blood cells; the latter function more like model cell systems than as real cells since they lack a nucleus and cannot reproduce. Much important information on mechanisms of photosensitized reactions and on how natural protective agents function has been obtained with such models. Carotenoids appear to have the property of protecting cellular and model systems from photosensitized oxidations. Bacteria which normally synthesize carotenoids are protected against photosensitized oxidation, in comparison to mutant strains which have lost the ability to synthesize these pigments. Similar observations are made in liposomes which have had carotenoid pigments introduced during their preparation. In both cases, the ability of the carotenoid pigments to protect is a function of the length of the polyene chain. This seems to be related to the triplet energy level, which in turn must be lower than the energy difference between ground state oxygen and singlet oxygen. Carotenoids are generally considered to protect largely by quenching singlet oxygen. However, caution must be exercised in interpreting experiments in which one observes carotenoid protection since there are now several examples of carotenoid pigments apparently exercising a protective function in systems which do not involve the production of singlet oxygen. Therefore the observance of protection does not necessarily define the mechanism of damage (see references 35, 36).

SUMMARY

Considerable discussion then followed, which cannot be detailed here, involving members of the Round Table and the audience. There seemed to be a general feeling that, although much remains to be learned about the fundamental photochemistry of sensitizers and the mechanisms of the photosensitized degradation of biomolecules in homogeneous solution, much more effort should be expended in carrying over such information to model systems (micelles, liposomes, artificial bilayers, etc.). Further, many participants felt that much more effort should be devoted to photosensitized studies with cells, especially with mammalian and human cells in culture; both endogenous and exogenous sensitizers should be studied, and more studies of the mechanisms of protector action should be carried out. Some progress is presently being made in establishing the detailed molecular mechanisms of photosensitized reactions at the cellular level, especially effects on the plasma membrane and perhaps at the genetic sites. Most of the generally-used sensitizers (except furocoumarins), absorb visible light and have generally been studied using light in this wavelength range. However, many sensitizing molecules absorb strongly in the UV-A range; radiation of this type is present in fairly high intensity in our environment. Because of this, more effort should be devoted to UV-A studies involving both endogenous and exogenous sensitizers. Little is known about the genetics of sensitivity and resistance of cells to photosensitized attack; further work should be done in this area. Finally, more effort should be made to carry over basic information obtained at the molecular and cellular levels to a better understanding of clinical sensitized phototherapy and to the development of improved techniques in this area.

REFERENCES

Space does not permit a detailed listing of pertinent references; hopefully, those given below will provide a useful introduction to the literature on photosensitization at the molecular and cellular levels.

1. J. D. Spikes, Photodynamic action, Photophysiology 3:33 (1968);
    J. D. Spikes, Photosensitization, p. 87 in "The Science of Photobiology," K. C. Smith, ed., Plenum Press, New York (1977);
    J. D. Spikes, Photodynamic reactions in photomedicine, in "Photomedicine: Light in Disease and Therapy," J. D. Regan, ed., Plenum Press, New York (in press, 1980).
2. C. S. Foote, Photosensitized oxidations and singlet oxygen: consequences in biological systems, p. 64 in "Free Radicals in Biology," W. A. Pryor, ed., Academic Press, New York (1976).

3. G. Rodighiero and F. Dal'Acqua, Biochemical and medical aspects of psoralens, Photochem. Photobiol. 24:647 (1976).
4. I. Rosenthal, E. Ben-Hur, A Prager and E. Riklis, Photochemical reactions of chlorpromazine; chemical and biochemical implications, Photochem. Photobiol. 28:591 (1978).
5. I. Kochevar and A. A. Lamola, Chlorpromazine and protriptyline photooxicity: photosensitized, oxygen independent red cell hemolysis. Photochem. Photobiol. 29:791 (1979).
6. G. Jori and J. D. Spikes, Photosensitized oxidations in complex biological structures, in "Oxygen and Oxy-Radicals in Chemistry and Biology," M. A. J. Rodgers and E. L. Powers, eds., Academic Press, New York (in press, 1981).
7. A. W. Girotti, S. Lyman and M. R. Deziel, Methylene blue-sensitized photooxidation of hemoglobin: Evidence for cross-link formation. Photochem. Photobiol. 29:1119 (1979).
8. Hélène, C., Photosensitized cross-linking of proteins to nucleic acids, p. 149 in "Aging, Carcinogenesis and Radiation Biology," K. C. Smith, ed., Plenum Press, New York (1976).
9. I. E. Kochevar, Photoallergic responses to chemicals, Photochem. Photobiol. 30:437 (1979).
10. N. I. Krinsky, Singlet oxygen in biological systems, Trends Biochem. Sci., February (1977).
11. L. I. Grossweiner, Flash photolysis work in photobiology, Photophysiology 5:1 (1970).
12. W. Poppe and L. I. Grossweiner, Photodynamic sensitization by 8-methoxypsoralen via the singlet oxygen mechanism, Photochem. Photobiol. 22:217 (1975).
13. E.-R. Lochmann and A. Micheler, Binding of organic dyes to nucleic acids and the photodynamic effect, p. 223 in "Physico-Chemical Properties of Nucleic Acids," Vol. 1, J. Duchesne, ed., Academic Press, New York (1973).
14. O. Bagno, J. C. Soulignac and J. Joussot-Dubien, pH dependence of sensitized photooxidation in micellar anionic and cationic surfactants using thiazine dyes, Photochem. Photobiol. 29:1079 (1979).
15. C. Scofienza, A. Van de Vorst and G. Jori, Type I and Type II mechanisms in the photooxidation of L-tryptophan and tryptamine sensitized by hematoporphyrin in the presence and absence of sodium dodecyl sulphate micelles, Photochem. Photobiol. 31:351 (1980).
16. B. A. Lindig and M. A. J. Rodgers, Yearly Review: Molecular excited states in micellar systems, Photochem. Photobiol. 31:617 (1980).
17. B. R. Ryman and D. A. Tyrrell, Liposomes-methodology and applications, Front. Biol. 48:549 (1979).
18. S. M. Anderson, N. I. Krinsky, M. J. Stone and D. C. Clagett, Effect of singlet oxygen quenchers on oxidative damage to liposomes initiated by photosensitization or by radiofrequency discharge, Photochem. Photobiol. 20:65 (1974).

19. E. S. Copeland, C. R. Alving and M. M. Grenan, Light-induced leakage of spin label marker from liposomes in the presence of phototoxic phenothiazines, Photochem. Photobiol. 24:41 (1976).
20. M. Delmelle, Retinal sensitized photodynamic damage to liposomes, Photochem. Photobiol. 28:357 (1978).
21. K. Suwa, T. Kimura and A. P. Schapp, Reaction of singlet oxygen with cholesterol in liposomal membranes. Effect of membrane fluidity on the photooxidation of cholesterol, Photochem. Photobiol. 28:469 (1978).
22. T. Ito, Cellular and subcellular mechanisms of photodynamic action: the $^1O_2$ hypothesis as a driving force in recent research, Photochem. Photobiol. 28:493 (1978).
23. A. A. Lamola, Photodegradation of biomembranes, p. 53 in "Research in Photobiology," A. Castellani, ed., Plenum Press (1977).
24. M. T. Sa E Melo, D. Averbeck, R. V. Bensasson, E. J. Land and C. Salet, Some furocoumarins and analogs: comparison of triplet properties in solution with photobiological activities in yeast, Photochem. Photobiol. 30:645 (1979).
25. R. V. Bensasson, E. J. Land, C. Salet, R. W. Sloper and T. G. Truscott, Psoralen phototherapy and the possible involvement of triplet excited states, p. 431 in "Radiation Biology and Chemistry; Research Developments," H. E. Edwards, S. Navaratnam, B. J. Parsons, and G. O. Phillips, eds., Elsevier Scientific Publishing Co., Amsterdam (1979).
26. R. V. Bensasson, C. Salet, E. J. Land and F. A. P. Rushton, Triplet excited state of the 4'5' photoadduct of psoralen and thymine, Photochem. Photobiol. 31:129 (1980).
27. H. Berg, F. A. Gollmick, H. Triebel, E. Bauer, G. Horn, J. Flemming and L. Kittler, Redox processes during photodynamic damage of DNA. I. Results obtained by several physico-chemical methods, Bioelectrochem. Bioenerget. 5:335 (1978).
28. H. Berg, Redox processes during photodynamic damage of DNA. II. A new model for electron exchange and strand breaking. Bioelectrochem. Bioenerget. 5:347 (1978).
29. M.-P. Pileni and R. Santus, On the photosensitizing properties of N-formylkynurenine and related compounds, Photochem. Photobiol. 28:525 (1978).
30. J. P. Pooler and D. P. Valenzo, The role of singlet oxygen in photooxidation of excitable cell membranes, Photochem. Photobiol. 30:581 (1979).
31. K. Kohn and D. Kessel, On the mode of cytotoxic action of photoactivated porphyrins, Biochem. Pharmacol. 28:2465 (1979).
32. A. A. Lamola and F. H. Doleiden, Cross-linking of membrane proteins and protoporphyrin-sensitized photohemolysis, Photochem. Photobiol. 31:597 (1980).
33. T. M. A. R. Dubbelman, A. F. P. M. deGoeij, and J. van Steveninck, Photodynamic effects of protoporphyrin on human erthrocytes. Nature of the cross-linking of membrane proteins. Biochim. Biophys. Acta 511:141 (1978).

34. A. W. Girotti, Porphyrin-sensitized photodamage in isolated membranes of human erythrocytes, Biochemistry 18:4403 (1979).
35. N. I. Krinsky, Non-photosynthetic function of carotenoids, Phil. Trans. Royal Soc. (London) B284:581 (1978).
36. N. I. Krinsky, Carotenoid protection against oxidation, Pure Appl. Chem. 51:649 (1979).

ROUND TABLE SUMMARY : PREBIOTIC PHOTOCHEMISTRY AND PHOTOCHEMICAL REACTIONS IN SPACE

François Raulin

LPCE, Université Paris Val-de-Marne
Avenue du Général de Gaulle
94000 Créteil, France

INTRODUCTION

The Round Table on prebiotic photochemistry was mainly devoted to the problem of the origin of Life and the possible role of light. Most of the questions concerning the origin of living systems on the Earth, involving photochemistry, were reviewed during the presentations by the speakers and the discussions which followed. Those topics were mainly, by logical order : organic photochemical synthesis in planetary atmosphere, prebiotic photochemical synthesis in aqueous media, model systems of photosynthesis, origin of chirality, and living systems as open systems.

CHEMICAL EVOLUTION

It is generally believed that Life appeared on the Earth after a long chemical evolution from non-living matter to living systems (1) : a so-called "Chemical Evolution".

One of the first phases of this evolution was the formation of organic monomers, such as amino-acids, heterocyclic bases or sugars, from the simple molecules present in the atmosphere. In fact, this initial phase is currently divided into two distinct steps. In a first step, the transformation of simple molecules in the atmospheric phase, under the influence of various sources of energy, including UV light, gave rise to volatile organic compounds with low molecular weight. Some of these compounds, called atmospheric precursors, have unsaturated carbon chains and activated chemical groups, such as activated nitriles or aldehydes, and they are very reactive. In a second step, the evolution of these organic precursors in aqueous solution, after dissolving in the water of the primitive oceans, eventually in the presence of solid phases such as clays, allowed the formation of monomers of biological interest.

The following phase of Chemical Evolution, would have been the accumulation of those monomers in the primitive oceans and their condensation into polymers of biological interest : protoproteins, for instance. Then, the differentiation, by formation of a separated phase, would have allowed the formation of coacervates, by putting together the various polymers. The last phase -but not the least- would have been the formation by feed-back processes of organized structures of increasing complexity, giving rise to autoreproductive systems, and to the first primitive living system.

The last phase of Chemical Evolution is still almost unknown, but the three others have been the subject of many experimental and theoretical works. This is particularly true for the initial phase.

LIGHT AND ORGANIC SYNTHESIS IN PLANETARY ATMOSPHERES

There are still a lot of controversies about the composition of the primitive atmosphere of the Earth. Without opening the problem here, it may be useful to remember that models of the primitive Earth's atmosphere vary from so-called reducing models such as CH_4-N_2 or NH_3-H_2O mixtures, to non-reducing models, such as CO_2-N_2-H_2O mixtures. In any case, there was probably no free oxygen.

Since the first, now classical, Miller's experiment in 1953 (2), hundreds of experiments of this type have been carried out, with various atmosphere models and with various sources of energy. However, only a few experimental works have been achieved on the formation of atmosphere precursors, using UV light. A critical review of these works (3), presented by F. Raulin, has been performed in order to compare the role of UV light and of electrical discharges, the two most important sources of energy on the primitive Earth.

The comparison between the results obtained from those two sources of energy indicates four main differences, as shown on table 1.

- Unlike electric discharges, UV light is not an efficient source for producing unsaturated carbon chains.

- UV light is efficient for producing nitriles in CH_4-NH_3 mixtures, when the molar ratio of NH_3 is very low, while electric discharges need a higher molar ratio of NH_3.

- UV light is not able to produce nitriles from CH_4-N_2 mixtures, while electric discharges produce important quantities of diversified nitriles from these mixtures.

Table 1 - Compared efficiency[a] of synthesis of organic compounds from reducing model atmospheres, using UV light and electric discharges (3).

Synthesized organics	Energy sources	
	UV light	Electric discharges
Hydrocarbons	-	+
Nitriles - From CH_4-NH_3 poor in NH_3	+	-
Nitriles - From CH_4-NH_3 rich in NH_3	-	+
Nitriles - From CH_4-N_2	-	+
Aldehydes	-	+

a : relative efficiency : high (+) or low (-).

- UV light is not efficient for producing aldehydes from CH_4-H_2O model atmospheres, while electric discharges seem to be able to produce them more efficiently.

The presently available experimental data lead to believe that, although UV light was a source of energy more abundant than electrical discharges, in the primitive Earth (as it is now), it has played a less important role in the synthesis of atmospheric organic precursors of bionomers.

The most important contribution of UV light to the evolution of the primitive Earth's atmosphere has surely been the more or less rapid transformation of a CH_4-N_2-H_2O atmosphere to a CO_2-CO-N_2-H_2O atmosphere.

PREBIOTIC PHOTOCHEMICAL SYNTHESIS IN AQUEOUS MEDIA

Nevertheless, it is clear that light from the sun is the most abundant energy source on our planet. Was its possible role in Chemical Evolution more important in the following step -the transformation of atmospheric precursors in water- than in the atmospheric syntheses ?

In his presentation on prebiotic photochemical synthesis in

aqueous media, Dr. M. Halmann, from the Weizmann Institute, Israel, reviewed some of the several reports on prebiotic synthesis in aqueous solution which have been successfully achieved with the help of UV light.

But, as noticed by Dr. A. Bossard (3,4), from the University of Paris-Val-de-Marne, France, it is easy to calculate that, for most of the plausible reducing models of the atmosphere of the primitive Earth, the absorption of far UV light from the sun must occur in the middle and upper atmosphere (i.e. : 10 km or more), that means at low pressure and far from any liquid water. Most of the experiments of prebiotic synthesis in aqueous media which use UV light as energy source are not right simulation experiments of the plausible conditions which were available at the periphery of the primitive Earth. It seems, effectively, very improbable that the far UV light which is used in these experiments has never reached the surface of the oceans of the primitive Earth.

M. Halmann, in his presentation, described the last results concerning experiments on the photo-assisted reduction of carbon dioxide and the formation of organic compounds containing 2 and 3 atoms of carbon.

Semi conducting solids, by irradiation with UV light, are able to provide reduction of CO_2 in aqueous solution into formic acid, formaldehyde, methanol or even methane (5). This photo-assisted reduction of CO_2 occurs, for instance, in the presence of Fe^{II} salts, with UV light. But it occurs also <u>with visible light</u>. In the presence of inorganic minerals of semi conducting properties (6), CO_2 in aqueous media, in absence of molecular oxygen, undergoes reduction photosensitized by visible light. This reduction is very efficient when the used mineral is nontronite, an iron-containing clay mineral. It produces mainly formaldehyde and methanol.

$$CO_2 + H_2O \xrightarrow{h\nu} H_2CO + O_2$$

$$2\ H_2CO \xrightarrow{h\nu} CHO-CH_2OH$$

$$CHO-CH_2OH + H_2CO \xrightarrow{h\nu} CHO=CH_2-CHO + H_2O$$

Fig. 1. Photosynthesis of Malonaldehyde from CO_2 (6).

Such a photo-assisted oxidation-reduction reaction may be one of the initial processes of producing molecular oxygen and energy rich organic compounds, such as aldehydes. It appears as a model of prebiological photosynthesis, able to provide photoinduced carbon fixation reaction, through the formaldehyde reaction, as shown in figure 1.

Effectively, formaldehyde produced in this reaction can give rise, by photosynthetic condensation, to glyoxal and malonaldehyde (Fig. 1). Those condensations were achieved by irradiating dilute aqueous solutions of formaldehyde with UV light in the absence of molecular oxygen (6). Malonaldehyde is an important organic compound in prebiotic chemistry. Its condensation with urea or with guanidines forms hydroxypyrimidines and amino-pyrimidines. It may be a plausible source of pyrimidine on the primitive Earth.

MODEL SYSTEMS OF PHOTOSYNTHESIS

Those experiments show photochemical reaction systems involving light induced electron transport. Such systems are important and they must be considered, in order to study the origin and evolution of photosynthetic transformation in the living systems.

Dr. M. Grätzel, from the Institut de Chimie-Physique, Ecole Polytechnique de Lausanne, Switzerland, gave a presentation on model systems of photosynthesis. In particular, he described in detail new model systems which allow the photoproduction of molecular hydrogen from water (7). In a model, platinum catalysts, replacing hydrogenase, are put in the presence of isolated chloroplast. Using water as electron source, methylviologen MV^{2+} as redox mediator and the glucose oxidase-catalase as O_2 scavenger system, the catalysts evolve H_2 at rates comparable to that obtained with the enzymes.

In the photosystem presented in detail by M. Grätzel, methylviologen, MV^{2+} is used with the ruthenium Tris (bipyridine) complex $Ru(bipy)_3^{2+}$, with visible light. One of the main important points of this model is the fact that, in the same system, formation of molecular oxygen and of molecular hydrogen is obtained from water (8), using two catalysts in separated phases : platinum on one side and ruthenium dioxide on the other side. As shown on figure 2, the coupling of the MV^{2+} cycle and of the $Ru(bipy)_3^{2+}$ cycle allows the continuous regeneration of the two redox systems, with the simultaneous decomposition of water into O_2 and H_2, through the photoassisted redox reaction :

$$Ru(bipy)_3^{2+} + MV^{2+} \xrightarrow{h\nu} Ru(bipy)_3^{3+} + MV^{+}$$

This is a nice example of a non biological photosystem. Studies of the effect of various parameters on the efficiency of H_2 production has been achieved. The quantum yields seem to be relatively high and this system appears as a very promising way in the modelization of biological photosynthesis.

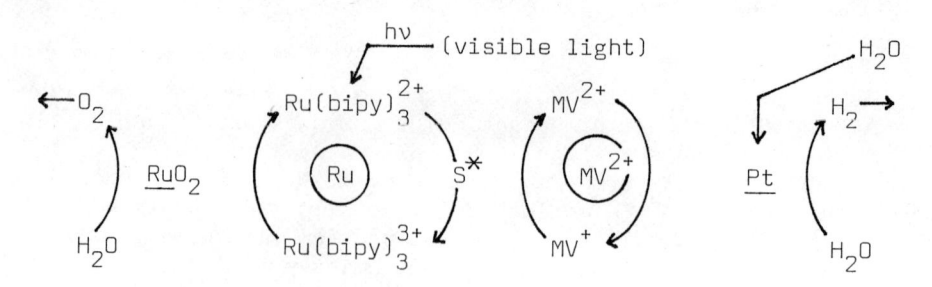

Fig. 2. An example of photoelectrolysis of water.

These artificial photosystems provide also very interesting examples of reactions involving photosensitized electron transfers, in the absence of living system. Some of the substances which are used in these experiments, as photoreceptors, may have been present in the early Earth's periphery. They could have played a role of primary photosensitizer, and they could have been involved in prebiotic photosynthetic systems. Thus, studies of artificial photosynthesis may give some important information on the chemical evolution of photosynthesis.

LIGHT AND THE ORIGIN OF CHIRALITY

Obviously, light has played a direct role in the electron transfer processes, during the several phases of Chemical Evolution, and it still plays such a role in the biological photosynthesis. Another possible role of light in Chemical Evolution is related to the origin of chirality in the building blocks of the living systems.

Table 2 - Chemical processes able to induce chirality.

Asymmetric synthesis	A → S, A → R
Kinetic resolution	R → R' R → S'
Racemic activation	R ⇌ S
Asymmetric conversion	R — R' ↕ S — S'

PREBIOTIC PHOTOCHEMISTRY

Dr. G. Balavoine, from the Laboratoire de Synthèse Asymétrique, Université de Paris Sud in Orsay, France, reviewed the different models involving light which can be, so far, pointed out to explain the origin of chirality in biological molecules.

Four different processes may be expected, which are able to induce the formation of chiral molecules during chemical transformations, as represented on table 2.

- Asymmetric synthesis of R and S from A, may allow the formation of optically active molecules from a non-chiral molecule A.

- Kinetic resolution can give rise from a racemic mixture (R + S) to a mixture of products (R' + S') which is not racemic, if the kinetic constants of the reactions R → R' and S → S' are not the same.

- Racemic activation related to the chemical equilibrium between two enantiomers R and S, may also produce an increase of the concentration of one enantiomer relatively to the other, from a mixture, initially racemic.

- Finally, asymmetric conversion of a racemic mixture (R + S), may give rise to a non racemic mixture (R' + S'), by coupling of kinetic resolution and racemic activity.

It has been shown that circularly polarized light can induce an optical activity from a racemic mixture, while destroying preferentially one of the two enantiomers of the initial mixture. For instance the photo-decomposition of camphor by polarized light can provide camphor with up to 35 % optical purity (9).

Circularly polarized light exists now : such a light is diffused, for instance, by Jupiter and other planets. It can be assumed that it existed also on the primitive Earth. Polarized light usually gives a small photoresolution between two enantiomers. However, the system does not need a very high optical purity, since, in an open system (and living systems are open systems) amplification processes (10,11) may occur which allow the total disappearance of one enantiomer, to the benefit of the other. With an appropriate chemical amplification system, the evolution toward such a final state may require only a very small excess of the other enantiomer, at the beginning.

LIFE : AN OPEN SYSTEM KEPT FAR FROM EQUILIBRIUM BY SOLAR LIGHT

The occurrence of amplification processes is one of the characteristics of the chemical open systems, which gave rise to the living systems on the Earth. But, as emphasized by Dr. P. Decker, from the Chemical Institute of Veterinary School in Hannover,

F.R.G., it is difficult to give a clear definition of Life. "How can we define a class, of which we know only one element ?". However, if we consider Life as a particular case of a more general system, then a definition is much easier to find.

For P. Decker, the general system is the "Bioid" : "the simplest chemical open system capable of Darwinian Evolution by mutation from one steady state into a more stable one". Life is just an example -a "nice one"- of bioid : it can exist in several steady states, it can concentrate and store information.

The biosphere represents a chemically reactive open system. This system is held far from equilibrium by the solar radiations. Probably, the first bioid system appeared on the Earth through self-organization processes of the available compounds into simple chemical cycles, involving autocatalytic feed-back reactions. With the help of the solar energy, by coupling with sensitized photoreactions, this system was able to increase continuously its information storage (12).

A nice model of photochemical reactions which could have been involved in those processes, is the formation of energy rich phosphate in a modified Fenton's reaction (13) : this formation can be easily photosensitized (12). An example is given on figure 3.

The coupling of solar energy, through sensitized photoreactions appears as the primary source of self-organization.

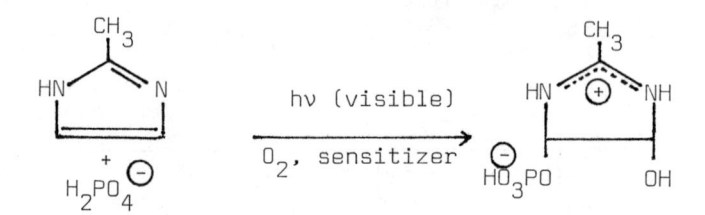

Fig. 3. An example of model photophosphorylation reaction using visible light (12).

CONCLUSIONS

Looking at the several questions which concern the problem of the origin of Life, on a photochemical point of view was very fruitful. This way of studying the different phases in the processes of Chemical evolution, clearly points out that we must consider the prebiotic systems which gave rise to the living systems, as open system, far from equilibrium, evolving under a continuous flux of energy.

Solar radiation, as the main energy source on the Earth, has played a crucial role in Chemical Evolution. It has probably not been directly involved into the formation of organic precursors in the atmospheric processes, unlike electrical discharges. However, it has surely been indirectly involved in those organic syntheses, since most of the energy of electrical discharges in the planetary atmosphere comes indirectly from the solar light.

In the complex processes which followed the evolution of the atmospheric organic molecules, after dissoluting in the aqueous media of the primitive oceans, light has played a direct role allowing the photo-assisted reactions to occur.

At all the stages of Chemical Evolution the conversion of photonic energy from the sun into chemical energy stored in organic compounds has occurred through the photoinduced chemical processes. Such processes may be considered as the ancestors of the present chemical processes which are involved in biological synthesis.

REFERENCES

1. See, for instance : A.I. Oparin, "Origin of Life", Mc Millan Co, N.Y. (1938) ; S.L. Miller and L.E. Orgel, "The Origin of Life on the Earth", Prentice Hall, New Jersey (1974) ; R. Buvet, "L'Origine des Etres Vivants et des Processus Biologiques", Masson et Cie, Paris (1974) ; S.W. Fox and K. Dose, "Molecular Evolution and the Origin of Life", Marcel Dekker, N.Y. (1977).
2. S.L. Miller, "A production of aminoacids under possible primitive Earth conditions", Science, 117:528 (1953).
3. A. Bossard, F. Raulin, D. Mourey and G. Toupance, "Organic chemical evolution of reducing model of the atmosphere of the primitive Earth. Role of UV light and electric discharges", submitted to J. Molec. Evolution (1980).
4. A. Bossard, "Rôle du rayonnement UV dans les synthèses prébiologiques", Thèse de Doctorat de 3ème Cycle, Spécialité Chimie-Physique, Université Paris VI (1979).
5. M. Halmann, "Photoelectrochemical reduction of aqueous carbon dioxide on p-type gallium phosphide in liquid junction solar cells", Nature, 275 (5676):115 (1978).
6. M. Halmann, B. Aurian-Blajeni and S. Bloch, "Photo-assisted carbon dioxide reduction and formation of two and three-carbon compounds". Proceedings of the 6th ICOL meeting, Jerusalem, June 1980, to be published.
7. J. Kiwi and M. Grätzel, "Hydrogen evolution from water induced by visible light mediated by redox catalysis", Nature, 281 (5733):657 (1979).
8. K. Kalyanasundaram, O. Micic, E. Pamauro and M. Grätzel, "Towards the construction of a complete cyclic water decomposition system, design and operation of an oxygen producing half cell", Helv. Chim. Acta, 62(7):2432 (1979).

9. G. Balavoine, A. Moradpour and H.B. Kagan, "Preparation of chiral compounds with high optical purity by irradiation with circularly polarized light, a model reaction for the prebiotic generation of optical activity", J. Amer. Chem. Soc., 96(16):5152 (1974).
10. F.G. Frank, "On spontaneous asymmetric synthesis", Biochim. Biophys. Acta, 11:459 (1953).
11. P. Decker, "The origin of molecular asymmetry through the amplification of stochastic information (noise) in Bioids, open systems which can exist in several steady states", J. Molec. Evolution, 4:49 (1974).
12. P. Decker, "Coupling to solar energy : sensitized photoreactions, the primary source of self-organization", Proceedings of the 6th ICOL meeting, Jerusalem, June 1980, to be published.
13. Oe. Saygin and P. Decker, "Formation of energy rich phosphate in Fenton's reaction", Proceedings of the 6th ICOL meeting, Jerusalem, June 1980, to be published.

BIOLUMINESCENCE AND ITS APPLICATIONS

A.M. Michelson

Institut de Biologie Physico-Chimique
Service de Biochimie-Physique
13, rue Pierre et Marie Curie - 75005 Paris

SUMMARY

VERS LUISANTS

Notre coeur a soif de tendresse,
Et nous aimons à pleine ivresse
Jusqu'à l'heure ou blanchit le jour ...
Cette lueur qui nous éclaire,
Diamant qui jamais ne s'altère,
C'est l'ardent flambeau de l'amour.
Et pendant que dans le ciel sombre
L'étoile glisse rayant l'ombre
De sa fine aigrette de feu,
On voit nos amoureuses flammes
Lumineux reflets de nos âmes,
Scintiller dans leur éclat bleu ...

DESPEYLOU

INTRODUCTION

The elucidation of the mechanism of firefly bioluminescence by McElroy in 1947, showing the involvement of ATP in the reaction, opened a new domain of biochemical analysis. Although analytical application for the estimation of ATP was at first very slow, in recent years, progress in the purification of luciferases and luciferins on a commercial scale (as opposed to purifications in the rather limited number of laboratories specialized in bioluminescence) coupled with an enormous progress in instrumentation for the measure of light output has led to an explosion of luminescent style analytical procedures. These can be used not only for the estimation of ATP, but for practically any biological molecule of interest, whether industrial, clinical or ecological.

The approaches used can be roughly classified into a number of sections based on different bioluminescent systems with their particular characteristics as shown below. Applications of chemiluminescence, particularly with respect to the oxidation of luminol, provides further possibilities.

1 - Measurements of ATP and of any enzyme which consumes or produces ATP, e.g. creatine kinase, using the firefly bioluminescent system.

2 - Measurement of NADH and NADPH or any enzymic system which consumes or produces these co-factors, as well as the substrates for these enzymes e.g. ethanol and alcohol dehydrogenase, using the bacterial bioluminescent system. In addition FMN, reduced FMN and any system in which FMN is involved e.g. production of FAD by the enzyme catalysed reaction of FMN with ATP, can also be estimated.

3 - Peroxidation systems e.g. measurement of peroxidase with pholad luciferin, or for determination of Fe^{2+} ions.

4 - Precharged bioluminescent systems such as Aequorin, for the estimation of Ca^{2+} ions.

5 - Use of bioluminescent systems e.g. firefly, or bacteria, to estimate levels of O_2.

6 - Extension to luminescent immunoassays in which a luciferase is coupled to a specific antibody or a luciferin is used to detect a

peroxidase-coupled antibody, with the various extrapolations and combinations inherent in this approach.

7 - Chemiluminescence generally involving oxidation of luminol to estimate H_2O_2, glucose, glucose oxidase (or other oxidases and their substrates) using peroxidase to catalyse the reaction, or for the estimation of superoxide radicals and superoxide dismutases.

Apart from direct assay applications it is likely than a number of physiological problems could be studied in vivo. To some extent this is already the case with certain bioluminescent organisms such as the firefly (for an excellent review of the physiological control and regulation of bioluminescence at the molecular level see ref. 1).

Among the various advantages of analytical measurement of bioluminescent and the generally less sensitive chemiluminescent (quantum yields in chemiluminescence are almost always considerably lower than enzyme catalyzed reactions) systems may be cited the extremely high sensitivity (indeed measurements with 100 cells are essentially routine and single cell experiments are not impossible), the rapidity of measurement (no lag phase in light emission, and modern instruments such as the Lumac Biocounter M 2010 allow up to 150 sample estimations per hour), the extremely wide linear range (5 to 7 decades is not unusual) and a relatively low cost. With respect to sensitivity, ATP (and hence any system involving ATP such as RNA polymerase) can be routinely measured at 10^{-17} moles per assay (even a mass produced comercial instrument such as the Lumac can be used to measure 0.2 pg ATP per µl, subfemtomole quantities) whereas NADH and NADPH systems can be estimated at 10^{-16} moles per assay (routinely picomole quantities with the above mentioned instrument).

Finally, one of the possible future developments could lie in the use of solid state bioluminescent systems using luminescent reagents coupled to a solid support to permit continuous estimation of different samples, or, by circulation in a closed circuit, fluctuations in a given system. This immobilisation can be achieved by covalent fixation of for example a given luciferase to a solid support[2], or simply by inclusion of the necessary components in a gelatine pill[3], or even simply to adsorb the luminescent system on a glass cover slip. With the astounding technical development of commercial instruments in recent years in which single photon counters are coupled to microprocessors or microcomputers with visual display

of the purpose built programme parameters and results (e.g.
Lumac Biocounter M 2010) it is now of historical interest to read
in a review[4] written only twenty years ago "The arrangement
should be set up in a room that can be completely darkened and
the observer should remain in darkness for not less than 30 minutes
before making the observation, in order for his eyes to be maximally sensitive. He should move his eyes from side to side and up
and down, because the rods, which are the more sensitive light
receptors in the retina, are not located in the center'".

As mentioned earlier, the applications of luminescent
systems are many and various if not unlimited. Industriel applications of luminescent analytical chemistry with respect to fermentation, brewery, dairy, food and agricultural industries ranging
from bacterial counts in process water to the number of somatic
cells in milk have been documented[5] as well as microbiological,
environmental (activated sludge) and medical (erythrocyte and
platelet viability) applications[5]. It is not the purpose here to review
estimations of malate, pyruvate, oxaloacetate and the corresponding
enzymes or the many other published applications, but rather to
present certain results obtained in the author's laboratory as examples of specific applications.

In 1887 Raphael Dubois defined two components extracted
from the luminescent clam Pholas dactylus, which gave rise to
light emission on mixing, as luciferin and luciferase[6]. This system
is thus closely linked with the first defined terminology. Pholas
dactylus is a mollusc which bores holes in relatively soft rocks
which at the approach of a predator ejects a jet of luminescent
material. This material, the luciferase and the luciferin, is localized in specialized cells. The system is somewhat special, if not
unique, in certain respects[7]. Thus both the luciferase (enzyme)
and the luciferin (substrate) are acidic glycoproteins. The luciferase has a molecular weight of 310 000 composed of two identical
subunits and contains 2 atoms of Cu^{2+}. The luciferin is a single
unit of molecular weight 35 000 containing three sulphate groups
and undoubtedly a small prosthetic group (as yet unidentified) which
is probably the true luciferin. Since rapid loss of activity occurs
at temperatures above 60° C, or by treatment with proteases, it
appears that the integrity of the protein is essential for bioluminescence. The λmax of emission due to oxidation of luciferin catalyzed
by luciferase is at about 500 nm (Fig. 1) with a quantum yield of
about 0.27.

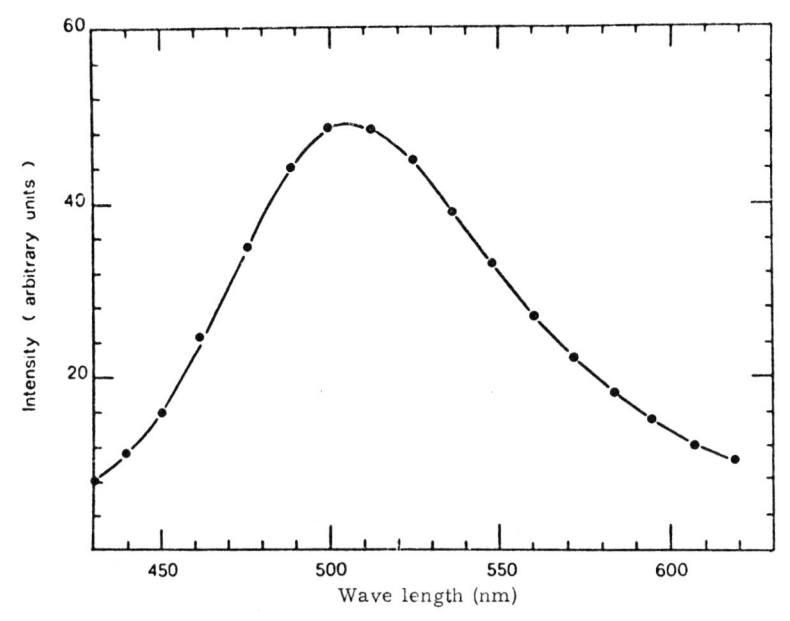

Fig. 1 - Bioluminescent emission spectrum of pholas luciferin plus luciferase.

Apart from oxidation by O_2 catalyzed by luciferase, Pholad luciferin is oxidized with concomittant light emission and always with the same spectral distribution, by a wide range of oxidants including all systems which produce the superoxide radical anion $O_2^{\cdot -}$. This luciferin can thus be used to measure Fe^{2+} ions (in presence of ferric ions) by injection of the sample into a solution of luciferin in phosphate buffer.

$$Fe^{2+} + O_2 \xrightarrow{phosphate} Fe^{3+} + O_2^{-}$$

$$luciferin + O_2^{-} \longrightarrow oxyluciferin + h\nu.$$

The sensitivity is such that about 0.1 nanomole quantities of Fe^{2+} can be conveniently measured in presence of Fe^{3+}, Ni^{2+}, Co^{2+} etc... in a 10 000 to 100 000 fold excess.

More importantly, Pholad luciferin is readily oxidized by horse radish peroxidase (also a glycoprotein), the latter acting as an oxidase and not a peroxidase with this substrate, since whereas molecular O_2 is necessary, H_2O_2 is not. The reaction occurs at physiological pH and can be used to estimate peroxidase with a sensitivity very much greater than that obtained using

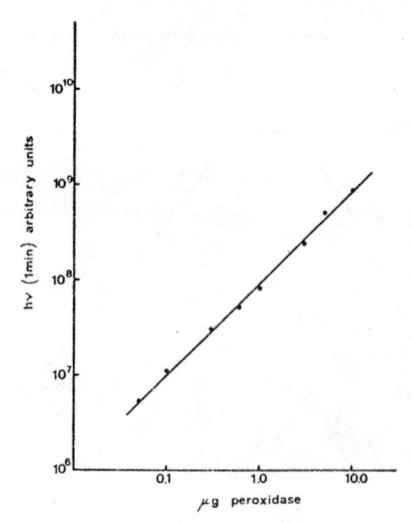

Fig. 2 - Dosage of peroxidase by oxidation of luminol in presence of 10^{-4}M H_2O_2.

peroxidation of luminol with H_2O_2 by peroxidase[8]. Estimation of peroxidase by the luminol technique is shown in Fig. 2 (10^{-5} M H_2O_2). It can be seen that the signal is linear between 0.1 to 10 µg of peroxidase per ml.

When Pholad luciferin is used (in absence of H_2O_2) linearity of the signal at constant peroxidase as a function of luciferin concentration is shown in Fig. 3a while the linearity with variation of peroxidase concentration at two different concentrations of luciferin is presented in Fig. 3b. Typical signals observed at 400 ng to 0.01 nanograms per ml using 6×10^{-8} M luciferin are shown in Fig 4 whereas linearity of the estimations are given in Fig. 5a. It can be seen that, at the highest concentrations, linearity is lost. This occurs because, at 400 ng of peroxidase, there are only six molecules of luciferin per molecule of peroxidase, and hence the steady-state oxidation (i.e. V_{max}) cannot be maintained, due to disappearance of substrate, as seen also in Fig. 6. Finally, linearity down to about 75 000 molecules per ml (10^{-4} femtomoles) is presented in Fig. 5b. In principle, using a 20 µl sample on a microscope cover slip, it should be possible to estimate precisely 750 molecules (10^{-6} femtomoles) of peroxidase.

While estimation of peroxidase is in itself of very limited interest this technique is readily applicable to the determination of

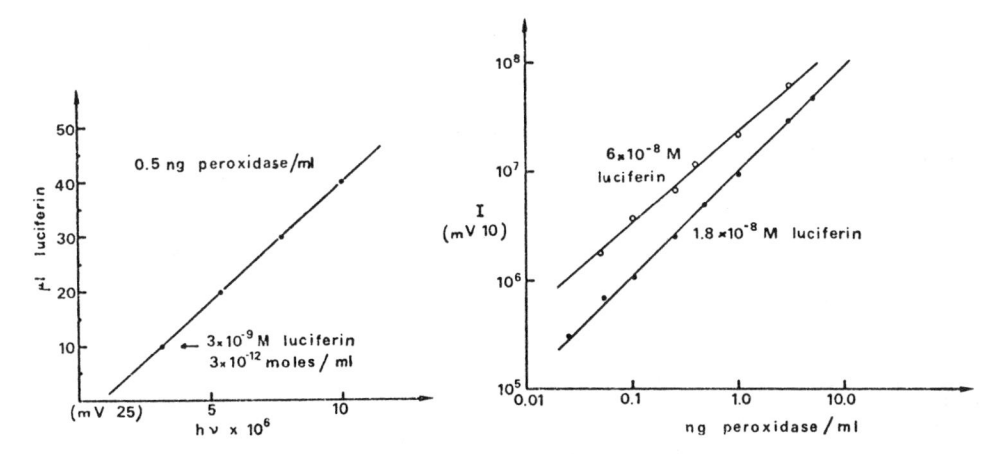

Fig. 3 a) - Linearity of light emission due to oxidation of pholad luciferin by horseradish peroxidase as function of the concentration of luciferin.
b) - Linearity of light emission at two different concentrations of pholad luciferin as a function of the concentration of peroxidase.

the number of antigenic sites on the outside of a cell membrane. We have used peroxidase-labeled anti-immunoglobulin antibodies to quantify immunoglobulins present on the surface of mouse lymphocytes. Antibodies were isolated and coupled with peroxidase. Control cells were incubated with the same enzyme-labeled antibodies diluted in antigen solution, i.e. normal mouse serum[8].

Spleen cell suspensions were prepared, fixed for 1 hr at 4°C with 4 % formaldehyde in 0.1 M phosphate buffer, pH 7.5 and then washed. Subsequent incubations of cell suspensions with enzyme-labeled antibodies, washings, and light emission was measured.

The results are shown in Fig. 6 and indicate, very approximately, about 1200 molecules of active peroxidase bound per cell. The approach can be increased in sensitivity by measuring the total light emitted within a given time, as shown in Table I.

Similar results were obtained with glucose oxidase coupled antibodies. Here the determination depends on measurement of the H_2O_2 produced by the oxidation of glucose catalyzed by enzyme. The H_2O_2 is then estimated by the oxidation of luminol catalyzed by peroxidase, after incubation of the antibody treated cells with glucose for 20 min at 25°C. The results are compared with a standard curve and indicate that about 5000 molecules of glucose oxidase are bound per cell (Fig. 7).

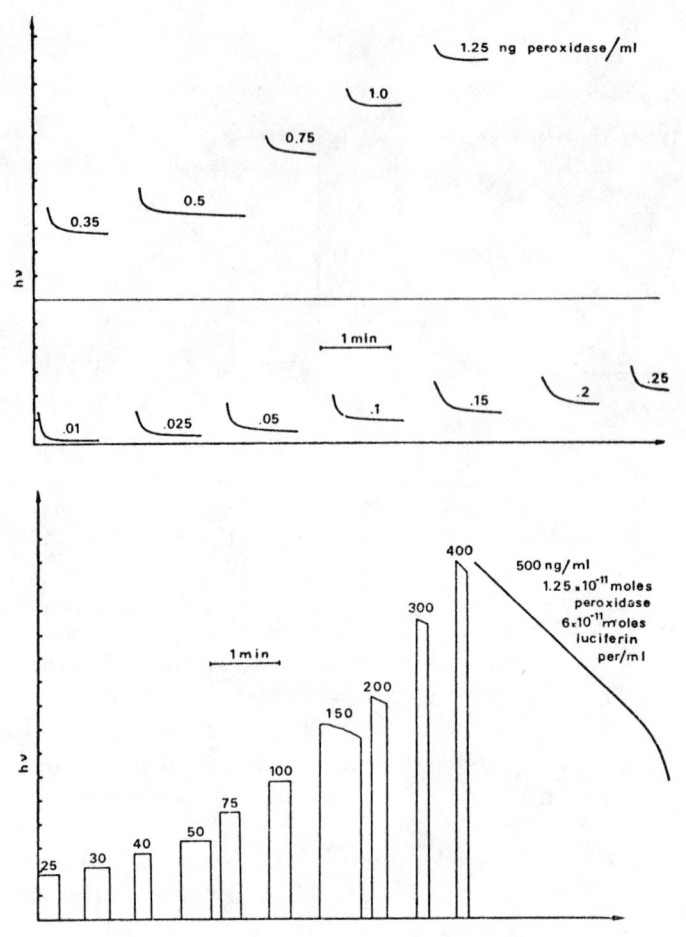

Fig. 4 - Light emission due to oxidation of pholad luciferin by different amounts of peroxidase.

      This technique can also be applied in a continuous fashion. Thus, cells were suspended in 0.1 M phosphate, pH 7.8 containing 10^{-4} M luminol, 3.3 µg of peroxidase/ml, and 0.33 mg of glucose/ml, and the light emitted due to peroxidase-catalyzed oxidation of luminol by the H_2O_2 produced by the glucose oxidase was measured as a steady-state output, and the values were compared with those obtained for known amounts of glucose oxidase. The results indicated approximately 3000 molecules of glucose oxidase bound per cell[8].

      Light emission techniques using oxidation of luminol have

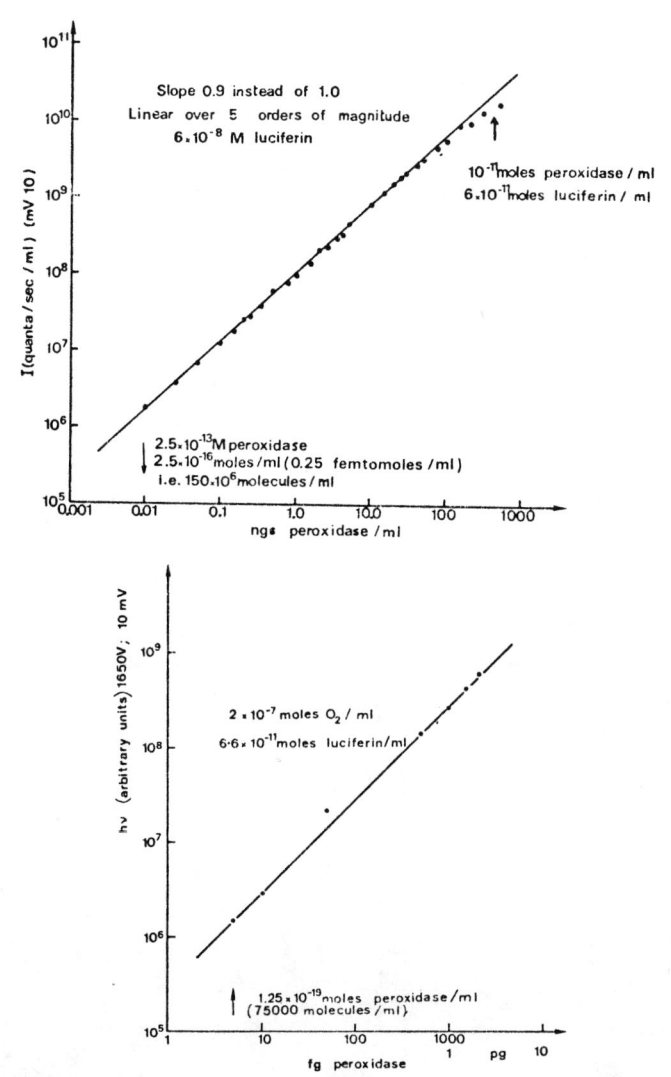

Fig. 5 a) - Dosage of peroxidase by oxidation of pholad luciferin.
b) - Femtogram dosage of peroxidase by oxidation of pholad luciferin.

been developed for microestimations of H_2O_2, O_2^-, glucose, glucose oxidase (or other oxidases and their substrates which produce H_2O_2), peroxidase, catalase, and superoxide dismutases. These methods are relatively sensitive, since estimations in the general range of 10^{-10} to 10^{-15} mol can be performed with precision and can be extended to enzyme-coupled antibodies in many cases. The general approach involves peroxidase-catalyzed oxidation of luminol by H_2O_2 and measurement of light emission (I_{max} and total light). Apart from the sensitivity obtained in static measurements

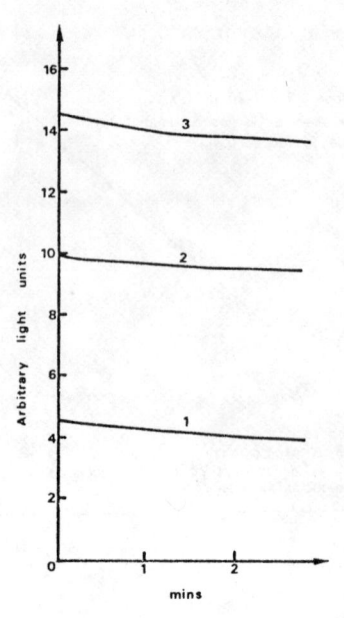

Fig. 6 - Light emission due to oxidation of pholad luciferin by mouse lymphocytes to which are attached peroxidase-bound antibodies: (1) background due to luciferin: (2) plus 1.5×10^6 cells; (3) plus 3×10^6 cells.

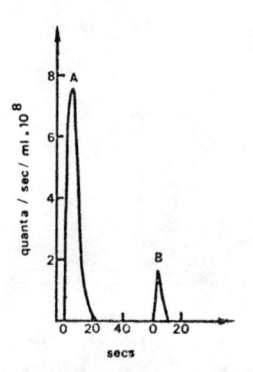

Fig. 7 - Light emission due to peroxidase-catalyzed oxidation of luminol by H_2O_2, produced after a 30-min. incubation at 25°C of mouse lymphocytes to which is attached a glucose oxidase antibody in glucose solution.
A. lymphocytes: B. control cells.

of a given component, the techniques can often be modified to follow continuous variations or repeated fluxes of an enzymic activity or product in vitro or in vivo (since the reagents are generally non toxic), for example, production of O_2^- by polymorphonuclear neu-

TABLE I

Arbitrary Light Units

| | Corrected for luciferin background ||
	I	L (2 min)
Luciferin		
+ 10 µl of cells	1.2	37
+ 50 µl of cells	5.5	151
+ 100 µl of cells	9.8	288

TABLE I - Cell suspension : 100 µl is equivalent to 3×10^6 cells. I corresponds to Vmax, whereas L (2 min) is the total amount of reaction during this time.

trophiles during phagocytosis, since the rate-limiting step is not emission (and measurement) of light.

Examples for the estimation of H_2O_2, glucose and glucose oxidase are shown in Fig. 8. This approach can be used for the estimation of any other oxidase which produces H_2O_2, since specificity is conferred by use of the specific substrate for the enzyme[9]. Catalase can also be estimated by the same techniques, using rate of disappearance of H_2O_2 (Fig. 9). This method is much more sensitive than any other yet described and indeed precise determinations are obtained with the equivalent of 0.02 µl of human blood[10]. Finally, use of the O_2^- induced chemiluminescence of luminol (using a steady state production of O_2^- by means of xanthine oxidase plus hypoxanthine) allows the precise determination of femtomole quantities (0.2 ng) of superoxide dismutase[7,10].

FUTURE DEVELOPMENTS

Coupling of any antibodies with a suitable luciferase (e.g. Cypridina luciferase), followed by reaction with the specific antigen and then reaction with the corresponding luciferin could give a convenient, rapid, highly sensitive general method for the estimation of an extremely wide range of materials, with a remarkable flexibility in application, using either direct measurement or competition

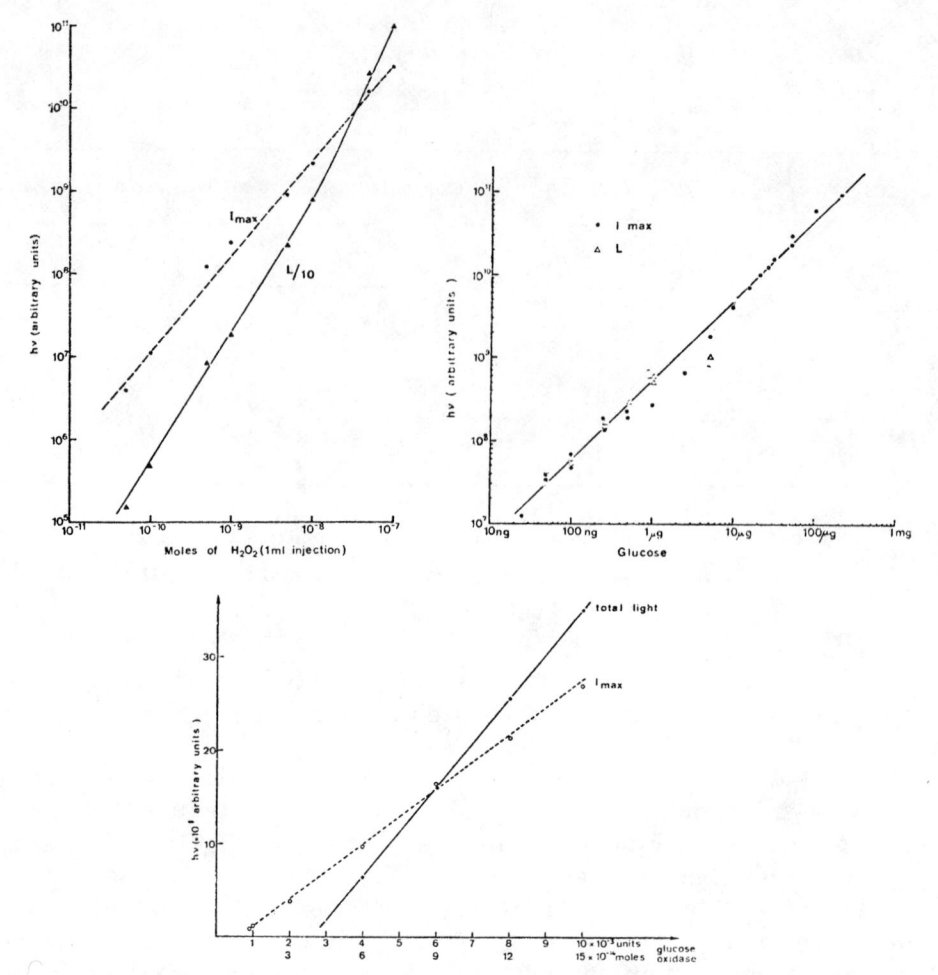

Fig. 8 a) - Light emission due to oxidation of luminol by horse radish peroxidase as a function of amount of H_2O_2. Imax ●——●; total light/ 10: ▲——▲.
b) - Estimation of glucose by oxidation of luminol by H_2O_2 produced by glucose oxidase. Imax ●——●; total light △——△.
c) - Estimation of glucose oxidase.

procedures. The luciferase itself provides an enzymic amplification if total light output over a given time period is measured. Such an approach is very useful for biopsy measurements since often 1 mg samples should be ample. Indeed, present techniques which are valuable for measurements of 100 cells could surely be extrapolated to single cell measurements.

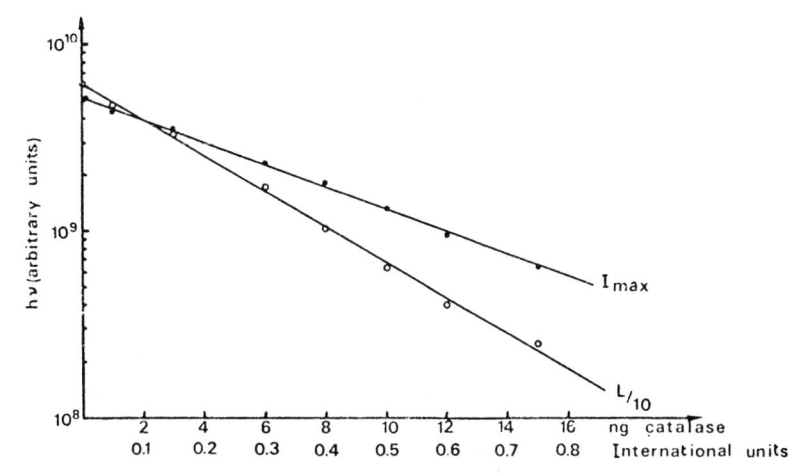

Fig. 9 - Calibration curves (Imax and total light, L) for the estimation of catalase activity.

In addition light amplifier techniques could be used in conjunction with the above approaches for the cellular localization (by direct image visualization) of a certain number of cell components or membrane antigens.

REFERENCES

1. J. P. Henry and A. M. Michelson, Photochem. Photobiol. 28 : 293 (1978)
2. B. F. Erlanger, M. F. Isambert and A. M. Michelson, Biochem. Biophys. Res. Communs 40 : 70 (1970).
3. V. Henri, C. R. Acad. Sci. Paris 142 : 97 (1906).
4. A. M. Chase "The Measurement of Luciferin and Luciferase" in "Methods in Biochemical Analysis" Ed. D. Glick, 8 : 61 (1960) Wiley, Interscience, New-York.
5. Lumac Documentation, Lumac B. V., Baanstraat 83, 6372 AD Schaesberg, The Netherlands.
6. R. Dubois C. R. Acad. Sci. Paris 105 : 690 (1887).
7. A. M. Michelson "Purification and Properties of Pholas dactylus Luciferin and Luciferase" in Methods in Enzymology 57 : 385 (1978) Academic Press, Inc. New-York.
8. K. Puget, A. M. Michelson and S. Avrameas, Analytical Biochem. 79 : 447 (1977).
9. K. Puget and A. M. Michelson, Biochimie 58 : 757 (1976).
10. A. M. Michelson, K. Puget, P. Durosay and J. C. Bonneau in "Superoxide and Superoxide dismutases" Academic

Press, London, 1977, Eds. A.M. Michelson, J.M. Mc Cord and I. Fridovich, pp 467-499.

THE TRENDS AND FUTURE OF PHOTOBIOLOGY : PHYSICAL AND
BIOPHYSICAL ASPECTS

J. W. Longworth

Biology Division
Oak Ridge National Laboratory*
Oak Ridge, Tennessee 37830

PROLOGUE

These remarks are my attempt to comment on symposium papers and contributed poster sessions presented during the congress in biophysical science. Biophysics is a diverse subject of studies, and my view of it is biased toward spectroscopy, as is photobiology. In addition, because I had quite recently listened to several of the symposium speakers at meetings in the United States, at this meeting I tended to listen to alternative talks; thus the lack of comment on some presentations. This is my way of pointing out that these comments are just that and are not a balanced review of even this Congress.

GENERAL IMPRESSIONS

My overriding impression at this Congress was the perceptive and critical application of many complex measurement methods which require the latest devices. Lasers can provide light in pulses of nanosecond or picosecond duration, and these light pulses can contain a large number of photons. Laser pulses of short duration can be bright, which can create complications — nonlinear dependencies on light intensity are present. Singlet exciton annihilations dominate picosecond laser excitation studies of fluorescence of chloroplasts. This requires the use of image intensifiers prior to the streak camera. Photoionization of indoles has been followed by the absorption of the solvated electron, which has a small molar

---

*Operated by Union Carbide Corporation under contract W-7405-eng-26 with the U. S. Department of Energy.

absorption coefficient. Hence intense nanosecond laser pulses are used to achieve adequate detection. Several facile paths of biphotonic photoionization are present, in addition to monophotonic ionization. These pathways can be disentangled by irradiating samples with two laser pulses of different color at various delays. Another property of laser light is its well-defined collimation so that it can be focussed into a small diameter. Flowing of a sample permits the study of the resonance Raman spectra of photochemical intermediates found in the initial 100 ns, the time taken to pass by the focus. Another strategy for studying the spectroscopy of intermediates is to combine the flowing sample with a picosecond pulsed mode-locked laser to extend the time interval into picosecond times and flow out the photolyzed sample prior to a subsequent actinic laser pulse.

FUTURE TERMS

During the meeting there was discussion of the clear feasibility of entering new measuring regimens. In time, femtosecond (10^{-15} s) pulses can be reached with mode-locked dye lasers, which is the time range for transfer of electronic energy between chlorophyll pigments in pigment—protein complexes. Attomolar (10^{-18}) quantities (picograms) can be measured by bioluminescence assays which, when combined with antibody interactions, offer similar advantages to radioimmune assays.

CURIOSITY UNANSWERED

My interest was drawn to an abstract from a Chinese delegate entitled "The microminiaturization of movie projectors," but this paper was not presented. The author described, in part, the advantage of using pulsed light sources (presumably in synchrony with the filmgate) with a resonant charging of the energy-storage component, and alluded to another principle which helps in size reduction. But the motive for the microminiaturization of the projector is enigmatic.

COMPLEX INSTRUMENTS FOR RAPID EVENTS

N. Geacintov (New York University) used several picosecond-wide pulses to excite chlorophyll fluorescence of chloroplasts. The fluorescence was amplified with an image intensifier, and the time dependence of the intensity was followed with a streak camera. The fluorescence of the reaction center, which is sensitized by the antenna chlorophyll pool, was delayed in time from the direct fluorescence of the antenna chlorophylls.

T. Bannister and S. Brody were the first to make attempts to detect the delay in the late 1950's. The delay and profile of the rise time of sensitized fluorescence contain a measure of the transfer time and are reminiscent of well-established measurements described in nuclear spectroscopy texts.

J. Breton (Saclay) presented a precis of the work of G. Paillotin and a synopsis of his own concepts on the organization of pigment molecules in thylakoid membranes. He suggested that the chlorophyll—protein binding pigment complex of Chlorobium represented a paradigm for chlorophyll—protein complexes. There would be a small number of pigment molecules per molecule, and each would have a specific orientation and distance to each other, largely different from random. Mode-locked dye laser pulses of several hundred femtoseconds can now be repetitively and reproducibly produced at suitable wavelengths for chloroplast fluorescence excitations. The rise-time of fluorescence anisotropy will contain information on the transfer rates and orientations of the chlorophylls.

VISION; SOMETHING NEW, SOMETHING OLD

A red fluorescence arising from metarhodopsin in insect eyes was reported by two separate groups (D. G. Stavenga, Groningen; and N. Francischini, Tubingen and Marseille). Fluorescence of metarhodopsin will offer a new measurement procedure and will be of particular interest if it responds to the Ca(II) binding protein involved in modulating a Ca(II) gradient.

More progress in defining the stereoisomeric state of bathorhodopsin was obtained through use of a synthetic retinal specifically deuterated to selectively perturb the resonance Raman spectrum in the stereoisomer "signature" region. An electronically perturbed all-trans stereoisomeric state was proposed for retinal in bathorhodopsin.

LEST WE FORGET

At every photobiology meeting there are reports of simple but readily overlooked effects based on the first law of photochemistry, that there must be absorption of light.

The lens of the human eye contains a yellow pigment which provides two benefits: there is a reduction in the chromatic aberration due to a reduction in the transmitted spectral interval, and the retina is protected from damage by UV radiation. Cataracts are treated by removal of the lens, which is replaced by an artificial lens made of materials which absorb little of the damaging radiation. S. Zigman and co-workers (Rochester) emphasized the importance of using pigmented lenses for aphakics.

J. Parrish (Boston) noted that human skin is more transparent to UV-B radiation when freshly washed. Chromatographic analyses showed that this was caused by the removal of urocanic acid. Urocanic acid absorbs UV-B radiation and is a metabolite of histidine, which accumulates in skin since the tissue cannot further metabolize it.

R. Santus (Paris) pointed out the presence of appreciable concentrations of pterine in the surface waters of tropical oceans. Pterine is an efficient former of singlet oxygen and thus could influence the ecology through appreciable photosensitized killings.

NEW ANALYTIC METHODS

P. Vigny (Paris) introduced the use of skin suction blisters as a noninvasive procedure to isolate interstitial fluid. The pharmacokinetics of fluorescent drugs could be readily followed because of the high sensitivity offered from single-photon counting. The variety of other high-sensitivity analytic procedures is obvious, and this simple technique is very attractive. J. Boyle (Manchester) described the isolation of a monoclonal hybridoma antibody with UV-irradiated DNA as the antigen. The ability to produce highly specific antibody in some quantity is crucial in developing high-sensitivity analyses for specific materials. The other component in the analytic procedure was discussed by A. M. Michelson (Paris). He emphasized that the high sensitivity of single-photon counting can be exploited by use of a purified luciferin—luciferase enzyme. Bioluminescence, or appropriate chemiluminescence reactions, can be combined with monoclonal antibodies to provide analytic methods comparable in sensitivity to radioimmune assay. The advantage of the bioluminescence methods is considerable because of our extreme caution in disposing of radioactive waste materials from laboratories.

LIGHT SOURCES

An intense UV-A light source which produced excellent recovery from psoriasis was reported by H. Lang (Jena). There were no side effects from this therapy, which required no prior medication. The spectral output from the lamp was presented, together with a photograph of the treatment bed. A filtered, medium-pressure mercury lamp is a possibility, where the filter suppresses both 313- and 365-nm mercury lines and transmits the intervening continuum and lines. T. Fitzpatrick (Boston) informed me that direct UV-A effects have been reported previously, but suffer from immediate regression. There were no detailed clinical data given in this poster.
I. Kochevar and L. Haber (New York) used a concave holographic grating and high-power xenon arc lamp to obtain intense light of moderate spectral purity and were able to measure the action spectra of a photoallergic response in patients.

TO BE OR NOT TO BE?

The number of synchrotron radiation sources available to photobiologists has increased, and will increase still more in the coming years. Though the unique spectral range of the radiation (0.1–100 nm) is not traditionally claimed by photobiologists, the radiation and light emitted from storage rings has properties of value in photobiology. J. Sutherland (Brookhaven) gave results obtained on Surf II (Gaithersberg) from a vacuum UV circular dichrometer. Excellent signal-to-noise ratios were achieved in several seconds of data collection, and the forthcoming storage ring NSLS-I at Brookhaven will allow facile studies. The pioneering studies of J. Brahms (Paris) on the vacuum UV-circular dichroism of proteins in the spectral interval 165–200 nm definitively demonstrate the utility of circular dichroism in defining the proportion of α-helix, β-sheet, β-bend, and unstructured conformers in proteins. After 25 years of intensive study, circular dichroism spectroscopy now appears realizable as envisaged.

The time properties of synchrotron radiation are of considerable interest, and several papers exploited ACO (Orsay), where a coincidence spectrometer is available to biophysicists. J. Balini (Orsay) and M. Daniels (Corvallis) measured the fluorescence decay at room temperature of polynucleotides and observed a long-lived component which is unexpected and of uncertain origin.

The use of synchronously pumped, mode-locked dye lasers to produce tunable light pulses of picosecond duration, at high repetition, provides considerable competition for storage-ring light sources. The cost is comparable to a beam line attachment to the ring, and the dye laser has the advantage of being in one's own laboratory and being accessible to use. Dye lasers can form pulses smaller than 10 ps and maintain stability for many hours of operation. The pulse width of storage rings is varied from the long 1.5-ns pulses of NSLS-I at Brookhaven and the 1 ns of ACO at Orsay to the 200 ps of SSRL at Stanford and the 100 ps of BESSY in Berlin. A. Miehe (Strasbourg) described a device which fully exploits the time range of dye laser pulses. A streak camera was synchronously triggered by a photodiode driving a constant fraction discriminator. The centroid of the trigger pulse was stable to ±1 ps, even though the pulse width is 200 ps. The repetitive sweeping permits averaging, and this allows quantitative measurements from a streak camera over a 40-ps interval.

INDOLE REVISITED

There was a group discussion on indole photoionization. R. Santus (Paris) had measured the photoelectron yield after laser pulsed excitation over a wide intensity range and temperature range. A marked temperature effect on electron yield was found. At low light intensities, agreement with the photoelectron yield measured with scavengers was noted (G. Kohler, Vienna). In contrast, G. Laustriat (Strasbourg), working with histidine as a scavenger, found much smaller photoelectron yields. L. Grossweiner (Chicago) presented his results on an enhancement of electron formation upon simultaneous exposure to UV radiation and green light (second and third harmonic of an Nd-YAG). It is not always made clear in published studies whether the green light has been suppressed with a color filter. Another difficulty is created by the low sensitivity of the detector and modest molar absorption of the solvated electron in the near-infrared, necessitating intense actinic pulses to achieve useful signal-to-noise ratios in data. Though the actinic pulse intensity may be changed over a wide range, this may be insufficient to disclose whether or not the mechanism leading to photoelectrons is caused by one or two photons. The scavenging results are obtained at sufficiently low intensities of UV radiation to be assured of monophotonic production of electrons from the singlet state. In addition, biphotonic electron production from both the excited singlet and triplet is possible. L. Lindqvist (Orsay) elaborated upon this by using two dye laser pulses of different colors which were separated by differing time intervals. He was able to demonstrate biphotonic photoelectron production from both triplet and excited singlet intermediaries, in addition to electron production directly from the excited singlet. The broad outline of photoionization from indole was described by the discussion group, and the conclusion was that care must be taken in interpreting published data. Large tasks remaining include determination of the role of temperature on each of the three channels of electron formation and the relationship to fluoresence yield and its temperature-dependent radiationless channel and to solvent dielectric and hydrogen bonding ability. Excitation of indole to its third electronic transition also enhances photoelectron yield and reduces fluorescence yield, suggesting a fourth channel of electron formation.

EPILOGUE

In many ways, the interest in photoelectron production by indole is a paradigm for biophysical photobiology. Indole is unusual in its electron structure in having two nearly degenerate electronic transitions. It interacts strongly with polar solvent, and more so in its excited state. It has the ability to eject electrons readily and has many channels of radiationless decay. It has appreciable yields of fluorescence, triplet formation, and phosphorescence and only a modest photochemical yield. Intensive studies over 30 years, with increasingly complex instrumentation, have not revealed a complete interpretation of its photophysics. Thus the use of indole as a spectroscopic probe, since it is naturally occurring, cannot be based on broadly defined paradigms but is restricted to special cases and to differences.

II - MUTAGENESIS, CARCINOGENESIS AND DNA REPAIR

CELL INACTIVATION AND MUTAGENESIS BY SOLAR ULTRAVIOLET RADIATION

Rex M. Tyrrell

Biophysics Institute, Federal University of Rio de Janeiro, Rio de Janeiro, R.J., 21941, Brazil

I. INTRODUCTION

The spectrum of solar radiation that reaches the earth's surface ranges from the highly energetic ultraviolet (UV) with wavelengths as short as 290-295 nm, through the near-UV region into the visible and infra-red regions. This presentation is concerned with the effects of solar-UV (290-380 nm), and where general terminology is inadequate, the specific wavelengths employed will be mentioned.

Despite clear epideminological evidence that sunlight is directly involved in the induction of basal and squamous cell carcinomas of the skin and is almost certainly a critical factor in provoking the more dangerous condition of malignant malanoma, the literature relating to genetic changes induced by sunlight and the consequent biological changes in simple systems, is extremely sparse. The main challenge in designing an experimental programme to help correct this deficit is how to obtain information that will be meaningful in attempts to predict the biological action of a polychromatic source such as sunlight whose incident energy changes rapidly not only in terms of dose-rate but also in spectral distribution. A comprehensive study of the action of solar radiation on a biological system may be divided into three distinct experimental phases. Firstly, information must be obtained concerning the action of individual monochromatic radiations, or mixtures of such, on the biological system for the parameter(s) in question. Secondly a suitable system of solar dosimetry must be exploited or developed. Finally, we will need studies with natural sunlight. While the first and third stages may be conducted simultaneously, precise and meaningful dosimetry is essential to obtain a useful correlation between laboratory and field measurements. The following is a description of experimental data and information that serve to illustrate this general approach. A large part of the data is concerned with inactivation and mutation of Escherichia coli and will be drawn from studies in our own laboratory. However, where possible we will place this information in the perspective of studies in other laboratories.

2. ACTION OF MONOCHROMATIC RADIATION OR MIXTURES OF WAVELENGTHS

2.1 - Methodology

The use of monochromatic radiation at selected wavelengths avoids the complication introduced by the lethal and mutagenic interactions which have been shown to occur over the solar-UV region (see below). This also allows precise and reproducible dosimetry and a ready comparison between laboratories that have adopted a similar protocol. Monochromatic radiation may be obtained using a high intensity light source in combination with a monochromator. Slit widths should be carefully selected to give the band-pass desired. Even so, additional filtration is usually necessary in order to eliminate scattered radiation, particularly at shorter wavelengths that fall outside of the chosen wavelength band. The cheapest system for reasonably accurate dosimetry is a radiometer with a uniform spectral response throughout the entire wavelength range being examined. All irradiations should be performed under controlled temperature (generally 0°C) and ambiental lighting (generally yellow) conditions.

The molecular and cellular action of solar-UV radiation has been reviewed (1, 2 and 3). The following description concerns the types of DNA damage induced and their susceptibility to repair and summarises the lethal and mutagenic action of solar-UV radiation in E. coli.

2.2 - Lesions induced

Several biological consequences of solar-UV almost certainly result from sites other than damage to DNA. For example, near-UV damage to the 4-thiouridine residues in transfer RNA is now believed to be the cause of near-UV induced growth delay (4, 5) and as mentioned later, this in turn may influence the mutation frequency. Near-UV also inhibits oxidative phosphorylation (6) and macromolecular synthesis. Recent studies have also implicated membranes, membrane binding proteins and membrane transport systems as sites of near-UV damage (see 2). However the nature of the chromophores involved in the various types of near-UV damage is not clear. Direct absorption, particularly by DNA and proteins, is probably important at the shorter wavelengths. However much of the action may be indirect and involve free-radical production and/or sensitized reactions involving many types of endogenous sensitizer molecules.

TABLE I

LESIONS INDUCED IN DNA BY SOLAR UV

	313 nm	334 nm	365 nm	405 nm
PYRIMIDINE DIMERS	○○○○○ ●●●●●	○○○○○	○○○	UNDETECTABLE
ENDONUCLEASE - SENSITIVE SITES	○○○○○	NOT MEASURED	○○○	NOT MEASURED
SINGLE - STRANDS BREAKS (alkaline - labile bonds)	○ ●	○	○○	○○○
THYMINE GLYCOLS	●●●●	LOW LEVELS	—	—
SPORE PHOTOPRODUCT - SPORES ONLY	○○○○○	NOT MEASURED	○○○○○	NOT MEASURED

Data from Tyrrell (1973), Tyrrell etal (1974), Cerutti and Netrawali (1979), Ley etal (1978), Tyrrell (1978) and Unpublished Observations (Ley; Tyrrell).

○ BACTERIAL DNA
● HUMAN CELL DNA

TABLE II

EVIDENCE FOR REPAIR OF NEAR-UV (365nm) DAMAGE

	REPAIR PATHWAY	EVIDENCE
BACTERIA	phr	SENSITIVE STRAINS PHOTOREACTIVABLE
	uvr	uvr STRAINS MORE SENSITIVE
	rec	rec STRAINS MUCH MORE SENSITIVE; FAST STRAND - BREAK REPAIR PATHWAY PARTIALLY DEFECTIVE IN rec STRAINS
	pol	pol STRAINS SLIGHTLY MORE SENSITIVE; FAST STRAND - BREAK REPAIR PATHWAY PARTIALLY DEFECTIVE IN pol STRAINS
PHAGE	v AND x	T4v AND T4x MUTANTS MORE SENSITIVE
	HOST-CELL REACTIVATION	PHAGE SLIGHTLY MORE SENSITIVE ON hcr HOST
	WEIGLE REACTIVATION	NO EVIDENCE FOR UV-ENHANCED REACTIVATION

The first table is a summary of the types of DNA lesions that have been shown to be induced by monochromatic wavelengths as long as 405 nm. The circles represent the predominant lesions at any particular wavelength and are not intended to represent a quantitative relationship between wavelengths. The table illustrates qualitatively that with increasing wavelength the relative levels of pyrimidine dimers or other recognizable base damage diminishes in relation to the level of strand breaks (alkali-labile lesions) induced. Thymine glycols, base damage thought to result from the indirect action of radiation, have been detected at 313 nm and at low levels after 334 nm in human cells (9). DNA-protein crosslinks have been reported after broadband near-UV treatment of mammalian DNA (11), but the wavelength dependence of the induction is unknown. Other base-damage has been reported (12) but not identified. However, it still appears that considering the total lesion induction by wavelengths in the most energetic part of the solar-UV spectrum i.e. 290-320 nm, the cyclobutane-type dimers (which are the predominant lesion induced by far-UV radiation) are still in the majority. This certainly does not mean that the non-dimer DNA lesions are not important. In fact, the relative biological importance of these lesions is not understood in repair proficient strains at any solar-UV wavelength despite quite reasonable quantitation, particularly at 365 nm. However, it is clear that the spectrum of solar-UV damage is quite distinct from that established in the shorter UV region.

2.3 - Lethal consequences and repair of solar-UV damage

Several laboratories have been concerned with the question as to whether near-UV damage is repairable. The second tables summarises the types of evidence obtained in bacteria and bacteriophage that illustrate that certain classes of damage induced by radiation at 365 nm, a wavelength in the middle of the solar-UV range, are susceptible to repair. The first evidence that 365 nm-induced pyrimidine dimers were repairable was obtained by R.B. Webb (13) who showed that certain repair-deficient mutants of E. coli damaged by 365 nm radiation are susceptible to photoreactivation, a repair process that is specific for pyrimidine dimers. A more general approach to determining whether near-UV damage is repairable has been to test the near-UV sensitivity of several mutants of bacteria and bacteriophage that have been shown to be sensitive to far-UV treatment. The 365 nm sensitivities of various E. coli mutants are shown in Fig. 1.

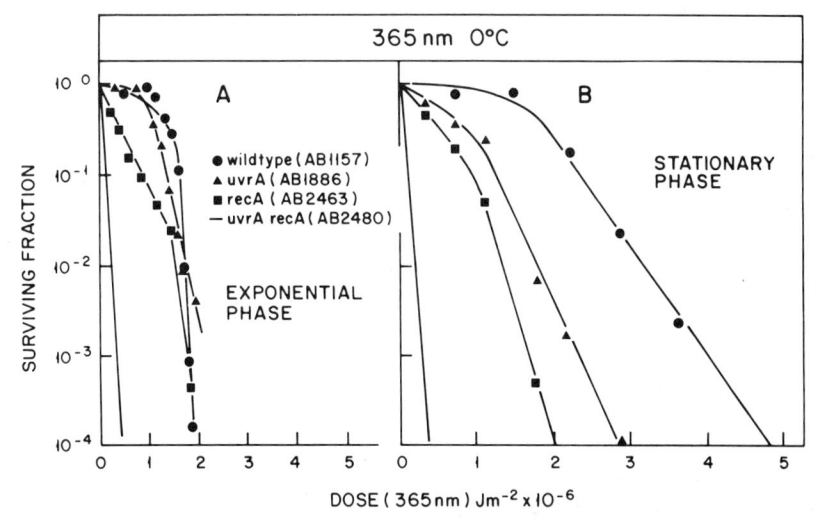

CELL INACTIVATION AND MUTAGENESIS BY U.V. RADIATION   159

Fig. 1 : A comparison of the sensitivity of repair proficient and repair defective strains of Escherichia coli to near-UV (365 nm) in exponential phase and stationary phase : A) exponential phase (adapted from ref. 14) ; B) stationary phase (adapted from ref. 2).

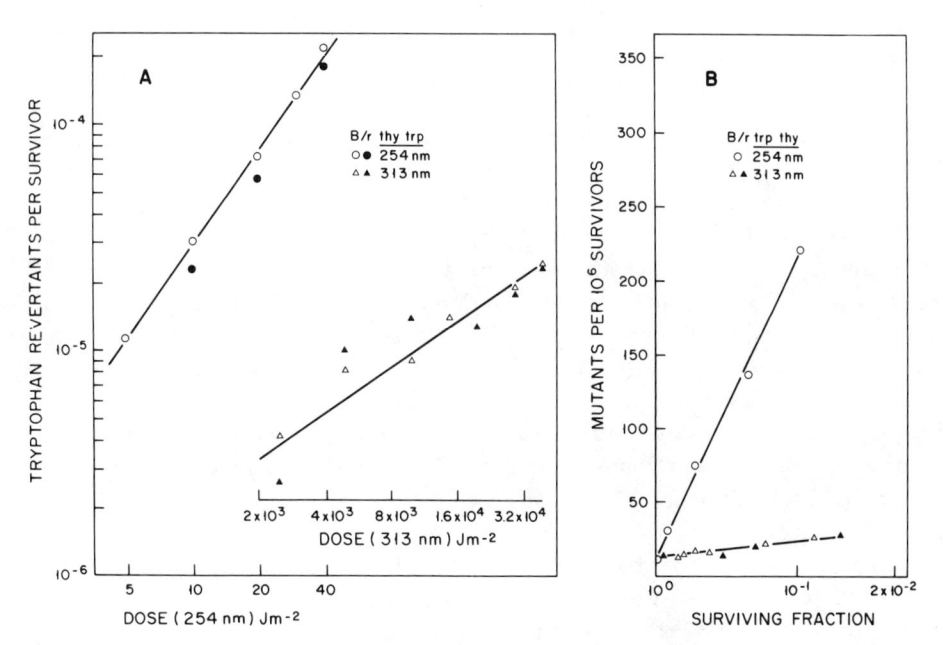

Fig. 2 : Mutation to tryptophan independence by UV radiations in the repair competent strain E. coli B/r trp thy : A) as a function of dose, B) as a function of surviving fraction (Redrawn, ref. 24)

From this data it is clear that strains defective in repair pathways controlled by the recA gene (exponential and stationary phase Fig. 1A and 1B) and by the uvrA gene (stationary phase Fig. 1B) are more sensitive than the repair competent strains to near-UV damage. When clone-forming ability is assayed on minimal medium plates, certain strains are very much more sensitive to near-UV than when plated on enriched media (K.C. Smith, personal communication ; our unpublished data). Under these conditions the recA mutant may exhibit near-UV sensitivity close to (our data) or identical with (15) the wild type and Tuveson and Jonas (15) have reported evidence that an additional gene product is involved in confering UV resistance to certain strains. However, more recent data (S.H. Moss and K.C. Smith, in preparation) has indicated that the additional sensitivity on minimal medium plates is largely due to salt sensitivity which, in turn, could reflect membrane damage. Mutants defective in polymerase 1 activity are also more sensitive than the repair competent strains to near-UV damage. In this particular case it has been shown that polA strains are unable to carry out a rapid repair of the near-UV induced breaks that has been observed in the repair competent strains (16). In further studies in our own laboratory (Miguel and Tyrrell, unpublished results) we have recognized at least 3 distinct pathways of strand-break repair, at least one of which is dependent upon the recA gene product but is not analogous to growth-medium dependent repair of ionizing radiation damage (17).

Repair of near-UV damage may also be tested by examining the sensitivity of bacteriophage mutants. Phage have the obvious advantage that they can be irradiated independently of the bacterial host. In this case, mutants defective in phage-controlled repair processes such as the v-gene and x-gene dependent repair systems in T4 coliphage (18) are found to be sensitive to near-UV relative to the wildtype strain (19). In fact, a larger sector of the 365 nm-induced damage is susceptible to the v-gene controlled process than damage induced by far-UV radiation. The situation is quite different when host-controlled processes are studied. There is only a small sector of excision-related host-cell reactivation (HCR) of near-UV damage of bacteriophage lambda. Particularly interesting is that the inducible weigle-reactivation process does not appear to work at all on near-UV lethal damage induced at long wavelengths. This observation may be particularly important to our mutation studies (see later) since this type of inducible repair is supposedly an error-prone repair process and a major source of mutation after both radiation and chemical damage.

2.4 - Damage to repair and lethal interactions

An additional factor that relates to repair must be considered when analysing the biological actions and interactions in this wavelength region. It's quite clear that the longer wavelengths of solar-UV also damage or disrupt virtually all the repair systems so far studied. This has been shown from both biological and physicochemical measurements and has been reviewed in detail elsewhere (3, 20). A biological manifestation of this damage to repair is that, to a greater or lesser extent, bacteria are sensitized to a spectrum of both physical (heat, far-UV and ionizing radiation) and chemical (methyl methane sulphonate, ethyl methane sulphonate) agents by prior exposure to near-UV radiation. The near-UV doses involved are quite large and we will return to this problem when considering the action of sunlight itself.

2.5 - Mutagenic action

With this background concerning the types of damage induced by near-UV and the lethal consequences we began to study the mutagenic action of near-UV and the possibility of mutagenic interactions. Reversion to tryptophan independence in auxotrophic repair proficient and excision deficient strains of Escherichia coli has been used as the model mutation system, although more recently we have tested additional markers. Most trp revertants after far-UV irradiation are suppressor mutations, with a minority of tru revertants (21). Mutations to tryptophan independence are detected by diluting and plating the treated cells onto media which contain only a low quantity of tryptophan (22). This media can be used for scoring both mutation and survival.

Mutation induction by both 254 nm and 313 nm radiations in the repair competent strain, E. coli B/r trp thy is shown in Fig. 2. The left hand panel shows induction of mutations as a function of the square of the dose. The well-known dose-squared relationship is followed at both wavelengths. A quite different aspect of this result is seen when mutants are plotted against surviving fraction (right hand panel). Since the abscissa is now proportional to the number of lethal events induced, it is evident that many fewer mutations are induced at 313 nm than at 254 nm per lethal hit. Although we are still unsure of the reason for this difference, several experiments suggest that it reflects the different spectrum of lesions induced by the two wavelengths. For example, an important class of lethal lesions induced at 313 nm may not be susceptible to error-prone repair. Alternatively, 313 nm induced lesions may lead to the selective inactivation of cells containing pre-mutagenic lesions. No mutations are induced in exponential phase cells at wavelengths longer than 313 nm in repair competent strains.

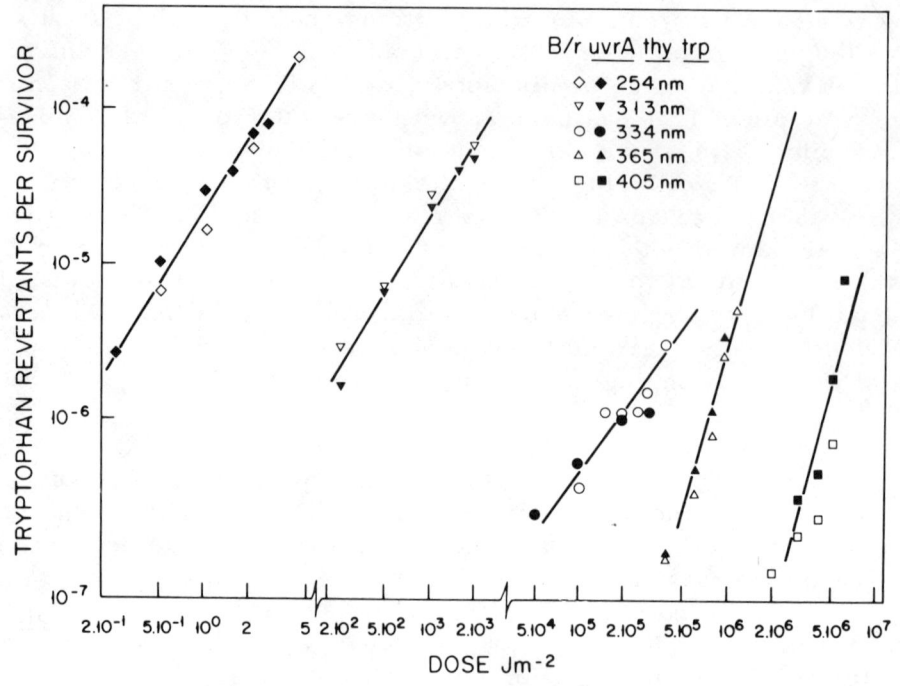

Fig. 3 : Mutation to tryptophan independence as a function of near-UV and visible dose in the excision deficient strain E. coli B/r trp thy uvrA. (redrawn from ref. 24)

In the excisionless strain, mutations are detected at wavelengths as long as 405 nm (Fig. 3), results which essentially confirm those of R.B. Webb (ref. 2, cited as preliminary observations) and, at least at 365 nm, we are in close quantitative agreement. The kinetics of mutation induction change with increasing wavelength from the dose squared response characteristic of the 254 nm induction curve to the situation at 365 nm and 405 nm where the mutation frequency increases exponentially with dose. Again it is interesting to consider mutation frequency as a function of lethal event (Fig. 4). The number of mutations per lethal event is again much lower at 313 nm than at 254 nm (a result confirmed independently in a K12 uvrA strain by S.H. Moss). At 365 nm, the contrast is even more marked (left hand panel, Fig. 4). In fact, it's clear from the curves in the right hand panel that a reduced mutation frequency per lethal event is characteristic of the longer wavelength radiations. Although this again may reflect the lesion distribution at these wavelengths, the mechanism of near-UV

CELL INACTIVATION AND MUTAGENESIS BY U.V. RADIATION   163

mutagenesis remains unclear.

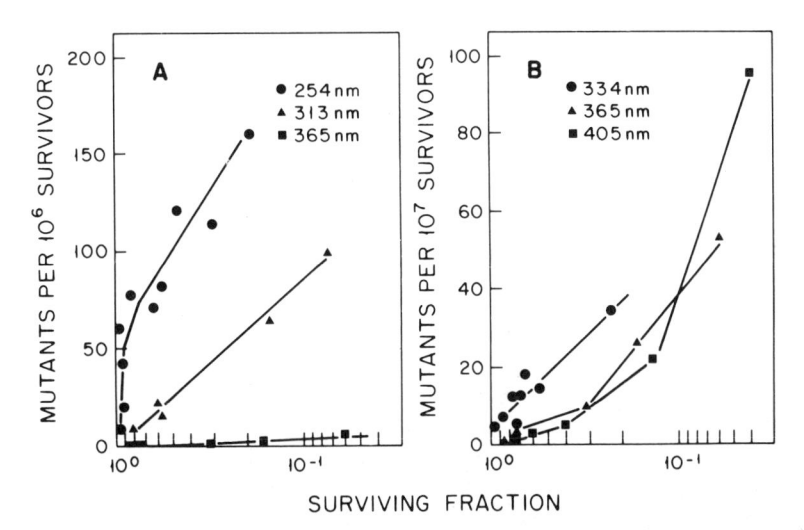

Fig. 4 : Mutation to tryptophan independence as a function of the fraction of the population surviving irradiation by various near-UV and a visible wavelength in the excision deficient strain E. coli B/r thy trp uvrA. (Redrawn from ref. 24).

2.6 - Mutagenic interactions

The most intriguing observation, at least in relation to mutagenesis by natural sunlight, was that pre-treatment of cells with near-UV radiation strongly reduced the frequency of mutations induced by the more energetic shorter wavelengths (24). Since radiation at 313 nm is dimer producing and a convenient emission line of mercury that lies in the mutagenic region of the solar spectrum, we later tested the modification of the mutagenic potential of both 254 nm and 313 nm radiation by pretreatment at the longer wavelengths (23).

Fig. 5 shows the mutagenic interaction between 334 nm radiation and these two shorter wavelengths. The upper curves show the surviving fraction after 334 nm radiation only and the additional lethal action after a fixed dose of either 10 Jm^{-2} at 254 nm or 10^4 Jm^{-2} at 313 nm. Most interesting is that when 254 nm or 313 nm mutation induction is followed as function of varying pretreatment doses at 334 nm, we see a rapid drop in mutation frequency at near-UV doses that are virtually non-lethal and show little or no lethal interaction. In this repair competent strain, the level of mutants falls to below 1 percent of the original induction level by a dose of 5×10^4 Jm^{-2} at 334 nm. There is no mutagenic interaction whatsoever when excision repair is non-functional (23). Similar experiments have been carried out at 3 longer wavelengths (365 nm, 405 nm and 434 nm) with almost identical results. At all wavelengths, the drop in mutation occurs at dose-levels very much lower than those where a lethal action or lethal interaction of the longer wavelength occurs.

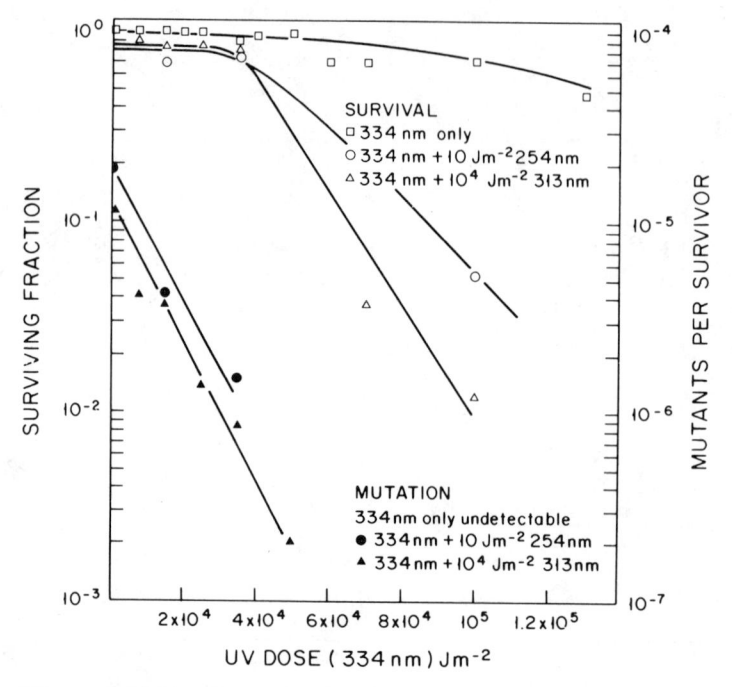

Fig. 5 : The modification of 254 nm and 313 nm mutation induction and survival by pre-treatment with near-UV (334 nm) radiation in the repair competent strain E. coli B/r thy trp. (redrawn from ref. 24).

In accordance with current models of far-UV mutagenesis (25) we may suggest that the dose-dependent mutation suppression reflects a progressive increase in the ratio of error-free to error-prone repair by near-UV. More specifically, these is evidence that the suppression is associated with a near-UV induced growth delay that allows more time for the constitutive primarily error-free pre-replication repair process to operate. This interpretation is consistent with the lack of suppression in the excisionless strain. The near-UV growth delay hypothesis is strongly supported by some earlier studies by Jagger et al. (26) who have measured growth delay and associated phenomena at various wavelengths. We have compared the relative efficiencies of various wavelengths in inducing growth and delay with the relative efficiencies of suppressing mutation induction (23). The relative doses required to reduce the percentage of the population not showing growth delay to 10 correspond reasonably well with the relative doses required to reduce the level of 254 nm induced mutants to 10 percent of the original value. They are clearly not related to the relative efficiencies of enhancing the lethality of various agents which in turn have been related to damage to repair. The actual near-UV doses required to cause growth delay and suppress mutation at any particular wavelength are also in close quantitative agreement. More important, these doses are sufficiently low that from these laboratory experiments and from spectroradiometric measurements of natural sunlight, we were able to predict that only a few minutes of noon sunlight should be sufficient to suppress mutation by the shorter wavelengths in sunlight itself.

3. SOLAR DOSIMETRY

As mentioned earlier, an essential phase in a study of the biological action of natural sunlight is the selection of a suitable dosimetry system. This is important not only for obtaining day-to-day reproducibility but also for correlating laboratory measurements with field measurements. Since the spectral distribution of sunlight, particularly at wavelengths shorter than 320 nm is critically dependent upon latitude, time of day, season and many other factors, it would be useful to take complete spectroradiometric measurements simultaneously with each biological measurement. An estimate of the biologically effective dose would require, in addition, a precise action spectrum for the organism and parameter under consideration. Aside from being complex, such a procedure would still not be valid where wavelength interactions occur.

A simpler approach has been to select a detecting system whose wavelength sensitivity corresponds to the action spectrum for the parameter in question and will therefore give a relative integral measurement of the effective dose or dose-rate. Such dosimeters fall into three classes:

a) _electronic_ In this case solar energy is converted by a photo-electric device to give a reading of dose-rate or integral dose on a meter. One such device designed to measure solar energy in "sun-burning units" employs a nickel oxide filter which transmits radiation in the 300 nm region. The filtered radiation then activates a magnesium tungstate phosphor which emits green light which may then be detected by a photoemissive tube or, when the system needs to be portable, by a photovoltaic cell (27).

b) _chemical_ Solar energy may also be used to drive a chemical reaction and obtain an integral measure of dose. One such system employs a photo-sensitive sulphonate film whose chemical modification is accompanied by a colour change. This change may then be measured directly by spectrophotometry. The main difficulty with such actinometric systems is to find a chemical whose light sensitivity as a function of wavelength corresponds to the biological effectiveness for the parameter in question.

c) _biological_ A UV sensitive bacterial spore isolated by Munakata (28) exhibits essentially exponential inactivation throughout the solar-UV range. Furthermore the relative sensitivity of the spore as a function of wavelength corresponds to the relative absorption of DNA, at least at wavelengths as long as 320 nm. Survival measurements after given exposure times will therefore give integral measurements of the relative energy absorbed by the DNA in a transparent system for any given exposure (10). The system, is extremely useful where DNA is believed to be the primary site of action and we have therefore used spore dosimetry for our studies of the lethal and mutagenic action of natural sunlight. Deviations from the result expected simply from DNA changes will give an indication that other targets are assuming importance. This system has recently been modified (V. Wang and C.S. Rupert, in preparation) for use with dry spores on cover slips. This extends considerably the manoevrability of the dosimeter.

4. ACTION OF SOLAR RADIATION

4.1 - Lethal action and interactions

The inactivation of colony-forming ability of several repair proficient and repair deficient strains of _Escherichia coli_ by natural sunlight has been observed. The measurements were made at ice temperature and arranged so that noon always fell in the centre of the total irradiation period. Knowing the solar exposure time required to inactivate the spores to any given survival level on the test day and using pre-determined spore survival curves at various short-UV wavelengths, it is possible to convert the solar dose to an equivalent dose at any chosen wavelength. When 254 nm is chosen, the pattern of solar inactivation resembles quite closely the pattern of inactivation by this non-solar UV wavelength in terms of relative rates. However, the bacteria are inactivated at rates at least 4 times faster than expected if 254 nm were a good model wavelength for extrapolation to sunlight studies.

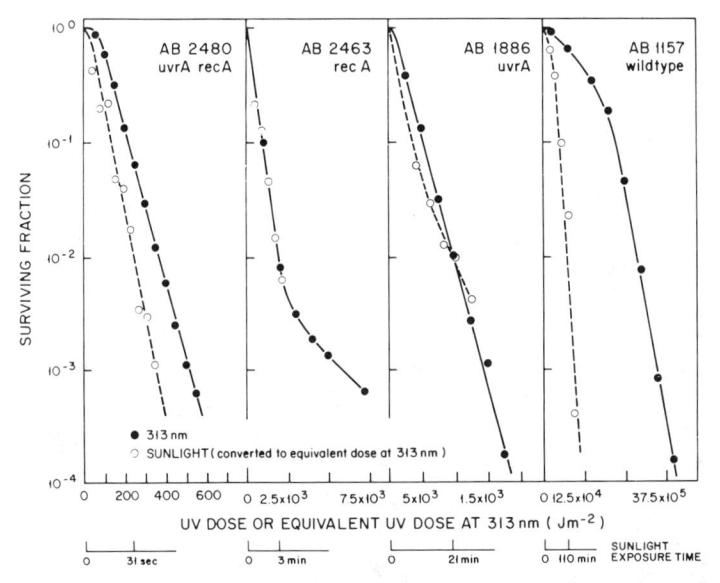

Fig. 6 : Inactivation of repair proficient or repair-deficient strains of E. coli by 313 nm or natural sunlight. The sunlight dose has been converted to an equivalent dose at 313 nm as described in the text. The actual solar exposure times to inactivate through two log cycles are shown on the bottom scale. (Data from Tyrrell and Souza-Neto, this laboratory).

When 313 nm is chosen as the standard wavelength, as illustrated in Fig. 6, both the relative rates of inactivation of the three sensitive strains (recA, uvrA and uvrArecA) and the equivalent sunlight dose at 313 nm correspond closely with the true inactivation curves at 313 nm. The repair competent strain is inactivated 3 or 4 times faster by a dose of sunlight equivalent to a similar dose at 313 nm. We believe that the most probable explanation is that the longer wavelengths in sunlight which are known to disrupt repair systems, sensitize the bacteria to the shorter DNA-damaging wavelengths by analogy with the results obtained with monochromatic radiations (see above). Under sunlight conditions where the ratio of longer to shorter wavelengths is higher (eg. lower zenith angles) the interactions may well be even more marked. Nevertheless, 313 nm appears to be a reasonable model wavelength under conditions where interactions are not observed.

In additional experiments the lethal interaction of sunlight with other agents has been tested by pre-treating repair proficient populations of bacteria with sunlight doses that inactivate only a few percent of the population and then exposing them to either far-UV, mild heat (52°C) or methyl methane sulphonate. Distinct positive interactions occur in all cases (unpublished results, this laboratory). Such a result would also be predicted from the lethal interactions observed with monochromatic radiations (see earlier).

4.2 - Mutagenic action and interactions

The results are particularly interesting when the parameter studied is mutation. Several years ago, some workers in England (29) reported that the sun did not induce mutations in repair competent bacteria. We were also unable to detect mutations in repair competent strains after a maximum solar exposure time of 20 min. Such an exposure is equivalent in terms of DNA damage to approximately $10^4 Jm^{-2}$ at 313 nm, a dose that would normally lead to a mutation frequency of 3×10^{-6} per survivor (20-30 mutants per plate). We therefore tested the possibility that some component of solar radiation itself is suppressing mutation.

Samples of sunlight-treated bacteria were exposed to a constant dose of far-UV (254 nm) radiation and the modification in mutation frequency was followed. The upper curves in Fig. 7 represent survival as a function of sunlight exposure either alone or including an additional exposure of $10 Jm^{-2}$ at 254 nm. There is very little lethal interaction in this range. However, the frequency of mutants induced by a constant exposure to $10 Jm^{-2}$ at 254 nm drops rapidly with solar exposure time to a few percent of the original frequency within a few minutes.

The two mutation curves on Fig. 7 merely compare experiments at 0°C and ambient temperature on days on which the solar dose-rate corresponds closely. In an additional experiment, it was observed that a 10 minutes exposure to sunlight reduced the level of mutants induced by subsequent exposure to a range of doses at either 254 nm or 313 nm below the level of detection. In several experiments, a window-glass filter that cut off radiations

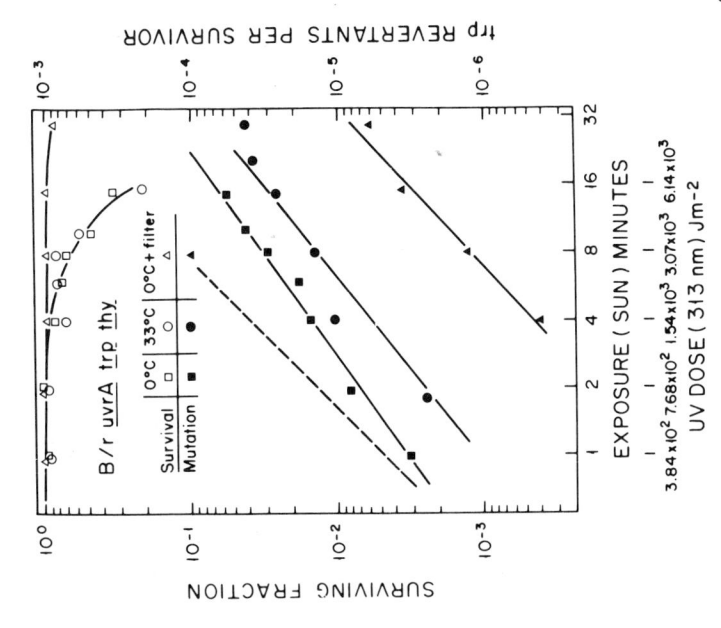

CELL INACTIVATION AND MUTAGENESIS BY U.V. RADIATION

Fig. 7 : Inactivation and mutation of E. coli B/r trp thy by sunlight and by mixtures of sunlight with a constant dose (10 Jm⁻²) of far-UV (254 nm). Solar irradiation at ice (0°C) or ambient (35°C) temperature. Rio de Janeiro, noon, June 1979. (from ref. 31).

Fig. 8 : Inactivation and mutation of E. coli B/r trp thy uvrA by sunlight after irradiation at 0°C, at ambient temperature and at 0°C employing a glass cut-off filter. Rio de Janeiro, noon, april-may 1979. The dotted line corresponds to the lower scale and represents mean mutation data for experiments carried out at 313 nm.

below 320 nm covered the quartz irradiation vessel. Little or no reduction was observed in the capacity of sunlight to suppress the induction of mutations by 254 nm. The filter reduced the DNA-damaging efficiency of sunlight by a factor of 3. This suggests that a major contribution to the mutation suppression is by wavelengths longer than 320 nm.

Results of similar experiments with an excisionless strain are shown in Fig. 8. As predicted from experiments with monochromatic radiation (see earlier), sunlight does not modify the frequency of induction of mutations by far-UV in this strain. However, sunlight alone now does induce a high level of mutation. By spore dosimetry we calculate that the dose-rate on the test day was equivalent to 6.4 Wm^{-2} at 313 nm. This value has been used to construct the lower dose-scale in Fig. 8 and a true mutation curve at 313 nm (dotted lines, mean of several experiments) has been included on the figure.

At ambient temperature, sunlight induces lower levels of mutation in the excisionless strain (Fig. 8) presumably because of simultaneous photoreactivation. When wavelengths below 320 nm are filtered out, the mutation rate is cut by a factor of 15. This probably reflects not only the elimination of a good fraction of the DNA-damaging wavelengths by the filter but also the longer time available for simultaneous photoreactivation to eliminate a much larger fraction of the damage.

The essential factor to emerge from these experiments is that the results of the sunlight studies confirm predictions from experiments with selected monochromatic wavelengths. Short exposures to sunlight can evidently provoke changes in the post-irradiation metabolism of cells that leads to a strong suppression of mutation induction in repair competent strains and may be responsible for the lack of mutation by sunlight itself. The observation that sunlight also provokes growth delays (30) lends support to the possibility that a similar mechanism for the mutation suppression may be operating after both sunlight and monochromatic radiations.

5. SUMMARY

An experimental approach to a study of the biological action of sunlight has been illustrated by a description of the lethal and mutagenic action of sunlight and it's component wavelengths. Experimental information has been collected using monochromatic radiation and mixtures of wavelengths and some of the predictions arising have been tested with natural sunlight. The spectrum of lesions induced in the solar-UV region is quite different from the far-UV region. This is reflected in results from mutation studies where many fewer mutations are induced than would be expected from the known levels of DNA damage, particularly at the longer wavelengths. The longer wavelengths of solar-UV also induce two important classes of metabolic alteration. In the low dose-range, there is inhibition of growth and macromolecular synthesis and in the higher dose range, disruption of repair systems. These two effects lead respectively to strong mutagenic interactions and strong lethal interactions between these wavelengths and other DNA-damaging agents. There is some evidence that lethal interactions occur between the longer and shorter wavelengths in sunlight itself in the higher dose range and that sunlight is also able to provoke mild lethal interactions with other DNA-damaging agents. On the other hand, mutational interactions occur in the low dose region and are therefore observed after low exposures to natural sunlight. Solar mutagenesis itself is almost certainly influenced by these phenomena. Furthermore laboratory measurements predict that low doses of natural sunlight may lead to both antagonistic and synergistic interactions with mutagenic chemicals according to the dose-range and specific chemical agent employed. The possibility of an analogous response in more complex cells cannot be overlooked.

ACKNOWLEDGEMENTS

This work was supported by the following Brazilian granting agencies : CNPq (National Research Council), CNEN (National Nuclear Energy Council), CEPG/UFRJ (University Council for Post-Graduate studies) and FINEP (B-76-79-074).

REFERENCES

1. Eisenstark, A. (1971) in Advances in Genetics, Vol. 12 (E. W. Caspari, ed.) pp. 167-198, Academic Press, New York.
2. Webb, R. B. (1977) in Photochemical and Photobiological Review, vol. 2 (K. C. Smith, ed.) pp. 169-261, Plenum Press, New York.
3. Tyrrell, R. M. (1978a) in International Symposium on Current Topics in Radiobiology and Photobiology (R. M. Tyrrell, ed.) pp. 73-80, Academia Brasileira de Ciências, Rio de Janeiro.
4. Ramabgadran, T. V. and J. Jagger (1976) Proc. Natl. Acad. Sci. US 73:59-63

5. Thomas, G. and A. Favre (1975) Biochem. Biophys. Res. Commun. 66:1454-1461.
6. Kashkett, E.R. and A.F. Brodie (1962) J. Bacteriol. 83:1094-1100.
7. Tyrrell, R.M. (1973) Photochem. Photobiol. 17:69-73.
8. Tyrrell, R.M., R.D. Ley and R.B. Webb (1974) Photochem. Photobiol. 20:395-398.
9. Cerutti, P. and Netrawali, M. (1979) in Proceedings of the Sixth International Congress of Radiation Research (ed. S. Okada, M. Inamura, T. Terashima and H. Yamaguchi) pp. 423-431, Tokyo, Japan.
10. Tyrrell, R.M. (1978d) Photochem. Photobiol. 27:571-579.
11. Bradley, M.O., Hsu, C. and C.C. Harris, Nature (in press).
12. Cabrera-Juarez, E. and J.K. Setlow (1977) Biochim. Biophys. Acta 475:315-322
13. Brown, M.S. and R.B. Webb (1972) Mutat. Res. 15:348-352.
14. Tyrrell, R.M. (1976) Photochem. Photobiol. 23:13-20.
15. Tuveson, R.W. and Jonas, R.B. (1979) Photochem. Photobiol. 30:667-677.
16. Ley, R.D., Sedita, B.A. and E. Boye (1978) Photochem. Photobiol. 27:323-329.
17. Town, C.D., R.C. Smith and H.S. Kaplan (1973) Cur. Topics Radiat. Res. Quart. 8:351-399.
18. Rupert, C.S. and W. Harm (1966) in Advances in Radiation Biology, Vol.2 (L.G. Augenstein, R. Mason and M. Zelle, ed.) pp. 1-81, Academic Press, New York.
19. Tyrrell, R.M. (1979a) Photochem. Photobiol. 29:963-970.
20. Tyrrell, R.M. (1978b) in Photochemical and Photobiological Reviews, vol.3 (K.C.Smith, ed.) pp.35-113, Plenum, New York.
21. Bridges, B.A., R.E. Dennis and R.J. Munson (1967) Genetics 57:897-908.
22. Demerec, M. and E. Cahn (1953) J. Bacteriol. 65:27-36.
23. Tyrrell, R.M. (1980) Photochem. Photobiol. 31:37-47.
24. Tyrrell, R.M. (1978c) Mutation Res. 52:25-35.
25. Witkin, E.M. (1976) Bacteriol. Rev. 40:869-907.
26. Jagger, J., W. Curtis-Wise and R. Stafford (1964) Photochem. Photobiol. 3:11-24.
27. Robertson, D.F. (1972) PhD thesis, Queensland University, Australia.
28. Munakata, N. (1974) J. Bacteriol. 120:59-65.
29. Ashwood-Smith, M.J., Copeland, J. and J. Wilcockson (1967). Nature (London) 214:33-35.
30. Jagger, J. (1975) Photochem. Photobiol. 22:67-70.
31. Tyrrell, R.M. (1979b) Biochem. Biophys. Res. Comm. 91:1406-1415.

PHOTOREACTIVATION OF PYRIMIDINE DIMERS GENERATED BY A

PHOTOSENSITIZED REACTION IN RNA OF INSECT EMBRYOS (SMITTIA SPEC.)

Klaus Kalthoff and Herbert Jäckle

Department of Zoology
University of Texas at Austin
Austin, Texas 78712

OUTLINE OF EARLY INSECT EMBRYOGENESIS

Insect eggs are large cells, ranging from a few tenths to several millimeters in length. The egg cell is surrounded by a vitelline membrane, and an outer shell, the chorion. The yolk-rich and opaque endoplasm contains glycoproteid spheres, lipid droplets, and glycogen particles, whereas the superficial periplasm is yolk-free. After fertilization or parthenogenetic activation, the nucleus undergoes a series of mitotic divisions within the yolk endoplasm. This period is referred to as intravitelline cleavage, although the egg cell is not cleaved. Rather, the embryo develops in a plasmodial state, containing eventually hundreds of energids, i.e. nuclei with jackets of cytoplasm. During nuclear migration stages, most energids move into the yolk-free periplasm, where they become enclosed by infoldings of the plasmalemma of the egg cell. The resulting blastoderm cells may be eventually separated from the yolk endoplasm, or may remain connected to it by cytoplasmic bridges. While cellularization is in progress, nuclear divisions may continue. Following this period of nuclear proliferation and the formation of an apparently homogeneous layer of blastoderm cells, regional differentiation begins. Many of the blastoderm cells build the originally unsegmented germ anlage while the remainder form the embryonic membranes. After gastrulation and segmentation, the embryo reaches the germ band stage which already reflects the basic organization of the larva. Recent descriptions of early Drosophila embryogenesis have been given by Fullilove and Jacobson (1978), Turner and Mahowald (1976), and Zalokar and Erk (1976). For reviews on insect development see Counce and Waddington (1972, 1973).

PHOTOREVERSIBLE UV-EFFECTS ON SMITTIA EMBRYOS

The experiments described below were carried out with early embryos of the chironomid midge Smittia spec.. Eggs of this species are extremely small, about 200 µm long and 90 µm wide, and have a perfectly transparent chorion. After UV-irradiation, three biological effects can be observed. (1) Upon selective irradiation of the anterior pole region, embryos develop with a distinctly abnormal segment pattern ("double abdomen"). (2) Exposure of entire embryos to UV causes arrest of development ("inactivation"). (3) The same treatment results in inhibition of protein synthesis. All of these UV-effects are photoreversible.

UV-Induction of Double Abdomens

UV-irradiation of anterior pole regions of Smittia embryos prior to blastoderm formation causes the development of embryos with a distinctly abnormal body pattern called "double abdomen" (Kalthoff and Sander, 1968). Such embryos have their head and thorax replaced by a mirror image of most of the abdomen (Fig. 1). The plane of symmetry varies from third thoracic to fifth abdominal segment. The supernumerary (anterior) abdomen arises from cells that would otherwise have formed head and thorax. Moreover, formation of this abdomen does not seem to require interaction between anterior and posterior halves of the

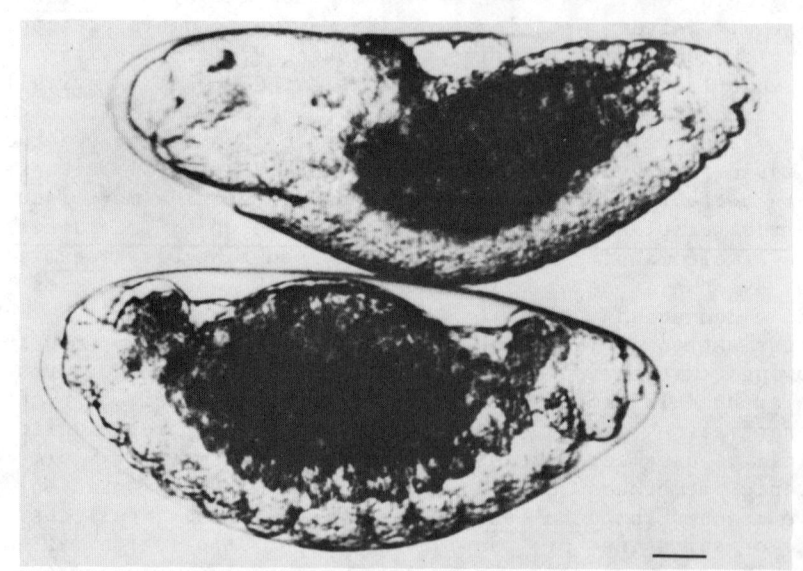

Fig. 1. Top: Smittia embryo with normal body segment pattern. Bottom: Embryo with aberrant pattern "double abdomen". Bar: 20 µm. Photograph: K. Sander. From Kalthoff (1970).

developing embryo (Sander, unpublished; Ritter, 1976; reviewed in Kalthoff, 1979). The development of the anterior abdomen is ascribed to inactivation of anterior morphogenetic determinants whose normal function is required for head and thorax to develop (Kalthoff, 1979). Exposure of UV-irradiated embryos to photoreactivating light seems to facilitate the repair of these components, with normal formation of head and thorax ensuing. The two alternate pathways of development can be discerned morphologically from the beginning of germ anlage formation (Kalthoff and Sander, 1968; Kalthoff, 1975a). However, the synthesis of certain "indicator" proteins in anterior or posterior halves of embryos during blastoderm stages foreshadows the formation of either head and thorax or an abdomen (Jäckle and Kalthoff, 1980a, 1981).

UV-Inactivation (Arrest of Development)

Irradiation of entire embryos, without shielding and with the long axis perpendicular to the UV-beam, causes arrest of development (Kalthoff, 1976). Embryos exposed to UV during intravitelline cleavage, i.e. with all somatic nuclei still in the yolk-rich endoplasm, typically die before blastoderm formation. Those which proceeded through this stage were scored as "normal" in our experiments if they formed pigmented eyes which become clearly visible in advanced stages of embryogenesis. Exposure of UV-irradiated embryos to near UV or blue light causes photoreactivation.

UV-effects on Protein Synthesis

UV-irradiation of Smittia embryos under conditions causing inactivation also has marked effects on protein synthesis (Jäckle and Kalthoff, 1980b). Application of UV-doses causing inactivation of 95% of the exposed embryos also reduces the in vivo rate of [^{35}S]-methionine incorporation into polypeptides to less than half of the normal rate. Two-dimensional gel electrophoresis reveals that synthesis of some proteins is reduced more than that of others, and that some new proteins appear in UV-irradiated embryos. Moreover, the stimulation of a messenger RNA-dependent reticulocyte cell-free system by total RNA, or polyadenylated RNA only, from Smittia embryos is reduced after UV-irradiation of the embryos in vivo (Fig. 2). Finally, the apparent degradation, during early embryogenesis, of maternal polyadenylated RNA is delayed after UV-irradiation. All of these effects, including UV-inhibition of messenger-RNA activity in vitro, are photoreversible. Since we do not know to which extent the UV-effects on protein synthesis causes inactivation of the embryos, we discuss the effects separately.

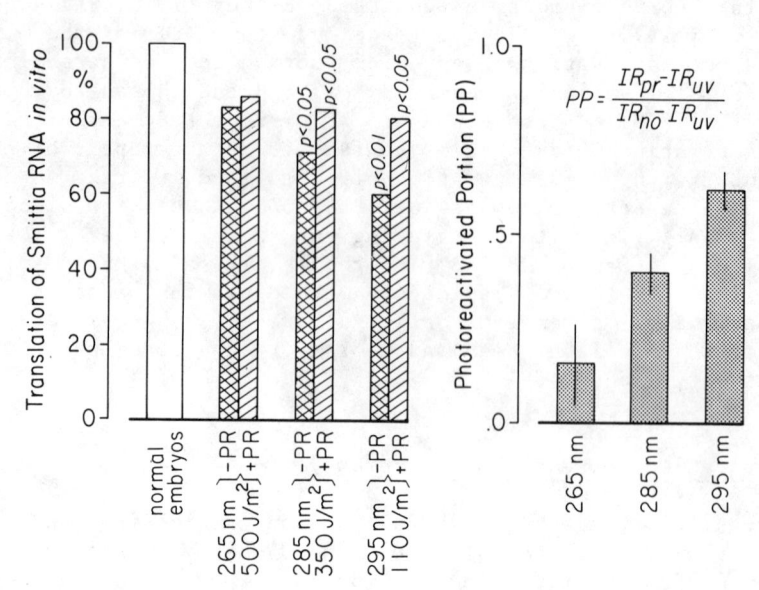

Fig. 2. Effects of UV-irradiation and photoreactivation on translation in vitro of total RNA extracted from Smittia embryos. Rate of [^{35}S]-methionine incorporation into polypeptides obtained with RNA from normal embryos was set to 100%. UV-irradiations with wavelengths and doses indicated were isoeffective with respect to inactivation of embryos (5% normal development). Statistical significance of differences between incorporation rates obtained with RNA from UV-irradiated versus normal embryos ($IR_{no} - IR_{uv}$), and from UV-irradiated versus photoreactivated embryos ($IR_{pr} - IR_{uv}$), is indicated on top of columns. Bars indicate standard deviations of photoreactivated portions obtained in three independent experiments. Data from Jäckle and Kalthoff (1980b).

SUBCELLULAR LOCALIZATION OF THE EFFECTIVE TARGETS OF UV

The effective targets for UV-induction of double abdomens are not associated with nuclei. This has been shown by irradiation of the anterior eighth of newly deposited eggs, whose anterior half is devoid of nuclei at this stage (Kalthoff, 1971). Embryos were also centrifuged with their long axis perpendicular to the direction of the centrifugal force, under conditions causing stratification of yolk components, mitochondria, and endoplasmic reticulum. The different layers were irradiated with a UV-microbeam. The highest double abdomen yields were obtained by irradiating a layer of cytoplasm that did not contain any of the stratified components, and no other conspicious organelles except ribosomes. Irradiation of this layer also led to significantly higher double abdomen yields than irradiating a corresponding zone in uncentrifuged embryos. We conclude that the effective targets are not directly stratified under the conditions used but only displaced or shielded by other components that become stratified (Kalthoff et al., 1977). The data suggest that the effective targets for UV-induction of double abdomens are localized in the cytoplasm and that they are smaller than mitochondria, i.e. the smallest components stratified under the conditions used.

The effective targets for UV-inactivation, as well as UV-inhibition of protein synthesis, also appear to be extranuclear, although the evidence is less clear cut. Embryos are particularly sensitive to UV-inactivation during intravitelline cleavage stage, when nuclei are shielded by yolk-rich cytoplasm absorbing more than 90% of the incident radiation (Kalthoff, 1973, 1976). Also, no damage to nuclei was reflected in the capacity for RNA synthesis by embryos irradiated during intravitelline cleavage. Those who survived until blastoderm stages (when RNA synthesis becomes prominent), showed reduced rates of protein synthesis while both uptake and incorporation of uridine were unaffected (Jäckle and Kalthoff, 1980b).

INVOLVEMENT OF DAMAGE TO RNA IN THE OBSERVED UV-EFFECTS

In all three UV-effects observed, damage to RNA seems to play a major role. UV-induction of double abdomens can be mimicked by application of RNase to the anterior pole of Smittia embryos (Kandler-Singer and Kalthoff, 1976). Also, the action spectrum for double abdomen induction by perpendicular irradiation of the anterior quarter during intravitelline cleavage has a distinct shoulder at 265 nm (Kalthoff, 1973). UV-irradiation under conditions leading to arrest of development and inhibition of protein synthesis also results in the formation of both RNA-protein crosslinks and of pyrimidine dimers in RNA (Jäckle and Kalthoff, 1978, 1979). Moreover, photoreactivating treatment of the embryos causes the disappearance of pyrimidine dimers from their RNA (Fig. 4). Finally, the translatability in vitro of RNA extracted from UV-irradiated embryos is reduced, and becomes partially restored after photoreactivation of the embryos (Fig. 2). These data suggest very strongly that UV-damage to RNA also accounts, at least in part, for the inhibition of protein synthesis in vivo.

PHOTOSENSITIZED FORMATION OF PYRIMIDINE DIMERS IN RNA IN VIVO

Despite the apparent role of RNA in the biological effects observed, RNA is not necessarily the primary absorber of UV in any of these cases. Rather, the wavelength-dependence of each of the biological effects indicates the involvement of a photosensitizer. Action spectra for double abdomen induction show a distinct peak at 285 nm (Kalthoff, 1973; Ripley and Kalthoff, 1980). The action spectrum for inactivation which shows a very sharp peak at 295 nm, does not even remotely resemble the absorption spectrum of nuclei acids (Fig. 3). This observation came as a surprise because the embryos could be photoreactivated very efficiently, and photoreactivation is usually associated with light-dependent repair of UV-damage to nucleic acids, in particular with the monomerization of UV-induced pyrimidine dimers (Harm, 1976; Gordon et al., 1976).

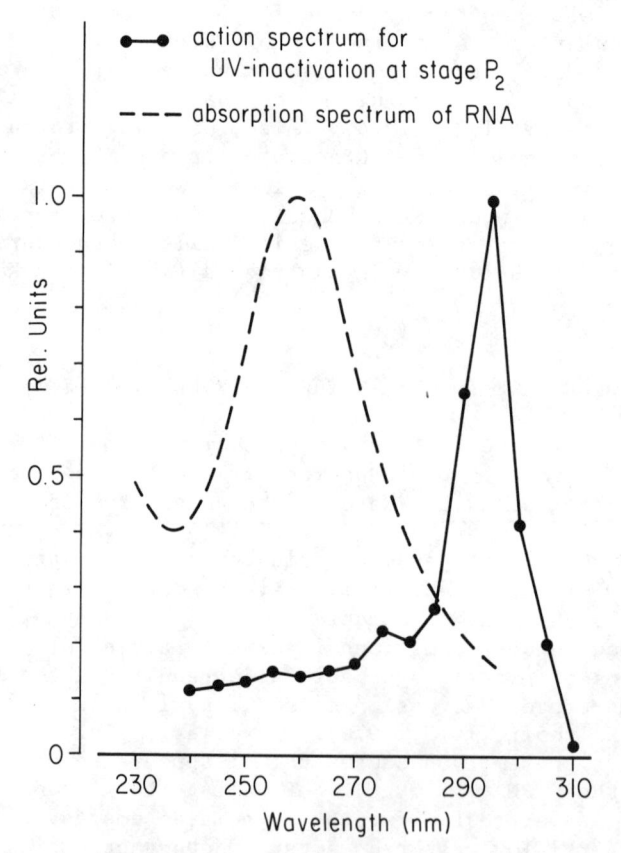

Fig. 3. Absorption spectrum of RNA and action spectrum of inactivation of Smittia embryos during intravitelline cleavage (stage P2). Data from Kalthoff (1976).

The involvement of photosensitizers in the generation of pyrimidine dimers in RNA of Smittia embryos in vivo was shown by UV-irradiation at different wavelengths, followed by RNA extraction, enzymatic or alkaline hydrolysis, and two-dimensional paper chromatography. Since these experiments were rather laborious, they were carried out only after irradiation at the following wavelengths and doses: 265 nm with 500 J/m^2, 285 nm with 350 J/m^2, 295 nm with 110 J/m^2. The wavelengths were selected because the action spectra for UV-induction of double abdomens, and for UV-inactivation, peaked at one or two of these;

TABLE I

Pyrimidine dimers in RNA irradiated in vitro, or extracted From Smittia embroys irradiated in vivo.

wavelength (nm)		% dimerized pyrimidines	incident UV-dose (J/m^2)	transmittance [@]	absorbed energy[!] (J/m^2)	% dimers per abs. 100 J/m^2
265	in vivo[&]	0.01	110	0.0000	92	0.01
		0.06	500	0.0000	420	0.01
265	in vitro[#]	0.20	100	0.0000	100	0.20
		0.38	200	0.0000	200	0.19
295	in vivo[&]	0.15	110	0.0001	95	0.16
295	in vitro[#]	0.026	100	0.2239	78	0.03
313	in vitro[*] 5% acetone	2.6	4,320	0.7499	1,080	0.24

[@] Transmittance of embryos was extrapolated from measurements using 5 μm layers of homogenate (Kalthoff, 1973, 1976).

[!] Absorbed energy is calculated from incident UV-dose by subtracting energy absorbed in egg shell, or transmitted (Kalthoff, 1973)

[#] [^{14}C]-uridine labeled transfer RNA from E. coli, RNA irradiated in distilled water

[&] [^{3}H]-uridine labeled total RNA from Smittia embryos, irradiated in vivo with long axis perpendicular to UV-beam

[*] [2-^{14}C]-uracil labeled ribosomal RNA from E. coli, RNA irradiated in 5% (v/v) acetone solution (data from Huang and Gordon, 1972)

the doses were adjusted to be isoeffective with respect to inactivation (5% survival). In addition, the combination 265 nm with 110 J/m^2 was used to test whether pyrimidine dimers were produced in a dose-dependent way. This was in fact observed (Fig. 4a,f). The most remarkable result, however, was that pyrimidine dimers were produced at least ten times more efficiently by 295 nm radiation as compared to 265 nm radiation. We have also compared the yields of pyrimidine dimers obtained after in vitro irradiation of [^{14}C]-labeled transfer RNA from E. coli to the yields determined in the (largely ribosomal) RNA extracted from Smittia embryos irradiated in vivo (see Table I). At 265 nm wavelength, the percentage of dimerized pyrimidines per 100 J/m^2 absorbed energy after irradiation in vivo was only about 5% of the corresponding yield in vitro. This can be ascribed to absorption, within the embryo, by components other than RNA and/or to the formation of RNA photoproducts other than pyrimidine dimers. Conversely, at 295 nm wavelength, the percentage of dimerized pyrimidines per 100 J/m^2 absorbed energy was about 5 times higher after irradiation in vivo as compared to irradiation in vitro. We conclude that the pyrimidine dimers resulting from in vivo irradiation at 295 nm are generated in a photosensitized reaction. The nature of the photosensitizing component(s) in the embryo are not known. Ketones might be considered as candidates since Huang and Gordon (1972) have shown that acetone is an efficient photosensitizer for generating pyrimidine dimers in RNA in vitro (last line in Table I).

LIGHT-DEPENDENT DISAPPEARANCE OF PYRIMIDINE DIMERS FROM RNA, CORRELATION WITH PHOTOREVERSAL OF BIOLOGICAL UV-EFFECTS IN SMITTIA EMBRYOS

The pyrimidine dimers generated by UV-irradiation of Smittia embryos disappeared upon subsequent exposure to light of longer wavelengths also causing biological reactivation (Fig. 4c,d). In addition, we have found that the translatability of total RNA, or polyadenlated RNA, from Smittia embryos in a cell-free system was reduced after UV-inactivation in vivo, and restored after photoreactivation of the embryos (Fig. 2). Moreover, the photoreactivable sector of biological inactivation with isoeffective doses at three UV-wavelengths is correlated with the percentage of pyrimidines dimerized under these conditions (Fig. 4). A corresponding quantitative correlation exists between the photoreactivated portion of UV-inhibition of protein synthesis in vivo and the percentage of dimerized pyrimidines (Fig. 1 in Jäckle and Kalthoff, 1980b).

The photoreactivable sectors of all three biological UV-effects observed are rather high after UV-irradiation at relatively long wavelengths. For double abdomen induction at

285 nm, the maximum value obtained was 0.87 (Ripley and Kalthoff, 1980). For UV-inactivation at 295 nm, it was 0.75 (Fig. 4; for complete data, see Kalthoff, 1976). For inhibition of protein synthesis in vivo by UV of 295 nm, photoreactivated portions of up to 0.86 were obtained (Jäckle and Kalthoff, 1980b). After UV-irradiation at 265 nm, the photoreactivable sectors were generally smaller, as fewer pyrimidine dimers were generated in the RNA of the embryos (Fig. 4, see also Kalthoff, 1976; Jäckle and Kalthoff, 1980b). Taken together, the data strongly suggest that pyrimidine dimers in RNA constitute the photorepairable fraction of the lesions causing the biological effects of UV-irradiation observed in Smittia embryos.

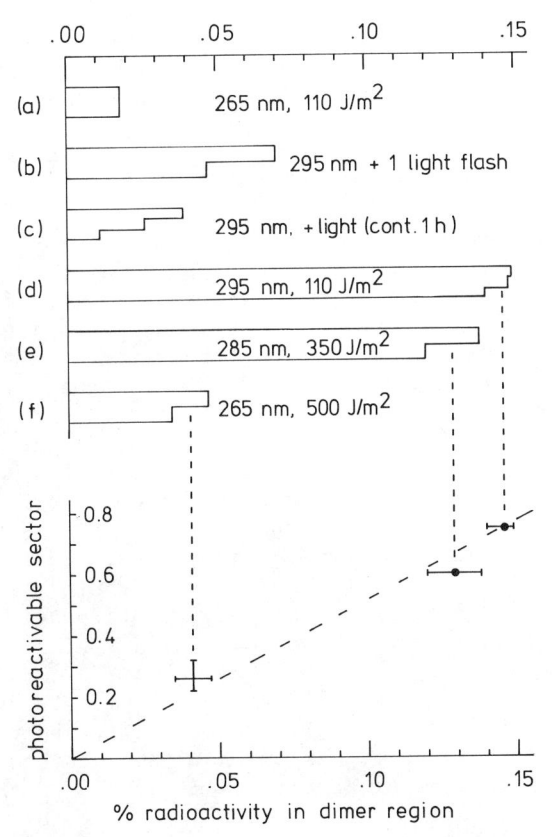

Fig. 4. Correlation between fraction of dimerized pyrimidines in RNA (columns and abscissa) and the photoreactivable sector of UV-inactivation of Smittia embryos. The UV-wavelengths and doses used in (d) to (f) were isoeffective with respect to inactivation. Note disappearance of pyrimidine dimers upon exposure of UV-irradiated embryos to continuous light (400 nm, 28,000 J/m²), or to a single flash of blue light (c,d). Data from Jäckle and Kalthoff (1978).

Photoproducts other than pyrimidine dimers, such as RNA-protein crosslinks (Jäckle and Kalthoff, 1979) or pyrimidine hydrates, might become more dominant lesions under irradiation conditions less conducive to generation of pyrimidine dimers in RNA. This seems to be the case when Smittia embryos are irradiated at 265 nm during intravitelline cleavage (Table I, Fig. 4). Similarly, Levin and Kozlova (1973, 1976, 1977) found that Drosophila eggs and early embryos could not be photoreactivated after UV-inactivation at 254 nm, but were photoreactivable after UV-irradiation at longer wavelengths (313 nm or 290-380 nm). On the other hand, insect eggs and early embryos have been irradiated at 254 nm in a way exposing (pro)nuclei, and were photoreactivable under these conditions (Von Borstel and Wolff, 1955; Amy, 1965, Ghelelovitch, 1966). Thus 254 nm radiation seems to generate a major fraction of photorepairable damage in nuclei (probably pyrimidine dimers in DNA) but not in the cytoplasmic RNA of insect embryos.

IS A "PHOTOREACTIVATING ENZYME" INVOLVED?

The photoenzymatic splitting of pyrimidine dimers is considered to be the predominant mechanism for photoreactivation (PR) of UV-irradiated DNA (Cook, 1970; Harm, 1976). However Hurter et al. (1974) have shown that UV-irradiated RNA from tobacco mosaic virus can also be photoreactivated in vitro with extracts from its host plant, Nicotiana tabacum. The active material in the extracts seems to be associated with protein, and the repaired lesions appear to be pyrimidine dimers in the viral RNA (see Gordon et al., 1976). Hurter et al. (1974) also reported that partially purified enzyme from yeast or pinto bean did not cause PR. Similarly, Rupert (1964) had found earlier that photoreactivating enzyme from yeast did not accept UV-irradiated RNA as substrate. So far, there is no direct evidence for the occurrence of any photoreactivating enzyme acting on RNA in animal cells.

Nevertheless, PR of the biological UV-effects observed in Smittia embryos shows several characteristics known from cases based upon photoenzymatic repair of DNA. The action spectrum of PR after UV-induction of double abdomens has a rather distinct peak at 440 nm, similar to action spectra for PR after UV-inactivation of Streptomyces griseus and Agmenellum (see Kalthoff, 1973). The action spectrum for PR after UV-inactivation of Smittia embryos has a broad maximum around 380 nm, similar to the action spectra for PR after UV-inactivation of E. coli and inhibition of green colony formation of Euglena gracilis (see Kalthoff, 1976). PR treatment of Smittia embryos is effective only after UV but not before. After UV-induction of double abdomens, PR shows dose rate saturation at 6 W/m^2 (Fig. 5), and is temperature dependent at this dose rate (Kalthoff, 1973). When flashes of blue light were used after UV-inactivation, the efficiency of PR increased with the length of the intervals between UV-irradiation and the first flash, and between successive flashes (Kalthoff et al., 1978). Taken together, the data suggest that PR of the observed UV-effects in Smittia embryos involves at least two mechanisms, both of which seem to include light-independent steps, which become rate-limiting under certain conditions. These steps should involve cofactors to account for the PR action spectra observed.

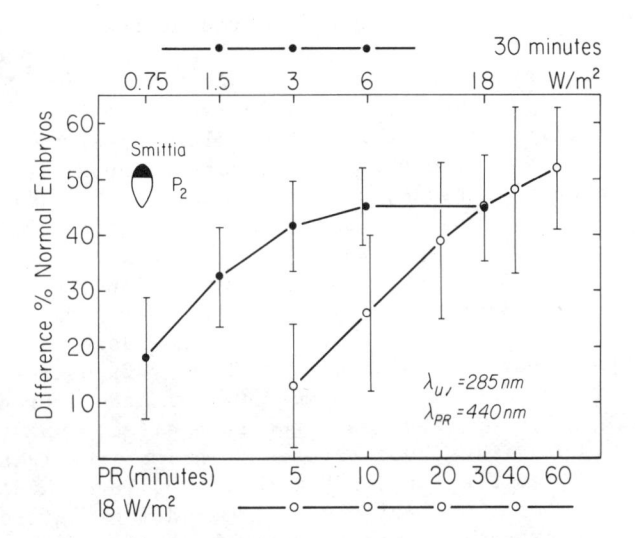

Fig. 5. Dose rate saturation of photoreversal after UV-induction of the aberrant segment pattern "double abdomen" in Smittia by irradiating the anterior quarter of embryos during intravitelline cleavage, using 285 nm wavelength. Photoreverting light (440 nm) was given either at a constant dose rate (18 W/m^2) for increasing periods of time (open circles, lower abscissa), or for a constant period of time (30 min) with increasing dose rates (filled circles, upper abscissa). Photoreversal was measured as the difference in the percentage of normal embryos, obtained with and without photoreversal. Bars represent confidence limits for p = 0.05. Data from Kalthoff et al. (1978).

The nature of the supposed cofactors is not known. It is tempting to speculate that they might be similar to the photoreactivating enzymes acting on UV-irradiated DNA which have been prepared from other organisms. However, this speculation leads to the somewhat puzzling conclusion that the supposed "PR enzyme" in Smittia embryos represents more than 1% of the total weight of non-yolk protein. This figure can be estimated on the basis of 7×10^9 dimers per embryo disappearing after one flash of blue light, and under the assumptions that (1) each "PR enzyme" molecule does not repair more than one dimer per flash ($t_{0.1}$= 1 msec), and that (2) the molecular weight of the supposed "PR enzyme" in embryos is about 35,000 daltons, like the one isolated from E. coli (see Kalthoff et al., 1978). Alternatively, one might assume that the photoreactivating cofactors thought to act on UV-irradiated RNA in Smittia embryos are smaller molecules and/or that they also serve other purposes.

COMPENSATION FOR SOLAR UV-DAMAGE BY SUNLIGHT OF LONGER WAVELENGTH

The spectrum of the sunlight received at the earth's surface overlaps, below 310 nm wavelength, with the action spectrum for inactivating Smittia embryos during intravitelline cleavage (Fig. 3). The egg shell and the jelly of the cluster provide little shielding from this part of the solar spectrum. Smittia females still do not hide their eggs, nor is deposition confined to certain circadian periods, at least not in the laboratory. Suspecting that the PR mechanisms found in Smittia embryos might help them to repair damage inflicted by solar UV, we have determined how well they survived under bright sunlight, and when exposed to solar UV alone. The latter was mimicked by 305 nm radiation of 5 nm band width and 0.28 W/m^2 dose rate obtained from our monochromator (Kalthoff, 1975b). Exposure to this "solar UV equivalent" inactivated 50% of the embryos within 12 minutes (Fig. 6). By contrast, the same effect was observed only after 3 hours in embryos exposed to bright sunlight (around noon, in June 74, at Freiburg, 270 m above sea level, 48°N). We conclude that nearly all of the solar UV-damage inflicted upon the embryos is constantly being repaired by their PR mechanisms utilizing sunlight of longer wavelengths. This would translate into a photoreactivable sector as high as 0.93.

Generally, protein synthesis in early embryos of insects and many other animals is supported by RNA supplies accumulated during oogenesis. This endowment seems to be necessary for many embryos to carry them over an initial period of rapid mitoses which restrict or even inhibit RNA synthesis. Messenger RNA and other RNA types stored during oogenesis thus represent a precious gift which, upon inactivation, cannot be reproduced immediately from the embryonic genome. Protective measures against UV-damage to RNA should therefore be expected in early embryos exposed to solar radiation. Light-dependent repair seems to be one of these measures.

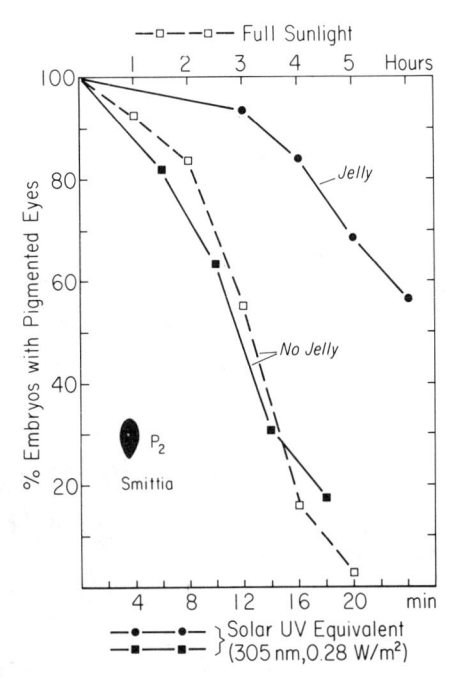

Fig. 6. Dose response curves for inactivation of Smittia embryos during intravitelline cleavage by full sunlight (open squares, upper abscissa), or by "solar UV equivalent" (305 nm, 0.28 W/m^2; filled symbols, lower abscissa). Embryos were left within the jelly of the clusters (circles) or released from the jelly (squares). Minimum number of embryos (experiments) per point = 400 (4). Data from Kalthoff (1975b).

ACKNOWLEDGEMENT

Our work has been supported by a research fellowship from the Deutsche Forschungsgemeinschaft to H.J., and by research grants to K.K. from the Deutsche Forschungsgemeinschaft, SFB 46, and the U.S. National Institutes of Health, AI 15046 TMP.

REFERENCES

Amy, R.L., 1965, Photorecovery from the effects of ultraviolet radiation in the developing Bracon embryo, Radiat. Res., 25:674-683.

Borstel, R.C. von and Wolff, S., 1955, Photoreactivation experiments on the nucleolus and cytoplasm of the Habrobracon egg, Proc. Natl. Acad. Sci. USA, 41:1004-1009.

Cook, J.br., 1970, Photoreactivation in animal cells, in: "Photophysiology", A.C. Giese, ed., Vol. V, pp. 191-233, Academic Press, London/New York.

Counce, S.J. and Waddington, C.H., 1972-73, "Developmental Systems: Insects", Academic Press, London/New York.

Fullilove, S.L. and Jacobson, A.G., 1978, Embryonic development:descritpive, in: "The Genetics and Biology of Drosophila", M. Ashburner and T.R.F. Wright, eds., Academic Press, London/New York.

Ghelelovitch, S., 1966, La sensibilitè des oeufs des la Drosophile (Drosophila melanogaster, Meig.) à l`action lètate des rayons ultraviolets. Evolution de la sensibilitè avec l`age de l`embryon, Int. J. Radiat. Biol.,11:255-271.

Gordon, M.P., Huang, C.W., and Hurter, J., 1976, Photochemistry and photobiology of ribonucleic acids, ribonucleoproteins, and RNA viruses. in: "Photochemistry and Photobiology of Nucleic Acids", S.Y. Wang, ed., Vol. II, pp. 265-308, Academic Press, London/New York.

Harm, H., 1976, Repair of UV-irradiated biological systems -- photoreactivation. in: "Photochemistry and Photobiology of Nucleic Acids", S.Y. Wang, ed., Vol. II, pp. 219-263, Academic Press, London/New York.

Huang, C.W. and Gordon, M.P., 1972, Photoreactivation of tobacco mosaic virus and potato virus X ribonucleic acids inactivated by acetone-sensitized photoreaction, Photochem. Photobiol., 15:493-501.

Hurter, J., Gordon, M.P., Kirwan, J.P., and McLaren, A.D., 1974, In vitro photoreactivation of ultraviolet inactivated ribonucleic acid from tobacco mosaic virus, Photochem. Photobiol., 19:185-190.

Jäckle, H., and Kalthoff, K., 1978, Photoreactivation of RNA in UV-irradiated insect eggs (Smittia spec., Chironomidae, Diptera) I. Photosensitized production and light-dependent disappearance of pyrimidine dimers, Photochem. Photobiol., 27:309-315.

Jäckle, H. and Kalthoff, K., 1979, Photosensitized formation of RNA-protein crosslinks in an insect egg (Smittia spec., Chironomidae, Diptera), Photochem. Photobiol., 29:1039-1040.

Jäckle, H. and Kalthoff, K., 1980a, Synthesis of a posterior indicator protein in normal embryos and double abdomens of

Smittia spec. (Chironomidae, Diptera), Proc. Natl. Acad. Sci.USA, (submitted).

Jäckle, H. and Kalthoff, K., 1980b, Photoreversible UV-inactivation of messenger RNA in an insect embryo (Smittia spec., Chironomidae, Diptera), Photochem. Photobiol., (in press).

Jäckle, H. and Kalthoff, K., 1981, Region-specific and stage-dependent synthesis of proteins during early insect embryogenesis (Smittia spec., Chironomidae, Diptera). II. Double cephalons and double abdomens., Develop. Biol., (submitted).

Kalthoff, K., 1970, Der Einfluss verschiedener Versuchsparameter auf die Häufigkeit der Missbildung Doppelabdomen in UV-bestrahlten Eiern von Smittia (Diptera, Chironomidae), Zool. Anz. Suppl., 33:59-65.

Kalthoff, K., 1971, Position of targets and period of competence for UV-induced of the malformation of "double abdomen" in the egg of Smittia spec. (Chironomidae, Diptera), Wilh. Roux`. Arch. Entwicklungsmech. 168:63-84.

Kalthoff, K., 1973, Action spectra for UV-induction and photoreversal of a switch in the developmental program of the egg of an insect (Smittia), Photochem. Photobiol., 18:355-364.

Kalthoff, K., 1975a, Smittia spec. (Diptera). Normale Embryonalentwicklung, Aberration des Segmentmusters nach UV-Bestrahlung, Encyclop. Cinematograph. (Göttingen) Film E2158/1974.

Kalthoff, K., 1975b, Compensation for solar uv-damage by solar radiation of longer wavelengths, Oecologia, 18:101-110.

Kalthoff, K., 1976, An unusual case of photoreactivation observed in an insect egg (Smittia spec., Chironomidae, Diptera), Photochem. Photobiol., 23:93-101.

Kalthoff, K., 1979, Analysis of a morphogenetic determinant in an insect embryo, in: "Determinants of Spatial Organization, S. Subtelny and I. Konigsberg, eds., pp. 97-126, Academic Press, London/New York.

Kalthoff, K., and Sander, K., 1968, Der Entwicklungsgang der Missbildung Doppel abdomen in partiell UV-bestrahlten Ei von Smittia parthenogenetica (Diptera, Chironomidae). Wilh. Roux`. Arch. Entwicklungsmech. Org. 161:129-146.

Kalthoff, K., Hanel, P., and Zissler, D., 1977, A morphogenetic determinant in the anterior pole of an insect egg (Smittia spec., Chironomidae, Diptera). Localization by combined centrifugation and UV-irradiation, Develop. Biol. 55:285-305.

Kalthoff, K., Urban, K., and Jäckle, H., 1978, Photoreactivation of RNA in UV-irradiated insect eggs (Smittia spec., Chironomidae, Diptera), II. Evidence for heterogeneous light-dependent repair activities, Photochem. Photobiol., 27:317-322.

Kandler-Singer, I., and Kalthoff, K., 1976, RNase sensitivity of an anterior morphogenetic determinant in an insect egg (Smittia spec., Chironomidae, Diptera), Proc. Natl. Acad. Sci. USA, 73:3739-3743.

Levin, V.L., and Kozlova, M.A., 1973, Photoreactivation of the Drosophila embryonic cells irradiated by ultraviolet with 254 and 313 nm wavelengths, Tsitologia (Leningrad), XV:1415-1420.

Levin, V.L., and Kozlova, M.A., 1976, Inheritance of photoreactivation ability of Drosophila zygotes, Genetica (Moscow), XII:105-108.

Levin, V.L., and Kozlova, M.A., 1977, Photoreactivation of cytoplasmic damage in Drosophila embryos irradiated at stages of meiosis and zygote by UV of different wavelengths, Tsitologia (Leningrad), XIX:434-440.

Ripley, S., and Kalthoff, K., 1980, Double abdomen induction with low UV-doses and virtually complete photoreversibility in embryo of Smittia spec., (Chironomidae, Diptera), Wilh. Roux`. Arch., (submitted).

Ritter, W., 1976, Fragmentierungs und Bestrahlungsversuche am Ei von Smittia spec. (Chironomidae, Diptera), Staatsexamensarbeit, Fakultät für Biologie der Universität Freiburg.

Rupert, C.br., 1964, Photoreactivation of ultraviolet damage, in: "Photophysiology", A.C. Giese, ed., Vol. II, pp. 283-327, Academic Press, London/New York.

Turner, F.R., and Mahowald, A.P., 1976, Scanning electron microscopy of Drosophila embryogenesis, 1. Structure of the egg envelopes and formation of the cellular blastoderm, Develop. Biol., 50:95-108.

Zalokar, M., and Erk, I., 1976, Division and migration of nuclei during early embryogenesis of Drosophila melanogaster, J. Microscopie. Biol. Cell, 25:97-106.

MOLECULAR ASPECTS OF ERROR PRONE REPAIR IN ESCHERICHIA COLI

Kevin McEntee

Department of Biochemistry
Stanford University School of Medicine
Stanford, CA 94305

ABSTRACT

Ultraviolet (UV) irradiation of Escherichia coli stimulates expression of a complex set of responses (SOS responses) that include mutagenesis, induction of prophage in lysogenic cells, inhibition of proper cell division (filamentation), and stimulation of reactivation and mutagenesis of DNA damaged phage (W-reactivation and W-mutagenesis, respectively).

These functions are coordinately regulated by the products of the recA and lexA genes acting as positive and negative control elements, respectively. The lexA protein appears to be a transcriptional repressor of some of the SOS functions and, upon UV treatment or other types of DNA damage, is cleaved into two fragments by an activated form of the recA protein. This activated recA protein cleaves other repressors of SOS genes or operons such as the phage λ repressor. Genetic evidence suggests several genes are directly repressed by lexA protein, such as those coding for the cell division inhibitor and the recA protein, while others, such as the umuC gene, are probably repressed by other recA protease-sensitive repressors distinct from lexA protein.

Error prone repair and other SOS-functions are expressed in tif-1 mutant cells at 42°C in the absence of DNA damage. This unusual recA mutation results in an altered recA protein capable of interacting with cellular components that are poorly recognized by the wild type (recA+) product. This interaction allows activation of the tif-1 protein under normal growth conditions causing expression of mutagenesis, prophage induction and other inducible functions without DNA damage.

In vitro replications of single stranded phage G4 DNA is blocked by UV irradiation, presumably at the site of the pyrimidine dimer. Neither the recA+ nor the tif-1 mutant protein by themselves permits DNA polymerase III holoenzyme to synthesize past a pyrimidine dimer on a single strand DNA template indicating that other proteins are required for "bypass synthesis."

INTRODUCTION

Faithful replication of DNA is a biological necessity for every organism. Genetic continuity of living systems demands that the copying of DNA during each cell division be performed with little or no error. Not surprisingly, replicative polymerases display great fidelity in copying DNA templates in vitro producing errors at frequencies of approximately 1 mismatch in 10^6 nucleotides. In addition to this faithful replication machinery, repair mechanisms contribute to the maintenance of the genetic program by repairing DNA damage in an 'error free' manner.

Although faithful replication is essential, it is also important to possess mechanisms whereby changes can be made in the genetic script - i.e. mutations - that permit variation of the organism as well as its adaptation to new environmental factors. In E. coli several mechanisms appear to contribute to a low level of mutagenesis and at least one such mechanism, designated 'error prone repair', produces mutations following DNA damage or blockage of DNA replication. Expression of error prone repair appears to be carefully controlled in the cell: it is normally repressed but can be 'induced' by DNA damage or DNA replication block. Error prone repair is only one of several inducible functions in E. coli that are expressed following these treatments. The induction of these diverse responses requires functional recA and lexA gene products.

In this paper, several aspects of error prone repair in E. coli are discussed, including current ideas about the regulation of this process and the enzymes involved in mutagenesis. The role of the recA protein in SOS responses in general and in error prone repair in particular is detailed.

Phenomenology of Error Prone Repair

In 1953, Weigle reported results of experiments with phage λ that demonstrated a surprising interaction between phage and host during the repair of UV damaged DNA. The survival of phage λ decreased with increasing UV dose as measured by the efficiency of plating on a bacterial host. However, UV irradiation of the bacterial cells prior to or at the time of infection with the damaged phage significantly stimulated phage survival. Moreover, a high frequency of the 'reactivated' phage contained mutations (1). The extents of reactivation and mutagenesis of UV damaged phage λ depended upon the UV dose to the phage and to the bacterial host and saturated at high UV doses to the bacteria.

Weigle obtained a similar result with irradiated λ and bacteria treated with nitrogen mustard, a potent DNA damaging agent. In the same set of experiments, visible light exposure of infected undamaged cells led to a significant increase in survival but this reactivation was not accompanied by mutagenesis. Furthermore, exposure of UV irradiated bacteria to visible light eliminated mutagenesis of the UV treated λ phage (1).

The demonstration that the primary lesion introduced into DNA by UV light is the intrastrand pyrimidine dimer suggested that these adducts are the relevant lesions for the DNA repair process operating on the λ and bacterial chromosomes. Enzymatic monomerization of pyrimidine dimers, a process called photoreactivation, requires visible light and is specific for pyrimidine dimers in DNA. The reversibility of mutagenesis by visible light during phage reactivation suggests that pyrimidine dimers are the relevant premutational lesions in DNA.

Genetics of the Error Prone Repair Process

Since these observations by Weigle, considerable genetic evidence has been obtained demonstrating a relationship between W-reactivation and W-mutagenesis of phage λ and bacterial repair and mutagenesis. The functions of the bacterial recA and lexA genes are required for bacterial mutagenesis (2, 3) and for W-reactivation and W-mutagenesis of phage λ (4, 5). RecA and lexA mutations sensitize cells to killing by UV irradiation but completely block UV mutagenesis of E. coli (6, 7). Furthermore, an unusual mutation in the recA gene, designated tif-1, increases both phage and bacterial mutagenesis at 42°C in the absence of DNA damage (8). Recently, Kato et al. isolated mutants of E. coli designated umuC, which were isolated based on their inability to be mutagenized by UV irradiation. In addition to preventing bacterial mutagenesis, the umuC defect blocks W-reactivation and W-mutagenesis of phage λ (9). Although the roles of the recA and lexA genes are complex and affect other aspects of DNA metabolism and cell regulation (see below), it is believed that their participation in phage and bacterial survival and mutagenesis reflects common or overlapping error prone repair pathways.

DNA damage produced by UV irradiation is corrected faithfully by photoreactivation, excision repair mechanisms dependent upon the uvrA, uvrB and uvrC gene products and by a recombinational 'tolerance' mechanism that exchanges dimers between partly replicated damaged DNA and intact regions of homologous chromosomes (for review see ref. 10). This latter process, termed post replication repair, depends upon the recA, recB, recC recL and recF functions (11, 12). The involvement of the recA gene in recombinational repair and in mutagenesis led to speculation that a minor recombination repair pathway requiring lexA gene function was mutagenic. This idea is made untenable by the results of Kato et al. (13) that demonstrate the occurrence of UV mutagenesis in strains that retain recA gene function but which are unable to homologously recombine ($<10^{-4}$ of wild type) due to mutations in the recF and recBC genes. Several other lines of evidence indicate that the mutagenic pathway is independent of recombination although both processes require the recA$^+$ gene function. These results suggest that recA protein participates in a nonrecombinational error prone repair process that does not require functions of the recB, recC or recF genes.

Two other genes have been implicated in error prone repair in E. coli. Johnson (14) identified a mutation, designated lexC113, that blocks UV mutagenesis and produces pleiotropic effects similar to lexA mutations but is genetically distinct from this locus. Recently, Meyer et al. (15) and Glassberg et al.(16) have provided evidence that the lexC gene is the structural gene for the helix destabilizing protein (or single stranded DNA binding protein, SSB) of E. coli. In addition to a proposed role in DNA repair and recombination, SSB is necessary for DNA replication in vivo and in vitro (17).

A preliminary report from Bridges and Mottershead (18) implicates the DNA polymerase III enzyme or holoenzyme in UV induced mutagenesis and reactivation. The involvement of some form of the replicative DNA polymerase in recA dependent repair is suggested by the work of Sedgwick (19). It is not clear in the latter case that this repair is mutagenic, however.

Requirements for W-reactivation and mutagenesis

Defais and coworkers, investigating the kinetics of the W-reactivation and W-mutagenic activities in E. coli, demonstrated that these processes are maximally expressed within 1 generation time following UV irradiation. The increase in these activities is followed by a decay over a comparable time period. The induction of W-reactivation and W-mutagenesis of phage λ show the same kinetics and both processes are abolished if, following UV irradiation, the cells are treated with chloramphenicol to prevent protein synthesis (20).

Witkin has demonstrated qualitatively similar kinetic effects for bacterial mutagenesis enhanced by tif-1 mutant gene expression

(21). Her results indicate that error prone repair capacity increases to a maximal level following temperature shift and, following a return to low temperature, decays over a similar time scale. Evidence from several laboratories indicates that protein synthesis is necessary for conversion of a premutational lesion (pyrimidine dimer) to a mutation. These results supported the proposal of Witkin (22) and Radman (23) that the error prone or mutagenic repair process is inducible in E. coli and requires new protein synthesis following UV irradiation.

Regulation of Error Prone Repair

According to the SOS hypothesis, error prone repair of DNA damage is one of several coordinately regulated functions expressed in response to DNA damage or blockage of DNA replication. Among the other SOS responses are (i) the induction of certain temperate prophages such as λ (24); (ii) inhibition of proper cell division resulting in filamentation (25); (iii) control of DNA degradation by the recBC enzyme (26), and (iv) induction of a 40,000 molecular weight protein, originally designated protein X and later shown to be the recA gene product (27, 28). Mutations in the recA or

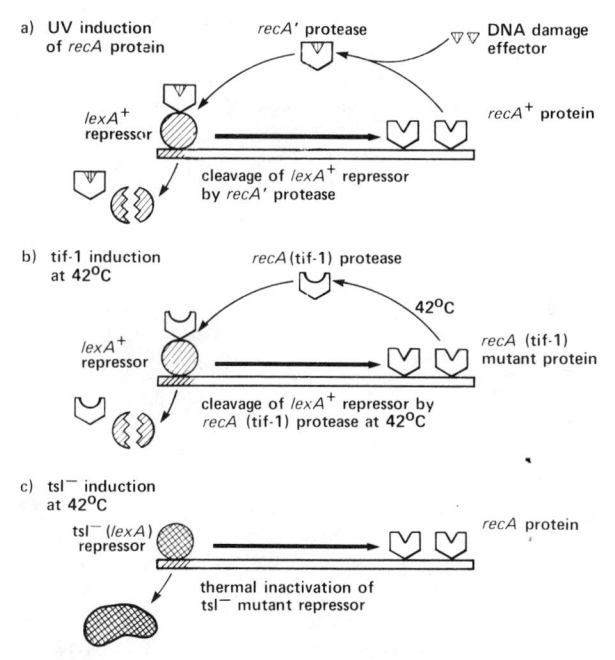

Fig. 1. Regulation of the recA gene in vivo. The regulatory features of this model are described in the text. The broad arrow represents increased transcription of the recA gene.

lexA genes block all of the SOS responses, including the induction of recA protein synthesis, which suggests a form of autoregulation for this gene. The tif-1 mutation results in expression of the SOS responses at 42°C in the absence of DNA damage. The tif-1 mutation has been shown to be a mutation in the recA structural gene which changes the isoelectric point of the protein (27, 28).

Based upon the results of Gudas and Pardee (29) and Gudas (30) concerning protein X expression in several mutant strains of E. coli, a regulatory model was proposed which accounted for several of the lexA and recA mutants and their effects upon recA protein synthesis (27, 28). The basic features of this model are shown in Figure 1 and are described below.

The recA gene is expressed at a low basal level in normally growing cells. The low level of recA gene expression is adequate for initiating homologous recombination but the amount of recA protein is insufficient for extensive repair of DNA required after UV irradiation or other types of DNA damage. The limited expression of the recA gene is due to repression by the lexA gene product. Introduction of DNA damage into the cell results in the generation of a "DNA damage effector" which interacts with the $recA_1$ protein and converts the recA protein to an activated form, $recA^1$. The biochemical nature of this "DNA damage effector" is not known. In vitro studies with purified recA protein (see below) suggest that a single-stranded polynucleotide can intereact with recA protein and, due to a conformational change in the polypeptide, induces new activities. Oishi and coworkers (31) have used a permeabilized cell system to study effectors of another SOS response, temperate prophage induction. Their results suggest that a dinucleotide is capable of eliciting prophage derepression in permeabilized cells. Taken together it appear that a likely effector of this system is some DNA degradation product or structure such as a gap produced during the replication of a region of DNA containing a pyrimidine dimer.

In this scheme the activated recA protein possesses anti-repressor activity. Specifically the activated recA is capable of specific proteolytic cleavage of the lexA protein, thereby removing this repressor and stimulating transcription of its own gene. The level of induction is estimated to be 10-50 fold and results in a large pool of recA protein which is available for DNA repair.

This regulatory loop accounts for expression of recA protein under a variety of conditions and defines the roles of the lexA protein as a negative control element and the recA protein as a positive control element in this circuit. Mutations in the lexA gene which block induction of the recA gene are explained by this model as altered repressor molecules which cannot be inactivated (cleaved) by recA protein. These noninducible alleles should, like the noninducible mutations in the λ repressor, be dominant to the wild type allele. Mount et al. have shown that the lexA⁻ mutations are dominant to the lexA⁺ allele in partial diploid strains (32). Mutations in the recA structural gene (recA⁻ alleles) block its own induction by producing an inactive form of the recA protein that is unable to interact with the effector or is altered in its ability to proteolytically cleave the lexA protein. RecA⁻ mutations of both types have been isolated and their properties characterized.

The unusual tif-1 allele of the recA gene is capable of inactivating the lexA protein in the absence of DNA damage at 42°C. This result suggests that the tif-1 mutation alters the recA protein in such a way that at high temperature (a) it requires no effector binding to alter its conformation or (b) it is altered in its cofactor requirements or its binding affinity in such a way that it recognizes structures or molecules that normally exist in the cell (such as near the growing fork in replication) and these can activate the tif-1 mutant protein but not the recA⁺ protein.

In studies of recA cleavage of the phage λ repressor in vitro, Roberts has reported that oligonucleotides serve as cofactors in the protease reaction (33). We have observed that short oligonucleotides (dT$_{12}$) are effective cofactors for the tif-1 mutant form of recA protein but are considerably less effective for the recA protein. In an analogous way, short oligonucleotides may be more effective as cofactors for cleavage of the lexA repressor by the tif-1 mutant protein than by the recA protein. Short regions of single stranded DNA, such as those in or near the growing fork of a replicating chromosome, might activate the tif-1 protein and not the recA protein. Completion of rounds of DNA replication would eliminate these single strand regions and no activation of tif-1 could result. Indeed, the dnaA mutation which permits completion of chromosome replication but blocks reinitiation suppresses tif-1 expression (34).

Additional mutations in the lexA gene such as tsl or spr are readily explained by this model as being altered repressor molecules that are temperature sensitive or partially defective, respectively. Thus, high levels of recA protein are expressed in either tsl or spr mutants because the repressor can be inactivated by a mechanism that does not require recA protein function . This thermal denaturation of lexA protein in tsl mutants results in overproduction of mutant as well as wild type recA proteins without invoking DNA damage.

Biochemical evidence that supports this regulatory scheme has come from several studies. McPartland et al (35) have examined recA gene transcription by RNA-DNA hybridization after DNA damaging treatment. Their results indicate that the rate of recA gene transcription increases more than 10-fold following UV treatment and this induction requires a recA$^+$ lexA$^+$ genotype.

The identification of the lexA protein by Little and Harper (36) and Brent and Ptashne (37) and the subsequent demonstration that this protein could be cleaved into two fragments in a system that depended upon recA protein, and ATP (or the analog ATPγS) (38), provides strong support for the idea that the primary event leading to recA gene derepression is proteolysis of the lexA coded repressor.

Brent and Ptashne (37) and Little et al (36) have also demonstrated another feature of lexA gene regulation which was not incorporated in the original recA regulation model. LexA protein negatively regulates its own expression, presumably at the level of transcription. Indeed the nucleotide sequence of control regions proximal to both recA and lexA genes are strikingly similar, suggesting that these nucleotide sequences are sites for interaction with a common repressor molecule (R. Brent, personal communication). The autoregulatory aspects of lexA repressor synthesis suggest a means by which cell, induced for SOS functions, can reestablish repression of the recA gene. UV irradiation elicits the production of an effector which we presume binds to recA protein and conformationally alters the enzyme to cleave the lexA protein. The cleavage of the lexA repressor derepresses synthesis of both the recA and the lexA gene product. The lexA protein will continue to be inactivated by activated recA protein (recA^1) as long as the effector is present or is capable of interacting with recA protein. The repair of the damaged sites in DNA removes the effector from the cell with the result that activated recA^1 protein is converted to a nonproteolytic form. This attenuation of recA protease permits synthesis of normal levels of the lexA repressor and blocks further recA gene transcription. Repression is thereby reestablished although the cells may continue to contain high levels of recA protein (in the nonactivated state). The recA protein, once made, is extremely stable and is diluted through cell division (unpublished results).

The analogy between the lexA protein and phage λ repressor extends to the interaction of these proteins with recA protein. Roberts and coworkers (38) first demonstrated cleavage of λ repressor in recA+ cells following UV irradiation or mitomycin C treatment. Using a filter binding assay that measured the amount of functional λ repressor in cell extracts, these investigators purified an activity capable of inactivating the repressor in an ATP dependent reaction. Craig and Roberts (39) demonstrated that the purified recA protein cleaved λ repressor in a polynucleotide dependent reaction. Thus λ repressor is a functional analog of the lexA repressor and presumably responds to the same primary induction mechanism following DNA damage. Furthermore, like λ repressor, lexA protein autoregulates its own synthesis to maintain a constant level of repressor in the cell.

Genetic evidence suggests that other genes are negatively regulated by the lexA protein. The temperature sensitive lexA allele, tsl, causes expression of some but not all SOS functions. At 42°C, tsl⁻ mutant cells form filaments and this filamentation is independent of the induction of recA protein in the cell (40). This result strongly suggests that the genes involved in blocking cell division are regulated (presumably repressed) by the lexA product. Expression of colicin E1 is induced in tsl mutant strains at 42°C in the absence of recA protein function (41) a result consistent with the notion that the plasmid uses the host lexA function for controlling expression of its genes.

Unlike filamentation, mutagenesis and prophage induction are not expressed in a tsl mutant at the nonpermissive temperature in a recA⁺ background. Formally, therefore, mutagenesis and prophage induction are indirectly regulated by the lexA repressor since a lexA gene is required for UV mutagenesis. We presume, therefore, that the genes involved in the productions of mutations are controlled by a separate regulatory element. The dependence on recA protein for expression of these mutagenic activities and the strong parallel with prophage induction suggest that the error prone repair gene(s) are negatively regulated by another repressor. The umuC mutation identified by Kato and Shinoura, presumably defines one of the enzymatic activities needed for error prone repair. However, the possibility remains that the umuC mutation is in the presumptive repressor of error prone genes and renders it resistant to inactivation by recA protein (analogous to the ind⁻ mutations in the λ repressor gene).

According to the regulatory model proposed above, high levels of recA protein are not necessary for inactivation of the lexA repressor, since in the induction of the recA gene only the basal level of recA protein suffices to cause derepression. Therefore, the rate limiting component for the UV induction of the SOS responses is the intracellular concentration of the effector, which directly determines the level of activated recA protein. Recently, Satta et al. (40) demonstrated that recA protein induction by DNA damage was selectively inhibited by low concentrations of rifampicin (approximately 4 μg/ml). These authors found that at these low drug concentrations, DNA was more extensively degraded than in the untreated cells, results interpreted to mean that high levels of recA protein were needed to protect DNA from degradation following damage. Baluch et al. (43) have presented evidence that prophage λ induction is not blocked by low concentration of rifampicin, consistent with the idea that recA protein overproduction is not a prerequisite for λ induction. Conflicting results have been reported by Kato and Shinoura, who find evidence that low rifampicin concentrations inhibit prophage induction and mutagenesis (44).

These and other data argue that UV stimulated expression of all SOS functions require a regulatory function of the recA protein. However, the DNA pairing and annealing activities of this protein are likely to be important in repair of DNA damage (45) and may be required to directly participate in both error free and error prone modes of DNA repair. At the present time, it is not known whether UV mutagenesis requires recA protein continuously or only for derepression of the mutagenic enzyme.

The original SOS hypothesis postulated that several genes would be coordinately controlled by the recA/lexA system. Kenyon and Walker (46) have employed a Mu-phage derivative to isolate fusions between the lacZ gene, the structural gene for beta-galactosidase, and genes that are induced by DNA damaging agents. Five distinct loci were identified in which the operon fusions caused UV inducibility of β-galactosidase. At least one of these loci has been tentatively identified by genetic mapping and phenotypic characterization of the insertion strain. The dinE insertion (damage inducible) is located within the uvrA gene, producing a UV-sensitive and HCR$^-$ phenotype. Induction of β-galactosidase in the din insertion strains by UV treatment or mitomycin C exposure was blocked by recA$^-$ or lexA$^-$ mutations. These results provide the first demonstration that enzymes involved in excision repair are regulated by the recA and lexA genes and are induced (approximately 5-fold) in response to DNA damage. Thus both error proof and error prone pathways for DNA repair are expressed at elevated levels in cells following UV damage. RecA and lexA proteins appear to participate in these pathways at least at the level of regulation. These results argue strongly for a role of the excision repair enzymes in W-reactivation of phage λ. Although uvrA$^-$ strains show W-reactivation, both the maximal level and optimal dose needed to elicit this repair are considerably reduced in comparison to uvrA$^+$ strains (4).

The quantitative requirement for recA protein in W-reactivation and W-mutagenesis of phage λ has been investigated by Martine Defais (personal communication). Her results indicate that by increasing the recA gene dosage in the cell (the recA gene being present on a multicopy colE1 recombinant plasmid), the level of W-reactivation of UV irradiated phage λ is increased approximately five fold or more over the the nonplasmid strain. This increase in survival of UV damaged phage is observed in the unirradiated bacterial strain containing the plasmid. However, despite this increase in phage survival, the frequency of mutants among the survivors is independent of the recA gene dosage. These preliminary results imply that the amount of recA protein is limiting in W-reactivation but another component is limiting in the accompanying W-mutagenesis.

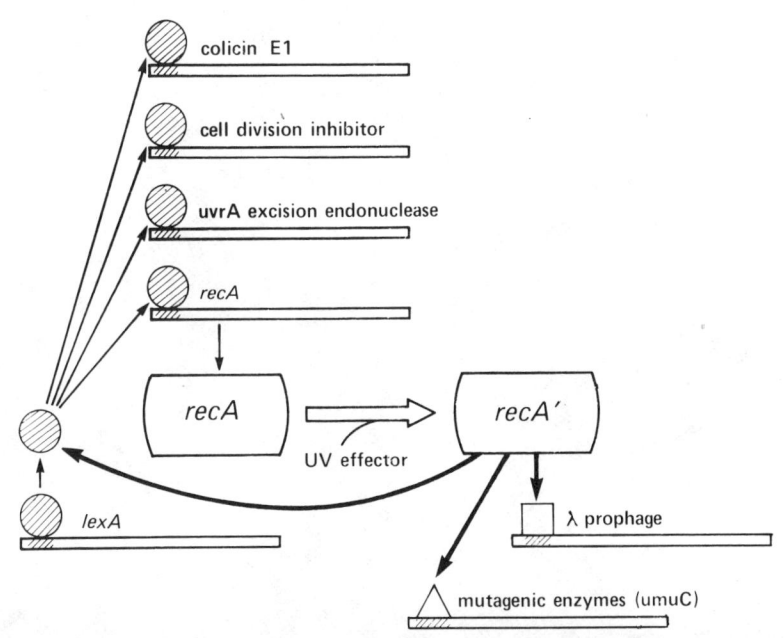

Fig. 2. Regulation of error prone repair and other inducible genes by recA and lexA proteins. The features of this model are discussed in the text. LexA protein directly represses the recA and uvrA genes, cell division inhibitor(s) and colE1. Both prophage λ and the enzymes needed for UV mutagenesis are negatively regulated by their own repressors. RecA protein is activated by binding an effector (RecA^1) proteolytically cleaves lexA protein as well as other repressors.

A summary of this regulatory network is shown in Figure 2. Much of the evidence in support of this scheme comes from genetic experiments. Efforts are currently underway to provide biochemical confirmation for these proposed regulatory roles of recA and lexA proteins.

Biochemical Mechanism of Error Prone Repair

DNA synthesis by any of the three DNA polymerases of E. coli is blocked at pyrimidine dimers in the DNA template. In the case of DNA polymerase I, Villani and coworkers (47) have demonstrated an 'idling' of the polI enzyme at UV lesions on primed single-stranded DNA in vitro: the polI enzyme inserts a base opposite the first pyrimidine in the dimer and this unpaired base is rapidly removed by the 3'-5' exonuclease of the polymerase (proof-reading activity). These investigators suggested that the generation of high concentration of deoxynucleoside monophosphates could eventually inhibit the 3'-5' exonuclease activity and permit insertion of nonpairing bases opposite the pyrimidine dimer. This 'bypass synthesis' would result in a misincorporated 'doublet' at the site of the dimer. Coulondre et al. (78) have

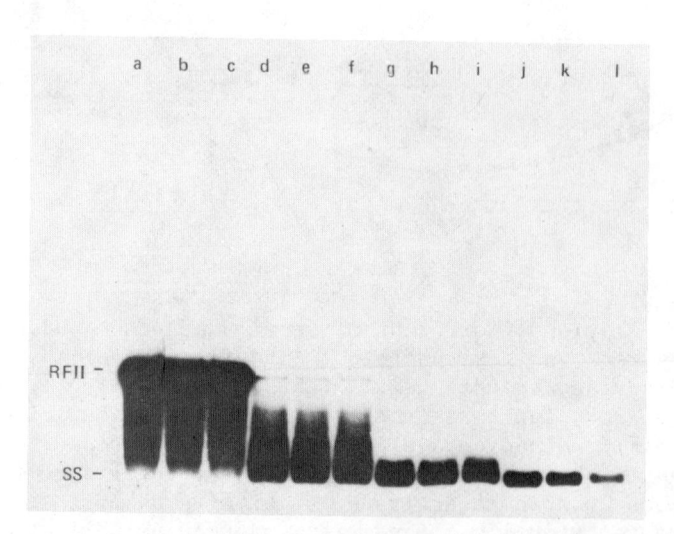

Fig. 3. Effect of UV irradiation on replication of G4 single stranded DNA. The condition for in vitro G4 DNA replication on UV irradiated DNA will be published elsewhere. Sample 1-3, no UV, no UV + recA protein, no UV + tif-1 protein; samples 4-6, 30 sec. UV, 30 sec. UV + recA protein, 30 sec. UV + tif-1 protein; 90 sec. UV + recA protein, 90 sec. UV + tif-1 protein; 300 sec. UV + recA protein, 300 sec. UV + tif-1 protein.

shown by sequence analysis that UV induced mutations in the lacI gene are frequently pairs of base changes in the DNA sequence corresponding to insertion of two bases opposite the dimer. These results provide the strongest evidence that bypass synthesis past pyrimidine dimers in a DNA template is a mechanism for producing mutations at the site of DNA lesions.

Reversal of the 3'-5' exonuclease activity of DNA polymerase has been proposed as a mechanism for insertion of bases opposite noncoding lesions. Schroeder (49) has suggested that the 3'-5' exonuclease function of replicative polymerases, which normally hydrolyzes single-stranded DNA from the 3' end, can in principle polymerize deoxynucleoside monophosphates in a template independent reaction. Such a mechanism would require a high dNMP concentration in the cell or in the microenvironment of the replication complex. However, following UV irradiation or other treatments that lead to mutagenesis, there is no indication that the dNMP pools in the cell increase.

Kato and Shinoura (44) suggested that association of DNA polymerase III or holoenzyme with an inducible factor alters this polymerase such that it is capable of synthesizing past pyrimidine dimers or other premutational lesion. Such a factor would alter base pairing discrimination as well as block 3'-5' exonuclease removal of the mispaired bases. This factor, presumed to be the umuC gene product, may be regulated by the recA protein as described earlier, or alternatively the umuC gene product may be made constitutively and act together with the recA protein in the error prone repair step. In either case the ability to perform bypass synthesis would be induced in cells that have sustained DNA damage.

We have performed experiments to determine whether recA protein (or tif-1 protein) promotes replication through a pyrimidine dimer. Using a reconstituted replication system for the conversion of single stranded phage G4 DNA to the replicative form (50), we have demonstrated that UV irradiation blocks replication of the single stranded template and the extent of replication is dose dependent (Figure 3). However, neither the addition of highly purified tif-1 mutant protein or recA+ protein alters the pattern of incorporation. We conclude that bypass synthesis must require other proteins in addition to or instead of recA protein. Experiments are in progress to demonstrate bypass synthesis in vitro using this reconstituted system.

Mutagenesis without DNA Damage

Treatments that damage DNA or block normal DNA replication are generally mutagenic-producing heritable alterations in the genetic script. The site of the DNA lesion is thought to be the eventual mutational site due to errors in the repair process. However, Ichikawa-Ryo and Kondo demonstrated mutagenesis in undamaged phage λ when infection is carried out in UV irradiated cells (51). This increased mutagenesis was considerably lower than the level seen in W-mutagenesis.

Thermal expression of the tif-1 mutation results in elevated mutagenesis at 42°C in the absence of DNA damage. Tif-1 expression also enhances mutagenesis resulting from UV irradiation and other damaging agents. These cases very likely represent a suboptimal expression of recA/lexA dependent error prone repair rather than a new mutagenic pathway. Perhaps structures that are generated during normal DNA replication can be acted on by the mutagenic enzymes, albeit less efficiently than damage containing structures. Inducible factors which relieve the strict base pairing requirements of the replicative polymerase need not act exclusively at sites of DNA damage although these sites may be more favorable kinetically. Such factors could promote a low level of "error prone replication" at sites where the polymerase "pauses" during elongation. The enzymatic requirements for this "untargeted" mutagenesis may not be identical to those for damage induced mutagenesis since the DNA substrates in the two situations are not identical. Nevertheless, according to this model, there is a fundamental relationship between targeted and untargeted mutagenesis.

ACKNOWLEDGEMENTS

I wish to thank Dr. I. Robert Lehman for his support, Martine Defais and Guiseppe Villani for thoughtful suggestions and useful information. K. McEntee was a recipient of an American Cancer Society Senior Fellowship.

REFERENCES

1. Weigle, J. J. (1953). Proc. Natl. Acad. Sci. USA 39, 628-636.
2. Kondo, S., Ichikawa, H., Iwo, K. and Kato, T. (1970) Genetics 66, 187-217.
3. Witkin, E. (1967). Brookhaven Symp. Biol. 20, 17-55.
4. Defais, M., Fauquet, R. Radman, M. and Errera, M. (1971). Virol. 43, 495-503.
5. Miura, A. and Tomizawa, J. (1968). Mol. Gen. Genet. 103, 1-10.
6. Kondo, S. (1969). Proceedings of the XIIth International Congress on Genetics, Vol. II.
7. Howard-Flanders, P., and Boyce, R. P. (1966). Radiat. Res. 6 (Suppl.) 156-184.
8. Castellazzi, M., George, J. and Buttin, G. (1972). Mol. Gen. Genet. 119, 139-152.

9. Kato, T. and Shinoura (1977) Mol. Gen. Genet. 156, 121-131.
10. Witkin, E. (1976). in Bacteriol. Rev. 40, 869-907.
11. Horii, Z. I. and Clark, A. J. (1973). J. Mol. Biol. 80, 327-344.
12. Rothman, R.H. and Clark, A. J. (1977). Mol. Gen. Genet. 155, 267-277.
13. Kato, T., Rothman, R. H. and Clark, A. J. (1977). Genetics 87, 1-18.
14. Johnson, B. F. (1977). Mol. Gen. Genet. 157, 91-97.
15. Meyer, R. R., Glassberg, J. Scott, J. V. and Kornberg, A. (1980). J. Biol. Chem. 255, 2897-2901.
16. Glassberg, J., Meyer, R.R. and Kornberg, A. (1979) J. Bacteriol. 140, 14-19.
17. Weiner, J. H., Bertsch, L. and Kornberg, A. (1975). J. Biol. Chem. 250, 1972-1980.
18. Bridges, B. A. and Mottershead, R. P. (1978). Mol. Gen. Gent. 162, 35-41.
19. Sedgwick, S. G. (1975). Proc. Natl. Acad. Sci. USA, 72, 2753-2757.
20. Defais, M. Caillet-Fauquet, P. Fox, M. S. and Radman, M. (1976). Mol. Gen. Genet. 148, 125-130.
21. Witkin, E. M. (1974). Proc. Natl. Acad. Sci. USA, 71, 1930-1934.
22. Witkin, E. M. (1973). Ultraviolet mutagenesis in bacteria: the inducible nature of error prone repair. An. Acad. Bras. Ciene., 45, 188-192.
23. Radman, M. (1975). SOS Repair Hypothesis phenomenology of an inducible DNA repair which is accompanied by mutagenesis p. 355-367. In P. Hanawalt and R. B. Setlow (eds.) Molecular Mechanisms for Repair of DNA part A Plenum Press. N.Y.
24. Lwoff, A., Simonovitch, L. and Kjeldgaard, N. (1950). Ann. Inst. Pasteur, Paris, 79, 815-859.
25. Witkin, E. M. (1967). Proc. Natl. Acad. Sci. USA, 57, 1275-1279.
26. Marsden, H. S., Pollard, E. C. Ginoza, W. and Randall, E. P. (1974). J. Bacteriol. 118, 465-470.
27. McEntee, K. (1977). Proc. Natl. Acad. Sci. USA 74, 5275-5279.
28. Gudas, L. S. and Mount, D. W. (1977). Proc. Natl. Acad. Sci. USA, 74, 5280-5284.
29. Gudas, L. J. and Pardee, A. B. (1975). Proc. Natl. Acad. Sci. USA 72, 2330-2334.
30. Gudas, L. J. (1976). J. Mol. Biol. 104, 567-587.
31. Oishi, M. Smith, C. L. and Friefeld, B. (1978). Cold Spring Harbor Symp. Quant. Biol. 43, 897-907.
32. Mount, D. W., Low, K. B. and Edmiston, S. J. (1972). J. Bacteriol. 112, 886-893.
33. Roberts, J. W., Robert, C. W. and Craig, N. L. (1978). Proc. Natl. Acad. Sci. USA, 75, 4714-4718.
34. D'Ari, R., George, J. and Huisman, O. (1979). J. Bacteriol. 140, 381-387.

35. McPartland, A. Green, L. and Echols, H. (1978). Transcriptional Regulation of the recA Region of E. coli. pp. 383-386. In DNA Repair Mechanisms (P. C. Hanawalt, E. C. Friedberg and C. F. Fox, ets.) Acad. Press, N.Y.
36. Little, J. W. and Harper, J. E. (1979). Proc. Natl. Acad. Sci. USA, 76, 6147-6151.
37. Brent, R. and Ptashne (1980). Proc. Natl. Acad. Sci. USA, 77, 1932-1936.
38. Little, J. W., Edmiston, S. H., Pacelli, L. Z. and Mount, D. W. (1980). Proc. Natl. Acad. Sci. USA, 77, 3225-3229.
39. Craig, N. L. and Roberts, J. W. (1980). Nature (London), 283, 26-30.
40. Huisman, O., D'Ari, R. and George, J. (1980). J. Bacteriol. 142, 819-828.
41. Hull, R. A. (1975). J. Bacteriol. 123, 775-776.
42. Satta, G., Gudas, L. S. and Pardee, A. B. (1979). Mol. Gen. Genet., 168, 69-80.
43. Baluck, J., Sussman, R. and Resnick, J. (1980). Mol. Gen. Genet., 178, 317-323.
44. Kato, T. and Shinoura, Y. (1978). XIV Int. Congr. of Genetics (in press).
45. Weinstock, G. M., McEntee, K. and Lehman, I. R. (1979). Proc. Natl. Acad. Sci. USA 76, 126-130.
46. Kenyon, C. J. and Walker, G.C. (1980). Proc. Natl. Acad. Sci. USA 77, 2819-2823.
47. Villani, G., Boiteux, S. and Radman, M. (1978). Proc. Natl. Acad. Sci. USA 75, 3037-3041.
48. Coulondre, C. and Miller J. H. (1977). J. Mol. Biol. 117, 577-606.
49. Schroeder, C. (1979). Eur. J. Biochem. 102, 291-296.
50. McMacken, R., Rowen, L. Kunihiro, U. and Kornberg, A. (1978). Priming of DNA Synthesis on Viral Single Stranded DNA in vitro in the Single Stranded DNA Phages (D. T. Denhardt, D. Dressler, D. S. Ray, eds.) pp. 273-285.
51. IchikawaRyo, H. and Kondo, S. (1975). J. Mol. Biol. 97, 77-92.

ROUND TABLE SUMMARY : GENETIC ENGINEERING AND DNA REPAIR

W. Dean Rupp

Yale University School of Medicine
New Haven, CT 06510, USA

INTRODUCTION

The last few years have seen the introduction and development of new and exciting methods for the cloning of specific genes. These approaches provide investigators with powerful new approaches for elucidating many aspects of gene structure and function. The availability of cloned genes aids in the identification of the gene products, provides greater amounts of the gene products for biochemical studies and can also provide important information about the control of expression of the cloned genes. At the present time, most of the work on cloned genes related to DNA repair has been done with E. coli, but we can anticipate that work in this area will soon include the cloning of DNA repair genes from many diverse sources including humans.

This paper summarizes recent studies from several laboratories in which DNA repair genes have been cloned and used for further studies. These genes include recA, lexA, uvrA, uvrB, uvrC and ssb of E. coli and the denV gene of bacteriophage T4.

CLONING OF THE recA GENE

The value of gene cloning in elucidating gene structure, function and regulation is clearly demonstrated in the recent and rapid progress with the recA gene of E. coli. In only a few years, we have seen the progression from not even knowing the identity of the recA gene product to having its entire sequence, knowing many details about the enzymatic properties of the purified protein, and understanding much about the molecular mechanism of its complex regulation.

A key event in the studies on the recA gene was the cloning by McEntee (1976) of the recA gene on λ bacteriophage. He and others (McEntee, 1977; Gudas and Mount, 1977; Emmerson and West, 1977; Little and Kleid, 1977) soon demonstrated that the recA protein was a polypeptide of about 40,000 molecular weight and was identical to protein X, a protein that had previously been shown to be induced after UV-irradiation (Gudas and Pardee, 1975; Witkin, 1976). Once the recA protein was identified, it was straightforward to purify it and to study the reactions it catalyzed. In order to facilitate the preparation of large quantities of recA protein, purifications were done from strains carrying the cloned recA gene either on λ bacteriophage or on a multicopy plasmid. These biochemical studies demonstrated the advantage of having a cloned gene because the presence of the recA gene on a transducing phage or on a multicopy plasmid produces a much higher yield of the recA protein than can be obtained from a normal cell in which the gene is present only on the chromosome.

Several groups have been studying the reactions of the recA protein on promoting the interactions of homologous DNA molecules, properties of the protein that are clearly relevant to its participation in recombination. The summary of these experiments will not be given here, but the reader can refer to recent publications from these laboratories to obtain the present understanding of these reactions and for references to the earlier studies carried out by these groups (Cassuto et al., 1980; Cunningham et al., 1980; McEntee et al., 1980).

Protease Activity of the recA Protein

Roberts and his colleagues (Craig and Roberts, 1980) have approached the study of the recA protein from a completely different direction. They were interested in the UV-induction of λ bacteriophage and observed that the λ repressor was cleaved during this process. They purified an activity that was responsible for the proteolytic cleavage of the λ repressor and found that it turned out to be identical to the recA protein. Even after extensive purification, this proteolytic activity is inseparable from the recA protein indicating that it is most likely an intrinsic activity of the recA polypeptide chain.

Physical Map and Nucleotide Sequence of the recA Gene

The availability of the recA gene on plasmids made it possible to determine the physical boundaries of the gene and to locate its promoter (Sancar and Rupp, 1979a; Ogawa et al., 1978). Once the boundaries of the gene were known, it was possible to determine the nucleotide sequence of the entire recA gene. This was accomplished independently in 2 laboratories on clones of the recA gene isolated from different strains of E. coli K12 (Horii et al., 1980; Sancar et al., 1980b). The results from both laboratories were in complete agreement on the nucleotide sequence for the structural gene.

Results in cell-free systems showed that the recA gene was transcribed very efficiently by RNA polymerase (Sancar and Rupp, 1979). The nucleotide sequence of the promoter is very close to the composite "ideal" promoter sequence obtained from a number of different promoters. There is also a good Shine-Dalgarno sequence near an initiation codon in the recA mRNA so that translation of the recA mRNA into protein should also be efficient. Thus, by just inspecting the DNA sequence, one can predict that a cell will rapidly synthesize recA protein in the absence of any negative control.

CONTROL OF EXPRESSION OF recA BY lexA

It has been clear for some time that the level of recA protein in cells is under regulation since it is normally present only at very low levels but after irradiation or other treatments that damage DNA it increases to high levels. The properties of lexA mutants led to the proposal that this gene product was a repressor of the recA gene (McEntee, 1977; Gudas and Mount, 1977; Emmerson and West, 1977; Craig and Roberts, 1980). In order to test this control model directly, Little and Harper (1979) and Brent and Ptashne (1980) cloned the lexA gene. Once the gene was cloned, they were able to identify a 24,000 dalton protein as the gene product. However, little protein was made even in cells with a multicopy plasmid carrying the lexA gene. In order to increase the yield, the lexA gene was fused to the lac promoter. With this strain, it was then much easier to obtain amounts of the lexA protein sufficient for various biochemical studies.

With the lexA protein available, various aspects of the model for the control of recA were then tested. Little (unpublished data) recently showed that the lexA protein binds specifically to the recA gene. If a protein binds tightly to a particular region of DNA, it protects that part of the DNA against nuclease degradation. Little showed that the protected region of DNA lies within the promoter, and specifically includes an inverted repeat of 22 basepairs that is in the promoter. This sequence with dyad symmetry had previously been noticed during the sequencing of the recA gene, and it had been suggested that this was a likely site for regulatory control of the recA gene (Sancar et al., 1980; Horii et al., 1980).

Little also showed that in a cell-free system, lexA protein prevented the transcription of the recA gene by RNA polymerase, but did not prevent the transcription of the amp gene present on the same plasmid DNA. These experiments provide convincing evidence that the lexA protein is the repressor of the recA gene since it acts by binding to the recA promoter thus preventing transcription by RNA polymerase.

This binding of the lexA protein to the recA operator satisfactorily explains the low expression of recA under normal conditions, but other information is required to understand how recA is switched on and expressed at high levels. The work of Little et al. (1980) demonstrates how this probably happens. They showed that the lexA protein is cleaved by the recA protein just as the λ repressor is cleaved. It thus seems that this is a common mechanism by which recA activates the expression of genes under recA control, and explains how it controls its own synthesis.

In studies with the lexA gene, it was found that synthesis of the lexA protein is under its own control (Little and Harper, 1979; Brent and Ptashne, 1980). While sequencing the control region of the lexA gene, Brent (unpublished data) observed several sequences in the lexA promoter region that were very similar to the inverted repeat present in the recA promoter and suggested that this might be a site for binding of lexA to its own promoter. Little showed directly that lexA protein binds to this part of the lexA gene and that the protected region is in fact the region that is similar to the inverted repeat in the recA promoter-operator. (This self-regulation is the reason that it was necessary to link the lexA structural gene to the lac promoter in order to obtain high yields of the lexA protein.)

CLONING OF THE uvr GENES AND CHARACTERIZATION OF THE GENE PRODUCTS

In order to identify gene products, increase the yield of the proteins and to understand their control, several labs have been active in cloning the uvr genes (Rupp et al., 1978; Pannekoek et al., 1978; Auerbach and Howard-Flanders, 1979). The presence of a DNA repair gene on a phage or on a plasmid is easily recognized because the introduction of a cloned gene confers UV-resistance to a strain with a mutation in that gene. Once such a clone is obtained, an important step is to identify the protein that is synthesized by the cloned gene. Several methods are available. The use of minicells is applicable if the gene is cloned on a plasmid. Minicells are formed in particular mutant strains in which the division and segregation is imperfect such that certain daughter "cells" (the minicells) are formed without any chromosomal DNA. The minicells can be separated from normal cells by differential centrifugation so that a fairly clean preparation can be obtained. When the original strain contains plasmids, these are found in the minicells and can support the synthesis of those proteins encoded by the plasmid DNA. With this system it can be determined what proteins are synthesized by the plasmid. A drawback to this system is that although the preparation of minicells is straightforward, it is time consuming and the yield is low.

Another approach is to use the cloned DNA as the template in a coupled in vitro protein synthesizing system. In principle, this is a good approach but in actual practice it is not easy to set up and often results in the formation of many protein fragments in addition to the intact polypeptides.

With λ bacteriophage, a useful approach is to infect heavily UV-irradiated cells with the unirradiated phage carrying the cloned gene. In this case, the high UV dose has made the cellular DNA inactive as a template for the synthesis of RNA and proteins so that the only proteins synthesized are those coded by the unirradiated bacteriophage that is added after irradiation. This method can be used to detect those proteins encoded by genes cloned on λ.

Use of Maxicells for Detecting Products of Genes Cloned on Plasmids

In our laboratory a new procedure was developed for specifically labeling plasmid proteins that is easier than the minicell method and is also more efficient in the incorporation of radioactive methionine into proteins (Sancar, Hack and Rupp, 1979). This approach, designated the maxicell procedure, involves irradiating a recA strain carrying the cloned gene on a plasmid with a low fluence of UV such that most of the plasmids in the cell, which are usually less than 1% the size of the chromosome, receive no UV lesions while many lesions are introduced into the chromosome. These cells undergo extensive DNA breakdown starting at the sites of lesions so that after a suitable period of incubation, the chromosomes are largely degraded while the plasmids persist and even replicate during this period. The addition of radioactive methionine to these "maxicells" then results in the specific labelling of those proteins encoded on the plasmid. This procedure has been of much value in our lab for the study of the uvrA (Sancar et al., 1980c) and uvrB (Sancar et al., 1980a) gene products and has been adopted by many other laboratories studying cloned genes.

Use of Transposons to Inactivate Cloned Genes

When a gene is cloned, it is common for adjacent genes to be present on the cloned fragment. Thus it is not always possible to identify a specific protein as the product of the cloned gene. Our approach was to inactivate the cloned gene by inserting a transposon into the middle of it and then using maxicells to detect which protein was lost. Using Tn1000, the gamma-delta sequence from the F factor, the uvrA and uvrB gene products were identified (Sancar and Rupp, 1979b; Sancar et al., 1980a and c). Furthermore, since insertion of Tn1000 into a gene causes premature termination of the polypeptide because of termination codons in the inserted sequence, it is possible to determine the direction of transcription and translation from the relation of the sizes of the truncated uvr polypeptides to the position of the insertions which were determined by restriction enzyme analysis of the plasmids containing Tn1000 inserted in the uvr genes.

Using these methods, the uvrA gene product was shown to be a protein of about 114,000 (Sancar et al., 1980c) while the size of the uvrB protein is about 84,000 (Sancar et al., 1980a).

Complementation Assay for uvr Proteins

Seeberg and his colleagues (Seeberg et al., 1976; Seeberg, 1978a and b) have developed an assay for the cutting of UV-irradiated DNA and have shown that the products of the three uvr genes, uvrA, uvrB and uvrC are all required. The development of this assay is particularly important because it provides a basis for the biochemical fractionation and purification of the various components.

Purification and Properties of uvr Proteins

Using the complementation assay, Seeberg (1978a, b and unpublished data) has partially purified the various components of the reaction. Although the proteins could be obtained from normal cells, it was found that the yields of the uvrA and uvrB proteins increased by about a factor of 10 when strains carrying plasmids with the cloned genes were used. This is particularly useful because the amounts of these proteins in a normal cell is very low. The uvrA and uvrB proteins have both been purified to near homogeneity using this approach. Seeberg estimates the size of the uvrA protein to be about 120,000 and the uvrB protein to be about 80,000. These estimates are thus in excellent agreement with those obtained in our laboratory using the completely different approach of identifying the gene products through the use of maxicells and insertional inactivation of the genes (Sancar et al., 1980 a and c).

It occurred to us that maxicells could be used to advantage in the purification of protein products of cloned genes. Since only the proteins synthesized from the plasmids are significantly labeled in maxicells, the total number of radioactive proteins is small, typically being less than 5. Thus it is usually easy to obtain radiochemically pure proteins after only 1 or 2 easy fractionations. Such preparations can be used to measure the interactions of the radioactive protein with other macromolecules. Although radiochemically pure, such preparations will generally contain many unlabeled proteins. However, these preparations are extremely useful for further purification because in subsequent fractionation steps, it is necessary to follow only the radioactivity to locate the protein that is being purified. Thus, the purification can be taken through one or many subsequent steps without the necessity to carry out laborious enzyme assays. This approach has worked well in our hands and the purifications to near homogeneity of both the uvrA and uvrB proteins have been accomplished without doing enzyme assays until the end of the purification (Kacinski and Rupp, unpublished data). In both cases, the purified proteins retained high activity in the complementation assay of Seeberg.

Properties of Purified uvrA and uvrB Proteins

Although the laboratories in Norway and at Yale have used very different approaches for purifying the uvrA and uvrB proteins, the characterization of the proteins in the 2 laboratories has resulted in excellent agreement about their properties (Seeberg, unpublished data; Kacinski and Rupp, unpublished data). As already pointed out, the size estimates agree. The uvrA protein binds to DNA with a somewhat greater affinity for single-stranded and UV-irradiated DNA than for double-stranded DNA. There is a DNA-stimulated ATPase activity that copurifies with the uvrA protein. The DNA stimulation is due to a decrease in the Km for ATP. The purified uvrA protein does not have any measureable nuclease or glycosylase activity.

Although purified uvrB protein is active in the complementation assay for the cutting of UV-irradiated DNA, no activities intrinsic to the isolated uvrB protein have yet been discovered. The purified uvrB protein does not bind to DNA and has no ATPase, nuclease or glycosylase activity.

Other Studies on Cloned uvr Genes

Pannekoek, van Sluis and their associates have also cloned the uvr genes. Pannekoek et al. (1978) originally cloned the uvrB gene by transferring the appropriate restriction fragment from a uvrB transducing phage to a plasmid vector. Subsequent studies have been done to show the orientation of transcription of the gene (clockwise on the E. coli genetic map), and also how the uvrB gene can be transcribed from an external promoter in those instances when the cloning procedure has removed the normal uvrB promoter from the structural gene (Pannekoek et al., 1979).

van Sluis (unpublished data) found a plasmid in the Clarke and Carbon (1977) collection that carried uvrC and then subcloned the fragment with the uvrC gene. From studies with this plasmid in minicells and maxicells, it was concluded that a protein of about 28,000 daltons is the uvrC gene product.

Use of Transposons to Clone Genes

Brandsma, van Sluis and van de Putte (unpublished data) have cloned the uvrA gene using an interesting approach that should be applicable to the cloning of mutant alleles as well as the wild-type gene. They inserted the Tn1 transposon carrying an antibiotic-resistance marker into a cloned chromosomal fragment adjacent to but not including the uvrA gene. The chromosomal segment with Tn1 was then incorporated into the chromosome by homologous recombination. By doing a partial restriction digest and selecting for a plasmid carrying Tn1, a large cloned fragment that carried the uvrA gene and several other genes was obtained. This method has the advantage that it can be used in those situations where there is already a plasmid available that carries a chromosomal segment near, but not including, the gene that is being cloned. Furthermore, since the selection is for antibiotic-resistance, this procedure can be used for the cloning of mutant genes where a direct selection is not practical.

Cloning on Bacteriophage λ

The bacteriophage λ has also been used successfully in the cloning of the uvr genes. In this procedure, random bacterial fragments are introduced into λ by in vitro DNA recombination. The phages with the cloned gene are then enriched by their ability to be host cell reactivated in a host strain that is mutated in one of the uvr genes. This approach has been used by Auerbach and Howard-Flanders (1979) to obtain the uvrA and uvrC genes and by Seeberg and Blingsmo (unpublished data) to obtain uvrC. Using the uvrC bacteriophage, Seeberg observed the synthesis of a protein of about 30,000 daltons that may be the uvrC gene product, an observation that agrees with the size of the uvrC protein as determined by van Sluis.

Regulatory Control of the uvr Genes

Because of previous difficulties in recognizing and assaying the uvr gene products, few details are known about the regulatory control of the uvr genes. However, recent data demonstrate regulation of the uvrA gene and the possible control of the uvrB gene as well.

uvrA. Kenyon and Walker (1980) demonstrated that fusion of the lacZ structural gene to certain chromosomal control regions resulted in the recA-lexA controlled inducibility of β-galactosidase and suggested that one of those sites was the uvrA gene. In other experiments (Kacinski, Seeberg, Sancar, Steinum and Rupp, unpublished), direct measurement of the uvrA protein showed that the uvrA gene is under recA-lexA control and that the lexA protein is probably also a repressor of the uvrA gene. The sequence of the uvrA control region is now being determined and it is expected that it will contain a sequence similar to that in the recA and lexA genes that has been demonstrated to be a binding site for the lexA protein.

uvrB. Although the sequence of the putative control region of the uvrB gene has been determined by Pannekoek et al. (unpublished) and by Sancar and Rupp (unpublished), details of the regulation (if it exists at all) remain to be elucidated. The possible role of recA and lexA in the control of uvrB is uncertain. Although the insertion of lacZ in or near uvrB results in the UV-inducibility of β-galactosidase (Schendel, unpublished data; Kenyon and Walker, unpublished data), we (Kacinski et al., unpublished) have been unable to observe an effect of lexA on the level of the uvrB protein. However, in determining the nucleotide sequence of the uvrB promoter, we (Sancar and Rupp, unpublished data) noticed a striking homology with the sequence in the recA promoter that binds the lexA protein, and anticipate that the lexA protein will also bind to the uvrB promoter.

Pannekoek, van Sluis, Brandsma, Dubbeld and Noordermeer (unpublished data) have cloned uvrB and uvrC on plasmids of different compatibilities so that both plasmids can exist together in the same cell. When maxicells of such a strain were examined, it was observed that the uvrB protein was expressed in much lower amounts than expected. This result led them to suggest that uvrB is regulated by uvrC, and specifically, that the uvrC protein might be a repressor that binds to a site in the uvrB promoter thus preventing transcription of the uvrB gene.

Although it is clearly too early to present a satisfactory scheme for the regulatory control of the uvrB gene, it is likely that there is negative control. A particularly exciting prospect is if uvrB turns out to be under dual control involving the recA-lexA pathway and also uvrC. Since the sequence of the uvrB control region has already been determined, rapid progress in understanding this possibly intricate regulation can be expected.

CLONING AND EXPRESSION OF BACTERIOPHAGE T4 denV GENE

The T4 denV gene codes for a protein involved in the repair of UV-irradiated DNA and complements E. coli mutants in the uvrA, uvrB, and uvrC genes. Lloyd and Hanawalt (unpublished data) constructed plasmids with properties indicating that they carry the denV gene. They used two diferent approaches. In the first, a specific Sal I restriction fragment known to contain the denV gene was isolated and inserted into pBR322. In the second procedure, a plasmid containing the denV gene was obtained by a shotgun approach in which all of the Xho I fragments of T4 were used as donor DNA and a recombinant plasmid complementing the UV-sensitivity of a E. coli uvrA recA strain was selected. The presence of either of these two plasmids in a E. coli uvrA recA strain caused the cells to become more resistant to UV, increased the host cell reactivation of bacteriophage λ and resulted in the disappearance of pyrimidine dimers (measured as T4 endo V sensitive sites). Unfortunately, these plasmids segregated out quite extensively making them difficult to study. Since these plasmids carried DNA fragments with several other T4 genes in addition to denV, it is possible that one of these other genes caused the observed instability and it is hoped that trimming off the excess DNA will result in a more stable recombinant plasmid carrying the denV gene.

ACKNOWLEDGMENTS

This paper summarizes data presented by the speakers at a Round Table sponsored by the C.N.R.S. during the Eighth International Congress on Photobiology at Strasbourg, France, July, 1980. Work done in the author's laboratory was supported by Grant CA 06519 from the U. S. Public Health Service.

REFERENCES

Auerbach, J. & Howard-Flanders, P. (1979). Molec. Gen. Genet. 168, 341-344.
Brent, R. & Ptashne, M. (1980). Proc. Natl. Acad. Sci. U.S. 77, 1932-1936.
Cassuto, E., West, S.C., Mursalim, J., Conlon, S. & Howard-Flanders, P. (1980). Proc. Natl. Acad. Sci. U.S. 77, 3962-3966.
Clarke, L., & Carbon, J. (1977). Cell, 9, 91-99.
Craig, N.L. & Roberts, J.W. (1980). Nature 238, 26-30.

Cunningham, R.P., DasGupta, C. Shibata, T., & Radding, C.M. (1980). Cell 20, 223-235.
Emmerson, P.T. & West, S.C. (1977). Molec. Gen. Genetics 155, 77-85.
Gudas, L.J. & Mount, D.W. (1977). Proc. Natl. Acad. Sci. U.S. 74, 5280-5284.
Gudas, L.J. & Pardee, A.B. (1975). Proc. Natl. Acad. Sci. U.S. 72, 2330-2334.
Horii, T., Ogawa, T. & Ogawa, H. (1980). Proc. Natl. Acad. Sci. U.S. 77, 313-317.
Kenyon, C.J. & Walker, G.C. (1980). Proc. Natl. Acad. Sci. U.S. 77, 2819-2823.
Little, J.W., Edmiston, S.H., Pacelli, L.Z. & Mount, D.W. (1980). Proc. Natl. Acad. Sci. U.S. 77, 3225-3229.
Little, J.W. & Harper, J.E. (1979). Proc. Natl. Acad. Sci. U.S. 76, 6147-6151.
Little, J.W. & Kleid, D.G. (1977). J. Biol. Chem. 252, 6251-6252.
McEntee, K. (1976). Virology 70, 221-222.
McEntee, K. (1977). Proc. Natl. Acad. Sci. U.S. 74, 5275-5270.
McEntee, K., Weinstock, G.M. & Lehman, I.R. (1980). Proc. Natl. Acad. Sci. U.S. 77, 857-861.
Ogawa, T., Wabiko, H., Tsurimoto, T., Horii, T., Masukata, H. & Ogawa, H. (1978). Cold Spring Harbor Symp. 43, 909-915.
Pannekoek, H., Noordermeer, I. A., & van de Putte, P. (1979). J. Bacteriol. 139, 54-63.
Pannekoek, H., Noordermeer, I. A., van Sluis, C. A. & van de Putte, P. (1978). J. Bacteriol. 133, 884-896.
Rupp, W. D., Sancar, A., Kennedy, W., Ayers, J., & Griswold, J. (1978). In: DNA Repair Mechanisms (P. C. Hanawalt, E. C. Friedberg & C. F. Fox, eds.), Academic Press, New York, pp. 229-235.
Sancar, A., , Clarke, N. D., Griswold, J., Kennedy, W. J. & Rupp, W. D. (1980a). J. Mol. Biol. In Press.
Sancar, A., Hack, A. M. & Rupp, W. D. (1979). J. Bacteriol. 137, 692-693.
Sancar, A. & Rupp, W.D. (1979a). Proc. Natl. Acad. Sci. U.S. 76, 3144-3148.
Sancar, A. & Rupp, W. D. (1979b). Biochem. Biophys. Res. Commun. 90, 123-129.
Sancar, A., Stachelek, C., Konigsberg, W. & Rupp, W.D. (1980b). Proc. Natl. Acad. Sci. U.S. 77, 2611-2615.
Sancar, A., Wharton, R. P., Seltzer, S., Kacinski, B. M., Clarke, N. D. & Rupp, W. D. (1980c). J. Mol. Biol. In Press.
Seeberg, E. (1978a). Proc. Natl. Acad. Sci, U.S.A. 75, 2569-2573.
Seeberg, E. (1978b). In: DNA Repair Mechanisms (P. C. Hanawalt, E. C. Friedberg & C. F. Fox, eds.), Academic Press, New York, pp. 225-228.
Seeberg, E., Nissen-Meyer, J. & Strike, P. (1976). Nature, 263, 524-526.
Witkin, E.M. (1976). Bacteriology Revs. 40, 869-907.

ASPECTS OF RADIATION-INDUCED MUTAGENESIS AND MALIGNANT

TRANSFORMATION

Ethel Moustacchi

Institut Curie, Biologie
Centre Universitaire, Bâtiment 110
91405   ORSAY   (France)

INTRODUCTION

Somatic cell mutation has long been considered as being involved in the carcinogenic process either of hereditary nature or induced by physical and chemical agents[1-3]. The mutation theory is essentially supported by a) The parallels between mutagenic and carcinogenic agents[4], b) The hereditary nature of certain tumors such as retinoblastoma, multiple polyposis of the colon and neurofibromatosis that are transmitted as autoromal dominant mutations, c) The association between cancer proneness and high induced mutability in vitro such as in Xeroderma pigmentosum[5-6].

The failure to demonstrate the presence of viral agents in a variety of spontaneous and induced tumors has provided an indirect argument in favour of the contribution of somatic mutations in the mechanism of cancer production. Recent developments, in viral tumorigenesis research, substantiate the unifying theories which have been proposed in the past[3].

The aim of this paper is to review the recent data on mammalian cell mutagenesis and malignant transformation induced by radiations which encourage or discourage the relationship between the two phenomena. I will limit myself to in vitro models, assuming with others that the analogies with the in vivo studies are sufficient[7] to justify the extrapolation from one situation to the other.

I) <u>The dose-effect relationships</u>.

A variety of mammalian cell culture systems are used to estimate mutation induction by radiation. The selective systems used often involve the purine and pyrimidine pathways. The quantity measured, as a function of dose, is the mutant frequency i.e. the number of mutant colonies divided by the total number of colony-forming cells in the culture. Mutants are often selected by their resistance to a toxic drug (thymidine analogues, hypoxanthine analogues, etc) whereas the total number of viable cells is scored in the absence of the drug. In contrast to the precision characteristic of microbial studies, variations in the response of mammalian cells have been encountered. Firstly, although most in vitro mutagenesis tests emphasize measurement of single gene mutations, this type of event may not predominate among mammalian cell mutants induced by radiations. Indeed large chromosomal defects such as translocations, deletions or aneuploidy are produced. When these changes affect functions essential to cell colony-forming ability, such mutants would escape detection and lead to an underestimation of the mutation frequencies. Secondly, quantitative mutagenesis in cultured somatic cells has, until recently, suffered from the variety of factors which influence the response. The problems inherent to the mutation systems include cell density, drug concentration, influence of media components and expression time. Recent technical advances have led to a relatively good agreement between quantitative results from different workers. A screening test capable of detecting chromosomal defects as well as single gene mutations has recently been proposed[8].

Tests for in vitro transformation focus on the changes in morphology and growth characteristics that accompany transformation of normal mammalian cells into malignant ones. Once such changes are recognized in culture, laboratory animals are injected with the transformed cell clones and production of tumors in vivo is determined. It should be noticed however that the two functional states -that is the transformed and the malignant phenotypes- are under separate genetic control in certain cell systems[9]. By making use of intraspecific human hybrids it was shown that malignancy, defined by the capacity of cells to produce tumors in a suitable host can be suppressed whereas many of the in vitro properties associated with transformation continue to be expressed. The quantitative assay of the rate of transformation is carried out by cloning the treated cells and scoring for colonies that have a transformed morphology, characterized by the piling up of cells and random orientation. The total number of colonies and the number of transformed colonies are obtained by direct count. Two of the best studied in vitro transformation tests use either a mouse cell line or hamster embryo cells[10-11]. It is only recently that a new experimental design has allowed the clear demonstration of neoplastic transformation with X-ray irradiation of human fibroblasts[12]. As with mutagenesis studies, cell density, expression time, etc, are essential for quantitative comparisons.

1) X-rays : Many of the reports dealing with mutation induction in cultured mammalian cells by X-rays have shown a non-linear response curve, with an initial low rate increasing steadily as the dose increases[13-16]. This appears to be the case for established rodent cell lines. However the validity of a non-linear mutation curve was questionned[17] and indeed when the mutation system circumvents some of the problems of mutation expression time, associated with the selection of mutants, linear relationships are found[18-20]. It may be significant that in contrast to the curvilinear increase in mutant frequency with dose observed for established cell lines, a linear increase is seen for human fibroblasts. Moreover when the differences in sensitivity to the inactivating effects of ionizing radiations are taken into account[19,20,8], i.e. when mutant frequency is plotted against the log of survival, the curves for different cell types are linear and demonstrate similar slopes (Fig. 1).

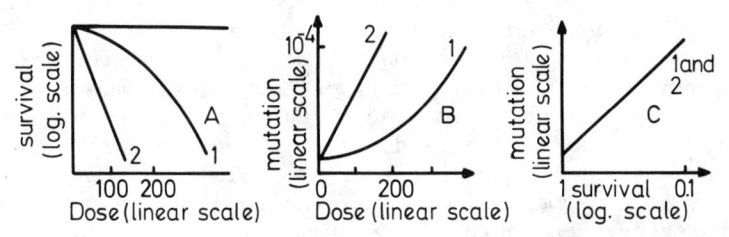

Fig. 1. General shape of dose-response relationships of X-ray-induced cell killing (A), of induced mutation frequency per survivors in cultured somatic cells as a function of dose (B) and of surviving fraction (C). Curve (1) refers to the type of data observed for survival and forward mutations from 8-Azaguanine or 6-Thioguanine sensitivity to resistance in Chinese Hamster cells[13,14,16,19]. Curve (2) corresponds to the type of data observed for survival and mutations to fructose utilization[18] or 6-Thioguanine resistance[17,19,20] in human diploid fibroblasts.

The dose-response data for transformation of either hamster embryo[10] or C3H mouse embryo (10T1/2)[11] cells by X-rays indicate that the percentage of cells transformed increases linearly by a factor of about 100 over the spontaneous background up to doses between 150 and 400 rad. A further increase in dose results in a plateau followed by a decrease in the proportion of cells transformed. Negligeabel cell killing occurs with doses up to 300 rad and it is only above this level that cell killing becomes exponential. In other words the transformation frequency declines in parallel with the decline in survival (Fig.2).

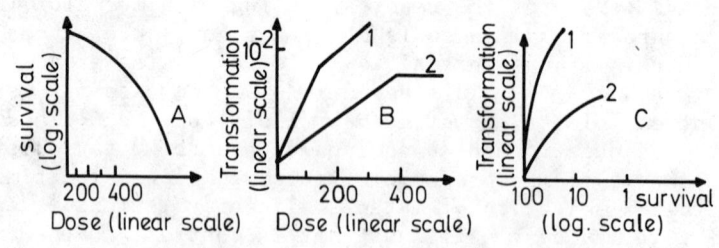

Fig. 2. General shape of dose response relationships of X-ray-induced cell killing (A), of induced transformation frequency in cultured rodent cells as a function of dose (B) and of surviving fraction (C). Curves are redrawn from Borek[10] (1) and Little[11] (2).

The comparison of figs. 1 and 2 shows that there are both qualitative and quantitative differences between mutation and transformation produced by ionizing radiations. In particular, induced frequencies of mutation were lower than transformation frequencies. The same conclusion was reached when mutagenic and carcinogenic compounds were compared for both these activities[21]. However it should be recognized that the two end-points were assayed separately, often on different cells and with different conditions of treatment. In view of the complexities of experimental factors involved in the two phenomena, there is an urgent need for experiments allowing the simultaneous assay of mutation and transformation. Some promising systems have recently been proposed[22].

Within the limits of our present knowledge it seems that the apparently large differences in X-ray dose-effect relationships between mutation and transformation are not a strong argument against the mutation theory for the oncogenic process. It is generally accepted that alterations in a single or limited number of genes represent at least one component of the steps leading to transformation. It is worth recalling that a) the transformation frequency is reduced to levels close to the mutation frequency when alteration of clonal morphology is not taken as the sole criterion of transformation, b) that a log-log plot of either mutants or transformants per survivors as a function of dose (data from the literature[10,18,19,20]) leads to single-hit or at most two-hit kinetics. For both mutation and transformation multi-hit models ($n > 2$) certainly do not fit with the data, c) since after exposure to very low doses (1 to 10 rads) transformation above the background level are detectable[10], the "transformation gene" is bound to represent either a hot spot (i.e. large target and/or unrepairable damage) or, as suggested[21], the other steps involved in the process are of the epigenetic type (derepression of vertically transmitted genetic sequences).

2) Ultraviolet (254 nm) : The survival curves found after exposure of both normal human fibroblasts[5,6,22] and rodent cells in culture are sigmoidal[23-29]. Generally, the induction of mutants is found to be a linear function of dose[6,25,27-29], at least around 20 J/m^2 (about 10 % survival). In some cases curvilinear responses have been described[5,24,26]. The end-points examined include resistance to azaguanine, 6-thioguanine, ouabain and diphteria toxin. These experiments involve either mutagenizing the cells, followed by replating for the drug challenge or an in situ treatment whereby surviving colonies rather than cells are challenged to locate the mutants . According to the method employed linear or non-linear curves were obtained under otherwise identical conditions[28]. In general, statistical analysis demonstrates that the linear model gives an excellent fit from about 1 to 10 J/m^2. With increasing dose, however, a decline in the rate of increase of mutation frequency leads to lack of fit of the linear model.

The dose-response relationship for the induction of transformation by UV[11,30] is similar in shape to that for X-rays[11], with a plateau level at about 7.5 J/m^2. As for X-rays, the steeply rising portion of the curves corresponds to the shoulder region of the survival curves. That UV is an effective transforming agent (with similar kinetics) has been confirmed with cells capable of dividing several times on soft agar[31,32].

The comparison of mutation and transformation data for UV indicates that both phenomena follow the same linear kinetics. Moreover when the two end-points are examined simultaneously on the same system[22] similar dose-response curves are obtained. In this case the transformation frequency remains as discussed above, higher than the mutation frequency. However, when one considers the small target size of the gene studied, the 3 to 10 fold difference observed[22] is relatively small. Moreover transformation is likely to be due to recessive mutation in view of the number of cell generations required for expression of mutants and transformants[22].

In summary, for UV light treatment it appears that there is a close correlation in response between in vitro mutagenesis and carcinogenesis.

II) <u>The role of repair processes.</u>

The role of DNA repair mechanisms in mutagenesis has been established in microorganisms and the concepts of "error-free" and "error-prone" repair differentiate those repair processes which restore damaged DNA to its original structure from those which generate mutations. In E.coli while the "error-free" DNA repair is constitutive, "error-prone" DNA repair is, at least partly, inducible.

1) **Comparison of normal and repair-deficient cells.**

a) <u>UV damaged DNA</u> : Evidence that repair processes are involved in mammalian cell mutagenesis essentially derive from experiments performed with cell strains obtained from human patients with the hereditary disorder Xeroderma pigmentosum[56,23] (XP). XP fibroblasts are more or less deficient, according to the complementation group involved, in either excision repair of UV-induced pyrimidine dimers or in the post-replication repair process ("variant") (for review see 33). Since XP patients often die of skin cancer on exposure to sunlight and XP cells in vitro show higher UV-induced mutation frequencies than normal cells at equal doses, the involvement of a repair mechanism in carcinogenesis and mutagenesis has been suggested. It is worth noticing that all the XP lines examined (groups A,C,D and variant) demonstrated, as a function of UV dose, higher frequencies of mutants than are produced in the normal type. This is true for a variety of end-points (resistance to azaguanine[5,34], thioguanine[23] and diphteria toxin[6]). However, when comparisons at equal survivals were made the induced mutation frequencies of normal and excision-deficient cells were equal[5,6], whereas the variant cells (normal for excision) still showed a higher frequency than normal fibroblasts[34,23]. The former results suggested that excision-repair in human cells is essentially "error-free" and that UV-induced mutants arise from another "error-prone" repair process. The latter result indicated that this process is not the one altered in variant cells ("post replication" repair). Indeed one has to assume that, at equal number of lethal events, the fidelity of the post replication repair in normal cells is even greater than in the excision-repair process.

In our present state of knowledge none of the repair-deficient mutants, characterized in mammalian cells, appear to be blocked in "error-prone" repair. The recent isolation of a variety of radiation-sensitive mammalian cell lines[35-38] is likely to accelerate the elucidation of the relationships between mutagenesis and DNA repair mechanisms.

UV-induced pyrimidine dimers in DNA appear to be involved in tumor production[39]. However, direct evidence of the involvment of DNA repair mechanisms in UV-carcinogenesis is still lacking. It has not been determined yet whether XP cells are more sensitive to UV-induced transformation in vitro. The extremely low transformation frequency of human fibroblasts at D_{37} by UV-irradiation was noted[22] but recent improvements[12] may help in the future.

b) <u>X-ray induced damage</u> : Ataxia telangectasia (AT) is another human genetic disorder which has been investigated because cells from such patients show a higher sensitivity to X-rays than normal cells. The nature of the molecular defect in AT cells is still questioned. Unlike XP cells. AT cells are either equally or less mutable than normal cells by X-rays and only about 10 % of AT patients develop neoplasms (for review see 33).

2) Delayed sub-cultures : Other attempts to demonstrate the possible role of repair in radiation-induced mutagenesis and transformation make use of the "liquid-holding recovery" procedure. Under non-growth conditions, a recovery of radiation-induced cell killing is observed with normal cells[40-43]. That repair processes contribute in eliminating potentially cytotoxic lesions, in quiescent conditions, is indicated by the lack of recovery in UV-treated XP group A[40,42] and in X-ray treated AT cells[40]. In terms of UV-induced mutation (X-ray mutagenesis was not studied) conflicting data are reported on the effect of holding normal cells. Holding of treated human fibroblasts either abolished induced mutants[42] over a range of doses or showed a dose dependance with an increase in mutant frequencies at high exposures[41]. Preliminary results with mouse cells treated with one UV dose confirm this last observation[43]. In an XP(A) line the mutation frequency did not change with time in confluence[42]. The message born out from such experiments is that repair processes are likely to take place. Whether they contribute in eliminating ("error-free") or in aggravating ("error-prone") mutagenic lesions is open to question.

Transformation experiments with UV[43] and X-rays[11] on mouse cells showed that within the initial few hours of holding there is a rise in transformant frequency followed by a decline to values below the 0 hour point. The functioning of an "error-prone" process followed by a second, slower, "error-free" repair mechanism may explain[11] this data.

It is clear that more experiments along these lines are needed before any parallels can be drawn between mutation and transformation, with respect to repair mechanisms.

3) Dose-fractionation : The strategy in split-dose experiments implies that lesions induced by the 1st dose may trigger a repair system which will favour the repair of damage induced by the 2nd dose. Indeed, mammalian cells in culture treated with fractionated doses of UV or X-rays are known to survive better than cells exposed to a single total dose. With UV radiation this recovery is less pronounced and is more cell-cycle-dependent than that observed after ionizing radiation.

In principle an "error-free" repair system will lead to a reduction in mutation and/or transformation frequencies when a split-dose regime is compared to a single dose equal to the sum of the two doses. The opposite is expected if the repair system is "error-prone". Abolition of such effects by protein synthesis inhibitors, given between the two doses, may be indicative of the inducibility of the putative repair system.

For γ-rays the mutagenic effects on Chinese hamster cells of fractionated doses, delivered at 6 hr intervals, were less than if the radiation was given in a single equal total dose[14,44]. Similarly, the fractionation of a UV dose reduced or did not change the mutation frequencies provided that the time interval between the two doses exceeded 3 hr[45,46]. However, at intervals of less than 2 hr a 3 fold increase in mutation frequency, with enhanced cytotoxicity, was found[45]. This may indicate that shortly after the 1st dose an "error-prone" process is involved, whereas with a longer time period between the two UV doses an "error-free" repair takes place.

Reports dealing with the induction of transformation by a fractionated X-ray dose (UV experiments are not yet available) have indicated that for low doses, delivered with a 5 hr interval, there was an increase in the proportion of clones transformed[10,47]. For higher doses and longer intervals there was a reduction in the proportion of clones transformed[11,47]. Obviously, more experiments are needed in order to define the role of time intervals, triggering doses and of defective repair systems. It should be kept in mind that actual in vitro mutagenesis and transformation split-dose data, leading to a reduction in frequencies, is difficult to reconcile with classical findings that X-ray split-doses enhance in vivo mutagenesis (mouse spermatogenia tested for dominant and recessive mutations on 13 different loci) and tumor production in treated animals.

An enhancement or speeding up of the nascent DNA chain elongation was observed in Chinese hamster cells[48] and in human fibroblasts[50] after a UV split-dose regime. (The interpretation of these results was questioned[51] but the criticism was overcome[49]). On the other hand, the rate of recovery of DNA semi-conservative synthesis was also found to be enhanced after split-dose UV-irradiation relative to a single total dose in normal human fibroblasts[52]. Although these two effects cannot be equated at the moment, since the increase in the rate of daughter-strand closure is persistant in XP cells[50] whereas the enhanced recovery of DNA semi-conservative synthesis is absent in XP cells[52], such experiments may provide a molecular clue to the split-dose effects.

In conclusion, the comparison of a variety of parameters on radiation-induced mutation and transformation do not provide definitive arguments in favour of the mutation theory of cancer. Often discrepancies appear to be related to technical problems which need to be overcome for suitable comparisons. The common features between the two processes are however sufficient to justify new experiments within this working frame.

1. T. Boveri, The origin of malignant tumors, Williams and Wilkins Co, Baltimore (1929).
2. A.G. Knudson, Mutation and Cancer : Statistical study of retinoblastoma, Proc. Natl. Acad. Sci. USA 68 : 820 (1971).
3. D.E. Comings, A general theory of carcinogenesis, Proc. Natl. Acad. Sci. USA 70 : 3324 (1973).
4. J. McCann, E. Choi, E. Yamasaki and B.N. Ames, Detection of carcinogens as mutagens in the Salmonella/microsome test : Assay of 300 chemicals, Proc. Natl. Acad. Sci. USA 72 : 5135 (1975).
5. V.M. Maher and J.J. McCormick, in "Biology of Radiation Carcinogenesis", J.M. Yuhas, R.W. Tennant and J.D. Regan, eds., Raven Press, New York, pp. 129-145 (1976).
6. T.W. Glover, C.C. Chang, J.E. Trosko and S.S.L. Li, Ultraviolet light induction of diphteria toxin-resistant mutants in normal and Xeroderma pigmentosum human fibroblasts, Proc. Natl. Acad. Sci. USA 76 : 3982 (1979).
7. J.C. Barrett and P.O.P. Ts'o, Evidence of the progressive nature of neoplasmic transformation in vitro, Proc. Natl. Acad. Sci. USA 75 : 3761 (1978).
8. C. Waldren, C. Jones and T.T. Puck, Measurement of mutagenesis in mammalian cells, Proc. Natl. Acad. Sci. USA 76 : 1358 (1979).
9. E.J. Stanbridge and J. Wilkinson, Analysis of malignancy in human cells : malignant and transformed phenotypes are under separate genetic control, Proc. Natl. Acad. Sci. USA 75 : 1466 (1978).
10. C. Borek, in "Biology of Radiation Carcinogenesis", J.M. Yuhas, R.W. Tennant and J.D. Regan, eds., Raven Press, New York, pp. 309-326 (1976).
11. J.B. Little, in "Origins of Human Cancer", H.H. Hialt, J.D. Watson and J.A. Winsten, eds., Cold Spring Harbor Conferences on Cell Proliferation, Vol. 4, Part B, p. 923 (1977).
12. C. Borek, X-ray-induced in vitro neoplastic transformation of human diploid cells, Nature 283 : 776 (1980).
13. E.H.Y. Chu, Mammalian cell genetics. III. Characterisation of X-rays-induced forward mutations in Chinese hamster cell cultures, Mutation Res. 11 : 23 (1971).
14. C.F. Arlett and J. Potter, Mutation to 8-azaguanine resistance induced by gamma-radiation in a Chinese hamster cell line, Mutation Res. 13 : 59 (1971).
15. R.J. Albertini and R. De Mars, Somatic cell mutation. Detection and quantification of X-ray induced mutation in cultured, diploid human fibroblasts, Mutation Res. 18 : 199 (1973).
16. H.J. Burki, Ionizing radiation-induced 6-thioguanine resistant clones in synchronous CHO cells, Radiation Res. 81 : 76 (1980).
17. J.W.I.M. Simons, Dose-response relationships for mutants in mammalian somatic cells in vitro, Mutation Res. 25 : 219

(1974).
18. R. Cox and W.K. Masson, X-ray dose response for mutation to fructose utilisation in cultured human diploid cells, Nature 252 : 308 (1974).
19. J. Thacker and R. Cox, Mutation induction and inactivation in mammalian cells exposed to ionising radiation, Nature 258 : 429 (1975).
20. R. Cox and W.K. Masson, X-ray-induced mutation to 6-thioguanine resistance in cultured human diploid fibroblasts, Mutation Res. 37 : 125 (1976).
21. S. Parodi and G. Brambilla, Relationships between mutation and transformation frequencies in mammalian cells treated in vitro with chemical carcinogens, Mutation Res. 47 : 53 (1977).
22. T. Kakunaga, J.D. Crow and C. Augl, in "Radiation Res. Proc. of Sixth Int. Cong. of Rad. Res.", S. Okada, M. Imamura, T. Terashima and H. Yamaguchi, eds., Topan Press, Tokyo, p. 589 (1979).
23. B.C. Myhr, D. Turnbull and J.A. DiPaolo, Ultraviolet mutagenesis of normal and Xeroderma pigmentosum variant human fibroblasts, Mutation Res. 62 : 341 (1979).
24. C.F. Arlett and S.A. Harcourt, Expression time and mutability in the estimation of induced mutation frequency following treatment of Chinese hamster cells by ultraviolet light, Mutation Res. 16 : 301 (1972).
25. A.W. Hsie, P.A. Brimer, T.J. Mitchell and D.G. Gosslu, The dose-response relationship for ultraviolet-light-induced mutations at the hypoxanthine-guanine phosphoribosyltransferase locus in Chinese hamster ovary cells, Somatic Cell Genet. 1 : 383 (1975).
26. C.C. Chang, J.E. Trosko and T. Akera, Characterization of ultraviolet-light-induced ouabain-resistant mutations in Chinese hamster cells, Mutation Res. 51 : 85 (1978).
27. J.E. Cleaver, Induction of thioguanine and ouabain-resistant mutants and single-strand breaks in the DNA of Chinese hamster ovary cells by ^{3}H-thymidine, Genetics 87 : 129 (1977).
28. M. Fox and S. McMillan, Evidence for the involvement of different repair mechanisms in mutagenesis and cell killing in V79 cells, in "DNA Repair Mechanisms", P.C. Hanawalt, C.F. Fox and E.C. Freidberg, eds., Academic Press, New York, p. 723 (1978).
29. H.J. Burki, C.K. Lam and R.D. Wood, UV-light-induced mutations in synchronous CHO cells, Mutation Res. 69 : 347 (1980).
30. J.A. DiPaolo and P.J. Donovan, In vitro morphologic transformation of Syrian Hamster cells by UV-irradiation is enhanced by X-irradiation and unaffected by chemical carcinogens, Int. J. Rad. Biol. 30 : 41 (1976).
31. Y. Ishii, J.A. Elliott, N.K. Mishra and M.W. Lieberman, Quantitative studies of transformation by chemical carcinogens

and ultraviolet radiation using a subclone of BHK21 clone 13 Syrian Hamster cells, Cancer Res. 37 : 2023 (1977).
32. T.J. Withrow, M.H. Lugo and M.J. Deunsey, Transformation of BALB 3T3 cells exposed to a germicidal UV lamp and a sun lamp, Photochem. Photobiol. 31 : 135 (1980).
33. E.C. Friedberg, U.K. Ehmann and J.I. Williams, in "Advances in Radiation Biology", J.T. Lett and H. Adler, eds., Academic Press, New York, Vol. 8 : 85 (1979).
34. V.M. Maher, L.M. Ouelette, R.D. Curren and J.J. McCormick, Frequency of ultraviolet light-induced mutations is higher in Xeroderma pigmentosum variant cells than in normal human cells, Nature 261 : 593 (1976).
35. T.D. Stamato and C.A. Waldren, Isolation of UV-sensitive variants of CHO-K1 by nylon cloth replica plating, Somatic Cell Genet. 3 : 431 (1977).
36. T. Shiomi and K. Sato, Isolation of UV-sensitive variants of human FL cells by a viral suicide method, Somatic Cell Genet. 5 : 193 (1979).
37. L.H. Thompson, J.S. Rubin, J.E. Cleaver, G.F. Whitmore and K. Brookman, A screening method for isolating DNA repair-deficient mutants of CHO cells, Somatic Cell Genet. 6 : 391 (1980).
38. D.B. Busch, J.E. Cleaver and D.A. Glaser, Large-scale isolation of UV-sensitive clones of CHO cells, Somatic Cell Genet. 6 : 407 (1980).
39. R.W. Hart, R.B. Setlow and A.D. Woodhead, Evidence that pyrimidine dimers can give rise to tumors, Proc. Natl. Acad. Sci. USA 74 : 5574 (1977).
40. R.R. Weichselbaum, J. Nove and J.B. Little, Deficient recovery from potentially lethal radiation damage in Ataxia telangiectasia and Xeroderma pigmentosum, Nature 271 : 261 (1978).
41. J.W. Simons, Development of a liquid-holding technique for the study of DNA-repair in human diploid fibroblasts, Mutation Res. 59 : 273 (1979).
42. V.M. Maher, D.J. Dorney, A.L. Mendrala, B. Konze-Thomas and J.J. McCormick, DNA excision-repair processes in human cells can eliminate the cytotoxic and mutagenic consequences of ultraviolet irradiation, Mutation Res. 62 : 311 (1979).
43. G.L. Chan, H. Nagasawa and J.B. Little, in "Radiation Res. Proc. of Sixth Int. Cong. of Rad. Res.", S. Okada, M. Imamura, T. Terashima and H. Yamaguchi, eds., Topan Press, Tokyo, p. 603 (1979).
44. J.C. Asquith, The effect of dose fractionation on γ-radiation induced mutations in mammalian cells, Mutation Res. 43 : 91 (1977).
45. W.G. Tilby and C. Heidelberger, Cytotoxicity and mutagenicity of ultraviolet irradiation as a function of the interval between split doses in cultured Chinese hamster cells, Mutation Res. 17 : 287 (1973).

46. C.C. Chang, S.M. D'Ambrosio, R. Schultz, J.E. Trosko and R.B. Setlow, Modification of UV-induced mutation frequencies in Chinese hamster cells by dose fractionation, cycloheximide and caffeine treatments, Mutation Res. 52 : 231 (1978).
47. R.M. Miller and E.C. Hall, X-ray dose fractionation and oncogenic transformation in cultured mouse embryo cells, Nature 272 : 58 (1978).
48. S.M. D'Ambrosio and R.B. Setlow, Enhancement of post replication repair in Chinese hamster cells, Proc. Natl. Acad. Sci. USA 73 : 2396 (1976).
49. S.M. D'Ambrosio, P.M. Aebersold and R.B. Setlow, Enhancement of post replication repair in ultraviolet-light-irradiated Chinese hamster cells by irradiation in G2 or S phase, Biophys. J. 23 : 71 (1978).
50. R.B. Setlow, F.E. Ahmed and E. Grist, in "Origins of Human Cancer", H.H. Hiatt, J.D. Watson and J.A. Winsten, eds., Cold Spring Harbor Conferences on Cell Proliferation, Vol. 4, Part B, p. 889 (1977).
51. R.B. Painter, Does ultraviolet light enhance post replication repair in mammalian cells ?, Nature 275 : 243 (1978).
52. E. Moustacchi, U.K. Ehmann and E.C. Friedberg, Defective recovery of semi-conservative DNA synthesis in Xeroderma pigmentosum cells following split-dose ultraviolet irradiation, Mutation Res. 62 : 159 (1979).

GENETIC ASPECTS OF REPAIR DEFICIENCY AND SKIN CANCER

Hiraku Takebe

Radiation Biology Center
Kyoto University
Kyoto 606, Japan

INTRODUCTION

There are several hereditary diseases known to be associated with DNA repair deficiency and high incidence of cancer. Among them, xeroderma pigmentosum (XP) has been most extensively studied clinically and experimentally. Many XP patients develop skin cancers presumably as a direct consequence of the repair deficiency. Since DNA repair is a genetically controlled function, the involvement of genetic factors in skin cancer in general is suggested. Genetic factors associated with the relationship between DNA repair and carcinogenesis could also enhance the incidence of cancers other than of skin. We examined 157 XP patients in Japan for their clinical and repair characteristics in order to investigate a possible relationship between level of DNA repair and clinical symptoms. Genetic heterogeneity has been demonstrated in XP by cell fusion techniques (11). Seven complementation groups (A through G) have emerged from those studies (1,4,6). We investigated the distribution of Japanese XP patients in these complementation groups. Characteristics of XP patients belonging to group F, so far found only in Japan, will be presented.

DNA REPAIR AND CLINICAL SYMPTOMS

Table 1 gives the age distribution, clinical characteristics including skin cancers, and DNA repair capacities of the cells cultured from patients as measured by the amounts of unscheduled DNA synthesis (UDS) within the first 3 hours after UV irradiation.

The ages represent the time of biopsy and, therefore, may not reflect the age of onset or time of first diagnosis.

Table 1. Age Distribution, Clinical Characteristics and DNA Repair of XP Patients in Japan

Ages	No.of Patients[1]	Mental Retardation & Neurolog.Abnorm. Yes	No	Unknown	UDS(% of Normal)[1] ≤ 5	5-30	30-60	$60\leq$
0-9	63(20)	37	3	23	55(18)	6(2)	1	1
10-19	35(22)	24	10	1	25(17)	4(3)	1(1)	5(1)
20-29	21(15)	3	15	3	4(4)	4(3)	3(1)	10(7)
30-39	11(6)	1	9	1	0	0	2(2)	9(4)
40-49	12(7)	0	11	1	0	3(1)	1(1)	8(5)
$50\leq$	15(11)	0	15	0	0	3(2)	1(1)	11(8)
Total	157(81)	65	63	29	84(39)	20(11)	9(6)	44(25)

1. number of patients with cancers in parenthesis.

The patients came from all parts of Japan, but not randomly, and approximately 60% of the total number examined comprise those from three cities (Osaka, Kobe and Tokyo). Table 1 may be summarized as follows.
(1) 40% are young patients, 0-9 years old, and most of them showed very low levels of UDS accompanied by neurological abnormalities.
(2) Approximately half of the patients have developed skin cancers.
(3) Low levels of UDS (below 5%) were not observed in patients over the age of 30 and with one exception they did not suffer from neurological abnormalities.
(4) More than 80% of the XP patients were young patients with low UDS levels and older patients (more than 20 years) with high UDS levels. Only a small number of patients with intermediate levels of UDS were observed.
(5) An inverse correlation between repair capacity measured by UDS and clinical symptoms appear to exist with only a few exceptions in agreement with the earlier findings by Bootsma (2).

The absence of cases with low repair levels (less than 5%) in the age group of 30 years or older suggests that low repair correlates with early death. Most of the patients with low UDS have developed severe symptoms with neurological abnormalities, and skin cancers have developed in almost all patients over 12 years old in this group, presumably due to their low repair capacity.
A few older patients with intermediate UDS levels (5-30%) were found. They may represent exception with regard to the relationship between UDS and clinical symptoms. Some of them belong to the complementation group F and will be discussed later.

GENETIC ASPECTS OF REPAIR DEFICIENCY AND SKIN CANCER

DISTRIBUTION OF PATIENTS OVER COMPLEMENTATION GROUP

Seven complementation groups have been found in the category of excision-deficient XP. In addition, a second caterogy of XP patients is recognized having normal levels of UDS (XP variants), whose cells show normal levels of UDS. Table 2 gives the distribution of these genetic groups in Japan in comparison with that in other countries.

Table 2.
Genetic Groups of XP Patients

Area	Complementation Group							Variant	Total
	A	B	C	D	E	F	G		
Japan	25	0	2	1	0	3	0	14[a]	45
Other Countries[b]	20	1	24	8	2	0	2	7	64
UDS (%) (UV)	0[c]<5	3<7	5<31	10<55	40<60	10<40	2<15	70<100	

a): Fujiwara (personal communication)
b): Bootsma and Kraemer (personal communication). They include patients from Europe, USA and the Middle East.
c): Several exceptions (up to 36% UDS) have been found in this group.

The distribution of the complementation groups in Japan seems to be different from that in other parts of the world.
Group C, the most frequent in other countries, is so far represented by only two patients in one family in Japan. Group D appears to be also rare in Japan. Group E was found only in the Netherlands in one family, group F only in Japan in two families and group G is represented by two families living in England. The group with very low repair as well as the group with relatively high repair levels in Table 1 appears to correspond to group A and XP variants respectively, being the most predominant groups in Japan.

CHARACTERISTICS OF GROUP F

The first group F patient was found in 1976 in Osaka, and was designated as XP230S. The patient was 45 years old, female, without neurological abnormalities and showed very mild symptoms and no cancer. In our routine UDS test her cells performed only 10% repair DNA synthesis compared with normal cells (1). However, the measurement of removal of T4 endonuclease susceptibility sites performed over a period of 24 hours suggested excision of UV-induced dimers to a larger extent than found in other XP cells with comparable UDS activity (3).

Also survival after UV exposure measured by colony forming ability and host-cell reactivation was found to be higher than with other XP cells with comparable UDS levels. Continuous measurement of UDS up to 24 hours after UV irradiation revealed that these group F cells had a slow but long-lasting excision repair, the final level reaching more than half that found in normal cells. These results suggest that the use of UDS tests, covering only the first 2 or 3 hours after UV exposure, as the only measure to evaluate the excision repair activity could be misleading.

HETEROZYGOTES AS A POSSIBLE HIGH CANCER RISK GROUP

In addition to skin cancers due to the action of UV, internal cancers may also be excepted in XP patients caused by chemical carcinogens to which XP cells have to shown to be hypersensitive in culture. We have little knowledge about internal cancers in XP patients. Kraemer collected 12 cases of XP patients bearing cancers of internal organs, but he did not estimate the frequency (5).

Swift et al.(8) reported that blood relatives of XP patients show a higher incidence of skin cancers than non-blood relatives. A similar observation was made in families with ataxia telangiectasia(7). The existence of genetically high cancer risk groups in the human population represented by heterozygotes of cancer-prone diseases was suggested by these findings. According to Hardy-Weinberg's law, a frequency of XP patients in Japan of 1 in 40.000 (9), would implicate the existence of 1% XP heterozygotes in the general population. A considerable fraction of skin cancers could then be attributed to XP heterozygosity. The contribution of the XP-trait to internal cancers may be less as is judged from the apparently low incidence of non-skin cancers in XP patients. However, the possibility exists that many XP patients, particularly in Japan, die before they develop cancers other than in the skin (10).

CONTRIBUTION OF SKIN COLOR

Another, and probably more important genetic factor in skin carcinogenesis is the skin color and its influence on the penetration of sunlight through epidermal cells. It is well known that people of Celtic origin show a high incidence of skin cancer, presumably related with their light skin color. Incidence of skin cancer in Japan is approximately 1/50 of that in caucasians at the same latitude. A genetically determined high risk group for skin cancer may be the result of a combination of XP heterozygocity and light skin color. This could be the case in white populations living near the equator for instance in Australia.

PREVENTION OF SKIN CANCER IN XP PATIENTS

The best and so far the only effective way to prevent XP patients from developing skin cancers is to avoid the exposure to sunlight. We have succeeded in many cases for the last 10 years in protecting them from over-exposure to sunlight by keeping them indoors and applying sun-screens on their skin. Although we have not followed these patients long enough to conclude how effective the protection is, the milder symptoms in the XP patients who have received good care in avoiding sun exposure are encouraging.

ACKNOWLEDGMENTS

The author wishes to express his thanks to colleagues who contributed data to Tables 1 and 2. The detailed report by all of the contributors will be published elsewhere. This work was supported by Grants-in-Aid from Ministry of Education, Science and Culture, Subsidy for Cancer Research from the Ministry of Health and Welfare and by The Princess Takamatsu Cancer Research Fund.

REFERENCES

1. Arase, S., Kozuka, T., Tanaka, K., Ikenaga, M. and Takebe, H. A sixth complementation group of xeroderma pigmentosum. Mutat.Res. 59: 143 (1979).
2. Bootsma, D. Defective DNA repair and cancer. in: "Research in Photobiology" (ed. A. Castellani, Plenum, New York, p.455 (1977).
3. Hayakawa, H., Ishizaki, K., Inoue, M., Yagi, T., Sekiguchi, M. and Takebe, H. Submitted to Mutation Research.
4. Keijzer, W., Jaspers, N.G.J., Abrahams, P.J., Taylor, A.M.R., Arlett, C.F., Zelle, B., Takebe, H., Kinmont, P.D.S. and Bootsma, D. A seventh complementation group in excision deficient xeroderma pigmentosum. Mutat. Res. 62:183 (1979)
5. Kraemer, K.H. Xeroderma pigmentosum. in: "Clinical Dermatology" (eds. D.J. Demis, R.L. Dobson and J. Mc Guire) Harper and Row, New York (in press).
6. Kraemer, K.H., de Weerd-Kastelein, E.A., Robbins, J.H., Keijzer, W., Barrett, S.F., Petinga, R.A., and Bootsma, D. Five complementation groups in xeroderma pigmentosum. Mutat. Res. 33:327 (1975).
7. Swift, M. and Chase, C. Cancer in families with xeroderma pigmentosum. J.Natl.Cancer Inst., 62:1415(1979).
8. Swift, M., Sholomon, L., Perry, M. and Chase, C. Malignant neoplasms in the families of patients with ataxia telangiectasia. Cancer Res. 36:209 (1976).
9. Takebe, H., Fujiwara, Y., Sasaki, M.D., Satoh, Y., Kozuka, P., Nikaido, O., Ishizaki, K., Arase, S. and Ikenaga, M. DNA repair and clinical characteristics of 96 xeroderma

pigmentosum patients in Japan. in: "DNA Repair Mechanisms" (eds. P.C. Hanawalt, E.C. Friedberg and C.F. Fox), Academic Press, New York, p.617 (1978).
10. Takebe, H. Xeroderma pigmentosum: DNA repair defects and skin cancer. GANN Monograph on Cancer Research, 24:103 (1979).
11. De Weerd-Kastelein, E.A., Keijzer, W. and Bootsma, D. Genetic heterogeneity of xeroderma pigmentosum demonstrated by somatic cell hybridization. Nature New Biol., 238:80 (1972).

IMMUNOLOGIC ASPECTS OF UV CARCINOGENESIS

Margaret L. Kripke

Cancer Biology Program

Frederick Cancer Research Center, Frederick, Maryland 21701

INTRODUCTION

I am most grateful to the Association Internationale de Photobiologie for permitting me to honor the memory of Dr. Edna Margaret Roe. Dr. Roe was interested in the broad application of radiation research to the solution of biological problems. In keeping with her interest, it seems quite appropriate to address one very specific aspect of this topic which concerns the use of UV radiation to investigate fundamental processes in carcinogenesis and immunology.

Although it has been known for nearly a century that UV radiation can induce skin cancer, it is only within the last few years that we have begun to realize that UV radiation also can affect certain immunologic processes. In addition to inducing neoplastic transformation of cells directly exposed to UV, irradiation of experimental animals with UV light also contributes to the pathogenesis of skin cancers through its effects on the immune system. This conclusion was reached as a consequence of an immunologic, rather than a photobiologic, approach to the problem of cancer. It originated from the discovery that UV-induced skin cancers in mice are highly antigenic, in that they induce a vigorous immune response in normal mice of the same inbred strain.

CHARACTERISTICS OF UV-INDUCED SUSCEPTIBILITY TO TRANSPLANTED TUMORS

The specific observation, upon which all subsequent findings are based, is that most skin cancers induced in mice by UV-B irradiation are so antigenic that they are rejected immunologically upon transplantation to normal, syngeneic recipients.[1,2] This unusual situation, which appears to be unique in chemical and radiation carcinogenesis, raised the question of how these tumors could escape immunologic destruction during their development in the primary host. The unexpected answer to this question turned out to be that UV radiation causes systemic immunologic alterations that prevent the rejection of these highly antigenic tumors.[3]

In spite of the fact that each UV-induced tumor is antigenically unique,[1,4] all of these tumors will grow progressively in UV-irradiated syngeneic hosts, even though they are immunologically rejected in unirradiated animals.[3,5] Recent experiments on the photobiological aspects of this susceptibility to transplanted tumors have demonstrated that the dose of UV radiation from FS40 sunlamps required to induce this susceptibility is quite low, relative to the carcinogenic dose. Susceptibility to the transplantation of UV-induced tumors is linearly related to the $\log_{10}$ dose of UV, and, unlike carcinogenesis, its induction exhibits reciprocity with regard to time, dose, and dose-rate of radiation. The active waveband in the sunlamps for the induction of tumor susceptibility in albino mice lies between 275 and 315 nm, based on experiments using plastic filters.[6,7]

MECHANISM OF TUMOR SUSCEPTIBILITY

The basis for tumor susceptibility in UV-irradiated hosts is immunologic in nature. *In vivo* lymphocyte transfer experiments demonstrated that UV irradiation results in the appearance of suppressor T lymphocytes in the spleen and lymph nodes of UV-treated mice.[8-10] These suppressor cells are part of a homeostatic system that normally regulates all immunologic responses. The suppressor cells induced by UV radiation in the absence of any other antigenic stimulus act to prevent the induction of an immune response against UV-induced skin cancers in the UV-irradiated host. They are specific, in that only the immune response against syngeneic UV-induced tumors is suppressed. The rejection of allogeneic tissues and chemically induced syngeneic tumors is unaffected by these suppressor cells.[8,9,11] However, this specificity is unusual, because each UV-induced tumor is antigenically distinct. Thus, the UV-irradiated host must contain many subpopulations of suppressor cells, each of which regulates the immune response to an individual tumor antigen; alternatively, the suppressor cells must regulate these individual immune responses through a common antigenic determinant, which is not readily detectable with other immunologic assays.

Based on our current understanding of immunologic processes, there must be an antigen (or antigens) produced by UV irradiation that is responsible for the occurrence of the UV-induced suppressor cells. Further, this antigenic stimulus must be related to the antigens that appear subsequently on UV-induced tumor cells, because it is the immune response against these tumor antigens that is abrogated by the UV-induced suppressor cells. Thus, in addition to causing neoplastic transformation, UV irradiation of the skin also produces new antigens that stimulate the production of specific suppressor cells.

SECOND IMMUNOLOGIC ALTERATION IN UV-IRRADIATED MICE

Insight into why the suppressor cell pathway is activated by UV-induced antigens, rather than an effector cell pathway, comes from recent studies on the immunologic capabilities of UV-irradiated mice. In order to determine whether the inability of UV-treated hosts to reject UV-induced tumors was caused by a specific immunologic deficiency or a generalized immunosuppression, we assessed the ability of UV-treated mice to make various immunologic reactions.[3,11-15] Most of the immunologic functions examined were found to be normal in UV-treated mice. These included skin and tumor allograft rejection, antibody formation against sheep erythrocytes, graft-versus-host reactivity, rejection of non-UV-induced syngeneic tumors, and various in vitro tests of lymphocyte and macrophage function. These results suggested that susceptibility to UV-induced tumors did not result from a generalized, nonspecific immunosuppression. However, in addition to their inability to reject UV-induced tumors, UV-irradiated mice were found to be immunologically deficient in two ways: they were unable to serve as recipients in a local graft-versus-host reaction and to develop a delayed hypersensitivity response against contact allergens.[12] An investigation of the mechanism of depressed contact hypersensitivity in UV-irradiated animals demonstrated that it was not caused by the inability of lymphocytes to recognize or respond to the contact allergen.[16] Rather, it suggested that an earlier step in the immune response to this antigen had been affected by UV radiation, probably at the level of processing or presentation of the antigen by macrophages, before the interaction of the antigen with lymphocytes.

Direct evidence for an alteration in the presentation of certain antigens in UV-irradiated mice has been obtained recently.[17,18] It is possible to induce contact hypersensitivity to certain antigens in mice by conjugating the antigen (i.e. trinitrophenyl, TNP) directly to splenic macrophages and using these antigen-presenting cells (APC) for immunization. Using this approach, it was found that the defective response in UV-irradiated mice could be bypassed by introducing the antigen on normal APC. Immunization of UV-treated mice with TNP-conjugated APC from normal donors resulted in a normal contact hypersensitivity response. Conversely, TNP-conjugated APC from UV-irradiated mice were unable to induce contact hypersensitivity, and, in addition, immunization of UV-treated mice with these cells resulted in the induction of antigen-specific T-suppressor cells that prevented immune responses to TNP.[17]

These findings suggest that this alteration in antigen presentation is the mechanism by which the suppressor cells for tumor rejection are formed in UV-irradiated mice. Presentation of a UV-induced antigen by these altered APC could lead to activation of the suppressor regulatory pathway, thereby generating T lymphocytes that specifically suppress the immune response to UV-induced antigens. For this reason, we are actively investigating the photobiologic characteristics and immunologic events associated with the UV-mediated suppression of contact hypersensitivity. It is likely that this suppression is a manifestation of the first step along the pathway leading to the failure of tumor rejection, and, hence, progressive tumor growth.

CONCLUSIONS

From the studies reviewed here, it is clear that the immune system is intimately involved in experimental photocarcinogenesis. Whether this will be true for human photocarcinogenesis as well remains to be determined. However, certain lines of evidence suggest that this may be a point worthy of consideration. First, it has become clear that one consequence of renal transplantation is an increased susceptibility to UV-induced skin cancer.[19-21] It is quite possible that this could be related to the long-term use of immunosuppressive therapy to prevent kidney rejection in these patients, although other explanations certainly can be considered as well. Second, there is some histologic evidence for an immunologic involvement in human skin cancer, even to the extent of suggesting a correlation between the pathogenesis of skin cancer and degree of lymphocytic infiltration.[22,23] Finally, there is a suggestion in the literature that local UV irradiation of human skin may depress the elicitation of delayed hypersensitivity.[24]

Whether or not these findings with UV-irradiated mice represent an appropriate model for human photocarcinogenesis, they have served to advance our understanding of certain biological processes and they provide a basis for further investigations into several aspects of immunology and carcinogenesis. Although the involvement of the immune system in this carcinogenesis model is perhaps extreme because of the peculiar immunologic consequences of UV-irradiation and the unusual antigenic characteristics of the tumors, it illustrates the extent to which immune processes can influence the course of tumor development. In addition, these studies clearly demonstrate that the complex immunologic regulatory pathways that have been described for immune reactions to foreign substances introduced from without, also control the immune responses to tumor-specific antigens that arise endogenously. This finding is important in attempting to develop immunologic approaches to alter cancer growth and progression.

The recent studies of the suppression of contact hypersensitivity by UV exposure have identified an important step in the induction of susceptibility to tumor growth by UV radiation. In addition, these studies have generated two findings of practical significance in basic immunology. It is now possible to induce the formation of specific T-suppressor cells to any contact allergen and, possibly, to certain soluble antigens as well, by introducing them at an appropriate time after a single UV exposure. This ability to induce a particular regulatory pathway at will should help to determine how regulatory pathways are activated during an immune response. This approach might eventually be exploited to induce suppressor cells that would prevent undesirable immunologic reactions, such as allograft rejection, allergies, and autoimmunity. Furthermore, the identification of an alteration in antigen-presenting cell function in UV-irradiated hosts provides a useful tool for inducing a selective immunologic deficiency in these animals. The fact that some immunologic responses proceed normally in UV-irradiated animals suggests that there are different pathways of antigen processing and presentation for different classes of antigens.

Finally, these studies provide the basis for further explorations of the relationship between neoplastic transformation and tumor antigenicity. Until the present time, the antigenicity of tumors induced by chemicals and radiation has been viewed as arising from random alterations in some cell-surface molecule. This is because each tumor has a different antigenic specificity, including those tumors induced by a single carcinogen in a single host. The finding that UV-irradiated mice can distinguish between UV-induced tumors and those induced by other carcinogens implies that antigenicity reflects the etiologic agent in some unknown fashion, and that antigenicity does not result from a random molecular alteration. The following findings support this possibility. 1) UV-irradiated mice can recognize UV-induced tumors of different histologic types, suggesting that recognition involves the etiologic agent, rather than the embryologic origin of the tissue. 2) Tumors induced by exposure of cells in culture to UV radiation seem to share the peculiar antigenic characteristics of tumors produced following *in vivo* irradiation. Preliminary experiments suggest that some BALB/c epidermal cells transformed *in vitro* by FS40 sunlamps are highly antigenic and may grow preferentially in UV-irradiated recipients (H. Ananthaswamy and M. Kripke, unpublished data). 3) Tumors induced in mice by psoralen and UV-A irradiation are not highly antigenic and do not grow preferentially in UV-B-irradiated hosts, suggesting that they do not belong to the same antigenic category as the UV-B-induced tumors (M. Kripke, W. Morison and J. Parrish, unpublished data).

If this hypothesis that antigenicity in some way reflects etiology is correct, the consequences could be of major epidemiologic importance. Because the UV-irradiated mouse can discriminate between UV-induced tumors and those induced by other agents, it might be possible to use this form of recognition to determine whether UV radiation was the agent that induced a particular tumor of unknown etiology. Whether tumors induced by other carcinogens also share some regulatory specificity that could be employed in a similar manner is unknown, but this possibility cannot be excluded on the basis of current information. Recently, we have had an opportunity to explore the use of this recognition system as an indication of tumor etiology. Some time ago, we discovered a malignant melanoma that had arisen on a C3H mouse. The animal had been given a few exposures to UV-B radiation and then treated for 2 years with croton oil.[25] A tissue culture line of the tumor was established and has been tested for growth and metastasis in UV-irradiated and normal recipients. It was found to grow preferentially in UV-irradiated animals (M. Kripke and I. J. Fidler, unpublished data). As this is the only C3H tumor that has exhibited this characteristic, aside from those induced by chronic UV-B irradiation, it seems quite likely that the UV radiation was of major etiologic significance in its induction.

If this approach can be generalized to tumors in other species and to tumors induced by other carcinogens, we can envision the development of a new approach to cancer epidemiology and the identification of etiologic agents in human cancer. Even if reality falls short of this goal, addressing the relationship between tumor antigenicity and neoplastic transformation using the tool of UV radiation should provide new insights into the carcinogenic process.

Research sponsored by National Cancer Institute under Contract No. N01-CO-75380 with Litton Bionetics, Inc.

REFERENCES

1. M. L. Kripke, Antigenicity of murine skin tumors induced by ultraviolet light, J. Natl. Cancer Inst. 53:1333 (1974).
2. M. L. Kripke, Latency, histology and antigenicity of tumors induced by ultraviolet light in three inbred mouse strains, Cancer Res. 37:1395 (1977).
3. M. L. Kripke and M. S. Fisher, Immunologic parameters of ultraviolet carcinogenesis, J. Natl. Cancer Inst. 57:211 (1976).
4. A. Graffi, K.-H. Horn and G. Pasternak, Antigenic properties of tumors induced by different chemical and physical agents, in: "Specific Tumor Antigens," R.J.C. Harris, ed., pp. 204-209, Munksgaard, Copenhagen (1967).

5. R. L. Daynes, C. W. Spellman, J. G. Woodward, and D. A. Stewart, Studies into the transplantation biology of ultraviolet light-induced tumors, Transplantation 23:343 (1977).
6. E. C. De Fabo, and M. L. Kripke, Dose-response characteristics of immunologic unresponsiveness to UV-induced tumors produced by UV irradiation of mice, Photochem. Photobiol. 30:385 (1979).
7. E. C. De Fabo, and M. L. Kripke, Wavelength dependence and dose-rate independence of UV radiation-induced immunologic unresponsiveness of mice to a UV-induced fibrosarcoma, Photochem. Photobiol. 32:183 (1980).
8. M. S. Fisher, and M. L. Kripke, Systemic alteration induced in mice by ultraviolet light irradiation and its relationship to ultraviolet carcinogenesis, Proc. Natl. Acad. Sci. USA 74:1688 (1977).
9. M. S. Fisher, and M. L. Kripke, Further studies on the tumor-specific suppressor cells induced by ultraviolet radiation, J. Immunol. 121:1139 (1978).
10. C. W. Spellman, and R. A. Daynes, Modification of immunologic potential by ultraviolet radiation. II. Generation of suppressor cells in short-term UV-irradiated mice, Transplantation 24:120 (1977).
11. R. T. Thorn, Specific inhibition of cytotoxic memory cells produced against UV-induced tumors in UV-irradiated mice, J. Immunol. 121:1920 (1978).
12. M. L. Kripke, J. S. Lofgreen, J. Beard, J. M. Jessup, and M. S. Fisher, In vivo immune responses of mice during carcinogenesis by ultraviolet radiation, J. Natl. Cancer Inst. 59:1227 (1977).
13. K. C. Norbury, M. L. Kripke, and M. B. Budmen, In vitro reactivity of macrophages and lymphocytes from UV-irradiated mice, J. Natl. Cancer Inst. 59:1231 (1977).
14. C. W. Spellman, J. G. Woodward, and R. A. Daynes, Modification of immunologic potential by ultraviolet radiation. I. Immune status of short-term UV-irradiated mice, Transplantation 24:112 (1977).
15. M. L. Kripke, R. M. Thorn, P. H. Lill, C. I. Civin, and N. H. Pazmiño, Further characterization of immunologic unresponsiveness induced in mice by UV radiation: Growth and induction of non-UV-induced tumors in UV-irradiated mice, Transplantation 28:212 (1979).
16. J. M. Jessup, N. Hanna, E. Palaszynski, and M. L. Kripke, Mechanisms of depressed reactivity to dinitrochlorobenzene and ultraviolet-induced tumors during ultraviolet carcinogenesis in BALB/c mice, Cell. Immunol. 38:105 (1978).
17. M. I. Greene, M. S. Sy, M. L. Kripke, and B. Benacerraf, Impairment of antigen-presenting cell function by ultraviolet radiation, Proc. Natl. Acad. Sci. USA 76:6591 (1979).

18. N. L. Letvin, M. I. Greene, B. Benacerraf, and R. N. Germain, Immunologic effects of whole body ultraviolet (UV) irradiation: Selective defect in splenic adherent cell function in vitro, Proc. Natl. Acad. Sci. USA 77:2881 (1980).
19. F. C. Koranda, E. C. Dehmel, G. Kahn, and I. Penn, Cutaneous complications in immunosuppressed renal homograft recipients, J. Am. Med. Assoc. 229:419 (1974).
20. V. Marshall, Premalignant and malignant skin tumours in immunosuppressed patients, Transplantation 17:272 (1974).
21. E. O. Hoxtell, J. S. Mandel, S. S. Murray, L. M. Schuman, and R. W. Goltz, Incidence of skin carcinoma after renal transplantation, Arch. Dermatol. 113:436 (1977).
22. A. L. Dellon, C. Potvin, P. R. Chretien, and C. N. Rogentine, The immunobiology of skin cancer, Plast. Reconstr. Surg. 55:341 (1975).
23. J. Viac, R. Bustamante, and J. Thivolet, Characterization of mononuclear cells in the inflammatory infiltrates of cutaneous tumors, Br. J. Dermatol. 97:1 (1977).
24. S. Horowitz, D. Cripps, and R. Hong, Selective T cell killing of human lymphocytes by ultraviolet radiation, Cell. Immunol. 14:80 (1974).
25. M. L. Kripke, Guest editorial: Speculations on the role of ultraviolet radiation in the development of malignant melanoma, J. Natl. Cancer Inst. 63:541 (1979).

THE TRENDS AND FUTURE OF PHOTOBIOLOGY: BIOCHEMICAL AND GENETIC ASPECTS*

Kendric C. Smith

Department of Radiology
Stanford University School of Medicine
Stanford, CA 94305

INTRODUCTION

Since the science of photobiology is in a rapid growth phase, it has been relatively easy to select areas of photobiology (biochemical and genetic) that will provide unusual opportunities for exciting research in the coming years. However, the future of the science of photobiology not only depends upon the continued production of "scientific breakthroughs", but also on the encouragement of young scientists to enter the field of photobiology. Accompanying this should be a commitment to educate photobiologists to be knowledgeable in all the areas of photobiology, not just in the area of their research specialty.

GENETIC APPROACH TO PHOTOBIOLOGY

One approach to solving problems in molecular biology is to isolate mutants that are deficient in the biological process under investigation. If all the mutants isolated are completely deficient in the process, then the process is probably accomplished by a single biochemical pathway. If several mutants are isolated that are only partially blocked in the process, then (assuming that the mutations are not leaky) there are probably several biochemical pathways to accomplish the process under investigation. This genetic approach can frequently give an overview of a biological process in a much shorter time than it would take to first isolate and characterize the

---

*Citations such as S19 and P20 refer to abstracts of papers presented at the 8th International Congress on Photobiology, Strasbourg, France, 1980.

different enzymes that might be involved. In addition, the availability of a mutant deficient in a given process is essential for the successful cloning of the gene of interest.

The availability of numerous mutants of Escherichia coli that are radiation sensitive, and therefore deficient in some aspect of the repair of deoxyribonucleic acid (DNA), was largely responsible for the rapid progress toward our current understanding of the molecular basis of DNA repair, and the recognition of the extreme complexity of the several DNA repair processes (Smith, 1979; Hanawalt et al., 1979).

A similar genetic approach is being pursued in the study of the metabolic disorders that render humans sensitive to sunlight and/or predispose them to the development of cancer (Friedberg et al., 1979). The first of these genetic diseases to be studied at the molecular level was xeroderma pigmentosum, a disease now well known to photobiologists, and one that was well reported at this Congress (S19, P155, 159, 160, 182, 281, 287, 325). Other human disorders described at this Congress that may be associated with a genetic defect in DNA repair are Gardner's syndrome (P282), Bloom's syndrome (P287), lupus erythematosus (P356), basal cell nevus syndrome (P156), and patients that suffer from actinic keratosis (P335, 336).

This genetic approach is now being pursued in the fields of photosynthesis (Sistrom, 1978), photomovement (P61), and chronobiology (Feldman and Atkinson, 1978). However, many other fields of photobiology might also progress faster if this genetic approach were used.

MUTAGENESIS

Although the field of mutagenesis is an old one (PS6; S16, 17; P268, 270, 274, 348, 349), a large increase in activity in this field should occur in the near future. The first reason is that many of the studies on the molecular basis of mutagenesis were published prior to the knowledge that mutagenesis is largely the result of errors made during the repair of damaged DNA (reviewed by Witkin, 1976). Even spontaneous mutagenesis in Escherichia coli is largely due to error-prone DNA repair mechanisms (Sargentini and Smith, manuscript in preparation), the damage in the DNA that is being repaired in an error-prone manner presumably arises as the result of normal cellular metabolism (see below; P257).

The second stimulant to research on the molecular basis of mutagenesis is the prediction that mutagenesis is the first step toward carcinogenesis (Trosko and Chang, 1978). The fact that most chemical carcinogens have now been shown to be mutagens (McCann and Ames, 1976) has led to the prediction that most cancer is caused by agents in the environment. However, one study on the mortality statistics for cancer in various countries has resulted in a markedly different conclusion (Totter, 1980). This study suggests that the spontaneous rate of carcinogenesis results from damage produced in DNA by radicals that result from the metabolism of oxygen, and only a small portion of cancer deaths arise from general pollution. This hypothesis of "normal" metabolic damage to DNA is consistent with the results for spontaneous mutagenesis in bacteria cited above. The mutagenesis field should be a lively one in the years to come.

RADIATION SYNERGISM AND ANTAGONISM

The interaction of two agents is considered to be synergistic if their interaction produces an effect greater than the sum of the effects of the two agents administered independently. If the interaction of the two agents is smaller than the sum of their independent action, the interaction is said to be antagonistic.

In his review on "Radiation Synergism and Antagonism", Tyrrell (1978) comments that "Despite rising concern for the possible detrimental effects of industrial by-products (chemicals, heat, and even radiation) cast into the environment, relatively little attention has been given to the enormous potential that these agents may have for interacting with each other and with natural environmental agents (including chemicals and sunlight) to cause profound biological changes. Thus, interaction studies should form a basic part of environmental control programs." I offer just a few examples to document the importance of this subject.

The classical examples of antagonistic responses come from the fields of photomorphogenesis (Pratt, 1979) and photoreactivation (Sutherland, 1977). The chromophore for photomorphogenesis, phytochrome, is converted into the active form by red light, and into the inactive form by far-red light. Thus, far-red light is antagonistic to the effect of red light.

UV radiation kills cells, but if these UV irradiated cells are exposed to a second irradiation with visible light, they are photoreactivated. The photoreactivated cells show a higher survival than cells not exposed to the second irradiation. Thus, visible light antagonizes the effects of the UV radiation.

Since sunlight contains both red and far-red light and both UV and visible light, cells in nature find themselves in an equilibrium condition that is determined by the relative intensities of the two antagonistic wave bands present in sunlight at any given time of day or year (S14).

Numerous synergistic interactions of radiation are known (Tyrrell, 1978; P270). In photosynthesis, the best known example of synergism is the Emmerson enhancement phenomenon (Myers, 1971). Most synergistic interactions, however, have been encountered in studies on the lethal and mutagenic effects of various types of radiation (Tyrrell, 1978).

However, any situation in which a given response in the presence of broad spectrum illumination differs from that predicted from its action spectrum will indicate the presence of synergism or antagonism. One example is the observation that a certain DNA repair deficient strain of bacteria was very easily killed by white fluorescent light, but an action spectrum demonstrated that this organism was not especially sensitive to any of the individual wavelengths present in white light (MacKay et al., 1976). Clearly, a synergistic interaction of the several wavelengths present in white light is suggested.

Since environmental lighting (both natural and artificial) is composed of many wavelengths of light, it is essential that all workers in photobiology be aware that the responses studied in the laboratory with monochromatic light may not adequately describe the response that their system will show in nature.

Furthermore, since the effects of radiation are known to interact nonadditively with heat and with chemicals (Tyrrell, 1978), all studies using broad band illumination (natural or artificial) must incorporate an awareness of the possibility of nonadditive effects modifying the experimental results.

IMMUNOLOGY IN PHOTOBIOLOGY

While some people are concerned over the immunological overkill in the literature (Waksman, 1980), the concepts and techniques of immunology are now catalyzing breakthroughs in several areas of photobiology.

Photocarcinogenesis

It was long a mystery as to why UV radiation-induced tumors (which are highly antigenic) could develop in UV irradiated hosts, but could not be easily transplanted to another member of the same inbred strain of animals. It is now known that the same radiation that produces the tumors also modifies the immunological response of the animal so that the tumor will grow at the site of the irradiation, or so that a tumor can be transplanted to a different site on a UV irradiated animal (Kripke, 1980; S18; P133, 134, 196, 199, 345 bis, 345 ter). Clearly, the immunological effects of light on man will be an exciting area of research in the years to come.

Immunological probes in photobiology

Antibodies can be made against specific biological molecules. These antibodies can then be used in very sensitive analytical procedures for these molecules. Radioimmune assays are widely used for detecting such substances as endorphins and hormones (Bizollon, 1979).

In photobiology, this technique is being used in several laboratories for the detection of specific types of radiation-produced damage in DNA, both in vivo and in vitro (Van Vunakis, 1980; P233, 269, 285), and for the detection of phytochrome (Pratt, 1979; P95). Other areas of photobiology should also consider using this sensitive assay procedure.

BIOCHEMICAL AND GENETIC ASPECTS OF PHOTOBIOLOGY

This antibody technique is useful not only for analytical determinations, but also for determining the precise location of the antibody attachment sites within cells or tissue sections (S13). For example, a fluorescent antibody against UV irradiated DNA could be used on tissue sections to determine how deeply UV radiation penetrates into skin in vivo.

PHOTORECEPTION (EXTRAOCULAR)

Numerous organisms have been shown to have photoreceptors that are not true eyes (Menaker, 1977; Wolken and Mogus, 1981). This extraocular photoreception has been studied both in invertebrates and in vertebrates, but few studies have been made on man. Even in the most thoroughly studied example of extraocular photoreception, that of testicular development in the house sparrow (Menaker, 1977), the molecular nature of the photoreceptor is not known.

The transduction of a light stimulus into a biological response requires that the light signal be amplified. The photoactivation of an enzyme provides an impressive system for light amplification that can directly affect cellular metabolism. A single photon can activate an enzyme, which can then process thousands of substrate molecules per minute. The products of the first enzymatic reaction may serve as effectors for another enzymatic reaction, and thus give rise to a second stage of amplification. The photoactivation of enzymes has been observed both in vivo and in vitro (Hug, 1978, 1981; S40, 45; P65). The photoactivation of enzymes is rapidly becoming a major subspeciality of photobiology.

Extraocular photoreception is a field of great importance, but an insufficient amount of research is currently being conducted in this field, and especially on man.

TRAINING NEW PHOTOBIOLOGISTS

Based upon my observations, the majority of the people who currently do research in photobiology have received their formal (or self) training only in the area of photobiology related to their research activities. Therefore, they know very little about the many other fields of photobiology. This is unfortunate because there are techniques and concepts that develop rapidly in one field of photobiology that might be equally useful in the other fields of photobiology. The science of photobiology would progress more rapidly with better cross-fertilization of the different sub-specialty areas.

I would wager that if most members of the several societies of photobiology that exist in the world were stopped on the street and asked "What is your profession?", very few would respond "I am a photobiologist". Rather, they would just mention the type of degree that they received in school.

The science of photobiology can only be considered to have reached a degree of maturity when courses that cover all areas of photobiology are taught throughout the world, and when members of scientific societies of photobiology finally list their profession as PHOTOBIOLOGY.

REFERENCES

Bizollon, Ch. A., 1979, Radioimmunology 1979, Elsevier/North-Holland Biomedical Press, Amsterdam.

Cilento, G., 1980, Photobiochemistry in the dark, in: Photochemical and Photobiological Reviews, Vol. 5 (K. C. Smith, ed.) pp. 199-228, Plenum Press, New York.

Feldman, J. F. and Atkinson, C. A., 1978, Genetic and physiological characteristics of a slow-growing circadian clock mutant of Neurospora crassa, Genetics 88:255-265.

Friedberg, E. C., Ehmann, U. K., and Williams, J. I., 1979, Human diseases associated with defective DNA repair, Adv. Radiat. Biol. 8:85-174.

Hanawalt, P. C., Cooper, P. K., Ganesan, A. K., and Smith, C. A., 1979, DNA repair in bacteria and mammalian cells, Ann. Rev. Biochem., 48:783-836.

Hug, D. H., 1978, The activation of enzymes with light, in: Photochemical and Photobiological Reviews, Vol. 3 (K. C. Smith, ed.) pp. 1-33, Plenum Press, New York.

Hug, D. H., 1981, Photoactivation of enzymes, in: Photochemical and Photobiological Reviews, Vol. 6 (K. C. Smith, ed.) In Press, Plenum Press, New York.

Kripke, M. L., 1980, Immunologic effects of UV radiation and their role in photocarcinogenesis, in: Photochemical and Photobiological Reviews, Vol. 5 (K. C. Smith, ed.) pp. 257-292, Plenum Press, New York.

Mackay, D., Eisenstark, A., Webb, R. B., and Brown, M. S., 1976, Action spectra for lethality in recombinationless strains of Salmonella typhimurium and Escherichia coli, Photochem. Photobiol. 24:337-343.

McCann J. and Ames, B. N., 1976, Detection of carcinogens as mutagens in the Salmonella/microsome test: Assay of 300 chemicals: Discussion, Proc. Natl. Acad. Sci. USA 73:950-954.

Menaker, M., 1977, Extraretinal photoreception, in: The Science of Photobiology, (K. C. Smith, ed.) pp. 227-240, Plenum Press, New York.

Myers, J., 1971, Enhancement studies in photosynthesis, Annu. Rev. Plant Physiol. 22:289-312.

Pratt, L. H., 1979, Phytochrome: function and properties, in: Photochemical and Photobiological Reviews, Vol. 4 (K. C. Smith, ed.) pp. 59-124, Plenum Press, New York.

Sistrom, W. R., 1978, Appendix: list of mutant strains, in: The Photosynthetic Bacteria (R. K. Clayton and W. R. Sistrom, ed.), pp. 927-934, Plenum Press, New York.

Smith, K. C., 1978, Multiple pathways of DNA repair in bacteria and their role in mutagenesis, Photochem. Photobiol. 28:121-129.

Sutherland, B. M., 1977, Introduction. Fundamentals of photoreactivation, Photochem. Photobiol. 25:413-414.

Totter, J. R., 1980, Spontaneous cancer and its possible relationship to oxygen metabolism, Proc. Natl. Acad. Sci. USA 77:1763-1767.

Trosko, J. E. and Chang, C., 1978, The role of mutagenesis in carcinogenesis, in: Photochemical and Photobiological Reviews, Vol. 3 (K. C. Smith, ed.) pp. 135-162, Plenum Press, New York.

Tyrrell, R. M., 1978, Radiation synergism and antagonism, in: Photochemical and Photobiological Reviews, Vol. 3 (K. C. Smith, ed.), pp. 35-113, Plenum Press, New York.

Van Vunakis, H., 1980, Immunological detection of radiation damage in DNA, in: Photochemical and Photobiological Reviews, Vol. 5 (K. C. Smith, ed.) pp. 293-311, Plenum Press, New York.

Waksman, B. H., 1980, Information overload in immunology: possible solutions to the problems of excessive publication. J. Immunol. 124:1009-1015.

Witkin, E. M., 1976, Ultraviolet mutagenesis and inducible DNA repair in Escherichia coli, Bacteriol. Rev. 40:869-907.

Wolken, J. J. and Mogus, M. A., 1981, Extraocular photoreception, in: Photochemical and Photobiological Reviews, Vol. 6 (K. C. Smith, ed.) In Press, Plenum Press, New York.

III - PHOTOMEDICINE

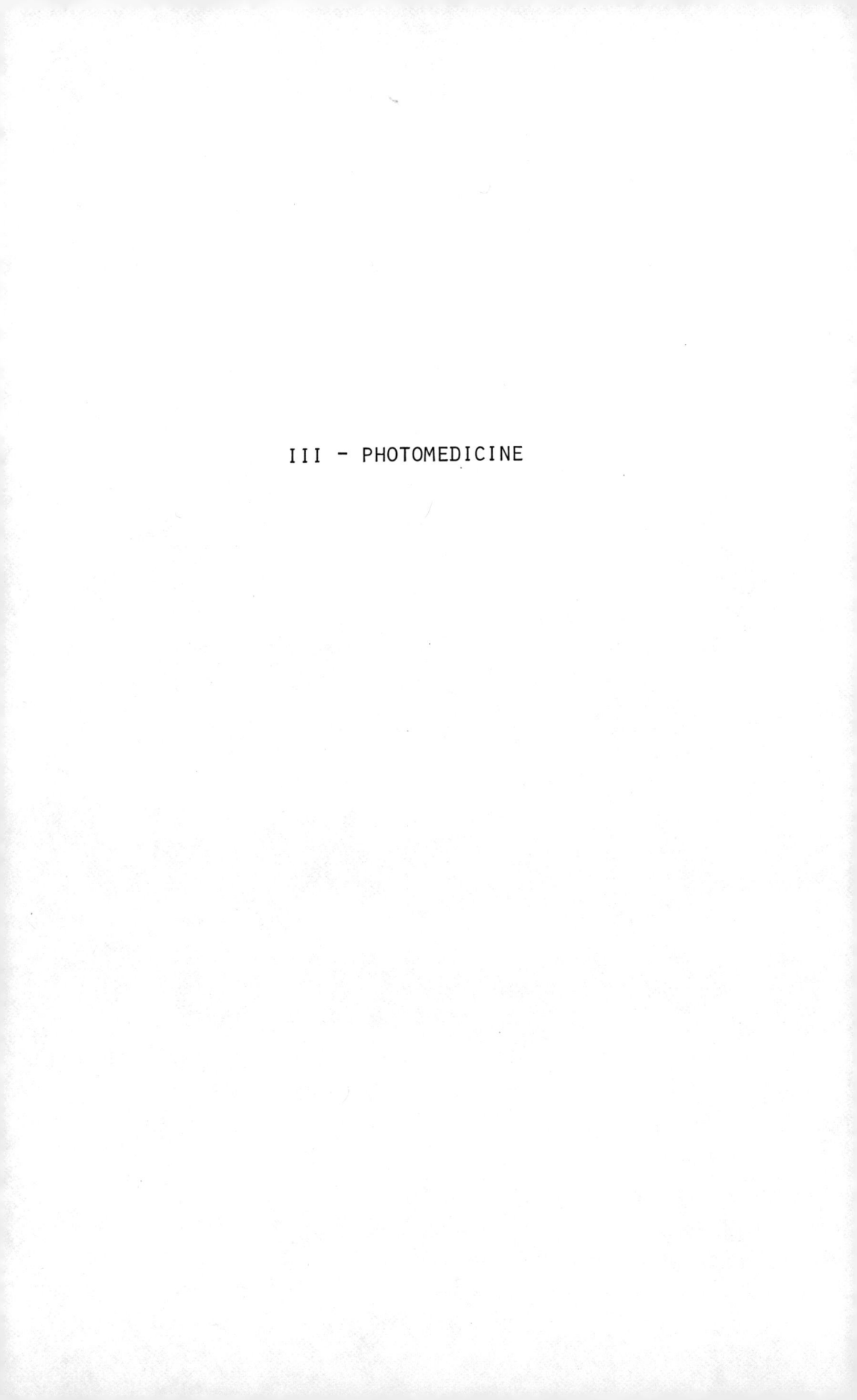

SKIN: STRUCTURE, NATURAL AND THERAPEUTICAL TARGETS OF ULTRAVIOLET RADIATION

Klaus Wolff

Dept. of Dermatology, University of Innsbruck
Innsbruck, Austria

Absorption of radiant energy by the skin initiates photochemical reactions which involve alterations of cellular metabolism, structure and function, changes of cellular kinetics, blood flow and pigmentation. Though most of these effects are due to the 290-320 nm (UV-B) or so-called sunburn spectrum there is now growing awareness that UV-A (320-400 nm) is also effective in inducing structural and functional changes in the skin.

A discussion of skin as a target organ of solar radiation will therefore first have to take into account the effects of different wavelengths at different energy levels. It will secondly have to consider that skin is a non-uniform tissue composed of different components of physical and chemical properties. In order to elicit a relevant biological effect radiant energy has to interact with living tissue for which transmission through the horny layer is important. In Caucasian skin, a substantial amount of incident radiation in the UV-B region is absorbed in the stratum corneum and only about 10% of this radiation penetrates to the upper dermis. By contrast, 50% of UV-A is transmitted to the dermis (1,2). Thirdly, due to the heterogeneity of the different skin compartments UV radiation elicits reactions in the non-vascular epidermis which are quite different from those in the dermis which has a rich supply of blood vessels. On the other hand, though different, these tissue compartments do not represent isolated systems but rather are interdependent, continuously interacting with each other. Fourthly and lastly, various autoprotective mechanisms make the skin vary in its susceptibility to the effects of equally energetic doses of UV-light. The thickness of the stratum corneum or the state of pigmentation and other as yet unidentified factors are responsible for the phenomenon that a defined dose of UV light will produce or fail to produce certain biological effects.

Erythema is the macroscopically recognizable and measurable feature of UV-injury to the skin; depending on the dose employed it may be accompanied by other signs of inflammation such as warmth, pain, swelling and may lead to blistering. The erythema action spectrum, which is a plot of erythemogenic effectiveness against wavelength is still not quite agreed upon in every detail but the most erythemogenic wave-band of sunlight is approximately at 308 nm (2). Delayed erythema to UV-B appears 2 to 6 hours and peaks 10 to 24 hours after irradiation. UV-C and UV-A also result in erythema, but the time course, clinical appearance and energy requirements for these erythemas are different (3). UV-B is approximately one thousand times more erythemogenic than UV-A (3,4); comparing UV-B with UV-C the degree of erythema following exposure to multiples of the minimal erythemogenic dose (MED) is greater for the longer wavelengths (5). Exposure to sub-erythemogenic doses of UV-B and UV-A produces erythematous reactions and this occurs when UV-A and UV-B are given in any order within a given period. The question whether UV-A radiation simply adds to or augments the UV-B erythema - in other words, whether this represents photo-addition (6) or photo-augmentation (7) - is not resolved.

Erythema represents a vascular response manifesting as vasodilatation and increased vascular permeability which, in brisk reactions in experimental animals can be demonstrated by the intramural deposition and leaking into the perivascular space of tracers introduced into the circulation. This raises the question whether the vascular wall proper is a target for UV radiation - particularly since UV-B and UV-C have been shown to have a direct vasodilator action on isolated arterioles in vitro (8) - or whether the vascular response is a secondary phenomenon. As indicated above, both UV-B and UV-A are transmitted to the papillary body and could thus directly interact with the vascular endothelium. This would suggest that a "direct hit" mechanism leads to UV erythema; an alternative mechanism, described by the diffusion theory, would involve the release of low molecular weight substances from other targets of UV-injury, particularly from UV-damaged epidermal cells; these substances would diffuse from the epidermis into the dermis to act on blood vessels (9). The fact that UV-B erythema is delayed and that, as will be shown later, epidermal changes due to UV-B precede those in the dermis would be in favour of the diffusion theory. The nature of the mediator(s) responsible is unknown but there is considerable evidence implicating prostaglandins in UV-B erythema (10) but not in erythema due to UV-A (11,12).

On the other hand, a direct interaction of light (or UV) and vascular endothelium has been shown to occur under certain circumstances, for instance when a photosensitizer, such as protoporphyrin, is present in the circulation; this is the case in the human disease erythropoietic protoporphyria (EPP) (13) or in experimental protoporphyria in the mouse (14). Here a photochemical reaction leads to the acute destruction of endothelial cells, sparing the epidermis and all other tissue components. Endothelial cells are vacuolized and disintegrate and thus lead to a spilling of vascular contents, plasma, cell debris and red blood cells, into the perivascular tissue. Recent evidence indicates that this phenomenon is due to the activation of the complement cascade (15) and it is interesting that whereas it results in the destruction of the endothelial cells of blood capillaries, it leaves the endothelia of lymphatics unchanged (14). Multiple phototoxic episodes of this type, each followed by regeneration of endothelial cells and the deposition of concentric vascular basement membranes, eventually lead to the pathognomonic appearance of vessels in chronically irradiated skin both in the protoporphyric mouse model and in EPP (16). Since a similar, though less pronounced pathology of vessels can be seen in skin chronically exposed to light (and UV) alone and since UV-A alone also induces similar, though less marked acute vascular changes (17), a similar pathogenesis may contribute to some of the dermal and vascular changes in chronic actinically damaged skin.

The vascular system is obviously not always and perhaps only exceptionally the primary target of UV damage. It thus appears appropriate to take a closer look at the tissue compartment which is hit by UV first, the epidermis.

This tissue is a stratified squamous epithelium in which germinative cells in the basal layers divide and randomly move towards the body surface. During this transepidermal migration, epidermal cells differentiate, eventually forming a unique protective layer on the surface of the skin, the stratum corneum (18). Light impinging on the skin is reflected, partially absorbed and scattered by the horny layer. When reaching the viable portions of the epidermis, it is aimed in many directions and is continuously scattered during its passage through successive cell layers. Different layers of skin are thus exposed to different quanta of radiant energy of different wavelengths.

The epidermis is not a uniform tissue, but consists of three different populations of cells, which are all affected in a different way by ultraviolet light: (i) keratinocytes which keratinize and constitute the majority of the epidermal cell population; (ii) melanocytes, which produce melanin with its unique photoprotective properties and (iii) Langerhans cells which are mesenchymal cells, constitute only 3-4% of the total epidermal cell population and represent the most peripheral outpost of the cellular immune system.

The earliest pathologic change observed in keratinocytes of skin irradiated with UV is intra- and intercellular edema; the former is usually perinuclear and often accompanied by the formation of large vacuoles which probably represent maximally distended endoplasmic reticulum. More pronounced changes include nucleolysis, disruption of cellular organelles and cytolysis. It appears important to emphasize that (i) these changes are dose dependent, i.e. after higher doses of radiant energy cytolysis will be apparent whereas lower doses lead only to intracellular edema. (ii) Cellular changes, though occurring preferentially in the upper spinous layers, are not uniformly present within the entire epidermis, but rather focal, seemingly affecting individual cells in a random manner. This is particularly evident with minor alterations occurring after low dose irradiation and it thus appears that some cells are more susceptible to damage than others. (iii) Identical changes occur after erythemogenic doses of UV-B and UV-A and even after phototoxic reactions due to psoralens and UV-A. They are thus not specific for irradiation with a particular wave-band and represent a secondary event, a final common pathway of tissue damage. The only difference between UV-B, UV-A and phototoxic reactions are different time courses and a different quantitative expression of the changes; alterations due to UV-B appear earlier and are more marked than those elicited by UV-A. (iv) Minimal changes of the type described occur also after irradiation with sub-erythemogenic doses of the various UV wavelengths, thus suggesting that damage to the epidermis occurs before and without a clinically visible erythematous response.

All this also holds for the classical histological hallmark of UV damage to the skin, the sunburn cell. This represents a brightly eosinophilic, shrunken epidermal cell with a pyknotic nucleus which appears to arise in a random fashion within the upper Malpighian layers after UV irradiation. At the ultrastructural level the cells show an aggregation of tonofilaments in the perinuclear region which leads to a progressive transformation of the entire cytoplasm into one densely packed fibrillar mass. These cells have been described as "dyskeratotic" but this term is a misnomer, for the observed phenomenon has nothing to do with faulty keratinization but rather represents massive dehydration and precipitation and condensation of pre-existing structural proteins. The term apoptosis, as employed by Kerr et al (19), which signifies a process of controlled cell death, therefore appears more appropriate.

The generation of sunburn cells is dose-dependent; time course studies by Woodcock and Magnus (20) have shown that in mice irradiated with 300 nm most sunburn cells appear within the first 24 hours after exposure; sunburn cells are not specific for any given wavelength for, as has been shown by Rosario and co-workers (21), they can be induced, though in a quantitatively different manner, with UV-A, UV-B and UV-C and with the combined action of psoralen and UV-A.

It is evident that the described epidermal changes represent secondary phenomena and therefore do not provide a clue as to the primary target of UV damage. According to their absorption spectra DNA, proteins, urocanic acid, melanin and polyunsaturated fatty acids of phospholipids could all play a role in the initial absorption of UV-energy in the epidermis (2). Lipid peroxidation by UV has been shown to occur and the membrane systems of the epidermis thus represent a potential primary target for UV rradiation. Candidates are the lamellae of Odland bodies (membrane coating granules), small vesicular structures in the upper epidermis (22), which after UV irradiation are transferred at a rate faster than normal into the intercellular space (23). Wolff-Schreiner(22) has postulated that they are the site of the UV-mediated conversion of provitamin D to vitamin D. Another potential candidate as UV target are lysosomes which occur in the epidermis (24); lysosomes are known to be lysed by UV light (25) and indeed, histochemical studies have indicated that UV irradiation of skin is followed by the release of hydrolytic enzymes from lysosomes; this has been considered the causal event for UV-induced cytoplasmic damage (26). After labelling epidermal lysosomes with a non-degradable tracer and exposing skin to UV light we have demonstrated that rupture of lysosomes and spilling of their contents into the cytoplasm indeed occurs but this phenomenon was observed only when considerable epidermal damage was already present; in fact, intact lysosomes were even seen in severely damaged sunburn cells, indicating that lysis of lysosomes is not the primary event in UV induced damage (27). This has recently been confirmed by biochemical methods (28). Other membranes, such as those of mitochondria are also affected by UV light; damaged mitochondria are sequestered into autolysosomes and are then digested (24) but again, this phenomenon follows the other changes already described and represents a repair process.

DNA is the major known biologic chromophore in the UV range and since it is present in high amounts in the epidermis it appears to be the biologically most important target for UV radiation (29). Indeed, pyrimidine dimers are produced in high yield after exposure of the epidermis to UV (30), but other photoproducts of DNA are also formed. Autoradiographic, biochemical and metaphase arresting techniques in mice have shown that DNA, RNA and protein synthesis and mitosis are inhibited within the first hour post irradiation and that this inhibition persists for up to 6 hours (31). As shown by Epstein and associates (32) for UV-injury in man recovery of DNA synthesis takes place within 24 hours and is followed by an accelerated activity for 48-72 hours which may persist longer in some species, eventually leading to epidermal hyperplasia. Among the better studied DNA injuries is thymine dimer formation (30) which has also been investigated by autoradiographic techniques to show unscheduled DNA repair activity within epidermal nuclei (32). Immediately after UV-B irradiation sparse labelling of the nuclei representing unscheduled DNA synthesis (UDS) is present within the epidermis and these lesions are detectable far earlier than any other changes described so far. Hönigsmann and co-workers (33) have recently shown that UDS occurs not only after UV-B but also after equally erythemogenic doses of monochromatic UV-A though quantitatively at a lower level. Since these DNA repair mechanisms may be error-prone, high doses of UV-A given over prolonged periods of time may be similarly hazardous as UV-B irradiation.

In summary then, the primary target of UV irradiation of keratinocytes is unknown, but although UV-energy may be absorbed by many chromophores at various levels of the skin, most of the available evidence suggests that nuclear DNA of the epidermis is the main and probably the initial target; however, UV damage is complex and a multitude of mechanisms may be operative, activating various cascades of events which eventually result in erythema or other manifestations of UV injury.

At this point, I would like to change the subject from injury to protection and to point out that the most significant available natural barrier to sunlight and thus UV radiation is melanin pigmentation. A suntan starts 48-72 hours after irradiation, it reaches a peak several days afterwards and gradually subsides over the next months. This delayed pigmentation is due to new pigment formation (melanogenesis) and the cellular site is a dendritic cell of neuroectodermal origin located within the epidermis at the dermal-epidermal interface, the melanocyte (34). This cell is uniquely equipped to produce the pigment, melanin, which represents a complex copolymer of black or brown indole-containing eumelanins and yellow or reddish sulphur-containing pheomelanins which arise through a common metabolic pathway in which dopaquinone is the key intermediate (35). Melanin is deposited in specific subcellular organelles with a distinctive protein substructure, the melanosome (36). The enzyme tyrosinase which utilizes tyrosine and dopa as the initial steps of melanin formation, is a unique property of the melanocyte (34) and can biochemically and cytochemically (at the ultrastructural level) first be demonstrated,as we and others (37) have shown, in the rough endoplasmic reticulum and the GERL of melanocytes from where it is transferred to melanosomes. It is here that the activated enzyme leads to the deposition of melanin pigment on the inner lamellae of these structures (38). Fully melanized melanosomes are then transferred via the dendritic processes of melanocytes into keratinocytes by means of a cytophagocytic process termed pigment donation (38). As shown by Fitzpatrick and Breathnach (39) a melanocyte is therefore associated functionally with a number of keratinocytes that surround it (epidermal melanin unit) so that the entire epidermis can be envisaged as a mosaic of multiple such units. Within the keratinocyte melanosomes are either singly or as multiples surrounded by a membrane (melanosome complex) into which acid hydrolases are incorporated. Melanosome complexes thus belong to the lysosome system and melanosomes are here subject to degradation (40). That degradation of melanosomes indeed occurs was shown by us when we incorporated a non-degradable tracer into such complexes and followed their fate during the transepidermal migration of the recipient keratinocytes (41). Within the keratinocyte melanin is thus present both in particulate form as melanosome and, since melanosomes but not melanin are degraded, as non-particulate amorphous melanin. Both are incorporated into the stratum corneum as the cell keratinizes. Melanin considerably reduces the transmittance of the epidermis and exerts its photoprotective role by (i) absorption of radiation whereby melanin acts as a neutral density filter, (ii) by scattering of radiation and (iii) by the trapping of free radicals generated in the tissue by UV irradiation (4, 34). However, it has also been postulated that tyrosine metabolites are cytotoxic and that the melanocyte is increasingly vulnerable under conditions of accelerated melanin synthesis (42).

Ultraviolet light activates melanocytes and melanogenesis but the mechanisms which regulate these events are unknown. The action spectrum falls into the sunburn range, but UV-A can also induce considerable pigmentation if enough energy is applied (3). Hyperpigmentation is associated with an activation of inactive melanocytes and an increase in melanocyte proliferation as has been shown for instance, by Rosdahl (43). Melanocyte hypertrophy and increased arborization occurs, there is activation of tyrosinase and synthesis of melanosomes followed by an increased melanization of melanosomes and an increased rate of melanosome transfer to keratinocytes.

Distinct from and preceding this delayed tanning response of melanogenesis is another phenomenon termed immediate pigment darkening (IPD). This type of pigmentation becomes apparent immediately after commencing irradiation, reaches its peak at the end of the exposure and gradually fades again over the next 6-8 hours (3). It is thus reversible and also with respect to color, which is greyish, not identical with delayed pigmentation. A prerequisite for this phenomenon is previous pigmentation, i.e. it is particularly pronounced in dark skinned or previously tanned individuals, and it also occurs when excised skin is exposed to long wave UV irradiation in vitro. IPD is not elicited by UV-B-, but by UV-A and possibly visible light (3). The underlying mechanism is unknown but most probably IPD is due to changes in preexisting melanin for which an oxidation of a monomeric unit via a semiquinone free radical state has been proposed (34). Another possible mechanism postulated by Jimbow and associates (44), would be a movement of microfilaments and melanosomes from the perinuclear cytoplasm to the dendrites of melanocytes; this would be analogous to what is known from the darkening of skin in amphibians, but definite proof that melanosome dispersion occurs in mammalian melanocytes has not yet been provided. IPD does not protect from UV radiation and its physiologic role is thus unknown.

Having discussed some effects of UV on keratinocytes and melanocytes, I finally wish to turn to the third cell population of the the epidermis, the Langerhans cell (LC), which in the past years has proved to be perhaps the most fascinating cell in skin research. LC are suprabasal dendritic cells, which can be demonstrated histochemically because of species-specific marker enzymes, such as ATPase, on their cytomembranes. Ultrastructurally, these cells have no tonofilaments and desmosomes and exhibit a specific cytoplasmic marker, the Langerhans cell granule (45). For decades, their function was unknown, but recent studies (46-53), primarily by my associate Georg Stingl and co-workers (46-48,53) have established that they originate from a bone-marrow-derived precursor cell (52) and that they are the only epidermal cells which express Fc-IgG receptors, receptors for the third component of complement and, above all, synthesize and express Ia antigens (46,47,50,51). Functionally, they thus belong to those cells which in vitro act as potent stimulators of antigen-specific and allogeneic T cell activation and indeed, Stingl and co-workers have shown that LC present antigen to immune T cells and serve as stimulator cells in the allogeneic mixed leukocyte reaction (48). LC are therefore now considered the most peripheral outpost of the cellular immune system (53) and this helps understanding their ubiquitous and superficial distribution within the epidermis as a system of recognition and first line defense.

Within the epidermis LC are accessible to UV light and the question therefore arises whether the immune functions of LC can be influenced by UV irradiation.

There have been several observations that UV irradiation of skin in vivo modulates the immune response. UV-B irradiation has been found to decrease cutaneous delayed-type hypersensitivity to dinitrochlorobenzene (DNCB) in DNCB-immune guinea pigs (54) and to intradermally injected antigens in man (55); to suppress the induction of contact hypersensitivity to trinitrochlorobenzene (TNCB) (56) and to abrogate the antigen-presenting cell function of splenic adherent cell populations in mice (57). Kripke and associates have shown that UV irradiated mice are unable to reject UV-induced tumours which are highly antigenic and are thus rejected by normal syngeneic animals and that this effect most probably is due to the generation of specific suppressor cells (58-61). Finally, Toews et al (62, 63) have demonstrated that UV-B irradiation of mouse skin induces specific unresponsiveness towards dinitrofluorobenzene

(DNFB) and have claimed that it depletes mouse skin of LC (employing the ATPase technique for LC demonstration).

Indeed, in a series of experiments carried out independently by our group it was shown that in UV-irradiated skin, ATPase-positive LC seemingly disappear from the epidermis (64); more importantly, this was also the case when LC were visualized, by immunofluorescence, through the Ia antigens on their membranes. Much to our surprise, however, Langerhans cells were still present in the epidermis when studied by electron microscopy, although their total population was reduced by approximately 30% and a small percentage of Langerhans cells exhibited profound morphological alterations, such as fibrillar condensation (similar to the changes seen in sunburn cells) or frank cytolysis (64,65). These changes occurred with doses of UV which left the other epidermal cell populations completely unaltered. These findings suggest that UV exerts a differentiated, non-uniform effect on LC surface markers and on the cells themselves and that it either deletes or modulates Ia antigens on the LC membrane (65).

Modulation of Ia antigens should have a profound effect on the function of these cells and that this is the case was shown in another series of experiments (66) in which epidermal cell suspensions were examined with regard to their capacity to stimulate allogeneic T cell proliferation in vitro: the epidermal cell-lymphocyte reaction (which reflects the allogeneic T cell stimulatory capacity of LC) was abrogated by UV irradiation indicating that LC had been deprived of this function; lastly we have shown that UV irradiation in vivo depletes Langerhans cells of their antigen-presenting capacity, i.e. abrogates their ability to serve as stimulators in antigen-specific T cell activation (67).

These results indicate that functionally the Langerhans cell is the most UV-sensitive cellular component of the epidermis and provide an explanation for some of the previous observations, referred to above, showing altered immune responsiveness of skin after UV exposure. A working hypothesis can thus be advanced (Fig.1): as suggested by Claman et al (68), the afferent limb of contact sensitivity is activated when an antigen couples to the Ia-bearing LC. This complex stimulates precursor T cells of delayed hypersensitivity to proliferate but it also stimulates auxiliary T suppressors. When Langerhans cells are exposed to UV (Fig. 1) Ia is modulated and the stimulatory activity of the antigen-Ia complex (of LC) on precursor T cells of delayed hypersensitivity is abrogated. Stimulation of T suppressors either by the Langerhans cells or the antigen would supervene and this would result in a tolerogenic effect. It is obvious that this has important implications not only for the presentation of haptens in contact hypersensitivity, but also for the recognition and presentation of viral antigens, tumour-associated antigens and alloantigens in skin graft rejection.

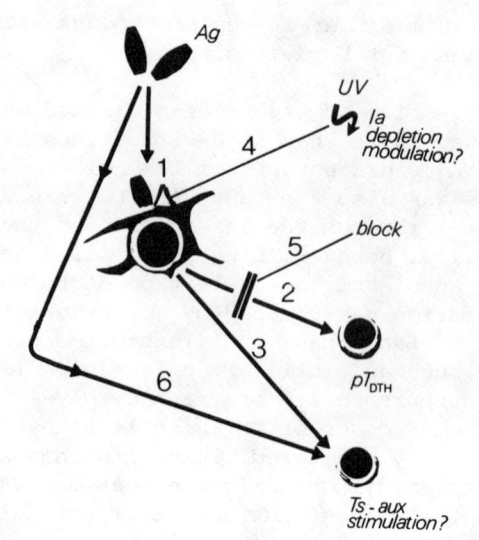

Fig. 1. When an antigen (Ag) is presented to skin it couples to the Ia (triangle) bearing Langerhans cell (1); this complex stimulates (2) precursor T cells of delayed hypersensitivity (pT-DTH) but auxiliary T suppressor cells (Ts-aux) are also stimulated (3). When skin is irradiated with UV the Ia on Langerhans cells is modulated (4) and the stimulatory signal of the antigen-Ia complex is abrogated (5). Stimulators of T suppressors by the Langerhans cell (3) or the antigen (6) can supervene and result in a tolerogenic effect.

REFERENCES

1. B.E. Johnson, F. Daniels, and I.A. Magnus, Response of human skin to ultraviolet light, Photophysiology 4:139 (1968).
2. H. Epstein, Photomedicine, in: "The Science of Photobiology," K.C. Smith, ed., Plenum Press, New York (1977).
3. J.A. Parrish, R.R. Anderson, F. Urbach, and D. Pitts, UVA: Biological effects of ultraviolet radiation, Plenum Press, New York (1978).
4. J.A. Parrish, H.A.D. White, and M.A. Pathak, Photomedicine, in: "Dermatology in General Medicine," T.B. Fitzpatrick, A.Z. Eisen, K. Wolff, I.M. Freedberg, and F.K. Austen, eds., McGraw Hill, New York (1979).
5. K.W. Hausser, and W. Vahle, Die Abhängigkeit des Lichterythems und der Pigmentbildung von der Schwingungszahl (Wellenlänge) der erregenden Strahlung, Strahlentherapie 13: 41 (1922).
6. C.Y. Ying, J. A. Parrish, and M.A. Pathak, Additive erythemogenic effects of middle (280-320 nm) and long (320-400 nm) wave ultraviolet light, J. Invest. Dermatol. 63: 273 (1974).

7. I. Willis, A. Kligman, and J.H. Epstein, Effects of long ultraviolet rays on human skin: photoprotective or photoaugmentative? J. Invest. Dermatol. 59: 416 (1972).
8. W.M. Sams, and R.K. Winkelmann, The effect of ultraviolet light on isolated cutaneous blood vessels, J. Invest. Dermatol. 53: 79 (1969).
9. J.C. van der Leun, Ultraviolet erythema. A study of diffusion processes in human skin. PhD. thesis, Utrecht (1966).
10. D.S. Snyder, and W.H. Eaglstein, Intradermal anti-prostaglandin agents and sunburn. J. Invest. Dermatol. 62:47 (1974).
11. W.L. Morison, B.S. Paul, and J.A. Parrish, The effect of indomethacin on long wave ultraviolet-induced delayed erythema. J. Invest. Dermatol. 68: 120 (1977).
12. A.P. Warin, The ultraviolet erythemas in man, Brit. J. Dermatol. 98: 473 (1978).
13. F. Gschnait, K. Wolff, and K. Konrad, Erythropoietic protoporphyria: submicroscopic events during the acute photosensitivity flare, Brit. J. Dermatol. 92: 545 (1975).
14. K. Konrad, H. Hönigsmann, F. Gschnait, and K. Wolff, Mouse model for erythropoietic protoporphyria. II. Cellular and subcellular events in the photosensitivity flare of the skin, J. Invest. Dermatol. 65: 300 (1975).
15. I. Gigli, personal communication.
16. H. Hönigsmann, F. Gschnait, K. Konrad, G. Stingl, and K. Wolff, Mouse model for protoporphyria. III. Experimental production of chronic erythropoetic protoporphyria-like lesions. J. Invest. Dermatol. 66: 188 (1976).
17. M. Kumakiri, K. Hashimoto, and I. Willis, Biologic changes due to longwave ultraviolet irradiation on human skin. An ultrastructural study, J. Invest. Dermatol. 69: 392 (1977).
18. A.S. Breathnach, and K. Wolff, Structure and development of skin, in: "Dermatology in General Medicine," T.B. Fitzpatrick, A.Z. Eisen, K. Wolff, I.M. Freedberg, and F.K. Austen, eds., McGraw Hill, New York (1979).
19. J.F.R. Kerr, A.H. Wyllie, and A.R. Currie, Apoptosis. A basic biological phenomenon with wide ranging implications in tissue kinetics, Brit. J. Cancer 26: 239 (1972).
20. A. Woodcock, and I.A. Magnus, The sunburn cell in mouse skin: preliminary quantitative studies on its production, Brit. J. Dermatol. 95: 459 (1976).
21. R. Rosario, G. Mark, J.A. Parrish, and M.C. Mihm, Histological changes produced in skin by equally erythemogenic doses of UV-A, UV-B, UV-C and UV-A with psoralens, Brit. J. Dermatol. 101: 299 (1979).
22. E.C. Wolff-Schreiner, Ultrastructural cytochemistry of the epidermis, Int. J. Dermatol. 16: 77 (1977).
23. G.F. Wilgram, R.L. Kidd, W.S. Krawczyk, and P.L. Cole, Sunburn effect on keratinosomes, Arch. Dermatol. 101: 505 (1970).

24. K. Wolff, and E. Schreiner, Epidermal lysosomes. Electronmicroscopic and electronmicroscopic-cytochemical studies. Arch. Dermatol. 101: 276 (1970).
25. G. Weissman, and H.B. Fell, The effect of hydrocortisone on the response of fetal rat skin in culture to ultraviolet irradiation, J. Exp. Med. 116: 365 (1962).
26. B.E. Johnson, and F. Daniels, Lysosomes and the reactions of skin to ultraviolet radiation, J. Invest. Dermatol. 53: 85 (1969).
27. H. Hönigsmann, K. Wolff, and K. Konrad, Epidermal lysosomes and ultraviolet light, J. Invest. Dermatol. 63: 337 (1974).
28. G. Volden, Acid hydrolases in blister fluid. Influence of ultraviolet light, Brit. J. Dermatol. 99: 53 (1978).
29. R.M. Tyrell, Molecular aspects of the interaction of near-UV radiation with living matter, in: "Proceedings of the International Symposium on Current Topics in Radiobiology and Photobiology, Rio de Janeiro 1977," R.M. Tyrell, ed., Academia Brasileira de Ciências, Rio de Janeiro (1978).
30. M.A. Pathak, D.M. Krämer, and U. Güngerich, Formation of thymine dimers in mammalian skin by ultraviolet radiation in vivo. Photochem. Photobiol. 15: 177 (1972).
31. J.H. Epstein, K. Fukuyama, and K. Fye, Effects of ultraviolet radiation on the mitotic cycle and DNA, RNA and protein synthesis in mammalian epidermis in vivo, Photochem. Photobiol. 12: 57 (1970).
32. W.L. Epstein, K. Fukuyama, and J.H. Epstein, Early effects of ultraviolet light on DNA synthesis in human skin in vivo. Arch. Dermatol. 100: 84 (1969).
33. H. Hönigsmann, K. Jaenicke, W. Brenner, W. Rauschmeier, F. Gschnait, and J.A. Parrish, Kinetics of thymine dimer repair in normal human skin after single and combined doses of UV-A, UV-B and PUVA, J. Invest. Dermatol. 74: 458 (Abst) (1980)
34. T.B. Fitzpatrick, G. Szabo, M. Seiji, and W.C. Quevedo, Biology of the melanin pigmentary system, in: "Dermatology in General Medicine," T.B. Fitzpatrick, A.Z. Eisen, K. Wolff, I.M. Freedberg, and F.K. Austen, eds., McGraw Hill, New York (1979).
35. G. Prota, Recent advances in the chemistry of melanogenesis in mammals., J. Invest. Dermatol. 75: 122 (1980).
36. M. Seiji, T.B. Fitzpatrick, and M.S.C. Birbeck, The melanosome: a distinct subcellular particle of mammalian melanocytes and the site of melanogenesis, J.Invest. Dermatol. 36: 243 (1961).
37. A.B. Novikoff, A. Albala, and L. Biempica, Ultrastructure and cytochemical observations on B-16 and Harding-Passey mouse melanomas - the origin of premelanosomes and compound melanosomes, J. Histochem. 16: 299 (1968).
38. K. Jimbow, and A. Kukita, Fine structure of pigment granules, in: "Biology of Normal and Abnormal Melanocytes," T. Kawamura, T.B. Fitzpatrick, and M. Seiji, eds., University of Tokyo Press, Tokyo (1971).

39. T.B. Fitzpatrick, and A.S. Breathnach, Das epidermale Melanin-Einheit-System, Dermatol. Wochenschr. 147: 484 (1963).
40. K. Wolff, and H. Hönigsmann, Are melanosome complexes lysosomes? J. Invest. Dermatol. 59: 170 (1972).
41. K. Wolff, Melanocyte-keratinocyte interactions in vivo: the fate of melanosomes, Yale J. Biol. Med. 46: 384 (1973).
42. J.M. Pawelik, and A.B. Lerner, 5,6-Dihydroxyindole is a melanin precursor showing potent cytotoxicity, Nature 278: 627 (1978)
43. K.I. Rosdahl, Melanocyte mitosis in UV-B-irradiated mouse skin, Acta dermatovenereol. 58: 217 (1978).
44. K. Jimbow, M.A. Pathak, and T.B. Fitzpatrick, Effect of ultraviolet on the distribution pattern of microfilaments and microtubules and on the nucleus in human melanocytes, Yale J. Biol. Med. 46: 411 (1973).
45. K. Wolff, The Langerhans cell, in: "Current Problems in Dermatology IV," J.W.H. Mali, ed., Karger, Basel-New York (1972).
46. G. Stingl, E.C. Wolff-Schreiner, W.J. Pichler, F. Gschnait, W. Knapp, and K. Wolff, Epidermal Langerhans cells bear Fc and C3 receptors, Nature 268: 245 (1977).
47. G. Stingl, S.I. Katz, E.M. Shevach, E.C. Wolff-Schreiner, and I. Green, Detection of Ia antigens on Langerhans cells in guinea pig skin, J. Immunol. 120: 570 (1978).
48. G. Stingl, S.I. Katz, L, Clement, I. Green, and E.M. Shevach, Immunological functions of Ia-bearing epidermal Langerhans cells, J. Immunol. 121: 2005 (1978).
49. I. Silberberg-Sinakin, R.L. Baer, and G.J. Thorbecke, Langerhans cells. A review of their nature with emphasis on their immunologic functions, Prog. Allerg. 24: 268 (1978).
50. G. Rowden, M.G. Lewis, and A.K. Sullivan, Ia antigen expression on human epidermal Langerhans cells, Nature 268: 247 (1977).
51. L. Klareskog, U. Malmnäs-Tjernlund, U. Forsum, and P.A. Peterson, Epidermal Langerhans cells express Ia antigens, Nature 268: 248 (1977).
52. S.I. Katz, K. Tamaki, and D. Sachs, Epidermal Langerhans cells are derived from cells which originate in bone marrow, Nature 282: 324 (1979)
53. G. Stingl, New aspects of Langerhans cell function, Int. J. Dermatol. 19: 189 (1980).
54. J. Haniszko, and R.R. Suskind, The effect of ultraviolet radiation on experimental cutaneous sensitization in guinea pigs, J. Invest. Dermatol. 40: 183 (1963).
55. S. Horowitz, D. Cripps, and R. Hong, Selective T cell killing of human lymphocytes by ultraviolet light, Cell Immunol. 14: 80 (1974).
56. F.P. Noonan, M.L. Kripke, and E.C. DeFabo, Suppression of contact hypersensitivity in mice with UV-radiation, in: "Book of Abstracts," Amer. Soc. Photobiol. Ann. Meeting, Colorado Springs (1980).

57. N.L. Letvin, M.I. Green, B. Benaceraff, and R.N. Germain, Immunologic effects of whole-body ultraviolet-irradiation: Selective defect in splenic adherent cell function in vitro, Proc. Natl. Acad. Sci. 77: 2881 (1980).
58. M.L. Kripke, and M.S. Fisher, Immunologic parameters of ultraviolet carcinogenesis., J. Natl. Cancer Inst. 53: 1333 (1974).
59. M.L. Kripke, and M.S. Fisher, Immunologic parameters of ultraviolet carcinogenesis, J. Natl. Cancer Inst. 57: 211 (1976).
60. M.S. Fisher, and M.L. Kripke, Systemic alteration induced in mice by ultraviolet light irradiation and its relationship to ultraviolet carcinogenesis, Proc. Natl. Acad. Sci. USA 74: 1688 (1977).
61. M.S. Fisher, and M.L. Kripke, Further studies on the tumor-specific suppressor cells induced by ultraviolet irradiation, J. Immunol. 121: 1139 (1978).
62. G.B. Toews, P.R. Bergstresser, J.W. Streilein, and S. Sullivan, Epidermal Langerhans cell density determines whether contact hypersensitivity or unresponsiveness follows skin painting with DNFB, J. Immunol. 124: 445 (1980).
63. J.W. Streilein, G. Toews, J.N. Gilliam, and P.R. Bergstresser, Tolerance or hypersensitivity to 2,4-dinitro-1-fluorobenzene; The role of Langerhans cell density within the epidermis, J. Invest. Dermatol. 74: 319 (1980).
64. W. Aberer, G. Schuler, G. Stingl, H. Hönigsmann, and K. Wolff, Effects of UV-light on epidermal Langerhans cells, J. Invest. Dermatol. 74: 458 (Abst) (1980).
65. W. Aberer, G. Schuler, G. Stingl, H. Hönigsmann, and K. Wolff, UV-induced damage of epidermal Langerhans cells. 8th Int. Congress Photobiol. Strasbourg, Poster-Abstract No. 133
66. W. Aberer, L.A. Gazze, G. Stingl, and K. Wolff, UV-induced abrogation of the epidermal cell-lymphocyte reaction. 8th Int. Congress Photobiol. Strasbourg, Poster-Abstract No.134.
67. G. Stingl, L.A. Gazze, W. Aberer, and K. Wolff, Abrogation of antigen-specific T cell stimulation of Langerhans cells by ultraviolet light. In preparation.
68. H.N. Claman, S.D. Miller, M.-S. Sy, and J.W. Moorhead, Suppressive mechanisms involving sensitization and tolerance in contact allergy, Immunol. Rev. 50: 105 (1980).

PHOTOCHEMICAL REACTIONS OF FUROCOUMARINS

Francesco Dall'Acqua

Institute of Pharmaceutical Chemistry
Padua University
Padova - ITALY

SUMMARY

The recent line of research in the field of furocoumarins, developed for obtaining new agents able to photosensitize biological substrates only through monofunctional photolesions to DNA, is discussed. The interactions between a group of methylderivatives of angelicin, recently prepared, and DNA both in the ground and in the excited state are referred. These new monofunctional furocoumarins appear interesting for studies in the field of molecular biology and seem to play the role of potential agents for the photochemotherapy of skin diseases characterized by cell hyperproliferation.

INTRODUCTION

Furocoumarins are tricyclic planar compounds that are able to intercalate in the ground state in duplex DNA [1], and by successive irradiation with UV-A light, to bind to the pyrimidine bases of the macromolecule, forming C_4-cycloadducts[2-4]. The most widely studied furocoumarins are psoralens, that is the linear isomers, which photobind to DNA both through mono- and bifunctional adducts (interstrand cross-linkages)[5]. Bifunctional adducts, even if from a quantitative point of view they represent only a minor product (between 3 and 20% of the total cycloadducts) of the photoreaction[6,7], from a photobiological point of view represent very important products, owing to their strong biological consequences [8].

In recent years a line of research in the field of furocoumarins has been developed with the aim of obtaining new DNA-mono-

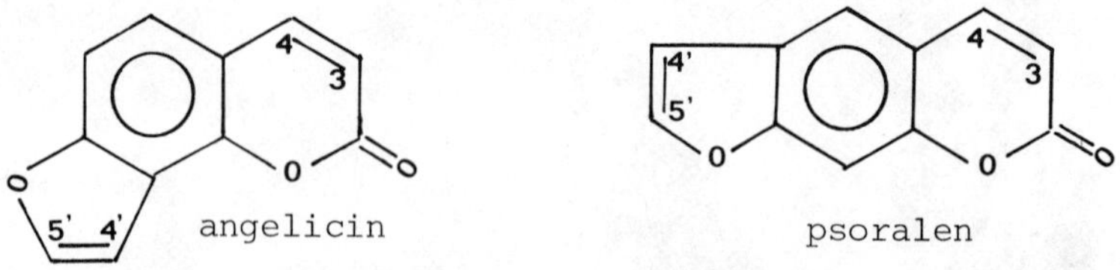

Fig. 1

functional reagents, which, maintaining the same selectivity of action as psoralens, should be able to induce photosensitization only by monofunctional lesions to DNA. This line has been stimulated both by some undesirable side effects of psoralens evidenced during the photochemotherapy of psoriasis and other skin diseases, and by the aim of enlarging the application of furocoumarins in the field of molecular biology, where effective pure DNA-monofunctional reagents are also requested in addition to bifunctional psoralens.

Some monofunctional reagents have been obtained from psoralens by cancelling the possibility of photoaddition at the level of one of the two photoreactive sites by introducing at this level either an encumbering group [9] or an electron withdrawing group [10]. In this case however the photobinding capacity of the resulting compound is generally lower than that of the parent psoralen. Other tricyclic structures (pyranocoumarins, benzodipyrones, difurobenzenes, etc.) have also been investigated [4].

A new series of DNA-monofunctional reagents have recently been obtained from angelicin. This compound, the parent angular furocoumarin, in spite of its bifunctional character (from a photochemical point of view two photoreactive sites, 3,4- and 4',5'-double bonds are present in its molecule, as in psoralens) behaves as pure monofunctional reagent towards native DNA. This behaviour is strictly connected with its angular structure and with the complex that it forms in the ground state with DNA undergoing intercalation between two base pairs of the macromolecule [11]. In fact when angelicin is intercalated in duplex DNA, only one of its two photoreactive sites can be aligned with the 5,6-double bond of one of the two pyrimidine bases involved in the intercalation: the photocycloaddition with pyrimidine bases can therefore lead to the formation of only monofunctional adducts [3,11,12].

By contrast, in the case of psoralen, both the two photoreactive sites (3,4- and 4',5'-) can be aligned with the two pyrimidines involved in the intercalation with the consequent possible formation of bifunctional (inter-strand cross-linkages) adducts [3,5,11].

Angelicin however is able to photobind to DNA only to a reduced extent and shows consequently a low photobiological activity [13].

In order to increase its photobinding capacity towards DNA one or more methyl- groups have been introduced in its molecule [14]; this chemical modification has been suggested by the previous observation that the introduction of methyl-groups in to the molecule of psoralen strongly increased its photobinding to DNA and its skin-photosensitizing activity [15,16].

So, various methyl- derivatives of angelicin have been prepared [12,14,17]. Their interactions with DNA both in the ground and in the excited state, as well as their photobiological properties have been studied. Their interactions with proteins have also been investigated.

Complexes in the ground state with DNA

The new methylangelicin derivatives have been able to form a complex with DNA in the dark as evidenced by the spectrophotometric or fluorimetric modifications observed in aqueous solutions of the compounds after addition of DNA, and by equilibrium dialysis experiments. The introduction of one, two or three methyl- groups in the angelicin molecule strongly increases the affinity towards DNA for the dark complex formation. This chemical modification in the molecule of angelicin increases the lipophilic character as shown by the marked decrease of the water solubility of its methyl derivatives; this fact should increase the affinity of these compounds for the

TABLE I. Binding parameters of the complexes, rate constants of the photoreactions with DNA and photobiological activity.

angelicin derivatives	water solubility ($\mu g/ml$)	K_a [a)	$1/n$ [a)	rate constant[6] min^{-1} x 10^2	relative photobiol. activity b) (ang = 100)
angelicin	20	560	0.063	1.1	100
4-methyl-	4	1400	0.076	1.6	105
5-methyl-	4.5	1560	0.071	3.4	274
5'-methyl-	12	1200	0.067	2.1	150
4,5'-dimethyl-	8	1450	0.094	4.0	230
5,5'-dimethyl-	3	1750	0.085	2.7	180
4,5,5'-trimethyl-	5	1200	0.075	0.8	low

a) determined according to Mc Ghee and von Hippel, J.Mol.Biol. 29:489 (1974).
b) in terms of capacity to inhibit DNA synthesis in Ehrlich ascites tumor cells[12].

internal lipophilic part of duplex DNA where intercalation occurs, with a consequent increase of the capacity to form the dark complex.

When in fact in the molecule of 4,5'-dimethylangelicin a polar group is introduced such as a hydroxymethyl- or a methoxymethyl- group, which slightly increases the water solubility, their affinity for DNA is slightly decreased, evidencing the role of the lipophilic character of the ligand for the intercalation [18].

On the other hand the affinity towards DNA is dramatically increased by the introduction into the molecule of a methylangelicin of side chain carrying a cationic head. When in fact an aminomethyl- group or a N,N-dimethylaminoethoxymethyl- group is introduced in the 4'-position of the 4,5'-dimethylangelicin the resulting compounds show a strongly increased capacity to form the intercalated complex with DNA (the association constant of the complexes is increased by one order of magnitude). This fact is due to the involvement of two types of interaction in the complex formation. Other than the intercalation of the tricyclic moiety between two base pairs of DNA also an external ionic binding between the cationic head and a phosphate group of DNA takes place [18,19,20].

The intercalation of angelicins in duplex DNA has been verified by flow dichroism studies. If a small ligand having a chromophore absorbing at longer wavelenghts than those of DNA undergoes intercalation in duplex DNA, a negative dichroism is observed in the correspondence of the absorption of the ligand similar to that peculiar to native DNA at 260 nm. The negative dichroism observed by flow dichroism measurements as well as the strong fluorescence quenching shown by the new angelicins in the presence of DNA are consistent with the intercalation of these compounds between two base pairs of DNA when complexed with the macromolecule.

This complex in the dark on the basis of its relatively low stability is not able to induce marked biological effects; a small

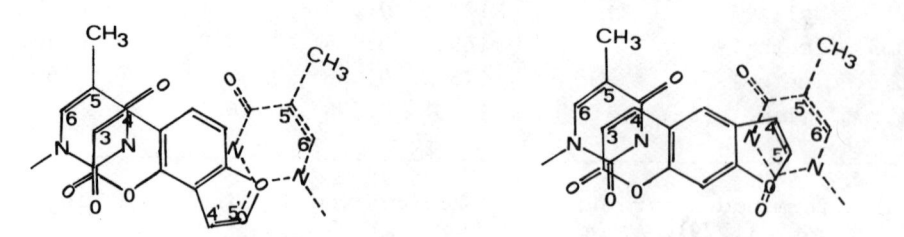

Fig. 2 - Representation of angelicin and psoralen intercalated.

PHOTOCHEMICAL REACTIONS OF FUROCOUMARINS

extent of frame shift mutations have been observed in the presence of relatively high amount of angelicins. Only in the case of the water soluble cationic derivatives of 4,5'-dimethylangelicin, in connection with the dramatically increased stability of their complexes with DNA some biological consequences can be observed [20] (e.g. inhibition of DNA synthesis in cells).

Photobinding capacity towards DNA

The schema of the photoreactions between angelicins and DNA is reported below. The system is rather complicated, however the factor that controls the photobinding to DNA is the amount of angelicin intercalated in duplex DNA; the complexed angelicin is in equilibrium with a certain amount of free angelicin dissolved in water; the latter can undergo photomodification during the irradiation; in this way a decrease of the free angelicin affects the extent of the complexed one. This effect is however very small, especially for the relatively short periods of irradiation because of the very low rate (K_3) of photoconversion of angelicins into their photodimers and photooxidation products.

The photocycloaddition reaction of the angelicins to DNA behaves as a pseudo first order reaction with respect to the complexed furocoumarins. The rate constant values are reported in Table I and II.

The amount of the various angelicins covalently photobound to DNA as a function of the time of irradiation was determined by radiochemical measurements on DNA irradiated in the presence of tritiated angelicins (see Fig.3 and Table I).

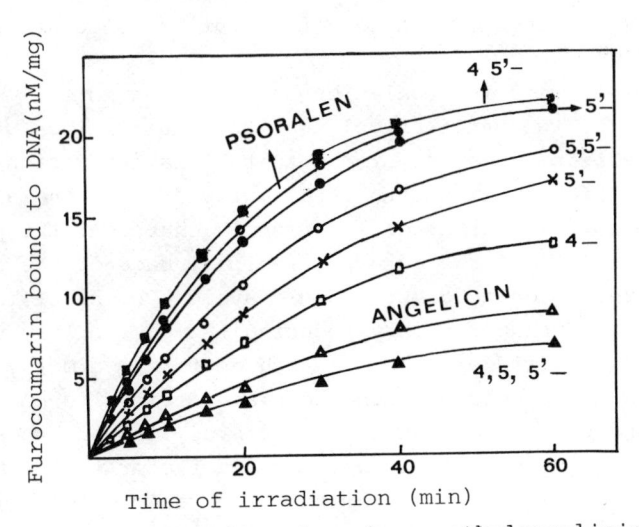

Fig. 3 -Photobinding capacity of variuos methylangelicins towards DNA; only the positions of the methyl-groups are indicated.

Three levels of photobinding capacity to DNA can be observed: the first, in the range of the photoreactivity of psoralen, includes 4,5'-dimethylangelicin and 5-methylangelicin. 5'-methylangelicin, 5,5'-dimethylangelicin and 4-methylangelicin can be included in the second while in the lowest 4,5,5'-trimethylangelicin shows a little lower photoreactivity than that of unsubstituted angelicin.

The increased photobinding to DNA observed in the methylangelicins in respect of the parent compound may be connected with various factors. The first is correlated with their increased affinity for DNA for the dark complex formation, which derives from their increased hydrophobic character. In this connection (see Table II) in the case of 4'-hydroxymethyl- and 4'-methoxymethyl- derivatives of 4,5'-dimethylangelicin, when the affinity for DNA for the complex formation decreases also photobinding to DNA decreases in an almost parallel way. So the increased capacity to form the dark complex in general favours the successive photoaddition in analogy with psoralens [1].

This general tendency however is not respected in the case of 4,5,5'-trimethylangelicin for which to an increased capacity to form the complex with DNA corresponds a reduced photobinding capacity. This behaviour may be explained in terms of geometry of intercalation; in other words a second important factor that affects photobinding to DNA depends on the position assumed by the angelicins when complexed between two base pairs of DNA. The three methyl-groups present in 4,5,5'-trimethylangelicin may play some role in terms of steric hindrance so that the position assumed by the intercalated trimethylangelicin does not favour the successive photocycloaddition.

The geometry of intercalation becomes even more important if we consider the two water soluble cationic derivatives of 4,5'-dimethylangelicin; when the bridge that links the cationic head is short such as for 4'-aminomethyl-derivative, while the affinity for DNA for the complex formation is strongly enhanced, the photobinding is very poor; in fact the external ionic linkage of the cationic head with the phosphate groups should avoid a complete intercalation of the tricyclic planar moiety reducing the possibility of cycloaddition. When the bridge is longer such as for 4'-dimethylaminoethoxymethyl-4,5'-dimethylangelicin the planar tricyclic moiety results more free and can assume, when intercalated, a position suitable for the successive photoaddition. A third factor that can affect photobinding to DNA depends from the electronic effect ($+I_s$) induced in the angelicin molecule by the introduction of the methyl-groups; this effect may be evaluated on the basis of the photophysical properties of the new angelicins; in this connection a lot of work has been done on various furocoumarins; for a review see ref.[4].

SCHEME OF THE PHOTOREACTION

$$\text{angelicin} + \text{DNA} \underset{}{\overset{K_a}{\rightleftarrows}} \text{complex} \underset{K_2}{\overset{K_1}{\underset{h\nu}{\nearrow\searrow}}} \begin{array}{l} 3,4\text{-non fluorescent cycloadducts} \\ 4',5'\text{-fluorescent cycloadducts} \end{array}$$

$$h\nu \downarrow K_3$$

photodimer and photo-oxidation products

Monofunctional photobinding to DNA

It is well known that the formation of bifunctional adducts (inter-strand cross-linkages) in DNA by psoralens induces peculiar properties in the macromolecule: cross-linked DNA, in fact, undergoes spontaneous renaturation after heat denaturation and its molecular weight, evaluated under irreversible denaturing conditions, is twice the weight of non cross-linked DNA. In DNA samples irradiated in the presence of the new angelicins no renaturation capacity and no increase of the MW, evaluated by hydroxylapatite chromatography and by alkaline sedimentation experiments respectively, have been observed[12].

TABLE II. Binding parameters of the complexes and rate constants of photoreactions

4,5'-dimethylangelicin derivatives	K_a a)	$1/n$ a)	rate constant[6] $\min^{-1} \times 10^2$	relative photobiol. activity, b)
parent compound	1400	0.094	4.0	230
4'-hydroxymethyl-	1250	0.087	3.1	270
4'-methoxymethyl-	700	0.050	1.6	110
4'-aminomethyl-	16300	0.255	0.8	25
4'-N,N-dimethylamino-ethoxymethyl-	80000	0.200	4.2	290

a) determined according to Mc Ghee and von Hippel, J.Mol.Biol. 29: 489 (1974).
b) in terms of capacity of inhibition of the DNA synthesis in Ehrlich ascites tumor cells [12].

Type of cycloaddition

In the monofunctional cycloaddition to pyrimidine bases of DNA only one of the two photoreactive sites of angelicins can be involved: therefore two different monoadducts can be formed: non fluorescent 3,4-cycloadducts and fluorescent 4',5'-cycloadducts; the violet fluorescence is referred to the excitation by UV-A light. (See scheme).

The fluorescence of DNA samples irradiated in the presence of the new angelicins has been determined after hydrolysis of the macromolecule; it has been observed that in almost all new angelicins their 4',5'-double bond is involved in the photocycloaddition to the pyrimidine bases of DNA. 4-methylangelicin and 5-methylangelicin show the higher amount of fluorescent adducts in respect of the total angelicin photobound to DNA.

Photobiological properties

The new angelicins did not show any photosensitizing activity on the skin of guinea pigs, even if tested in drastic conditions. This behaviour is to be connected with their inability to form interstrand cross-linkages; in fact, various other furocoumarins, such as 3-carbethoxypsoralen, 3-α,α-dimethylallylpsoralen, 7-methylallopsoralen, 4,7-dimethylallopsoralen or pyranocoumarins (xanthyletin), able to induce only monofunctional lesions to DNA, have been found completely unable to produce skin-photosensitization.

In spite of this fact, the new angelicins have shown a strong capacity to reduce the infectivity of T_2 phage and to inhibit the DNA and RNA synthesis in Ehrlich ascites tumor cells, when irradiated in their presence. Generally a good correlation has been observed between DNA photobinding capacity and photobiological activity (see Table I and II).

A strong inhibition of epidermal DNA synthesis in mice has been obtained after topical application or after oral administration of the new angelicins and irradiation with UV-A[20].

The mutagenicity of the new angelicins assayed in bacteria and chinese hamster cell in culture was generally much lower than that of psoralen derivatives. This results may be due to different enzymatic mechanisms of the DNA repair systems which remove the monoadducts and the inter-strand cross-linkages[20].

These compounds on the basis of their strong photobiological activity, their lacking of skin-photoxicity, their low mutagenic activity and their ability to inhibit DNA synthesis in the skin of mice, appear as possible agents alternative to psoralens for the photochemotherapy of psoriasis and other skin-diseases characterized by hyperproliferation[14].

Interactions of furocoumarins with proteins

In recent times the study of the interactions between furocoumarins and proteins have been investigated with the aim to clarify a possible role of these interactions in the mechanism of action of furocoumarins. In this connection the capacity of various furocoumarins to inactivate various enzymes under UV-A irradiation has been investigated .Poppe and Grossweiner, using 8-MOP observed a marked inactivation of lysozime and interpreted this effect in terms of singlet oxygen production by the furocoumarin[21]. Veronese et al. extending these experiments to various other enzymes evidenced that while lysozime was markedly inactivated other enzymes showed much less sensitive. The inactivation of enzymes,which is a parameter through which is possible to evidence a modification on proteins, is largely dependent from the furocoumarin structure. Among various furocoumarins, psoralen and 8-methylpsoralen appeared the most involved in enzymes inactivation [22].

In this line also covalent photobinding of various furocoumarins to proteins (serum albumin was generally used) was studied. Yoshikawa et al.[23] evidenced a covalent photobinding of 8-MOP to various proteins. Veronese et al. found that this photobinding is a general phenomenon and its extent depends upon the furocoumarin structure. This binding however occurred also by irradiating the furocoumarin alone and then adding it to the protein suggesting that the products of photomodification of furocoumarin are involved in the covalent binding to proteins. This binding was reduced in the presence of nitrogen. A very interesting correlation between the rate of photodegradation of the furocoumarins, the formation of covalent photobinding and the capacity to inactive the enzymes was observed [22].

The non covalent interaction between furocoumarins and proteins was also investigated. This aspect may be of interest in view of the pharmacokinetic properties of these compounds some of which are used systemically for the photochemotherapy of the psoriasis. A significant binding model has been observed using human and bovine serum albumin. The interaction is markedly affected by the protein structure (K_a, among 6 furocoumarins varies between 1.27×10^4 and 1.9×10^5). The binding is characterized by specificity since few binding sites (1.5-2) per protein molecule were found and small perturbances in the protein structure are accompanied by a decrease in the extent of the interaction. In the binding hydrophobic forces seem to be involved as shown by the correlation between lipophilic character of furocoumarins and extent of interaction[24].

REFERENCES

1. F.Dall'Acqua,M.Terboievich, S.Marciani, D.Vedaldi and M.Recher, Investigation on the dark interactions between furocoumarins and DNA,Chem.-Biol. Interactions,21:103(1978).
2. L.Musajo and G.Rodighiero, Mode of photosensitizing action of furocoumarins,in:"Photophysiology", vol.VII,C.Giese ed.,Academic Press,New York,pag.115(1972).
3. F.Dall'Acqua, New chemical aspects of the photoreaction between psoralen and DNA, in:"Research in Photobiology",A.Castellani ed., Plenum Press,New York,pag.245(1977).
4. P.-S.Song and K.J.Tapley Jr.,Photochemistry and photobiology of psoralens, Photochem. Photobiol.,29:1077(1979).
5. F.Dall'Acqua, S.Marciani and G.Rodighiero,Inter-strand cross-linkages occurring in the photoreaction between psoralen and DNA, FEBS letters,9:121(1970).
6. F.Dall'Acqua, S.Marciani, F.Zambon and G.Rodighiero,Kinetic analysis of the photoreaction(365 nm) between psoralen and DNA, Photochem. Photobiol.,29:489(1979).
7. F.Dall'Acqua, D.Vedaldi, F.Bordin and G.Rodighiero,New studies on the interaction between 8-MOP and DNA in vitro,J.Inv.Dermatol., 73:191(1979).
8. R.S.Cole,Inactivation of E.coli,F' episomes at transfer bacteriophage lambda by psoralen plus 360 nm light: significance of DNA cross-links,J.Bacteriol.,107:846(1971).
9. D.Vedaldi, F.Dall'Acqua, S.Caffieri and G.Rodighiero, 3-α,α-dimethylallylpsoralen: a linear furocoumarin forming mainly 4',5'-monofunctional adducts with DNA,Photochem.Photobiol.,29:227(1979).
10. D.Averback, E.Moustacchi and E.Bisagni,Biological effects and repair of damages photoinduced by a derivative of psoralen substituted at the 3,4-reaction site,Biochim.Biophys.Acta,518:464(1978).
11. F.Dall'Acqua, S.Marciani, L.Ciavatta and G.Rodighiero,Formation of inter-strand cross-linkings in the photoreactions between furocoumarins and DNA,Z.Naturforsch.,26b:561(1971).
12. F.Bordin, F.Carlassare, F.Baccichetti, A.Guiotto, P.Rodighiero,D. Vedaldi and F.Dall'Acqua,4,5'—dimethylangelicin:a new DNA-photobinding monofunctional agent,Photochem.Photobiol.,29:1063(1979).
13. F.Bordin, S.Marciani, F.Baccichetti, F.Dall'Acqua and G.Rodighiero,Studies on the photosensitizing properties of angelicin,an angular furocoumarin forming only monofunctional adducts with the pyrimidine bases of DNA,Ital.J.Biochem.,24:258(1975).
14. Italian patent application N.84134 A/79 of the august 20,1979, "Furocumarina per la fotochemioterapia della psoriasi e di altre malattie cutanee ad essa sensibili".

15. G.Caporale and A.M.Bareggi,Su alcuni metilpsoraleni,Gazz.Chim. Ital.(Rome) 98:444(1968).
16. F.Dall'Acqua, S.Marciani, D.Vedaldi and G.Rodighiero,Skin-photosensitization and cross-linking formation in native DNA by furocoumarins,Z.Naturforsch.,29c:635(1974).
17. F.Bordin, F.Carlassare, F.Baccichetti, A.Guiotto and P.Rodighiero,5-methylangelicin:a new highly photosensitizing angular furocoumarin,Experientia,35:1567(1979).
18. F.Dall'Acqua,D.Vedaldi,S.Caffieri,A.Guiotto,P.Rodighiero,F.Baccichetti,F.Carlassare and F.Bordin,New monofunctional reagents towards DNA as possible agents for the photochemotheraphy of the psoriasis:derivatives of 4,5'-dimethylangelicin,(1980), submitted for publication.
19. A.Guiotto,unpublished results.
20. F.Bordin,unpublished results.
21. W.Poppe and L.J.Grossweiner,Photodynamic sensitization by 8-MOP via the singlet oxygen mechanism,Photochem.Photobiol.,22:217 (1975).
22. F.M.Veronese, R.Bevilacqua and O.Schiavon, Interazioni farmaci-proteine.Indagine su furocumarine dotate di attività terapeutica,in Abstract Book of 1st Nat.Meeting Chim.Farm.,Soc.Chim. Ital.,Pisa(Italy),dec.13-15(1979).
23. K.Yoshikawa,N.Mori, S.Sakakibara,N.Mizuno and P.-S.Song,Photoconjugation of 8-MOP with proteins,Photochem.Photobiol.,29: 1127(1979).
24. F.M.Veronese, R.Bevilacqua, O.Schiavon and G.Rodighiero,Drug-protein interactions:plasma protein binding of furocoumarins, Il Farmaco, Ed.Sci., 34:716(1979).

A PHOTOCHEMICAL CHARACTERIZATION OF REACTIONS OF PSORALEN

DERIVATIVES WITH DNA

Stephen T. Isaacs, Corliss Chun, John E. Hyde,
Henry Rapoport and John E. Hearst

Department of Chemistry
University of California, Berkeley
Berkeley, California 94720

INTRODUCTION

Psoralens are nucleic acid photoreagents which form mono- and diadducts to pyrimidine bases upon irradiation with long wavelength ultraviolet light (Musajo et al., 1967). Diaddition results in the formation of interstrand bridges or crosslinks in nucleic acid double helices (Cole, 1970, 1971). Crosslink formation is a multistep process involving initial dark binding or intercalation of the psoralen, photoaddition of the psoralen to a pyrimidine base forming a monoadduct followed by conversion of the monoadduct to crosslink by a second photoaddition to a second pyrimidine base. Each of these steps may be investigated as a function of the type and position of the substituents of a given psoralen compound.

The two most widely used psoralen derivatives are 4,5',8-trimethylpsoralen (II) and 8-methoxypsoralen (III). In an earlier publication, we reported the synthesis and characterization of several new derivatives of 4,5',8-trimethylpsoralen (Isaacs et al., 1977). That study has now been expanded to include psoralen (I), 8-methoxypsoralen (III), derivatives of 8-methoxypsoralen (IV-XIII) and an additional trioxsalen derivative (XV).

The objective of the present study is to determine how substituent type and position effects the ability of a given psoralen to dark bind and photoreact with DNA. To this end the solubilities, DNA binding constants, photoaddition and photobreakdown rate constants and quantum yields for the various psoralen compounds were determined. The results of these experiments allow various structure-activity relationships to be established.

MATERIALS AND METHODS

Synthesis of the Psoralen Derivatives

8-Methoxypsoralen (III) was chloromethylated by modification of a reported procedure (Abouleze et al., 1973) to give 5-chloromethyl-8-methoxypsoralen (IV). Hydrolysis and methanolysis of IV gave the corresponding 5-hydroxymethyl-8-methoxypsoralen (V) and 5-methoxymethyl-8-methoxypsoralen (VI). In an attempt to prepare 5-aminomethyl-8-methoxypsoralen (VIII), the required Gabriel intermediate 5-N-phthalimidomethyl-8-methoxypsoralen (VII) was prepared by heating IV with potassium phthalimide in DMF. However, removal of the phthalyl moiety to give VIII did not occur with either hydrazine hydrate or concentrated HCl. Reaction of IV with pyridine at room temperature gave N-(5-methylene-8-methoxypsoralen)-pyridinium chloride (IX). Oxidation of alcohol V with γ-MnO$_2$ gave the corresponding aldehyde 5-formyl-8-methoxypsoralen (X). 5-Nitro-8-methoxypsoralen (XI) and 5-amino-8-methoxypsoralen (XII) were prepared by reported procedures (Brokke and Christensen, 1958). N-(4'-methylene-4,5',8-trimethylpsoralen)-pyridinium chloride (XV) was prepared by treating 4'-chloromethyl-4,5',8-trimethylpsoralen (XIV, Isaacs et al., 1977) with pyridine at room temperature. The synthetic scheme is shown in Figure 1.

To provide labeled psoralens for the characterization of the compounds, 4,5',8-trimethylpsoralen and psoralen were tritiated via T$_2$O exchange while tritiated 8-methoxypsoralen was prepared from 8-hydroxypsoralen (XIII) and tritiated methyl iodide. Labeled derivatives were prepared from the tritiated parent compounds. The details of this work are reported elsewhere (Isaacs et al., 1977, 1980).

5-Chloromethyl-8-methoxypsoralen (IV). 8-Methoxypsoralen (1655 mg, 7.7 mM) was dissolved in 100 ml glacial acetic acid by heating gently, then allowed to cool. An initial 5 ml portion of chloromethyl methyl ether was added followed by two additional 5 ml portions at 24 and 48 hr. After 72 hr, a precipitate was evident which increased following 8 hr cooling on ice. The crude product was collected by suction filtration and recrystallized from acetone to give pure IV (877 mg, 53%). It was necessary to immerse the

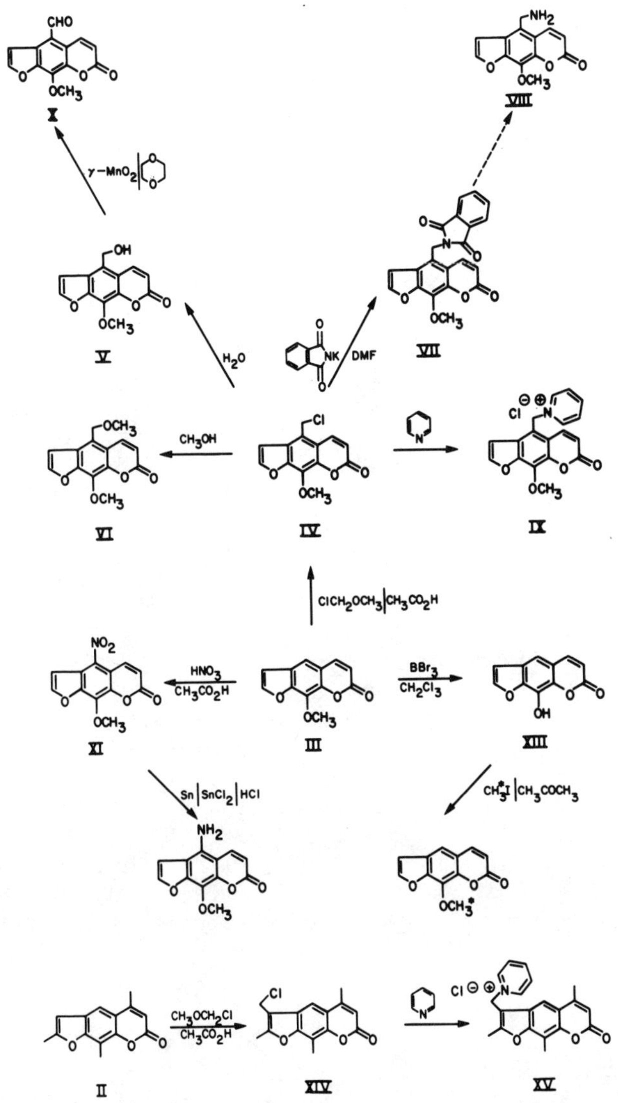

PHOTOCHEMICAL REACTIONS OF PSORALENS WITH DNA

Fig. 1.   Synthetic Paths to the Psoralen Derivatives.

recrystallization flask in ice prior to complete cooling to obtain crystalline compound; prolonged recrystallization at room temperature gave amorphous material. Aliquots of the reaction mixture taken at 24, 48, and 72 hr showed product formation to be (by NMR) 20, 90, and 100% complete. NMR (CDCl$_3$) δ4.4 (3H, s, OCH$_3$), 5.0 (2H, s, CH$_2$Cl), 6.4-6.6, d, C3-H), 7.0-7.1 (1H, d, C4'-H), 7.8-7.9 (1H, d, C5'-H), 8.1-8.3 (1H, d, C4-H); mass spectrum m/e (relative intensity) 230 (100), 264 (M+, 8.9), 266 (M+2, 2.4).

Analysis. Calculated for C$_{13}$H$_9$ClO$_4$: C, 59.0; H, 3.4; Cl, 13.4. Found: C, 58.9; H, 3.6; Cl, 13.3.

5-Hydroxymethyl-8-methoxypsoralen (V). 5-Chloromethyl-8-methoxypsoralen (IV, 118 mg, .45 mM) was refluxed in 20 ml distilled water for 2 hr (heterogeneous), after which no starting material could be detected by TLC (benzene/acetone 1:1). After cooling on ice for 2 hr, the product was collected by suction filtration (95 mg, 87%). Mass spectrum m/e (relative intensity) 246 (M+, 100). NMR (DMSO-d$_6$) δ4.1 (3H, s, OCH$_3$), 4.9 (2H, s, CH$_2$Cl), 6.3-6.5 (1H, d, C3-H), 7.2 (1H, d, C4'-H), 8.1 (1H, d, C5'-H), 8.3-8.5 (1H, d, C4-H).

Analysis. Calculated for C$_{13}$H$_{10}$O$_5$: C, 63.4; H, 4.1. Found: C, 63.2; H, 4.2.

5-Methoxymethyl-8-methoxypsoralen (VI). 5-Chloromethyl-8-methoxypsoralen (IV, 96 mg, .36 mM) was refluxed in 40 ml anhydrous methanol for 1.5 hr after which no starting material was detected by TLC (CHCl$_3$/CH$_3$OH 98:2). Most of the methanol was evaporated under reduced pressure, resulting in a slurry of white crystals. After cooling on ice (60'), the product was filtered off and air dried (82 mg, 87%): m.p. 168-169°C; NMR (CDCl$_3$) δ3.4 (3H, s, CH$_2$OCH$_3$), 4.4 (3H, s, OCH$_3$), 4.9 (2H, s, CH$_2$OCH$_3$), 6.4-6.6 (1H, d, C3-H), 7.0-7.1 (1H, d, C4'-H), 7.8-7.9 (1H, d, C5'-H), 8.2-8.4 (1H, d, C4-H); mass spectrum m/e (relative intensity) 260 (M+, 28).

Analysis. Calculated for C$_{14}$H$_{12}$O$_5$: C, 64.6; H, 4.6. Found: C, 64.4; H, 4.8.

5-N-Phthalimidomethyl-8-methoxypsoralen (VII). 5-Chloromethyl-8-methoxypsoralen (IV, 83 mg, .31 mM) was stirred into 50 ml dry (4A sieves) dimethylformamide at room temperature. Purified (Isaacs et al., 1977) potassium phthalimide (661 mg, 3.6 mM) was added in one portion and the yellow solution immersed in a preheated (100°C) oil bath and magnetically stirred. After 2 hr the reaction was complete as indicated by TLC (ET$_2$O). The solvent was removed under vacuum, giving a yellow paste which was transferred to a separatory funnel with 100 ml chloroform. The chloroform solution was washed with water (50 ml x 4), dried (MgSO$_4$) and evaporated, giving a white, amorphous solid. After drying (6 hr, 100°, 1 mm) the crude product was weighed (995 mg, 86%) and a small portion recrystallized from absolute ethanol for analysis (small white needles): NMR (CDCl$_3$) δ4.4 (3H, s, OCH$_3$), 5.3 (2H, s, C5-CH$_2$), 6.5-6.7 (1H, d, C3-H), 7.4-7.5 (1H, d, C4'-H), 7.8-8.0 (4H, m, phthal; 1H, d, C5'-H), 8.8-9.0 (1H, d, C4-H). Mass spectrum m/e (relative abundance) 375 (M+, 90).

Analysis. Calculated for C$_{21}$H$_{13}$NO$_6$: C, 67.2; H, 3.5; N, 3.7. Found: C, 66.6; H, 3.6; N, 3.7.

N-(5-methylene-8-methoxypsoralen)-pyridinium chloride. 5-Chloromethyl-8-methoxypsoralen (IV, 191 mg, .72 mm) was placed in 15 ml spectral grade pyridine and magnetically stirred at room temperature. Most of the solid dissolved after a few minutes. To effect complete solution, the mix was briefly heated on the hot plate resulting in a clear yellow solution. Almost immediately after clearing a granular appearing precipitate began to form. Stirring at room temperature was continued until no S.M. remained by TLC (CHCl$_3$/CH$_3$OH, 98:2). At 19 hr only a nonfluorescent spot at the origin was found (S.M. R$_f$ ~.6). The solid was filtered off, washed with 10 ml fresh pyridine, filter dried (1 hr) and vacuum dried (6 hr, 100°, 1 mm) to give 233 mg (94%) light tan powder. The product was freely soluble in water and was recrystallized from n-PrOH/H$_2$O. NMR (D$_2$O) δ3.4 (3H, s, OCH$_3$), 5.5 (2H, s, C5-CH$_2$), 5.6-5.7 (1H, d, C3-H), 6.3 (1H, d, C4'-H), 7.1-7.5 (4H, m, C5'-H, pyridine-H's), 7.8-8.2 (3H, m, C4-H, pyridine-H's).

Analysis (crude product): Calculated for C$_{18}$H$_{14}$ClNO$_4$: C, 62.9; H, 4.1; Cl, 10.3; N, 4.1. Found: C, 62.7; H, 4.2; Cl, 10.2, N, 3.9.

5-Formyl-8-methoxypsoralen (X). 5-Hydroxymethyl-8-methoxypsoralen (V, 205 mg, .83 mM), γ-MnO$_2$ (720 mg, .83 mM) and p-dioxane (10 ml) were heated in an oil bath at 70° while being continually stirred. After the third and sixth hour, additional 250 mg portions of γ-MnO$_2$ in dioxane (5-10 ml) were added. After 10 hr no S.M. remained by TLC (CHCl$_3$/CH$_3$OH, 98:2). The MnO$_2$ was removed by suction filtration, washed with chloroform (5 x 25 ml) and the solvents removed in vacuo. The residue was placed on an 8" silica gel column (60-200 mesh) and eluted with chloroform, the fractions containing product combined, dried (MgSO$_4$) and the solvent removed giving the crude aldehyde, which was recrystallized from absolute ethanol (71 mg, 35%). NMR (CDCl$_3$) δ4.5 (3H, s, OCH$_3$), 6.5-6.7 (1H, s, C3-H), 7.5 (1H, d, C4'-H), 7.8 (1H, d, C5'-H), 8.9-9.2 (1H, d, C4-H), 10.7 (1H, s, CHO).

Analysis. Calculated for C$_{13}$H$_8$O$_5$: C, 63.9; H, 3.3. Found: C, 63.8; H, 3.4.

5-Nitro-8-methoxypsoralen (XI). 8-Methoxypsoralen (III, 1490 mg, 6.9 mM) was dissolved in glacial acetic acid (18 ml) by warming, then cooled to room temperature. Concentrated nitric acid (10 ml) was slowly added while maintaining the reaction temperature at 25°. The mixture initially turned red, then began to solidify. After stirring 30 minutes, the reaction mixture was poured onto 32 g of ice. After the ice had melted, the amorphous yellow product was collected by suction filtration and sucked dry (1720 mg, 96%). The crude product was recrystallized from absolute ethanol giving thin, bright yellow needles: m.p. 235-238°C; NMR (CDCl$_3$) δ4.6 (3H, s, OCH$_3$), 6.6-6.8 (1H, d, C3-H), 7.4-7.5 (1H, d, C4'-H), 7.9-8.0 (1H, d, C5'-H), 8.7-8.9 (1H, d, C4-H). Mass spectrum m/e (relative abundance) 261 (M+, 21).

Analysis. Calculated for $C_{12}H_7NO_6$: C, 55.2; H, 2.7; N, 5.4. Found: C, 55.1; H, 2.8; N, 5.3.

5-Amino-8-methoxypsoralen (XII). 5-Nitro-8-methoxypsoralen (XI, 1.0 g, 3.8 mM) stannous chloride (2.0 g, 10.5 mM) and granular tin (2.0 g, 16.8 mM) were stirred at room temperature (cooling as required) in a mixture of concentrated HCl (12 ml)/ethanol (4 ml) for 24 hr. The material was then filtered, washed with sodium bicarbonate solution, and recrystallized from absolute ethanol to give thick yellow needles (465 mg, 53%), m.p. 244-245°; NMR (DMSO-d$_6$) δ4.9 (3H, s, OCH$_3$), 6.1-6.3 (1H, d, C3-H), 7.3-7.4 (1H, d, C4'-H), 7.9-8.0 (1H, d, C5'-H), 8.3-8.5 (1H, d, C4-H).

Analysis. Calculated for $C_{12}H_9NO_4$: C, 62.3; H, 3.9; N, 6.1. Found: C, 62.4; H, 4.0; N, 6.0.

N-(4'-methylene-4,5',8-trimethylpsoralen)pyridinium chloride. 4'-Chloromethyl-4,5',8-trimethylpsoralen (XIV, 120 mg, .43 mM) was placed in 10 ml pyridine (spectral grade) and stirred at room temperature, dissolving after several minutes to give a yellow solution. After 15-20 min a white granular solid had begun to precipitate from solution. TLC of the mixture (CHCl$_3$/CH$_3$OH, 98:2) showed the high R$_f$ S.M. plus a nonfluorescent origin spot (product). After 9 hr the amount of precipitate had increased. However, TLC showed starting material still to be present. Stirring was continued and TLC redone at 24 hr, showing a single spot at the origin. The mixture was filtered and the pinkish-white solid sucked dry on the Buchner funnel, followed by vacuum drying (2 hr, 100°, 1 mm) to give the crude product (132 mg, 86%). The product was freely soluble in water and was recrystallized from n-PrOH/H$_2$O. Additional material in the filtrate was not recovered. NMR (D$_2$O) δ1.7, 2.2, 2.7 (9H, 3s, C 4,5',8-methyls), 5.7 (1H, s, C3-H), 6.0 (2H, s, C4'-CH), 7.2 (1H, s, C5-H), 8.1-9.5 (5H, m, pyridine ring H's).

Analysis (crude product). Calculated for $C_{20}H_{18}ClNO_3$: C, 67.5; H, 5.1; Cl, 10.0; N, 3.9. Found: C, 67.2; H, 5.2; Cl, 10.0; N, 4.0.

Dark (Noncovalent) Binding of the Psoralen Compounds to DNA. Noncovalent binding of psoralen derivatives to DNA was determined by equilibrium dialysis. The dialysis system consisted of 50 µg/ml calf thymus DNA (Worthington), 10 mM Tris (pH 8.0), 1 mM EDTA in 10 mm dialysis tubing (Spectrapor 2) dialysing against 10 ml of the buffer. An equal concentration of ^{3}H-labeled psoralen was placed inside and outside the dialysis membrane, such that total drug was present at one drug to fifteen base pairs. After allowing the system to equilibrate at room temperature for 24-48 hr, the equilibrium concentrations of bound and free derivative were determined by the counts found inside and outside the membrane. From the optical density of the DNA solution and the specific activity of each psoralen compound, the binding constant of each drug was determined.

Determination of Quanta Absorbed by the Photoreaction System. Irradiation of all psoralen solutions was done in polypropylene Eppendorf tubes placed within a jacket of 38% cobaltous nitrate solution which served as a filter selecting for maximum transmittance at 365 nm from the two mercury vapor lamps (General Electric 400W) on either side of the sample chamber. Constant temperature of 21° was maintained by regulating the temperature of the cobaltous nitrate solution.

The quantity of long wave UV light absorbed by the reaction system of psoralen in buffer solution was determined by uranyl-oxalate actinometry (Noyes, 1948). The concentration of the actinometric solution was adjusted to yield the optical density of the standard 5.05×10^{-6} M psoralen solution at 365 nm. Thus by irradiation of the uranyl-oxalate solution under indentical conditions of photoreaction as the psoralen solutions, and by the known quantum yield of the well-studied actinometric reaction, the quanta absorbed by the psoralen reaction system could be determined. Since the presence or absence of DNA in solution with psoralen did not affect the optical density of the reaction system at 365 nm, it was assumed that free psoralen and DNA-bound psoralen do not significantly differ in quanta absorbed per molecule second. No adjustment was made for the differences in extinction coefficients between derivatives so the quantum yields are approximate.

Determination of Photoaddition Rates. A 1 ml solution containing 50 µg/ml calf thymus DNA in 10 mM Tris (pH 8.0), 1 mM EDTA (TE buffer) was allowed to equilibrate overnight at room temperature with the ^{3}H-labeled psoralen derivative present at the ratio of one drug per 15 base pairs of DNA. Timed aliquots were taken from the irradiated sample and noncovalently bound drug was removed from the DNA by three ethanol precipitations of the DNA using 0.25 M NaCl and 2 volumes of absolute ethanol. Optical density at 260 nm of the resuspended pellet was used to determine the DNA concentration, which along with the remaining counts yielded the covalent adducts per base pair.

Determination of Photobreakdown Rates. Timed aliquots were taken from a room temperature irradiation of a solution containing ^{3}H-labeled drug in TE buffer. Volumetric amounts were spotted on silica gel thin layer chromatography plates (Eastman "Chromagram" #13181) with fluorescent indicator. After development in the appropriate solvent the counts migrating as the nonirradiated drug standard were monitored and the first order decay behavior was plotted.

The solvent system giving best resolution of the charged derivatives PMT, AMT, PMX consisted of 25% 1 M ammonium acetate (pH 6.0), 75% ethanol (95%). For TMP and HMT chloroform was used, while all other derivatives could be developed in a solvent of 50% benzene, 50% acetone.

RESULTS

A tentative mechanism for the reaction of the psoralens with DNA has previously been presented (Isaacs et al., 1977). The fundamental steps involved are as follows:

$$P + S \underset{k_{-1}}{\overset{k_1}{\rightleftarrows}} PS \tag{1}$$

$$PS + h\nu \overset{k_2}{\rightarrow} A \tag{2}$$

$$P + h\nu \overset{k_3}{\rightarrow} B \tag{3}$$

In this model P is the psoralen compound in question, S is a psoralen binding site in the DNA, PS is the noncovalent intercalation complex between psoralen and the DNA, A refers to covalent adduct of the psoralen to the DNA, and B is photobreakdown of the psoralen.

Noncovalent Binding of the Psoralen Derivatives to DNA. The dissociation constant for the noncovalent binding of each psoralen compound to DNA is defined by the expression

$$K_D = \frac{k_{-1}}{k_1} = \frac{[P][S]}{[PS]}$$

where [P] is the concentration of free psoralen, [S] is the concentration of unoccupied binding sites where each base pair is taken to be a binding site, and [PS] is the concentration of bound sites.

By determining the dissociation constant of each psoralen compound at a constant ratio of one psoralen for every fifteen base pairs, the effects of the various substituents on the dark binding step of the mechanism are reflected by the difference in the binding constants of the compounds. At this psoralen-base pair ratio, the equilibrium measurement is made far enough from saturation of binding sites that site exclusion is not a factor.

Table I shows the extinction coefficients (H_2O), solubilities, dissociation constants and ratios of concentrations of occupied to unoccupied binding sites in saturated solutions for the psoralen compounds. Inspection of Table I reveals the following trends in the dark binding of the various psoralens to DNA. Both 4,5',8-trimethylpsoralen (TMP) and 8-methoxypsoralen (8-MOP) exhibit enhanced binding relative to unsubstituted psoralen. Further substitution of TMP and 8-MOP at the 4' and 5 positions respectively with hydroxymethyl and methoxymethyl moieties decreases dark binding with increasing substituent size. 4'-Methoxymethyl-4,5',8-trimethylpsoralen (MMT) and 4'-hydroxymethyl-4,5',8-trimethylpsoralen (HMT) have dissociation constants about four times and two times greater than TMP, while 5-methoxymethyl-8-methoxypsoralen (MMX) and 5-hydroxymethyl-8-methoxypsoralen (HMX) have dissociation constants eight times and four times greater than 8-MOP. The equilibrium binding of psoralen with DNA involves the intercalation of the planar photoreagent in the DNA helix. Presumably, steric constrains due to these 4' and 5 substituents reduce the ability of these derivatives of TMP and 8-MOP to intercalate. However, if positively charged substituents are placed at these same positions in TMP and 8-MOP, an enhancement in the dark binding occurs. A size effect is again apparent with the larger substituents decreasing the degree of binding enhancement observed. Thus while 4'-aminomethyl-4,5',8-trimethylpsoralen (AMT) has a dissociation constant 20 times smaller than does TMP, charged N-(4'-methylene-4,5',8-trimethylpsoralen)-pyridinium chloride (PMT) with the bulkier pyridinium methyl substituent has a dissociation constant only five times smaller than TMP.

By dividing the molar solubility of a given psoralen compound by its dissociation constant, the ratio of bound to free sites at equilibrium in a solution saturated with that psoralen is obtained. As shown in Table I, site saturation can only be achieved with those psoralens which combine high aqueous solubility with low DNA dissociation constants. Only the positively charged derivatives AMT, PMT and PMX exhibit these properties.

The compounds 5-nitro-8-methoxypsoralen and 5-amino-8-methoxypsoralen showed no detectable binding to DNA.

<u>Photoaddition of Trimethylpsoralens and 8-Methoxypsoralens to DNA.</u> A time course of covalent photoaddition for each derivative is shown in Figure 2. Although the derivatives display a wide range of final levels of photoaddition and initial kinetic behavior, each curve is consistent with the proposed model of psoralen reaction. In the initial stage of the reaction, increase in covalent adduct

Table I.

Compound Name	Abbrev.	Molec. Wt.	Extinction Coefficient $\varepsilon 250nm(\ell/mol^{-1}cm^{-1})$	Solubilities μg/ml	Solubilities mol/l	K_D DNA mol/l	[PS]/[S] Saturated Solution
4,5',8-Trimethylpsoralen (Trioxsalen)	TMP	228	1.5×10^4	0.6	2.6×10^{-6}	1.3×10^{-4}	.05
4'Aminomethyltrioxsalen	AMT	257	2.5×10^4	>10^4	3.4×10^{-2}	6.6×10^{-6}	5000
4'Hydroxymethyltrioxsalen	HMT	258	2.1×10^4	41	1.6×10^{-4}	2.9×10^{-4}	.55
4'Methoxymethyltrioxsalen	MMT	272	2.5×10^4	10	3.7×10^{-5}	5.4×10^{-4}	.39
N-(4'methylenetrioxsalen)-pyridinium chloride	PMT	356	2.5×10^4	>10^4	2.8×10^{-2}	2.6×10^{-5}	1080
Psoralen	PSOR	186		35	1.9×10^{-4}	4.0×10^{-3}	.05
Xanthotoxin (8-Methoxypsoralen)	8-MOP	216	2.1×10^4	38	1.8×10^{-4}	1.3×10^{-3}	.14
5-Hydroxymethylxanthotoxin	HMX	246	1.8×10^4	46	1.8×10^{-4}	5.0×10^{-3}	.04
5-Methoxymethylxanthotoxin	MMX	260	1.7×10^4	55	2.1×10^{-4}	1.0×10^{-2}	.02
N-(5-methylene-8-methoxy-psoralen)-pyridinium chloride	PMX	344	2.3×10^4	>10^4	2.9×10^{-2}	5.2×10^{-5}	560

is linear with time, when the concentration of equilibrium bound complex has not significantly changed:

$$\frac{d[A]}{dt} = k_2[PS] \quad \text{where} \quad [PS] \approx [PS]_0 .$$

During the progress of irradiation the rate of covalent adduct formation must decrease due to depletion by not only the addition but also the breakdown pathway until no further photoaddition is possible. At this point the curve is seen to remain at the final plateau level.

Relative initial rates of photoaddition to DNA distinguish the trimethylpsoralens from the 8-methoxypsoralens with the latter producing lower initial rates. While persistence of the linear addition region extends through 20 minutes of irradiation for the 8-methoxypsoralens, it was necessary to take time points from 0.25 to 2 minutes to observe the trimethylpsoralen addition before the plateau effect dominated.. Observations of the initial rates of photoaddition of TMP, and AMT, are consistent with that reported in Hyde and Hearst (1978), in that trimethylpsoralen reveals the greatest rate of covalent adduct formation and reaches the point of plateau first, while AMT displays a lower initial rate of addition but surpasses TMP in the final level of the plateau. (It should be noted that the TMP rate observed is that for concentrations of derivative and DNA one-half that of the others due to the limited solubility of TMP.)

Efficiency of Photoaddition to DNA. That the positively charged derivatives capable of the highest levels of equilibrium binding do not produce the highest rates of photoaddition to DNA is evidence of their differential abilities of forming the covalent adduct in the bound conformation. The relative efficiencies at which the bound conformation is converted to the covalent adduct can be expressed by the quantum yield for the photoaddition reaction:

$$\frac{(\frac{d[A]}{dt})_0}{[PS]_0 \gamma} = \phi_2 ,$$

derived by taking the observed initial rate of photoaddition, normalizing by the initial amount of equilibrium bound derivative and dividing by the determined number of quanta absorbed per second by each psoralen molecule. Under conditions of irradiation used, actinometry yielded $\gamma = .43$ quanta absorbed per second per molecule.

The three greatest quantum yields for covalent photoaddition of equilibrium bound complex to DNA are demonstrated, interestingly, by the simplest molecules (Table II, column 1). They occur in the order of trimethylpsoralen, the unmodified scaffold molecule psoralen,

Table II.

	ϕ_2	ϕ_3	ϕ_2/ϕ_3	Equilibrium Bound per 1000 b.p.	Plateau Bound per 1000 b.p.	Plateau of Photoaddition Equilibrium Bound
TMP	8.4×10^{-2}	4.7×10^{-2}	1.8	15	28	1.9
AMT	2.2×10^{-2}	2.2×10^{-2}	1.0	61	67	1.1
HMT	1.6×10^{-2}	4.9×10^{-3}	3.2	14	38	2.7
MMT	4.7×10^{-3}	1.6×10^{-3}	2.9	13	23	1.8
PMT	1.8×10^{-3}	3.1×10^{-4}	5.8	49	55	1.1
PSOR	3.0×10^{-2}	8.3×10^{-3}	3.6	1.2	9	7.5
8-MOP	1.3×10^{-2}	5.5×10^{-4}	24	3.7	(42)	(11.4)
HMX	1.4×10^{-3}	1.4×10^{-4}	10	1.0	(4)	(4.0)
MMX	1.2×10^{-3}	8.9×10^{-4}	1.3	0.5	(2)	(4.0)
PMX	6.2×10^{-4}	8.3×10^{-4}	0.74	39	(67)	(1.7)

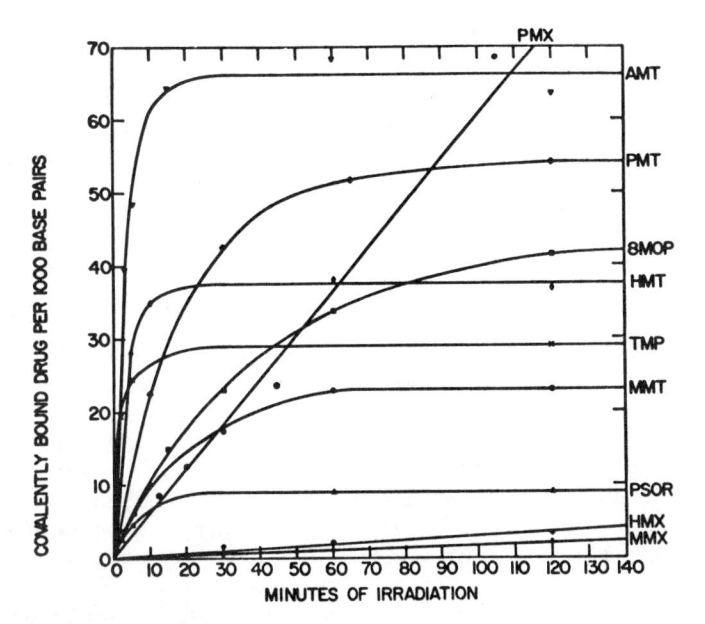

Fig. 2. Kinetics of photoaddition of trimethylpsoralens and 8-methoxypsoralens to DNA.
An initial psoralen to base pair ratio of 1:15 was used in all irradiations. Because of limited solubility, the trimethylpsoralen photoaddition was done with 0.5 OD DNA, while that of all other derivatives was done with 1.0 OD DNA.

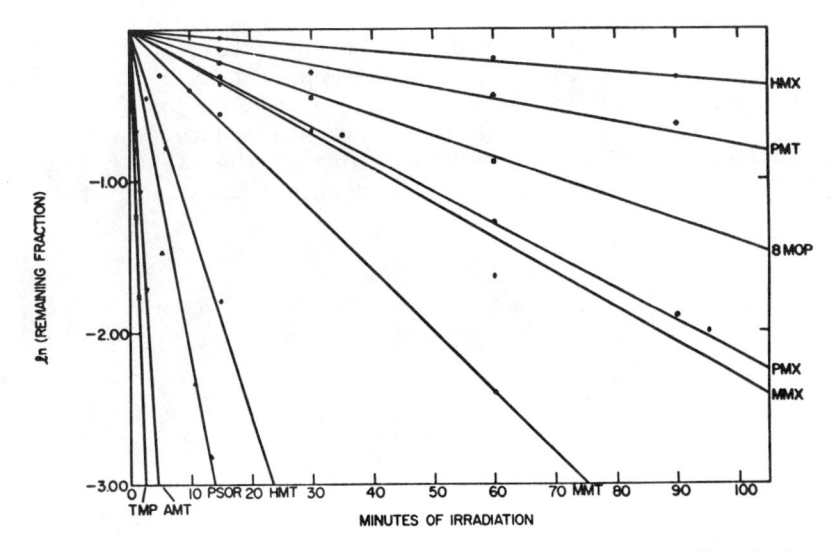

Fig. 3. First order kinetics of photobreakdown of trimethylpsoralens and 8-methoxypsoralens in aqueous solution.

and 8-methoxypsoralen. Relative to unmodified psoralen, the methoxy group at the 8-position detracts from efficiency of covalent addition, while the three methyl substitutions at the 4, 5, 8 positions actually enhance the photoaddition. Further modification upon either trimethylpsoralen or 8-methoxypsoralen framework decreases the efficiency of coversion to covalent adduct. In order of decreasing quantum yield of photoaddition, the substituent groups on trimethylpsoralen appear to be aminomethyl, hydroxymethyl, methoxymethyl, and pyridinium methyl. In the case of 8-methoxypsoralen, the order is consistent: hydroxymethyl, followed by methoxymethyl, then pyridinium methyl. Since this order reflects the size of the substituent groups, the trend suggests steric hindrance of bulky structures in the formation of covalent adduct from the equilibrium-formed complex, of either 4'-substituted trimethylpsoralens or 5-substituted 8-methoxypsoralens. Thus, although the charged pyridinium methyl group dramatically increased the equilibrium bound psoralen, the bulky substituent group may prevent efficient photoaddition by skewing the complex out of the most favorable configuration for photoreaction to form the covalent adduct.

Efficiency of Photobreakdown in Solution. Kinetic behavior beyond the linear phase of initial photoaddition also distinguishes the two groups of derivatives. Within one hour all the trimethylpsoralens can be seen to have reached a saturation or plateau level (Figure 2), as has the parent compound psoralen. In contrast, even at 120 minutes of irradiation, none of the 8-methoxypsoralen compounds has reached such a plateau level, indicating that a surviving pool of unreacted derivative is available for further bound complex formation and photoaddition.

The stability during irradiation of 8-methoxypsoralens relative to the trimethylpsoralens is readily demonstrated in the first order plot of photodecay of the free derivative in solution (Figure 3). Destruction of the ^{3}H-labeled derivative was monitored as stated in Materials and Methods. The extreme slope of the first order plot in the case of the trimethylpsoralen derivatives illustrates greater efficiency of photoreaction:

$$-\frac{d[P]}{dt} = k_3 [P]$$

$$\ln \frac{[P]}{[P]_0} = -k_3 t$$

$$-\frac{d(\ln \frac{[P]}{[P]_0})}{dt} = k_3$$

$$\frac{-d(\ln\frac{[P]}{[P]_0})/dt}{\gamma} = \phi_3$$

Efficiency of Photobreakdown in Solution. Again trimethylpsoralen and its aminomethyl derivative produce the two greatest rates of reaction (Table II, column 2). In this case, however, the reaction rate directly reflects quantum yield. (In fact, normalization of these observed rates by the very similar absorbtion coefficients of all the derivatives does not alter the order observed in the photodestruction rates.)

Unlike the quantum yield for photoaddition of the equilibrium bound complex, the quantum yield of photodestruction of free derivative in solution does not show strong correlation to substituent group size. However, the framework of trimethylpsoralen provides the more photoreactive structure for its derivatives, with the exception of pyridinium methyl, than the general 8-methoxypsoralen structure.

Effect of Greater Efficiency of Photoaddition than of Photobreakdown. It can be seen from the steps of the proposed model that the relationship of ϕ_2 to ϕ_3 determines whether the final level of covalent addition of derivative is greater than or less than the initial equilibrium bound level. Greater ϕ_2 would favor replenishment of equilibrium bound drug for further covalent addition, whereas greater ϕ_3 would favor emptying of previously complexed drug out into the pool of nonreacted molecules.

In general the derivatives display greater ϕ_2 than ϕ_3 and therefore covalent addition in excess of the equilibrium bound level is expected. Indeed for the more photoefficient trimethylpsoralens, and unmodified psoralen, all have, by two hours, reached plateau levels in excess of the initial equilibrium levels. For the more photostable 8-methoxypsoralen derivatives, by two hours each has surpassed its equilibrium level of binding.

It should be noted that in the case of a weaker equilibrium binder such as 8-methoxypsoralen or unmodified psoralen, the observed ratio of final covalent addition level to initial level bound by equilibrium may be dramatically greater than the ratio observed for a strong binder such as the pyridinium methyl derivative. This reflects not just a difference in the ϕ_2/ϕ_3 ratio value but also the availability of a larger pool of unreacted drug in the case of the weaker equilibrium binder.

Although PMX, like the other derivatives, displays a greater covalent addition level than the initial dark-bound level, its ϕ_2/ϕ_3 ratio is unexpectedly low. The cause for this inconsistency remains unclear at this time.

Table III. Relative quantum yield for photoaddition of the trimethylpsoralens and 8-methoxypsoralens to DNA.

x =	-H	-CH$_2$NH$_2$	-CH$_2$OH	-CH$_2$OCH$_3$	-CH$_2$-N⟨⎯⟩
4'-x-TMP	1.0	0.26	0.19	0.05	0.02
5-x-8 MOP	1.0	--	0.11	0.09	0.05

DISCUSSION

Among the trimethylpsoralen and 8-methoxypsoralen derivatives presented here were examples of extreme photosensitivity and photostability. With a procedure such as that outlined above, however, it becomes possible to relate the expected photoaddition result to the initial equilibrium of the derivative with DNA. The equilibrium constant, the quantum yield of photoaddition, and the quantum yield of photodestruction can be used as parameters by which future psoralen structures may be screened for desired kinetic behavior.

REFERENCES

Aboulezz, A. F., El-Attar, A. A. and El-Sockary, M. A., 1973, Acta Chimica Academiae Scientiarum Hungaricae, Tomus, 77(2):205-10.
Brokke, M. E. and Christensen, B. E., 1958, J. Org. Chem., 23:589.
Cole, R. S., 1970, Biochim. Biophys. Acta, 217:30-39.
Hyde, J. E. and Hearst, J. E., 1978, Biochemistry, 17:1251-57.
Isaacs, S. T., Shen, C.-K. J., Hearst, J. E., and Rapoport, H., 1977, Biochemistry, 16:1058-64.
Isaacs, S. T., Hearst, J. E., and Rapoport, H., 1980, in preparation.
Musajo, L., Bordin, F., Caporale, G., Marciani, S., and Rigatti, G., 1967, Photochem. Photobiol., 6:711-19.
Noyes, W. A., Jr., and Boekelheide, V., 1948, Photochemical Reactions, in: "Technique of Organic Chemistry, Vol. II", Weissberger, A., ed.

PHOTOBIOLOGY OF FUROCOUMARINS

D. Averbeck

Institut Curie, Section de Biologie,
26, Rue d'Ulm, 75231 Paris Cedex 05, France

INTRODUCTION

The photobiology of fucoroumarins has been covered by a number of reviews (1-4). In recent years we observed a rapid evolution in the knowledge on naturally occurring or synthetic furocoumarins which in the presence of near ultraviolet light or UVA, i.e. radiation in the range of 320 to 380 nm, exhibit photosensitizing properties in various biological systems (1-4). The reaction scheme for furocoumarins with nucleic acids, especially with DNA, is well established (1, 4, 5, 6). It involves the formation of complexes between furocoumarins and DNA in the dark, and in the presence of 365-nm radiation the induction of cyclobutane type of mono-additions of pyrimidine bases in DNA involving the 3,4 or the 4',5' double bond of the furocoumarin molecule. Some of the 4',5' cycloadditions may photoreact further forming interstrand DNA cross-links (1, 4-7).

Some aspects concerning the kinetics of the photoreaction (8), the involvement of singlet and triplet excited states (4, 9-12) and reactions of furocoumarins with RNA (4), amino acids (9) and proteins (13, 14), and the production of singlet oxygen (4, 15), were studied in some detail. In parallel new photoreactive furocoumarins were developed for the use in molecular biology as proves for the structure of DNA, RNA and chromatin by Hearst and co-workers (see Ref. 4). In recent photobiological studies research work was initiated with emphasis on the response of pro- and eucaryotic cells to treatments with furocoumarins and 365-nm radiation. The aim was to elucidate further the repair mechanisms in bacteriophage (16-18), in bacteria (19-21), in yeast (22-30) and in mammalian cells (31-33); the induction of mutations in bacteria (20, 21), in yeast

(25-27, 30, 34) and in mammalian cells (35, 36); and the induction of recombination in bacteria (37) and in yeast (25) after photosensitization by furocoumarins. Interesting investigations were also made on the metabolism of therapeutically active furocoumarins (38-40), on the genetic (21, 25, 34-36, 41-45) and carcinogenic risk (46-49) of furocoumarins, and on the use of a mono-functional furocoumarin in photochemotherapy (47, 50).

The present report describes some of the new research trends indicated by recent work on the genetic effects of mono and bi-functional furocoumarins using the genetically well defined eucaryotic unicellular system of the yeast Saccharomyces cerevisiae.

MATERIALS AND METHODS

We employed the materials and methods as described previously (34, 51). Mono- and bi-functional furocoumarins were always used in combination with 365-nm radiation (dose rate 0.96 $kJm^{-2}min^{-1}$, unless specified). The term "dose" is the same as that defined by the first Int. Congress of Photobiology in 1954 (see Ref. 52).

RESULTS AND DISCUSSION

Evidence for Differential Effects of Mono- and Bi-functional Furocoumarins

A reasonably good correlation has been observed between the capability to photo-induce DNA cross-links and the photobiological activity of furocoumarins (1, 5, 53, 54). However, certain furocoumarins exhibit a high photoreactivity with DNA and photosensitizing activity in spite of a low cross-linking capacity (2, 5) suggesting that also the induced mono-additions may contribute to the photobiological effects.

In bacteria, the repair of furocoumarin induced lesions is well documented (53, 59, 60). In order to determine the respective roles of the two types of lesions induced by furocoumarins in eucaryotic cells we used mono-functional furocoumarins, such as 3-Carbethoxypsoralen (3-CPs, Ref. 22) and angelicin (55, 56), unable to form cross-links, and bi-functional furocoumarins, such as psoralen and 8-methoxypsoralen (8-MOP) (1, 5, 55), able to induce mono- and bi-additions in DNA. Although in certain conditions angelicin may form a small amount of cross-links (57, 58) it may be still considered as a mono-functional compound. In the eucaryotic system of yeast, it was demonstrated that the interaction of different repair pathways differed for bi-functional (61) and monofunctional furocoumarins (22). Recent studies using mutants of yeast deficient in recombinational pathways (28) and newly isolated pso mutants (sensitive to the photo-addition of psoralens) (29) affecting also error prone repair (30) demonstrate that the repair of furocoumarin induced DNA cross-links depends at least on three distinct repair pathways: excision-resynthesis, error prone pathway and recombination pathway. Furthermore, it was shown in yeast that the repair of furocoumarin induced lesions depends on the growth phases (62). Generally, in yeast, lesions induced by bi-functional furocoumarins are less efficiently repaired than lesions induced by monofunctional furocoumarins (22, 25, 26, 63). This result appears to be in accord with recent findings in cultured mammalian cells (31, 33).

In view of the different repair capacities for mono- and bi-additions the study of genetic effects of mono- and bi-functional furocoumarins was of a particular interest. Bi-functional furocoumarins were shown to be mutagenic in several pro- and eucaryotic systems (3, 4). After treatments with different mono- and bi-functional furocoumarins we measured the induction of nuclear genetic events in haploid and diploid yeast (25, 34, 63). Per unit dose of 365-nm radiation bi-functional furocoumarins were always more effective than the mono-functional furocoumarins (25, 34, 63). This was also true when the results were expressed as a function of survival. The results are in accord with the idea that the repair of lesions photo-induced by bi-functional furocoumarins is more error prone than the photo-addition by mono-functional furocoumarins (34). The results obtained in yeast differ from those obtained in bacteria (20) where mono-additions are suspected to be responsible for the induction of the major fraction of mutations (20). Per viable cell, the induction of mitotic crossing-over in diploid yeast was more efficient after treatments with bi-functional than after treatments with mono-functional furocoumarins. No such differences appear to exist for the induction of gene conversion (25).

It can be noted that the potent mutagenic activity of bi-functional furocoumarins seems to correlate with a clear carcinogenic activity (in mice) (46, 47, 64), whereas mono-functional furocoumarins did not produce any detectable carcinogenic effects [47]. This relation holds for the induction of mutations as well as for crossing-over, but not for gene conversion. Thus, the idea may be put forward that carcinogenicity is more related to the induction of mutations and inter-genic mitotic recombination than to intra-genic mitotic recombination.

Mono- and bi-functional furocoumarins may be characterized by their respective activity on the induction of cytoplasmic "petite" mutations (i.e. respiratory deficient mutants, rho$^-$) in yeast indicating the induction of mitochondrial damage (including DNA). At comparable survival levels mono-functional furocoumarins were always much more efficient inducers of rho$^-$ mutants than bi-functional compounds (63, 22, 25, 51). The high efficiency for the rho$^-$ induction observed for mono-functional furocoumarins is likely to be

TABLE 1. PHOTOBIOLOGICAL PROPERTIES OF 8-METHOXYPSORALEN AND 3-CARBETHOXYPSORALEN

	8-METHOXYPSORALEN (BI-FUNCTIONAL)	3-CARBETHOXYPSORALEN (MONO-FUNCTIONAL)
PHOTOADDITION TO NUCLEIC ACIDS	+	+
FORMATION OF DNA INTER-STRAND CROSS-LINKS	+	-
INDUCTION OF CELL KILLING	+++	+
INDUCTION OF MUTATIONS	+++	+
INDUCTION OF MITOTIC RECOMBINATION: CROSSING-OVER	+++	+++
GENE CONVERSION	+	+++
INDUCTION OF MITOCHONDRIAL DAMAGE	+++	-
CARCINOGENIC EFFECTS IN MICE	+	-
INDUCTION OF PIGMENTATION	++	-
INDUCTION OF DIRECT PHOTOTOXICITY (SKIN ERYTHEMA)	++	++(TO BE CON-FIRMED)
THERAPEUTIC ACTIVITY ON HUMAN SKIN (PSORIASIS)		

(- ACTIVITY ABSENT, + ACTIVITY PRESENT, ++ MEDIUM ACTIVITY, +++ HIGH ACTIVITY)

due to an efficient repair of mono-additions induced in nuclear DNA (resulting in increased survival) accompanied by a low repair activity for furocoumarin induced lesions in mitochondrial DNA (63, 25). The clear differences observed in the genetic effects of mono- and bi-functional furocoumarins on the nuclear and cytoplasmic level facilitate the characterization of new photoreactive furocoumarins.

The Example of 3-Carbethoxypsoralen (3-CPs), a Mono-Functional Photoreactive Furocoumarin

In 1974, 3-CPs and related compounds were synthesized using a new method (65). Although 3-CPs was found to lack photosensitizing ability on guinea pig skin (2) studies in yeast demonstrated that 3-CPs exhibits an appreciable photoreactivity in respect to the induction of lethal events, cytoplasmic "petite" mutations and nuclear reversion. Table 1 summarizes the photobiological properties of 3-CPs as compared to 8-MOP. Investigations using biochemical and biophysical methods (22) revealed that 3-CPs is capable of inducing photo-additions to nucleic acids, however, no cross-linking capacity could be detected (this finding was recently confirmed, Ref. 68, 69).

In comparison to the photobiological activity of the bi-functional furocoumarin 8-MOP the activity of 3-CPs was characterized as that of a mono-functional compound showing relatively small effects on cell killing and a reduced effect on the induction of nuclear mutations and mitotic crossing-over, but a high efficiency on the induction of cytoplasmic "petite" mutations (mitochondrial damage) per viable yeast cell (22, 66, 67, 51, 34, 25).

After it was shown that in contrast to 8-MOP 3-CPs is lacking detectable carcinogenic effects in mice (50, 47), a preliminary clinical study was undertaken on psoriatic patients using topical treatments with 3-CPs (47, 50). The results clearly indicate a therapeutic activity of 3-CPs on psoriatic lesions which appears to be comparable to 8-MOP (when applied topically) (47). No evidence for hyperpigmentation or immediate phototoxic effects (erythema) as seen after treatments with 8-MOP could be demonstrated with 3-CPs. Obviously, hyperpigmentation or formation of erythema are not necessarily associated with therapeutic activity. Since 8-MOP is known to exert carcinogenic effects (46-49) the use of less mutagenic and non carcinogenic mono-functional furocoumarins like 3-CPs represents an advantage.

Mono-Functional Furocoumarins Exhibiting Photobiological Properties Comparable to 3-CPs

Recent work on psoralen derivatives (51) showed that molecules substituted at the 4',5' double bond need generally higher doses of

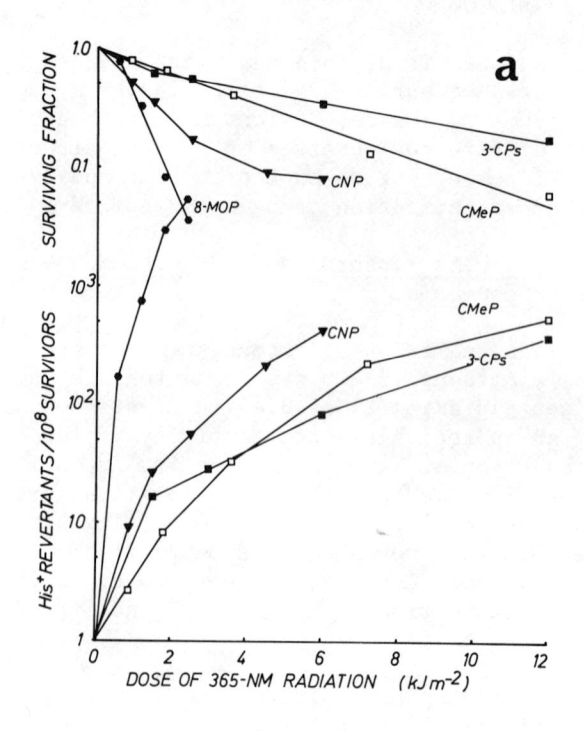

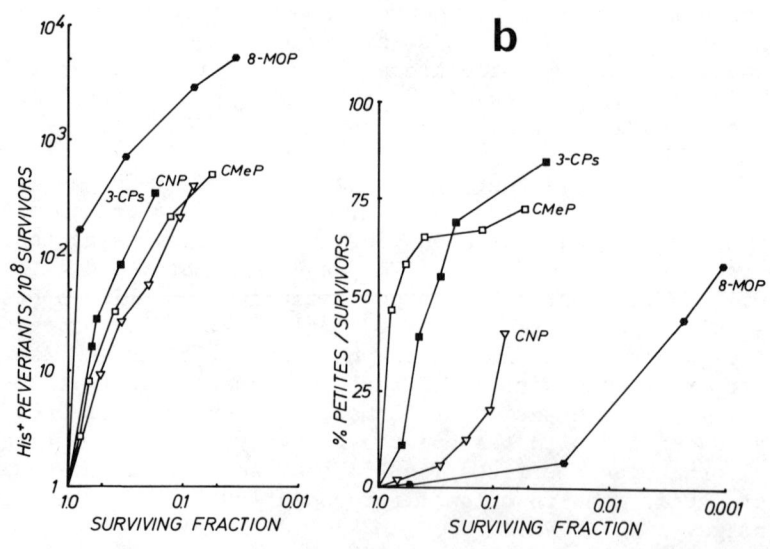

Fig. 1. The induction of lethal effects, nuclear reversions (His⁺) and cytoplasmic "petite" mutations in the haploid yeast strain N123 using two new psoralen derivatives CNP and CMeP, as well as 8-MOP and 3-CPs for comparison. Figure 1a above, shows the results as a function of dose; Fig. 1b, above 2nd row, shows results as a function of survival.

365 nm radiation for equal biological effects (in yeast) than molecules substituted at the 3,4 double bond. Based on their photobiological properties we were able to classify the new compounds as mono- or bi-functional compounds. Figure 1 shows survival curves and the induction of nuclear mutations (as a function of dose) (Fig. 1a) and the induction of nuclear and cytoplasmic mutations (as a function of survival) (Fig. 1b) obtained in the haploid yeast strain N123 after treatment with 8-MOP (bi-functional), 3-CPs (mono-functional) and two new psoralen derivatives 3-Cyanopsoralen (CNP) and 3-Carbomethoxypsoralen (CMeP) at a concentration of $5 \cdot 10^{-5}$ M. As a function of dose, CNP exhibits a higher activity than 3-CPs, whereas CMeP approaches the activity of 3-CPs. CNP and CMeP clearly show a lower activity on the induction of nuclear revertants (Fig. 1b, left panel) and a higher activity on the induction of cytoplasmic "petite" mutations (Fig. 1b, right panel) than the bi-functional furocoumarin 8-MOP. Thus, CNP and CMeP may be classified as monofunctional furocoumarins.

Recent studies indicate that CNP and CMeP are unable to produce inter-strand cross-links in calfs thymus DNA in vitro (70). In analogy to 3-CPs, for new photobiologically active mono-functional furocoumarins such as CNP and CMeP, and for the newly developed 4',5'-dimethylangelicin (71) a possible therapeutic activity merits to be checked.

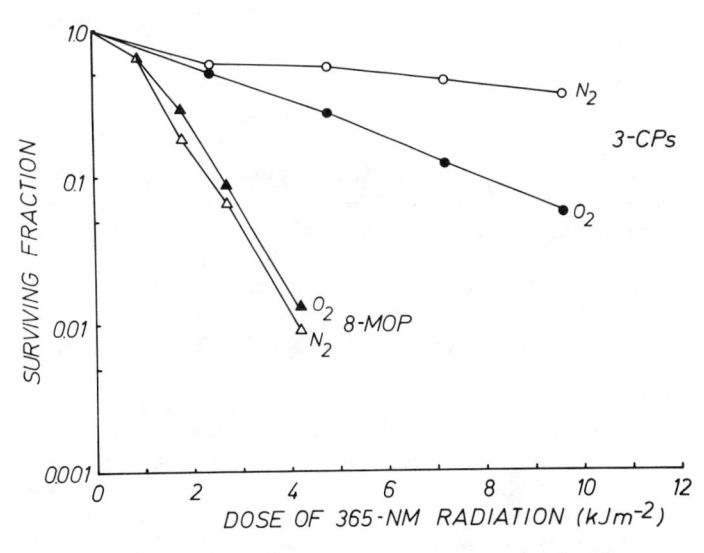

Fig. 2. The induction of lethal effects in haploid yeast (N123) after treatments with 8-MOP or 3-CPs ($5 \cdot 10^{-5}$M) in the presence (O_2) and absence (N_2) of oxygen as a function of 365-nm radiation.

Photosensitization in the Presence and Absence of Oxygen

The photoreactions of furocoumarins have been generally considered as being independent of molecular oxygen (1, 72, 73). For instance, the lethal effect of 8-MOP in bacteria was shown to be oxygen independent (74). Recently, the hypothesis has been put forward that an energy transfer from the psoralen triplets to O_2, generating singlet oxygen, may take place during photosensitization by furocoumarins (4). Indeed, furocoumarins such as 8-MOP are able to produce singlet oxygen in vitro (4, 15).

We verified in our experimental system (yeast) whether the photobiological effects of 8-MOP and 3-CPs were different in the presence and in the absence of oxygen. As seen in Fig. 2, we observed for the induction of lethal effects in yeast a sensitization in the presence of oxygen when the cells were treated with 3-CPs. No sensitization occurred in the case of 8-MOP. This may be related to the different properties of the triplet excited states of the molecules (76). Since also for 5,7-dimethoxycoumarin an oxygen-dependent photodynamic action may exist (75), an oxygen effect may be expected also for other furocoumarins. In addition, oxygen-dependent reactions may have some bearing on the therapeutic effects of furocoumarins. Photophysical studies may help to elucidate the underlying reaction mechanisms.

The Effect of Furocoumarins in the Presence of Low Dose-Rate 365-nm Radiation

An interesting approach to study the repair of furocoumarin induced lesions are investigations using high (HDR) (above 50 $kJm^{-2} h^{-1}$) and low (LDR) (below 1 $kJm^{-2}h^{-1}$) dose rates of 365-nm radiation (23, 27). Using 8-MOP a high resistance to cell killing was observed in haploid and diploid repair-proficient yeast cells at LDR (23, 27). The effect of LDR was totally suppressed in cells treated at 5°C (27) and, partially, in cells treated in the presence of cycloheximide, an inhibitor of protein synthesis. These findings suggest that active metabolic functions are involved in the LDR effect. The use of repair deficient mutants revealed that the LDR effect depends on excision-resynthesis as well as on other repair functions (27).

The question arises whether treatments at LDR favor an error prone mutagenic repair. It was shown that in the presence of 8-MOP the frequency of induced nuclear mutations in haploid yeast was lower at LDR than at HDR (27). Such a difference was not seen in excision defective mutants. The present study extends these results to the induction of nuclear and cytoplasmic genetic events in diploid yeast (D_7, Zimmermann). After treatments with 8-MOP, concomitant with a striking increase in survival, there is a marked decrease in the induction of mitotic intragenic recombination in conditions of LDR (see Fig. 3). The same is true for the induction of reversions and mitotic crossing-over (data not shown). Such an LDR effect is absent in the sensitive diploid mutant D_{7-6}(rad6) (Zimmermann) indicating that an error prone repair pathway (rad6) is involved (77). In an earlier study (27) it was shown that also the induction of cytoplasmic "petite" mutations is affected by treatments at LDR. In the strain D_7 a much higher induction of cytoplasmic "petite" mutations is observed at LDR than at HDR when the induction is expressed as a function of survival. The induction at LDR is as high as that observed with 3-CPs (mono-functional) at HDR. All these data put together suggest that treatments at LDR give rise to an efficient repair of 8-MOP induced mono-additions and DNA cross-links. This repair is very limited when furocoumarins of higher cross-linking capacity (5-MOP, Ref. 5) are used.

Study on the Photobiological Activity of 4,8-dimethyl,5'carboxypsoralen (DMeCP), a Metabolite of 4,5',8-Trimethylpsoralen (TMeP)

Photo-therapeutically active furocoumarins such as TMeP and 8-MOP are rapidly metabolized in the liver of treated patients (38, 40). Using one of the major metabolites (DMeCP) of TMeP (kindly

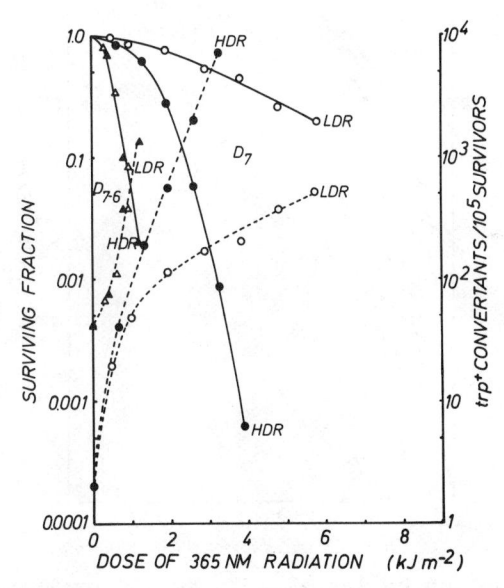

Fig. 3. The induction of lethal effects and gene conversion in the diploid strain D_7(Rad$^+$) and the repair deficient derivative D_{7-6}(rad6) after treatment with 8-MOP ($5 \cdot 10^{-5}$ M) and 365-nm radiation at high (HDR) (closed symbols) and low (LDR) (open symbols) dose rate.

provided by Dr. M.A. Pathak, Boston) we were able to show that it is a photobiologically active compound.

Figure 4 shows that in haploid yeast TMeP is the most effective compound followed by 5-MOP, 8-MOP, 3-CPs and DMeCP. The lethal doses for 37% survival are 1.3, 2.2, 12 and 22 times higher for 5-MOP, 8-MOP, 3-CPs and DMeCP, respectively, than for TMeP. In contrast to our results, 5-MOP was shown to be less active than 8-MOP in Escherichia coli and Chinese Hamster Cells (21) using different concentrations than ours ($5 \cdot 10^{-5}$ M.) When the above data are plotted as a function of survival DMeCP shows the same high mutagenic activity as the bi-functional furocoumarins (TMeP, 5-MOP, 8-MOP). Thus, DMeCP must be considered as a bi-functional compound. This holds with the cross-

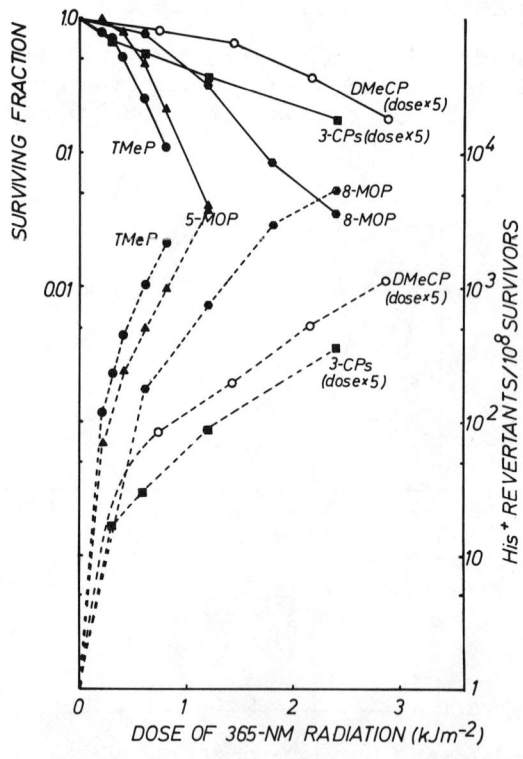

Fig. 4. The induction of lethal effects and nuclear reversion (His$^+$) in the haploid yeast strain N123 after treatments with $5 \cdot 10^{-5}$ M 4,5',8-Trimethylpsoralen (TMeP), 5-Methoxypsoralen (5-MOP) 8-Methoxypsoralen (8-MOP), 3-Carbethoxypsoralen (3-CPs) and the metabolite of TMeP 4,8-dimethyl, 5'carboxypsoralen (DMeCP) as a function of 365-nm radiation.

linking capacity reported earlier (38). These results clearly indicate that the metabolite of TMeP is photobiologically active and may (when present at high concentrations) contribute to the therapeutic effects observed with TMeP.

CONCLUSION

In the research work reported above some new trends were recognized that concern the photobiological studies with furocoumarins. These include: the genetic approach to the repair of furocoumarin induced lesions combined with biochemical analysis, studies on the relationship between cellular damage and genetic (and carcinogenic) effects induced by furocoumarins, studies on the involvement of oxygen-dependent reactions in the photobiological and phototherapeutic effects of furocoumarins, studies on the activity of metabolites and the development of photoreactive mono-functional furocoumarins as therapeutically useful drugs showing less mutagenic (and carcinogenic) effects than bi-functional furocoumarins.

REFERENCES

1. L. Musajo and G. Rodighiero, Photophysiology, Vol. VII, Ed. A. Giese, Academic Press, New York and London, pp. 115-147 (1972).
2. M. A. Pathak, D. M. Krämer, and T. B. Fitzpatrick, in: Sunlight and Man, Eds. M. A. Pathak, L. C. Harber, M. Seiji, A. Kukita, University of Tokyo Press, pp. 335-368 (1974).
3. B. R. Scott, M. A. Pathak, and G. R. Mohn, Mutation Res. 39: 29-74 (1976).
4. P.-S. Song and K. J. Tapley, Jr., Photochem. Photobiol. 29: 1177-1197 (1979).
5. L. Musajo, G. Rodighiero, G. Caporale, F. Dall'Acqua, S. Marciani, F. Bordin, F. Baccichetti, and R. Bevilacqua, in: Sunlight and Man, Eds. M. A. Pathak, L. C. Harber, J. Seiji, A. Kukita, University of Tokyo Press, pp. 369-387 (1974).
6. F. Dall'Acqua, in: Research in Photobiology, Ed. A. Castellani, Plenum Press, pp. 245-255 (1977).
7. F. Dall'Acqua, D. Vedaldi, and M. Recher, Photochem. Photobiol. 27:33-36 (1978).
8. F. Dall'Acqua, S. Marciano Magno, F. Zambon and G. Rodighiero, Photochem. Photobiol. 29:489-495 (1979).
9. R. V. Bensasson, E. J. Land, and C. Salet, Photochem. Photobiol. 27:273-280 (1978).
10. M. T. Sae Melo, D. Averbeck, R. V. Bensasson, E. J. Land, and C. Salet, Photochem. Photobiol. 30:645-651 (1979).
11. P. C. Beaumont, B. J. Parsons, G. O. Phillips, and J. C. Allen, Biochim. Biophys. Acta 562:214-221 (1979).
12. R. V. Bensasson, C. Salet, E. J. Land, and F. A. P. Rushton, Photochem. Photobiol. 31:129-133 (1980).

13. K. Yoshikawa, N. Mori, S. Sakakibara, N. Mizuno, and P.-S. Song, Photochem. Photobiol. 29:1127-1133 (1979).
14. S. Lerman, J. Megan, and I. Willis, Photochem. Photobiol. 31: 235-242 (1980).
15. N. J. De Mol and G. M. J. Beijersbergen van Henegouwen, Photochem. Photobiol. 30:331-335 (1979).
16. B. R. Zerler and S. Wallace, Photochem. Photobiol. 30:413-416 (1979).
17. F. Baccichetti, F. Bordin, and F. Carlassare, Exparientia 35: 183-184 (1979).
18. H. Fujita, H. Sano, and K. Suzuki, Tokai J. Exp. Clin. Med. 3: 35-42 (1978).
19. B. A. Bridges and R. P. Mottershead, J. Bacteriol. 139:454-459 (1979).
20. T. Seki, K. Nozu and S. Kondo, Photochem. Photobiol. 27:19-24 (1978).
21. M. J. Ashwood-Smith, G. A. Poulton, M. Barker, and M. Mildenberger, Nature 285:407-409 (1980).
22. D. Averbeck, E. Moustacchi, and E. Bisagni, Biochim. Biophys. Acta 518:464-481 (1978).
23. D. Averbeck and S. Averbeck, Mutation Res. 50:195-206 (1978).
24. I. V. Federova, Genetika, tom XIV, 11:1884-1891 (1978).
25. D. Averbeck and E. Moustacchi, Mutation Res. 68:133-148 (1979).
26. E. L. Grant, R. C. von Borstel, and M. J. Ashwood-Smith, Environm. Mutagenesis 1:55-63 (1979).
27. D. Averbeck and S. Averbeck, in: Studies in physical and theoretical chemistry 6, Radiation Biology and Chemistry, Research Developments, Eds. H. E. Edwards et al., Elsevier Scientif. Publ. Comp. Amsterdam, Oxford, New York, pp. 453-466 (1979).
28. J. A. P. Henriques and E. Moustacchi, Photochem. Photobiol. 31:555-562 (1980).
29. J. A. P. Henriques and E. Moustacchi, Genetics, June issue, in press (1980).
30. C. Cassier, R. Chanet, J. A. P. Henriques, and E. Moustacchi, Genetics, submitted paper (1980).
31. J. Coppey, D. Averbeck and G. Moreno, Photochem. Photobiol. 29:797-801 (1979).
32. S. Nocentini, Biochim. Biophys. Acta 521:160-168 (1978).
33. J. Kaye, C. A. Smith, and P. C. Hanawalt, Cancer Res. 40:696-702 (1980).
34. D. Averbeck and E. Moustacchi, Photochem. Photobiol. 31:475-478 (1980).
35. P. M. Burger and J. W. I. M. Simons, Bull. Cancer (Masson, Paris) 65:281-282 (1978).
36. P. M. Burger and J. W. I. M. Simons, Mutations Res. 60:381-389 (1979).
37. E. Cassuto, J. Mursalim, and P. Howard-Flanders, Proc. Natl. Acad. Sci. USA 75:620-624 (1978).

38. B. B. Mandula, M. A. Pathak, and G. Dudek, Science N. Y. 193:1131-1134 (1976).
39. V. Busch, J. Schmid, F. W. Koss, H. Zipp, and A. Zimmer, Arch. Dermatol. Res. 262:255-265 (1978).
40. B. B. Mandula and M. A. Pathak, Biochem. Pharmacol. 28:127-132 (1979).
41. B. A. Bridges, Hum. Genet. 49:91-96 (1979).
42. B. A. Bridges, Clin. Exp. Dermatol. 3:349-353 (1978).
43. D. Averbeck, Bull. Cancer (Masson, Paris) 67:245-254 (1980).
44. A. L. Gaynor and D. M. Carter, J. Invest. Dermatol. 71:257-259 (1978).
45. B. Lambert, M. Morad, A. Bredberg, G. Swanbeck, and M. Thyresson-Hök, Acta Dermatovener (Stockholm) 58:13-16 (1978).
46. D. D. Grube, R. D. Ley, and R. J. M. Fry, Photochem. Photobiol. 25:269-276 (1977).
47. L. Dubertret, D. Averbeck, F. Zajdela, E. Bisagni, E. Moustacchi, R. Touraine, and R. Latarjet, Brit. J. Dermatol. 101: 379-389 (1979).
48. R. S. Stern, L. A. Thibodeau, R. A. Kleinerman, J. A. Parrish, and T. B. Fitzpatrick, New England J. Med. 300:809-813 (1979).
49. B.A. Bridges and G. Strauss, Nature 283:523-524 (1980).
50. L. Dubertret, D. Averbeck, and E. Bisagni, C. R. Acad. Sci. Paris, T 288, D, 975-977 (1979).
51. D. Averbeck, E. Bisagni, J. P. Marquet, P. Vigny, and F. Gaboriau, Photochem. Photobiol. 30:547-556 (1979).
52. C. S. Rupert and R. Latarjet, Photochem. Photobiol. 28:3-5 (1978).
53. R. S. Cole, J. Bacteriol. 107:846-852 (1971).
54. F. Dall'Acqua, S. Marciani, D. Vedaldi, G. Rodighiero, Biochim. Biophys. Acta, 353:267-273 (1974).
55. F. Dall'Acqua, S. Marciani, L. Ciavatta, and G. Rodighiero, Z. Naturforsch. 26b:561-569 (1971).
56. F. Bordin, S. Marciani, F. Baccichetti, F. Dall'Acqua, and G. Rodighiero, Ital. J. Biochem., 24:258-266 (1975).
57. J. W. Lown and S. K. Sim, Bioorg. Chemistry 7:85-95 (1978).
58. L. Kittler, Z. Hradecna, and J. Sühnel, Biochim. Biophys. Acta 607:215-220 (1980).
59. R. S. Cole, D. Levitan, and R. R. Sinden, J. Mol. Biol. 103: 39-59 (1976).
60. R. R. Sinden and R. S. Cole, J. Bacteriol. 136:538-547 (1978).
61. D. Averbeck and E. Moustacchi, Biochim. Biophys. Acta 395: 393-404 (1975).
62. D. Averbeck, P. Chandra, and R. K. Biswas, Rad. Environm. Biophys. 12:241-252 (1975).
63. J. A. P. Henriques, R. Chanet, D. Averbeck, and E. Moustacchi, Molec. Gen. Genetics 158:63-72 (1977).
64. A. C. Griffin, R. E. Hakim, and J. Knox, J. Invest. Dermatol. 31:289-295 (1958).
65. P. Queval, and E. Bisagni, Eur. J. Med. Chem. 9:335-340 (1974).

66. D. Averbeck, VII. Int. Congress on Photobiology, 29 Aug.-3 Sept, Rome, book of abstracts, p. 137 (1976).
67. G. Rodighiero and F. Dall'Acqua, Photochem. Photobiol. 24:647-653 (1976).
68. F. Dall'Acqua, personal communication.
69. N. Magana-Schwencke, D. Averbeck, J. A. P. Henriques, and E. Moustacchi, C. R. Acad. Sci. Paris t.291:207-210 (1980).
70. F. Gaboriau (unpublished data).
71. F. Bordin, F. Baccichetti, and F. Carlassare, Z. Naturforsch. 33c:296-298 (1978).
72. E.-R. Lochmann and A. Micheler, in: Physico-chemical properties of nucleic acids, Vol. I, Ed. J. Duchesne, Academic Press, New York, pp. 223-267 (1973).
73. G. Löber and L. Kittler, Photochem. Photobiol. 25:215-233, (1977).
74. E. L. Oginsky, G. S. Green, D. G. Griffith, and W. L. Fowlks, J. Bacteriol. 78:491-493 (1976).
75. M. L. Harter, I. C. Felkner, and P.-S. Song, Photochem. Photobiol. 24:491-493 (1976).
76. E. J. Land, R. Bensasson, J. C. Ronfard-Haret, unpubl. results.
77. E. Moustacchi, R. Chanet, and M. Heude, in: Research in Photobiology, Ed. A. Castellani, Plenum Press, New York and London, pp. 197-206 (1977).

ACKNOWLEDGEMENTS

This work was supported by INSERM ATP 67-78-99 N° 25, CNRS RCP RCP 80572 and EURATOM 155-BIOF. It was carried out in collaboration with Drs. E. Moustacchi, E. Bisagni, L. Dubertret, F. Zajdela, T. Land, R. Bensasson, P. Vigny, F. Gaboriau, and M. A. Pathak. We thank Dr. R. Latarjet for his interest in our work. We are grateful to Madame S. Averbeck-Entéric for the excellent technical assistance.

PHOTOCHEMOTHERAPY WITH FUROCOUMARINS (PSORALENS)

Herbert Hönigsmann

Department of Dermatology,
University of Innsbruck,
Anichstr. 35, A-6020 Innsbruck, Austria

INTRODUCTION

Photochemotherapy is the combined use of a photosensitizing chemical compound with non-ionizing electromagnetic radiation to bring about a therapeutically beneficial result not produced by the drug or radiation alone. The drug may be applied topically or administered systemically to reach the skin via the circulation and is subsequently activated by irradiation with appropriate wavelengths.

This communication specifically refers to oral psoralen photochemotherapy which is based on the interaction of 8-methoxypsoralen (8-MOP) and long-wave ultraviolet (UV-A) radiation. This particular form of photochemotherapy, acronymously termed PUVA (psoralen and UV-A), has profoundly influenced dermatologic therapy and specifically the management of psoriasis.

Although the use of natural psoralens and sunlight as a remedy for vitiligo dates back to the ancient Egyptian and Indian healers, it was not until the early seventies when the introduction of photochemotherapy opened a new era in the treatment of severe and disabling psoriasis (1). From several treatment protocols* employing topical and/or systemic administration of 8-MOP or trimethylpsoralen, oral 8-MOP photochemotherapy emerged as the superior form (1,3,4).

The dramatic effectiveness of PUVA in producing and maintaining clinical remissions of psoriasis has been widely documented and confirmed by two cooperative clinical studies both in the United States and Europe (5,6). Today, the clinical possibilities and limitations

---
* for reference see ref. 2

of PUVA are well defined as are the guidelines for proper dosimetry and optimum management of the patients (7). The success of PUVA in treating psoriasis has stimulated the interest in other issues such as the treatment of disorders other than psoriasis, improvement of existing treatment protocols, the problems raised by maintenance therapy and the question of potential long-term risks and adverse reactions.

FUNDAMENTALS OF THE TREATMENT

The rationale of photochemotherapy of psoriasis is to induce remissions by repeated, controlled phototoxic reactions. Experimental evidence has shown that these reactions inhibit epidermal DNA synthesis, and thus may suppress the proliferative stimulus of psoriatic cells (8,9). Upon exposure to UV-A 8-MOP forms monoadducts and interstrand crosslinks with cellular DNA (10) and it is commonly assumed that the therapeutic effect of psoralens is largely related to the formation of crosslinks (1,3,11). However, in recent experiments, two separate groups could not detect crosslinks in epidermal DNA with oral psoralens administered in therapeutic doses (12,13). This suggests that crosslinking may not be the underlying mechanism in the treatment of psoriasis.

METHODS AND TREATMENT PROTOCOLS

The limiting parameters in photochemotherapy are erythema and pigmentation. Intense erythemas are undesirable side effects and careful dosimetry is essential to avoid such reactions. Pigmentation is an important factor as it increases the tolerance of skin to UV radiation and thus decreases the therapeutic effect. As a consequence UV doses have to be gradually increased during therapy to maintain a fairly constant therapeutic response. Since both the capacity to develop erythema and pigmentation vary considerably between individuals it is necessary to determine the individual sensitivity and predict the responses to PUVA treatments. To this end two different approaches are currently in use:

Skin Typing(5)

Patients are classified according to ethnic extraction and sunburn history and UV-A doses are delivered according to an empirical scheme based on this classification (Table 1).

Phototoxicity testing (7)

UV-A dosimetry is based on the determination of the individual minimal phototoxicity dose (MPD) by phototoxicity testing. The MPD informs about the patients threshold sensitivity and is therefore the most reliable and safest dose to start treatment. By using the MPD both over- and undertreatment can be avoided.

Table 1. Recommended UV-A Exposure per Skin Type*

Skin type	History	Recommended Joules/cm²
I	Always burn, never tan	0.5
II	Always burn, but sometimes tan	1.0
III	Sometimes burn, but always tan	1.5
IV	Never burn, always tan	2.0

Skin type	Physical examination	Recommended Joules/cm²
V	Moderately pigmented	2.5
VI	Blacks	3.0

* adapted from (14)

The current principle of the treatment is to hold the dose of the drug constant (0.6 - 0.8 mg 8-MOP/kg body weight) and to vary the dose of UV-A depending on the sensitivity of the patient. Two major treatment schedules of PUVA have evolved and have been employed in two cooperative clinical trials (5,6). The main difference between these two schedules are: (i) UV-A dosimetry is based on skin typing in the US multicenter trial (5), while the European regimen (6) requires phototoxicity testing and (ii) the European protocol is more aggressive in the clearing phase employing higher initial doses of UV-A and more treatments per week (4 times a week) than the US regimen (2-3 times a week), the rationale being to clear the patient before intense pigmentation develops as this may require earlier and higher UV-A dose increments. The European protocol (6) is more complicated than the US regimen but, because of phototoxicity testing it allows an individualized, i.e. optimal dosimetry for each patient.

CLINICAL RESULTS

Clearing Phase

Practically all forms of psoriasis respond to PUVA and this includes eruptive-guttate and chronic plaque type psoriasis, exudative and erythrodermic forms (1,3-6) as well as pustular psoriasis (15).

Table 2. Results of Initial Treatment
(Austrian PUVA study Innsbruck-Vienna)*

Number of patients	572
Patients cleared	534 (93%)
Treatment failures	37 (6.5%)
Number of exposures required for clearing	14.7 ± 8.3
Duration of treatment (days)	30.4 ± 26.6
Total UV-A-dose (J/cm^2)	78.7 ± 88.7

* adapted from (2)

A better than 90% clearing rate was achieved in the Austrian series comprising 572 patients who have been followed up for more than 5 years (2). As shown in table 2 a mean of 15 treatments and 30 days was required to achieve complete clearing. The total cumulative UV-A dose was 79 J/cm^2, but the large standard deviation points to the great variation in the individual dose requirements.

The results reported by the two multicenter PUVA trials are essentially similar. In the US study (5) (1139 patients) 88.2% of the patients showed complete clearing after 25 exposures (mean) and 12 weeks (mean) (Table 3). However, the mean total cumulative UV-A dose (245 J/cm^2) needed for this result was relatively high ranging from 191 J/cm^2 for skin type I patients to 296 J/cm^2 for skin type IV patients.

Table 3. Results of Initial Treatment
(US Cooperative Clinical Trial)*

Number of patients	1139
Patients cleared	1005 (88.2%)
Treatment failures	33 (2.9%)
Drop-outs	101 (8.9%)
Number of exposures required for clearing (mean)	25.2
Duration of treatment (weeks) required for clearing (mean)	12.7
Total cumulative UV-A dose (J/cm^2) required for clearing (mean)	245

* adapted from (5)

Table 4. Results of Initial Treatment
(European PUVA study)*

Number of patients	2995
Patients cleared (90-100%)	2658 (88.8%)
Patients cleared (50-90%)	170 (5.7%)
Treatment failures	139 (4.6%)
Patients worsened	28 (0.9%)
Number of exposures required for clearing (mean)	19.0
Duration of treatment (weeks) required for clearing (mean)	5.7
Total cumulative UV-A dose (J/cm^2) required for clearing (mean)	103.3

* adapted from (6)

In the European PUVA study (6) comprising 2995 patients complete clearing was achieved in 88.8% of the patients, but only 19 exposures (mean) and 6 weeks were required. More importantly the mean total cumulative dose required for clearing was 103 J/cm^2 (Table 4) for all skin types.

Thus, albeit a much more conservative and cautious approach was employed for the US cooperative study (5), both with regard to the initial UV-A dose and the dose increments during therapy, the mean cumulative UV energy requirements were considerably higher than with the more aggressive European regimen (6). Moreover, the number of treatments and the treatment time was much higher in the US study. It is obvious that an aggressive, but individualized treatment saves time and reduces the total UV-A energy load as compared with a more rigid dosimetry schedule which permits only minimal adjustments according to the individual patient's needs.

Maintenance phase

Various maintenance regimens have been tested to establish a treatment schedule which keeps psoriasis in remission for prolonged periods of time (5,6,16). To date the best results have been achieved by an arbitrary schedule in which the frequency of PUVA treatments is adjusted to the patients condition (16,17). A significantly higher recurrence rate was observed in both, the US study (5) and the Austrian series (16) in patients in whom no maintenance treatment had been performed. However, PUVA maintenance does have three major disadvantages: (i) Using a rigid, long-term therapy schedule patients with a stable remission may be overtreated as there exists no parameter to estimate how long maintenance will have to be continued in an individual patient to induce a long-lasting remission. (ii) Continuous maintenance is time-consuming for both the patient and the therapy center and may limit the admission of new patients. (iii) Most importantly long-term risks of PUVA are presumably related to the total cumulative phototoxic doses delivered to the skin during continuous or long-term maintenance therapy.

By analysis of our patient population it turned out that most of the recurrences occur during the first two months after termination of the clearing phase and if patients are not subjected to maintenance after clearing, only 11% are still free of disease one year later (18). On the other hand, maintenance therapy can be stopped after 2 to 3 months in those patients who have had no recurrence at this time and the percentage of patients who are still clear after 9 months will be 68% (18). Consequently, we now subject all patients to maintenance treatment for the first two months after clearing and discontinue PUVA therapy in those patients who are still in remission at this time. Preliminary follow-up observations indicate that the majority of these patients remain free of disease for at least 6 - 12 months (2).

This approach prevents overtreatment of patients who do not require continuous maintenance and thus reduces considerably the cumulative UV dose.

OTHER DERMATOSES RESPONSIVE TO PUVA

Besides psoriasis the most important indication for PUVA is cutaneous T cell lymphoma (mycosis fungoides, Sezary syndrome). Excellent results have been obtained for the early eczematous (T_o) and plaque stages (T_1, T_2) of the disease*, when the malignant T cell infiltrate is still confined to the epidermis and the superficial dermis (19-22). Thus, the therapeutic effect parallels the depth of penetration of UV-A. Surprisingly even the flat tumors of early tumor stage (T_3) show a slow, but definite response (20,21). Whether the natural course of mycosis fungoides with progression to later tumorous stages and stages with systemic involvement can be altered or perhaps aborted by PUVA will be determined by long-term observation only.

PUVA may clear mycosis fungoides by exerting a direct cytotoxic effect on the dermal infiltrate. Apparently this effect is not specific for T cells as is indicated by the observation that other cutaneous infiltrates as in B-cell leukaemias (23) and in cutaneous

---

* staging according to Edelson, J Amer Acad Dermatol 2: 89 (1980)

mastocytosis (urticaria pigmentosa) (24) also show regression when treated with PUVA.

Other dermatoses which respond to PUVA include atopic dermatitis (25,26), lichen planus (27) and pityriasis lichenoides (27). The mechanism by which PUVA is operative in these disorders in which epidermal hyperproliferation has no or only secondary significance is not fully understood. Several studies have shown recently that photochemotherapy (as well as UV-B phototherapy) may affect the immune system (28). PUVA decreases leukotactic activity in psoriasis, it suppresses contact sensitivity reactions, it possibly induces the generation of suppressor T cells and it may affect white blood cells circulating in the superficial capillaries*. One thus might speculate that PUVA is effective by altering certain immune mechanisms in diseases which are thought to be related to immunologic phenomena. The protective action of PUVA in polymorphous light eruption (29,30) may therefore be not only a consequence of increased melanogenesis and thus represent photoprotection, but may also be mediated by immunological effects.

SIDE EFFECTS AND LONG TERM RISKS

Short-term side effects are related to overdosage of UV-A (pruritus, erythema, blistering) or intolerance of the drug (nausea, headache, dizziness) (1,3,4). They are usually mild, reversible and can be avoided mostly by careful dosimetry (7). However, experimental evidence suggests that photochemotherapy may carry the risk of long-term side effects as cataractogenesis and carcinogenesis*.

Cataracts

8-MOP has been detected in the lenses of rats and it has been shown that photoproducts are formed with lens protein upon UV-A exposure. These photoproducts may remain permanently and accumulate within the lens. Cataracts have been produced in laboratory animals after very high doses of 8-MOP and subsequent UV-A exposure but not with doses comparable with those used in therapy*. To date there is no evidence of PUVA-induced cataracts in man but it is advisable that the patients wear appropriate UV-A-opaque glasses during the entire period of increased photosensitivity after 8-MOP ingestion.

Carcinogenesis

The induction or promotion of skin cancers represents the area of greatest concern. The interaction of psoralens and cellular DNA leads to the formation of DNA-psoralen photoadducts which have been

---

* for reference see ref. 2

shown to be mutagenic in bacterial systems. Chromosomal aberrations and an increased sister chromatid exchange have been observed in lymphocytes exposed to 8-MOP and UV-A in vitro, in a dose-dependent fashion, but not in PUVA patients in vivo. Under certain extreme experimental conditions tumors have been produced in laboratory animals. However, patients treated for vitiligo with trimethylpsoralen and sunlight for many years have not revealed an increased rate of skin tumors*.

Recently two large clinical studies (31,32) have demonstrated an increased incidence of non-melanoma skin tumors in patients subjected to long-term PUVA therapy if certain risk factors were present. These risk factors were previous treatment with ionizing radiation (31) and/or arsenic (32) and a history of previous skin carcinoma. On the other hand, if patients with risk factors were excluded from consideration the risk of tumor formation equalled that of a normal control population. Unfortunately, both studies have a significant draw-back in that appropriate control populations, (e.g. untreated psoriatics, psoriatics with risk factors but without PUVA treatment etc.) were not studied. It still will remain to be determined whether the incidence of skin cancer is increased after prolonged PUVA treatment of patients without risk factors and whether limits can be defined within which PUVA can be considered relatively safe.

IMPROVEMENTS AND PERSPECTIVES

As the long-term hazards of photochemotherapy are still unknown new approaches should aim at a reduction of the frequency of treatments and, concomitantly, a reduction of the total UV-A dose required for treatment.

Combination with Other Treatments

The effectiveness of PUVA is markedly increased by a combination with topical corticosteroids (33,34). However, one study suggested an increased recurrence rate with the combined treatment (33). An enhanced therapeutic effect was found with anthralin and PUVA (33), but yet there exists no information on a larger series of patients and on the recurrence rates.

Aromatic Retinoid (Chemophotochemotherapy)

A definite improvement has been obtained by the combination of PUVA with an orally administered aromatic retinoid (Ro-10 9359, Hoffmann-La Roche) (35,36). If the retinoid is given 1 week before PUVA therapy and is continued until clearing, the response rate of psoriatic lesions is accelerated, the number of treatments is re-

---

* for reference see ref. 2

duced by one third and, most importantly, the total cumulative UV-A dose is cut by more than half of what would be required for conventional PUVA. The mechanisms by which the aromatic retinoid improves photochemotherapy are unknown.

As remissions are achieved with lower final UV-A doses the dose requirements per exposure for maintenance treatment remain also at a lower range. Thus potential long-term hazards related to the total dose of UV-A applied over prolonged periods of time may be reduced. Standard PUVA maintenance without the retinoid is performed after the clearing phase and follow-up studies for a period of up to 3 years indicate that the recurrence rate is not higher than in patients treated by PUVA alone.

The only disadvantages of this treatment known at present are several unpleasant, but reversible short-term side effects (e.g. cheilitis, dry rhinitis). The possible teratogenic property of retinoids limits this therapy to a selected patient group.

5-methoxypsoralen (5-MOP, Bergapten)

5-MOP has been shown to be an excellent drug for photochemotherapy (37). The clearing rates are comparable with those of 8-MOP-PUVA, but 5-MOP-PUVA produces less pruritus and does not induce nausea. Moreover, 5-MOP has the advantage of being barely erythemogenic and thus psoriatic plaques can be treated without the development of erythema and this increases the safety margins of PUVA. However, the total UV-A dose requirements are not reduced as compared to 8-MOP-PUVA (37).

3-carbethoxypsoralen (3-CEP)

A new synthetic furocoumarin compound, 3-carbethoxypsoralen has recently been tested for antipsoriatic activity after topical application followed by UV-A irradiation (38). Preliminary results seem to promise a therapeutic efficacy comparable with 8-MOP. Systemic administration has not been tried yet. The possible advantages of this compound are obvious: (i) it forms only monoadducts with DNA, (ii) it is non-erythemogenic and (iii) it is non-carcinogenic in mice. Further studies are needed to show whether 3-CEP will be an ideal alternative to replace 8-MOP in photochemotherapy.

FUTURE ASPECTS

The success of photochemotherapy of psoriasis has undoubtedly stimulated the interest in such treatment modalities and has initiated a tremendous amount of basic research work in cutaneous photobiology. Light technology and measurements are constantly improving and laser techniques will probably find their way into photochemotherapy. Ultimate targets will perhaps be a new generation of photosensitizers which might act in a more selective way on certain cells or structures without being harmful to others.

REFERENCES

1. J.A. Parrish, T.B. Fitzpatrick, L. Tanenbaum and M.A. Pathak, Photochemotherapy of psoriasis with oral methoxsalen and long wave ultraviolet light, N. Engl. J. Med. 291: 1207 (1974).
2. K. Wolff and H. Hönigsmann, Clinical aspects of photochemotherapy, Int. Encyclopedia of Pharmacology and Therapeutics, in press (1980).
3. K. Wolff, H. Hönigsmann, F. Gschnait and K. Konrad, Photochemotherapie bei Psoriasis. Klinische Erfahrungen bei 152 Patienten, Dtsch. Med. Wschr. 100: 2471 (1975).
4. K. Wolff, T.B. Fitzpatrick, J.A. Parrish, F. Gschnait, B. Gilchrest, H. Hönigsmann, M.A. Pathak and L. Tanenbaum, Photochemotherapy with orally administered methoxsalen, Arch. Dermatol. 112: 943 (1976).
5. J.W. Melski, L. Tanenbaum, J.A. Parrish, T.B. Fitzpatrick, H.L. Bleich et al, Oral methoxsalen photochemotherapy for the treatment of psoriasis: a cooperative clinical trial, J. Invest. Dermatol. 68: 328 (1977).
6. T. Henseler, E. Christophers, K. Wolff, H. Hönigsmann et al, PUVA-study: results of oral photochemotherapy for psoriasis conducted in 17 European centers. In preparation (1980).
7. K. Wolff, F. Gschnait, H. Hönigsmann, K. Konrad, J.A. Parrish and T.B. Fitzpatrick, Phototesting and dosimetry for photochemotherapy, Brit. J. Dermatol. 96: 1 (1977).
8. J.F. Walter, J.J. Voorhees, W.H. Kelsey and E.A. Duell, Psoralen plus black light inhibits epidermal DNA synthesis, Arch. Dermatol. 107: 861 (1973).
9. P.O. Fritsch, F. Gschnait, G. Kaaserer, W. Brenner, S. Chaikittisilpa, H. Hönigsmann and K. Wolff, PUVA suppresses the proliferative stimulus produced by stripping on hairless mice, J. Invest. Dermatol. 73: 188 (1979).
10. G. Rodighiero and F. Dall'Acqua, Biochemical and medical aspects of psoralens, Photochem. Photobiol. 24: 647 (1976).
11. M.A. Pathak, D.M. Krämer and T.B. Fitzpatrick, Photobiology and photochemistry of furocoumarins (psoralens), in "Sunlight and Man: Normal and Abnormal Photobiologic Responses," M.A. Pathak, L.C. Harber, M. Seiji and A. Kukita, eds., T.B. Fitzpatrick, consulting ed., Tokyo University Press, Tokyo (1974).
12. T. Cech, M.A. Pathak and R.K. Biswas, An electron microscopic study of the photochemical cross-linking of DNA in guinea pig epidermis by psoralen derivatives, Biochim. Biophys. Acta 562: 342 (1979).

13. D. Lerche, J. Søndergaard, S. Wadskov, V. Leick and V. Bohr, DNA interstrand crosslinks visualized by electron microscopy in PUVA-treated psoriasis, Acta Dermatovener. 59: 15 (1979).
14. Current Status of Oral PUVA Therapy for Psoriasis, J. Amer. Acad. Dermatol. 1: 106 (1979).
15. H. Hönigsmann, F. Gschnait, K. Konrad and K. Wolff, Photochemotherapy for pustular psoriasis (von Zumbusch), Brit. J. Dermatol. 97: 119 (1977).
16. F. Gschnait, Orale Photochemotherapie, Wien. Klin. Wschr. 89, Suppl. 75: 1 (1977).
17. K. Wolff, F. Gschnait, H. Hönigsmann, K. Konrad, G. Stingl, E. Wolff-Schreiner and P. Fritsch, Oral photochemotherapy. Results, follow-up and pathology, in: "Psoriasis," E.M. Farber and A.J. Cox, eds, Yorke Medical Books, New York (1977).
18. K. Wolff, PUVA 1979, Klinik und Praxis, in: "Fortschritte der praktischen Dermatologie und Venerologie," Bd. 9, O. Braun-Falco and H.H. Wolff, eds, Springer Verlag, Berlin-Heidelberg (1979).
19. B.A. Gilchrest, J.A. Parrish, L. Tanenbaum, H.A. Haynes and T.B. Fitzpatrick, Oral methoxsalen photochemotherapy of mycosis fungoides, Cancer 38: 683 (1976).
20. H. Hönigsmann, K. Konrad, F. Gschnait and K. Wolff, Photochemotherapy of mycosis fungoides, in "Book of Abstracts, VIIth International Congress of Photobiology," Rome (1976).
21. K. Konrad, F. Gschnait, H. Hönigsmann and K. Wolff, Photochemotherapie bei Mykosis fungoides, Hautarzt 29: 191 (1978).
22. C. Hofmann, G. Burg, G. Plewig and O. Braun-Falco, Photochemotherapie cutaner Lymphome. Orale und lokale 8-MOP-UVA-therapie, Dtsch. Med. Wschr. 102: 675 (1977).
23. K. Wolff, Photochemotherapie cutaner Lymphome, Hautarzt 29, Suppl. III: 75 (1978).
24. E. Christophers, H. Hönigsmann, K. Wolff and A. Langner, PUVA treatment of urticaria pigmentosa. Brit. J. Dermatol. 98: 701 (1978).
25. F. Gschnait, H. Hönigsmann, K. Konrad, P. Fritsch and K. Wolff, Photochemotherapie (PUVA) bei Neurodermitis. Z. Hautkr. 52: 1219 (1977).
26. W.L. Morison, J.A. Parrish, T.B. Fitzpatrick, Oral psoralen photochemotherapy of atopic eczema, Brit. J. Dermatol. 98: 25 (1978).
27. W. Brenner, F. Gschnait, H. Hönigsmann and P. Fritsch, Erprobung von PUVA bei verschiedenen Dermatosen, Hautarzt 29: 541 (1978).
28. W.L. Morison, J.A. Parrish, J.H. Epstein, Photoimmunology, Arch. Dermatol. 115: 350 (1979).
29. F. Gschnait, H. Hönigsmann, W. Brenner, P. Fritsch and K. Wolff, Induction of UV light tolerance by PUVA in patients with polymorphous light eruption, Brit. J. Dermatol. 99: 293 (1978).

30. J.A. Parrish, M.J. LeVine, W.L. Morison, E. Gonzalez and T.B. Fitzpatrick, Comparison of PUVA and beta-carotene in the treatment of polymorphous light eruption, Brit. J. Dermatol. 100: 187 (1979).
31. R.S. Stern, L.A. Thibodeau, A.B. Kleinerman, J.A. Parrish, T.B. Fitzpatrick et al, Risk of cutaneous carcinoma in patients treated with oral methoxsalen photochemotherapy for psoriasis, N. Engl. J. Med. 300: 809 (1979).
32. H. Hönigsmann, K. Wolff, F. Gschnait, W. Brenner and E. Jaschke, Keratoses and non-melanoma skin tumours in long-term photochemotherapy (PUVA), J. Amer. Acad. Dermatol. in press (1980).
33. W.L. Morison, J.A. Parrish and T.B. Fitzpatrick, Controlled study of PUVA and adjunctive topical therapy in the management of psoriasis. Brit. J. Dermatol. 98: 125 (1978).
34. M. Schmoll, T. Henseler and E. Christophers, Evaluation of PUVA, topical corticosteroids and the combination of both in the treatment of psoriasis, Brit. J. Dermatol. 99: 693 (1978).
35. P.O. Fritsch, H. Hönigsmann, E. Jaschke and K. Wolff, Augmentation of oral methoxsalen-photochemotherapy with an oral retinoic acid derivative. J. Invest. Dermatol. 70: 178 (1978).
36. P. Fritsch, H. Hönigsmann, E. Jaschke and K. Wolff, Photochemotherapie bei Psoriasis: Steigerung der Wirksamkeit durch ein orales aromatisches Retinoid. Klinische Erfahrungen bei 134 Patienten. Dtsch. Med. Wschr. 103: 1731 (1978).
37. H. Hönigsmann, E. Jaschke, F. Gschnait, W. Brenner, P.O. Fritsch and K. Wolff, 5-methoxypsoralen (Bergapten) in photochemotherapy of psoriasis. Brit. J. Dermatol. 101: 369 (1979).
38. L. Dubertret, D. Averbeck, F. Zajdela, E. Bisagni, E. Moustacchi, R. Touraine and R. Latarjet, Photochemotherapy (PUVA) of psoriasis using 3-carbethoxypsoralen, a non-carcinogenic compound in mice, Brit. J. Dermatol. 101: 379 (1979).

ADVANCES IN PHOTOTHERAPY OF SKIN DISEASES

John A. Parrish

Department of Dermatology
Harvard Medical School
Massachusetts General Hospital
Boston, Massachusetts 02114

   Most of the ingredients of successful phototherapy were discovered over the past 60 years by clinicians using a trial and error approach and making observations with critical and prepared minds (1). Clinicians have the advantage of knowing the nature of the problem (in this case skin disease). They also have the stimulant of being forced to act - they are repeatedly faced with fellow humans suffering the signs and symptoms of disease. But hurried and repeated encounters with the patient as a whole organism often fills the time and mind of the clinician allowing insufficient attention to pathophysiology of disease and mechanism of therapy. What is new in phototherapy of skin diseases is that mechanisms and parameters of treatment are being studied. Effectiveness is being confirmed and documented in controlled studies. Bilateral comparison techniques are being used to identify and quantify active ingredients of multiparameter therapy. Phototherapy is being clarified and refined.

   It is fair to say that one stimulus for renewed interest in phototherapy of skin diseases has been the demonstrated effectiveness of oral psoralen photochemotherapy (2). There has been increased respect for nonionizing electromagnetic radiation as a therapeutic tool. There are several reasons why PUVA may prove to have an important positive influence on phototherapy. PUVA was developed at a time when it was possible to utilize newly developed exposure sources which were spectrally specific and of high intensity. PUVA was defined at a time when fairly accurate radiometry was easily available and relatively accurate dosimetry could be achieved. Exposure doses were given in energy per unit area and not in minutes of exposure to unmetered sources. In the presence of photoactive psoralens, the inflammation response resulting from exposure of normal skin to UVA has a steep dose-response curve making attention to details of dosimetry an important aspect of therapy. PUVA was new enough that fewer assumptions were made about dosimetry, side effects, and long-term risks. It was therefore studied carefully. In fact the absence of "newness" in part impedes progress in the study of phototherapy. Study is slowed or discouraged not so much by prejudice about effectiveness or absence of effectiveness, as by undocumented assumptions about how best to administer phototherapy. Although a much older treatment, phototherapy lags behind PUVA in taking advantage of the new capabilities of electro-optics and photomedicine. We do not yet know the limits or best techniques for phototherapy of skin diseases, but there has been progress in exploring and defining the potential. That progress is the topic of this chapter.

Ultraviolet phototherapy and heliotherapy are used for a variety of common and uncommon skin diseases. In many of these disorders the clinical improvement is not impressive. Some of the diseases are self-limiting and many of the disorders treated with phototherapy have not been studied by controlled studies using bilateral comparison techniques. These factors make it difficult to evaluate the effectiveness of phototherapy. Eczema, pityriasis rosea, parapsoriasis, and other diseases are occasionally improved by repeated exposures to sun or to artificial sources. Paradoxically, certain photodermatoses can be improved by controlled exposures to ultraviolet or visible radiation. Repeated exposures may cause clinical symptoms in these light-sensitive disorders, but eventually the threshold to radiation is raised so that patients can better tolerate sun exposures. Such treatments may deplete mediators, induce melanogenesis, injure cells involved in the pathogenesis of the photosensitivity, or act by other unknown mechanisms. Acne is often treated with ultraviolet radiation from sun or artificial sources. There is considerable debate whether ultraviolet radiation or heliotherapy diminishes the disease process or simply masks lesions by inducing redness and subsequent tan. Some dermatologists find that phototherapy of acne has no effect.

Certain forms of itching can be improved by phototherapy. Persons with chronic renal failure are often bothered by severe itching. The exact cause of this pruritus is not known but it is hypothesized that toxic metabolites normally cleared by the kidneys can accumulate in the skin where they affect nerve endings resulting in itching. Dryness of the skin, poor general health, malaise and depression magnify the itching so that it becomes a significant problem. In 1975, Saltzer (3) reported that seven of eight uremic patients were less pruritic within a month of beginning ultraviolet treatment. Gilchrest et al. (4) confirmed this observation by showing that repeated total body erythemogenic exposures to UVB led to relief of itching in persons with renal failure maintained with hemodialysis therapy. Black patients have also been effectively treated with UVB (5). The possibility of a major placebo effect was minimized by treating additional patients with total-body exposure in a UVA unit without benefit. Because the MED to UVA is 1000 times greater than that for UVB and the UVA intensity was only 10 times that of UVB, the UVA exposure doses were therefore markedly suberythemogenic and were considered placebo therapy. Response to therapy is not influenced by frequency of exposure. Response occurs sooner in patients treated more frequently. In patients treated with one to three treatments per week improvement usually follows four to six UVB exposures. Treatment to one half of the body leads to bilateral improvement (6). This observation is consistent with the possibility that phototherapy inactivates a circulating substance or substances present in uremic patients which is responsible for the pruritus experienced by many patients with chronic renal failure. Because pruritus is difficult to evaluate and quantify and some patients are not benefited, it has been claimed that observed therapeutic response is only subjective or is caused by a counter irritant effect of sunburn. However, if chronically ill uremic patients believe they have less itching and feel more comfortable, the treatment is worthwhile.

The skin disease most often treated by phototherapy is psoriasis. Because the therapeutic effect in this disorder is often quite striking, phototherapy of psoriasis has been the object of many basic and clinical studies and will be discussed in detail in this chapter. The kinds of advances made in the therapy of this disorder may improve phototherapy of other skin diseases. We will first examine some of the assumptions made about the mechanisms of phototherapy of psoriasis.

Psoriasis is a disease of unknown etiology characterized by increased epidermal cell proliferation. It is a common chronic skin disease that affects from 1% to 3% of people to a variable extent. The tendency to have psoriasis is inherited. Onset is often prior to adult life but psoriasis may begin at any age from birth to old age. Once the disease becomes manifest, its course is unpredictable. The individual lesions of psoriasis are raised, red, circumscribed scaling plaques which tend to occur symmetrically over the body with a predilection for bony prominences such as elbows and knees. Lesions may, however, occur at any site. Microscopically there is a marked thickening of the epidermis with a regular elongation of rete ridges and consequent elongation of dermal papillae which contains dilated capillaries. There is an increased number of dividing cells in the lower layers of the epidermis, a general acceleration of the epidermal cell cycle and alterations of cell differentiation. This results in the formation of scales as the outer layers of the epidermis are manufactured rapidly and abnormally.

PHOTOTHERAPY OF PSORIASIS

A host of chemical, metabolic and structural changes occur when living cells absorb ultraviolet radiation. It is assumed that the variety of radiation induced cellular alterations and cascade of tissue reactors known to occur in normal skin also occurs in diseased skin and that qualitative and quantitative differences in response to radiation may lead to improvement of diseased skin without irreversible or unacceptable damage to normal skin. Therapeutic effect may result because abnormal cells are more sensitive to radiation than are normal cells. Acute and chronic phototoxic effects on normal skin and blood are the limiting factors in phototherapy. While most attention has been placed on delayed erythema of normal skin, it is important to remember that this is only one manifestation of ultraviolet effects. Absence of erythema does not mean that no effects have occurred. DNA damage, dyskeratotic cells, and melanogenesis occur after suberythemogenic exposure doses of UV and other effects are also likely to occur.

Repeated episodes of phototoxicity induced by a variety of wavebands leads to improvement of psoriasis.(Table). The precise mechanisms are unknown and may vary with the waveband used. After 10 to 35 treatments the skin feels and looks normal; the disease process may stay in remission for days to years but almost always eventually recurs. UV exposure of skin of animals and humans causes a transient decrease in DNA, RNA and protein synthesis. It is assumed that ultraviolet radiation of psoriasis also causes decrease in DNA synthesis which influences the return of more normal cell kinetics. Medications known to be effective in treating psoriasis are those which suppress mitoses, DNA synthesis or cell proliferation. It is not known if psoriatic cells are more sensitive to ultraviolet damage because they are more metabolically active or are replicating more rapidly. It is not known if the therapeutic effect results because ultraviolet radiation decreases macromolecular synthesis in all psoriatic cells or if there is selective inhibition or killing of a smaller population of highly proliferative cells. When normal skin is exposed to erythemogenic ultraviolet radiation, the transient decrease in DNA synthesis is followed by a rebound hyperproliferative state which can lead to thickening of the epidermis or to desquamation. This proliferogenic stimulus, most marked by shorter ultraviolet wavelengths, may also occur in psoriatic tissue; it may act to normalize differentiation or it may be an undesirable antitherapeutic influence to be overcome by other more therapeutic ultraviolet effects.

TABLE

Examples of Effective Forms of Psoriasis Phototherapy
With and Without Adjunctive Therapeutic Agents*

Repeated Phototoxicity is Therapeutic for Psoriasis:

   Solar Therapy (9)

   Goeckerman (Tar-UVB) (25-28,31-33,35)

   UVB (7-11,32,39,40)

   Tar-UVA (36)

   Lubricant-UVB (32,39,40)

   UVA (15)

   Topical Psoralen-UVA

   PUVA (Oral Psoralen-UVA) (10,11)

*Numbers refer to references in this chapter

There are a variety of other mechanisms by which ultraviolet radiation may benefit psoriasis. Ultraviolet radiation may alter the recruitment of cells from a resting phase to a proliferative phase, alter differentiation, or act as gene regulator or deregulator. Phototherapy may compromise specific cells such as lymphocytes or polymorphonuclear leukocytes which are necessary for the pathophysiology of the disease. Metabolites or mediators which are necessary to maintain the hyperproliferative state may be photochemically altered. More than one mechanism may be involved. The gradual continuous decrease in the thickness and scaling of a plaque and the complete remission of disease after cessation of treatment may not necessarily result from the same molecular mechanism. The net effect of each episode of phototoxicity may be to decrease macromolecular synthesis in abnormal keratinocytes. The ultraviolet-induced remission may result from an effect on blood vessels, white blood cells, or on keratinocyte gene regulation. All of the treatments by some cumulative effect act in concert to re-establish normal kinetics and differentiation.

ACTION SPECTRUM

When considering the spectral sensitivity of skin, broadband exposure devices which are essentially UVB sources improve psoriasis when erythemogenic doses are used (7,8,9,10,11). High pressure mercury lamps, mercury xenon sources, fluorescent "sunlamps" and sunlight are all moderately effective if repeated erythemogenic exposure doses are used. With all of these sources, however, total clearing of psoriasis without the use of adjunctive therapeutic agents is not regularly achieved.

Within the range of 254 nm to 313 nm the action spectrum for clearing of psoriasis with repeated doses of UVR has been determined and compared to the erythema action spectrum of adjacent unaffected skin (12). 253.7 nm radiation was obtained from a low pressure mercury source filtered by a cyanine liquid filter which decreases stray light to less than 10^{-5}. Radiation at 280, 290, 295, 300, and 313 nm was obtained with a 5000 W mercury-xenon arc and a holographic grating monochromator (4 nm HBW). At each of these wavebands the minimal exposure dose required to induce erythema on uninvolved normal skin was determined. Separate 5 cm^2 sites within psoriatic plaques were then exposed to doses ranging from 0.1 to 2.5 times MED at each wavelength daily (6 times/week) for 4 weeks.

For all wavelengths shorter than 295 nm, no improvement occurred at any site at any exposure dose. At 254 nm and 280 nm repeated daily exposures to 20 and 50 times MED caused no change in psoriasis, establishing that at shorter wavelengths the action spectrum for induction of delayed erythema in normal skin is quite different from the action spectrum for phototherapy of psoriasis. The shorter wavelength is far more erythemogenic than therapeutic; its presence in treatment sources may unnecessarily limit the exposure doses used during phototherapy.

The marked decrease in effectiveness of wavelengths shorter than 295 nm may be explained by optical and structural differences between psoriatic and normal skin. In psoriasis the proliferative compartment at the bottom of the epidermis is thicker and its average depth is deeper than in normal skin. There is also an increased thickness of stratum corneum and an increased scale. Epidermal proteins absorb wavelengths shorter than approximately 290 nm, and optical scattering also increases at shorter wavelengths. In general, transmission decreases exponentially with increased thickness of an optical barrier, and scattering is inversely related exponentially to wavelength. Transmission of wavelengths shorter than 290 nm to the proliferating epidermal cells in psoriasis may therefore be markedly reduced when compared with normal skin.

Within the UVB region the threshold dose for induction of delayed erythema of normal skin and the minimal daily dose required for therapeutic response are quite similar. Studies with monochromatic radiation showed that MED doses of 295 nm radiation improved most patients. At 300 and 305 nm, sites repeatedly exposed to doses less than one MED may clear completely. 313 nm sites also may heal psoriasis at less than MED doses but exposure of 1.5 MED and greater may cause edema and increase in induration and scaling (12).

The action spectra for delayed erythema of normal skin and of therapy of psoriasis may also be closely aligned in the UVA region. UVA is markedly less erythemogenic than is UVB requiring as much as 1000 times more energy to cause redness in normal skin. At these longer wavelengths therapeutic effects also require 1000 times greater doses. Fischer (13) investigated the therapeutic effects of defined narrow wavebands at longer wavelengths by irradiating circular areas within psoriatic plaques. Using appropriate filters spectral lines were isolated and different experimental sites were exposed to 313 nm band (one-half, one and two times predetermined MED of adjacent normal skin), 334 nm plus 365 nm band (30 J/cm^2), 365 nm band (30 J/cm^2) and 404 nm band (30 J/cm^2). At these exposure doses the 313 nm band was the most effective and degree of improvement correlated with increasing exposure dose. The MED to these longer wavelengths was not measured; it is estimated that, with the mercury source at 334 or 365 nm, 30 J/cm^2 is near or below the MED for most light-skinned Caucasians. No significant improvement was seen in any site treated with 405 nm radiation.

Young and van der Leun (14) found that 14 daily exposures to 7-14 J/cm^2 of UVA had no effect on psoriasis. They used a high pressure mercury arc filtered through its own glass envelope plus an additional 3 mm of window glass to eliminate UVB radiation. No erythema was reported to occur on normal skin. Fisher (13) found 30 J/cm^2 of primarily 365 nm radiation to be ineffective in most test sites. Normal skin was not irradiated but this dose may be minimally erythemogenic to some but not all light Caucasians. By analyzing the studies done with monochromatic radiation it is possible to propose an action spectrum for phototherapy of psoriasis (see Figure 1).

Parrish (15), using larger exposure doses of 50 to 300 J/cm^2, found broadband UVA (320-400 nm) to be effective in clearing psoriasis from small exposure sites. Comparative studies showed these doses of UVA to be as effective as doses of UVB 1000 times smaller. MED to UVA with the same exposure sources was 20 to 100 J/cm^2. This observation is of interest because it does suggest that within portions of the UVA band the action spectrum for therapeutic effect is similar to the action spectrum for delayed erythema of normal skin, but the use of such large exposure doses with conventional sources is impractical and possibly unsafe for whole-body treatments.

Figure 1 Dashed line   = human erythema action spectrum
Solid line   = action spectrum for phototherapy of psoriasis based on several studies
Closed circles = reciprocal of lowest daily dose which clears psoriasis (295, 300, 305, 313 nm from ref# 12, 365 from ref# 15).
Open circles = reciprocal of daily exposure doses tried and found not to be effective (P from Parrish (12); F from Fischer (13); Y from Young & van der Leun (14)). Action spectrum for phototherapy must fall below these points.

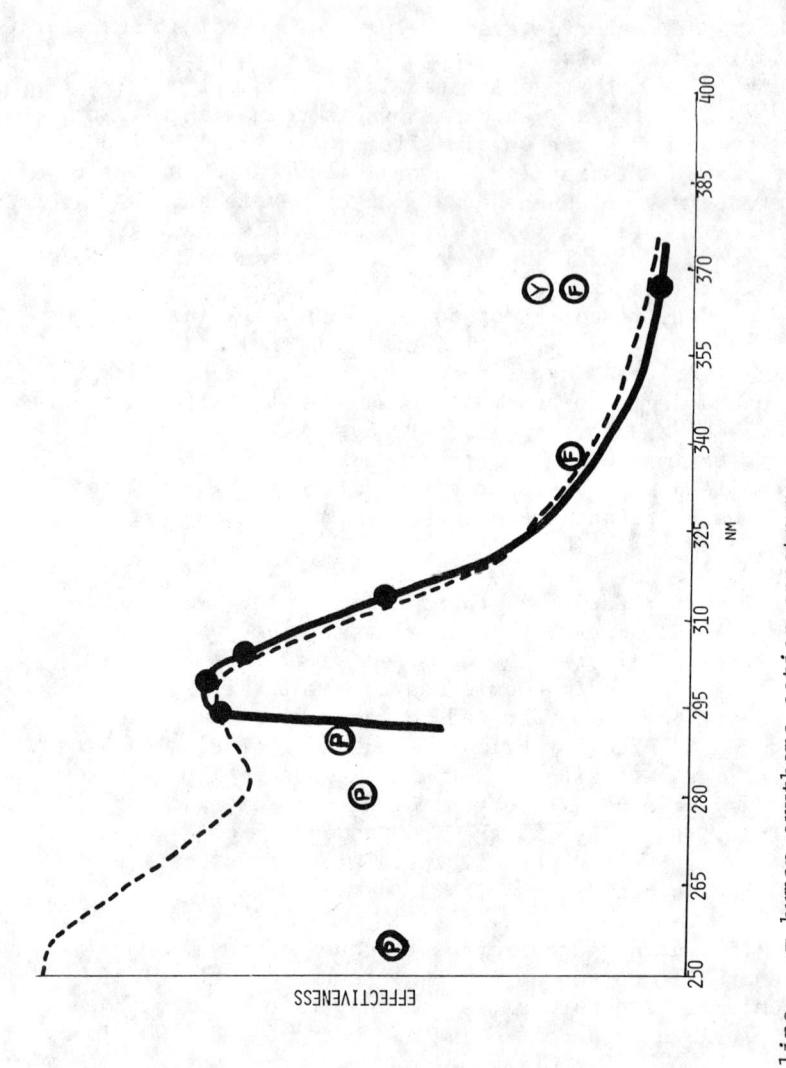

Polychromatic radiation with maximum output between 300 and 325 nm is claimed to be a practical and effective treatment. This treatment has been referred to as "selective ultraviolet phototherapy" (SUP) because it decreases the more erythemogenic shorter wavelength ultraviolet component present in many artificial ultraviolet sources, is therefore more like solar radiation (16) and may utilize radiation which is transmitted deeper into the psoriatic tissue (17). These spectral properties are consistent with the observations cited above. Ultraviolet radiation of wavelengths less than 290 is more erythemogenic than therapeutic (12). Longer wavelengths (>325 nm) are therapeutic but require massive doses (15). Longer wavelengths, however, penetrate more deeply into tissue and may be more likely to affect abnormal blood vessels and cellular infiltrates important to the pathophysiology of psoriasis. Compared to wavelengths longer that 310 nm ultraviolet radiation of 290-310 nm requires lower exposure doses to be both erythemogenic and therapeutic. The region between 300 and 320 nm may therefore represent an important "compromise region" (15) for phototherapy of psoriasis. In patients treated with erythemogenic exposure doses results with SUP have been reported to be good (18), better than more conventional broad spectrum ultraviolet treatment (16,19,20) and, in selected cases, as effective as PUVA (17). Other studies have found PUVA to be a better treatment for more difficult cases of psoriasis (21). Because PUVA is activated primarily by UVA, its better results may be partially explained by optical arguments related to a greater portion of photochemistry occurring deeper within the tissue (17).

A natural spectral shift toward longer wavelengths may partially explain why heliotherapy at the Dead Sea appears to be more effective than solar phototherapy at other locations (22,23, 24). Because the Dead Sea is 1200 feet below sea level, solar radiation must travel through a longer column of atmosphere. Due to the increase path length through air, shorter wavelengths are preferentially scattered and absorbed. This results in a net spectral shift to longer wavelengths at ground level. This effect is small and one might suppose that other mechanisms such as the allegedly higher aerosol content near the Dead Sea accounts for some of any biologically significant shift in the UVA/UVB ratio. It is also probable that soaking in the sea water alters optical properties of psoriasis in a way which increases the effectiveness of phototherapy.

Further narrowing of the wavebands used for phototherapy may result in increase in safety and effectiveness. Preliminary studies with monochromators show that complete clearing of small areas of psoriasis may be achieved with repeated exposure doses slightly less than the MED of normal skin (12). In order to clear psoriasis with broadband sources it is usually necessary to increase the exposure dose as the UV tolerance of normal skin increases; doses of five to ten times the original MED are often used toward the end of a course of therapy. This suggests that narrow waveband phototherapy at wavebands centering between 300 and 310 nm may have a wide safety margin when compared with broadband UVB sources. It is possible that certain wavebands have a net proliferogenic influence on the abnormal cells within psoriasis plaques. Parrish (15) observed that psoriasis initially cleared by 305 nm radiation relapsed when subsequently exposed to broad spectrum UVB (FS40 Westinghouse fluorescent bulbs). Further study of such waveband interactions is required.

ADJUNCTIVE AGENTS

Ultraviolet therapy of psoriasis often includes the topical use of tar or related topical compounds. Goeckerman first reported the beneficial effects of application of tar and exposure of the skin to ultraviolet radiation in 1925 (25). After evaluating other potential photosensitizing agents, he settled on crude coal tar which he believed would enhance the therapeutic effects of ultraviolet radiation (26). For more than 50 years this combination has been the customary treatment for generalized psoriasis and its effectiveness is without question (27,28). Until recent years, the mess, stain and offensive odor of tar made it necessary to hospitalize patients for this treatment. The current trend is to use ambulatory care facilities or day care units to provide tar and ultraviolet therapy.

The mechanism of topical tar as adjunctive agent for UV treatment of psoriasis and the exact role played by each therapeutic component are not known. Tar has been used since ancient times to treat skin diseases. It has been estimated that over 10,000 different components make up crude coal tar; most of these have not been identified. Tar alone has some definite but modest therapeutic effect on psoriasis (8,29,30), but less so than ultraviolet radiation (8,31,32). In some studies, ultraviolet radiation and tar used in combination appear to be more effective than either alone (31,32,33).

Topical crude coal tar, tar pitch and several commercially available tars are phototoxic, i.e. these compounds reduce the amount of ultraviolet necessary to induce injury of normal skin as manifest by delayed erythema. The action spectrum for this effect appears to be within the UVA. The photosensitizing capacity for most therapeutically used tars is not great; the threshold UVA dose necessary to induce delayed erythema is reduced at most by a factor of two (34).

It was assumed for decades that chemical photosensitization was an important aspect of tar plus ultraviolet radiation therapy but this is now in question. The treatment sources commonly used with topical tar include both UVA and UVB. However, the spectral power distribution of the sources and the spectral sensitivity of skin is such that phototoxicity (manifest by delayed erythema) occurs primarily from the UVB component of the sources. Considering the effects on normal skin, exposure is limited by UVB induced effects long before the sources have delivered enough UVA to induce tar photosensitization. Figure 2 is a diagrammatic conception of this dose relationship. Two studies compared the effectiveness of UVA plus tar with that of UVB plus tar in psoriasis therapy and found the UVB tar combination therapy to be more effective (8,35). However, the exposure doses of UVA in both these studies were insufficient to elicit tar photosensitization.

When using exposure doses adequate to cause cutaneous phototoxicity (manifest by delayed erythema), Parrish et al. (36) found UVA and UVB to be equally therapeutic to tar-treated psoriatic lesions. It was felt that the UVB therapeutic effect was independent of tar photosensitization because the exposure dose of UVB required to cause delayed erythema of normal skin was the same with or without the tar. Special sources, filters and long exposure times were necessary in order to obtain photosensitizing doses of UVA without inducing phototoxicity from the much more erythemogenic UVB present in most ultraviolet treatment sources. Tar-UVA phototoxicity was shown by the fact that less UVA was required to elicit delayed erythema in the presence of tar than when skin was irradiated without tar application. The dose of UVA required to benefit tar-treated skin was large enough to be impractical with currently available sources.

There is another problem with the use of tar photosensitization to treat psoriasis. After application of crude coal tar or tar products to the skin, subsequent exposure to UVA may lead to an unpleasant, burning, or painful sensation that has been referred to as the "smarting reaction." Smarting begins relatively abruptly at some point during UVA exposure, is relieved simply by discontinuing the exposure, and may return when exposure is resumed. Smarting usually occurs at doses of UVA slightly less than that needed to produce delayed erythema in tar-treated skin, but considerable individual variation exists. The action spectrum appears to include UVA but may extend into visible wavelengths. The mechanism of smarting is not known. Cutaneous nerve endings may be directly affected by some photochemical event, or they may be indirectly affected via damage of other cells or components in the skin.

Therefore, considering clinical studies, the spectral power distribution of treatment sources most often used and the

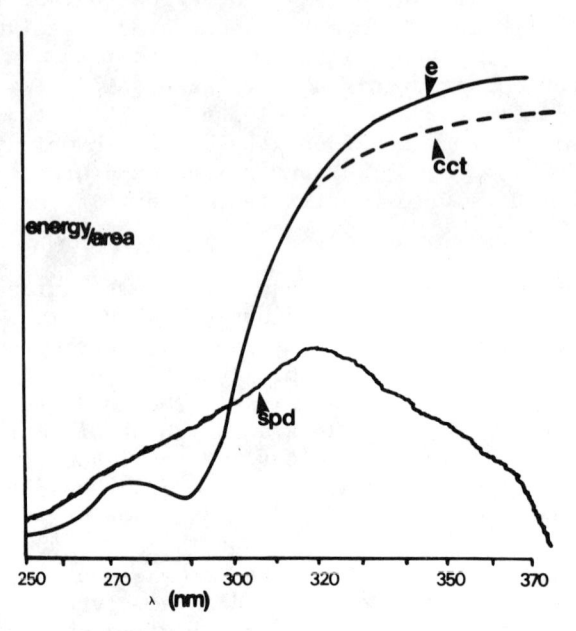

● = THRESHOLD FOR DELAYED ERYTHEMA
CCT = THRESHOLD FOR PHOTOTOXICITY OF TAR TREATED SKIN PI = 2
SPD = SPECTRAL POWER DISTRIBUTION OF PHOTOTHERAPY SOURCE

Figure 2
Diagramatic consideration of the threshold for delayed erythema without (e) and with (cct) crude coal tar. Energy/area is plotted on a log scale. Within the UVA waveband the application of tar lowers the threshold for delayed erythema by a factor of two (phototoxic index = 2). A typical UVB source (spd) may have a considerable portion of its energy in the UVA range but still threshold for erythema is exceeded in the UVB range before threshold is exceeded in UVA even in the presence of crude coal tar.

erythema action spectrum of tar-treated skin, it appears that the major therapeutic component of the popularly-used tar and ultraviolet radiation treatment is probably UVB. Other factors, such as tar, hospitalization, and patient-physician and patient-nurse interactions, also play an important role. Patient motivation is necessary for tolerance of the inconvience and messiness of the treatment and is also a component of the therapy. The vehicles used to apply the tar and the ointments used to treat scaling, dryness, and itching may also be therapeutic.

Tar is usually applied at a 1% to 6% concentration in an ointment or lubricant base. The vehicle used to apply tar may enhance the therapeutic effect of ultraviolet radiation. In fact, LeVine et al. (32) found that the lubricant vehicle when used in conjunction with repeated daily erythemogenic exposures to UVB was as effective as tar and ultraviolet therapy. It has been shown that hydrated petrolatum and other hydrophobic, nonvolatile substances alter the surface of psoriasis lesions so that less ultraviolet radiation is back-scattered from the surface of the scales (37). This reduced remittance is broad spectrum and does not appear to be due to ultraviolet absorption by the vehicle but is best explained by an immediate alteration of the optical properties of psoriatic plaques caused by matching of refractive index between the applied material and the numerous superficial flakes of stratum corneum. The optical effects may be cosmetic as well as therapeutic; less visible light is back-scattered and the lesions appear to be less silvery or white and more the color of normal skin. These optical effects probably account for a portion of the increased effectiveness of ultraviolet radiation when used in combination with topical tars; the vehicle used to apply the tar enhances the effect of ultraviolet radiation by increasing transmission to the proliferating cells of the epidermis. Because this effect does not occur to the same extent in normal skin (37,38), selective increased ultraviolet penetration into psoriasis occurs. Tar may actually absorb enough UV radiation to decrease effectiveness of phototherapy but if thin layers and low concentrations are used this effect will be small and may be offset by the therapeutic effects of tar alone and the optical changes due to the vehicle in which the tar is applied. Lubricants or tar vehicles may also facilitate the removal of scale.

Outpatient studies have shown that topical application of white petrolatum and subsequent exposure to erythemogenic doses of UVB either 3 times per week (39) or 5 times per week (40) leads to clearing of psoriasis vulgaris. When using broad spectrum UVB it appears that erythemogenic doses are required. Some investigators have claimed to achieve clearing with suberythemogenic exposures (41) to broadband sources but the studies need confirmation with careful dosimetry and larger numbers of patients.

Ultraviolet phototherapy has also been used in combination with other topical agents such as liquor carbonis detergens (42-44) and with dithranol (8,45,46). The relative effectiveness of ultraviolet radiation reported in these studies is variable and seems to depend in part on the ultraviolet doses used, the duration of the study, the enthusiasm of the investigator, and the thoroughness of the patient. The therapeutic effects of dithranol appear least affected by ultraviolet radiation. The combination of ultraviolet phototherapy with orally administered retinoic acid derivatives appears to offer the short-term advantages of inducing remission with less number of phototherapy treatments and with less side effects than seen with the oral agent alone (47,48). Duration of remissions and long-term risks require further study.

SUMMARY

Successful phototherapy of psoriasis can be achieved by a variety of measures all of which induce phototoxic injury to cells. Repeated exposures are necessary. To date, all successful forms of phototherapy and photochemotherapy including tar and ultraviolet combinations (49,50) are known to induce damage to DNA. In normal skin, UV is known to cause transient decrease in DNA synthesis. At wavelengths longer than 290 nm the therapeutic action spectrum appears similar to the action spectrum for delayed erythema. However, at wavelengths shorter than 290 nm the two spectra begin to separate so that greater multiples of MED must be to achieve therapeutic effects. Wavelengths shorter than 290 nm become much more efficient in producing erythema of normal skin but less therapeutic: at 254 nm daily exposure to doses of 50 times MED does not seem to benefit psoriasis. This decreased therapeutic effectiveness is probably related to high optical density and increased scatter by the abnormal stratum corneum. The chromophore for therapeutic effect in psoriasis is unknown but it is most likely that photons must reach the proliferative compartment in order to be therapeutic. The major portion of the proliferative compartment is deep within the thickened psoriatic epidermis and these are more likely to be reached by electromagnetic radiation of longer wavelengths which are known to penetrate more deeply into tissue. It is also possible that radiation of blood vessels or other dermal structures is necessary for improvement or important in the induction of remission. Tar often has a minimal but definite therapeutic effect on psoriasis and this effect may add to benefit derived from phototherapy. Topical agents may be used to alter the optical properties of the surface to selectively increase transmission of ultraviolet into psoriatic plaque while less effectively altering the optics of normal skin. The long-term effects of aggressive erythemogenic phototherapy are not certain. Long-term tar and ultraviolet phototherapy may increase the incidence of skin cancer in psoriatics (51); both agents are well-known carcinogens.

There are several means by which phototherapy of psoriasis could be made safer and more effective. Methods of selectively increasing UV transmission into psoriatic tissue include refractive index matching, alteration of the optical properties of the surface of psoriasis, mechanical removal of scale, and selective elution of UV-absorbing substances from the abnormal tissue. Waveband selection may further improve effectiveness and reduce risks. Possibilities include the use of monochromatic radiation or sources with spectral power distribution matching phototherapy action spectrum. It may prove beneficial to alter the spectral power distribution of the source as therapy progresses. Intelligent use of waveband interactions requires more study.

REFERENCES

1. J.A. Parrish, Phototherapy of psoriasis and other cutaneous disorders, in: "Photomedicine," JD Regan, JA Parrish, eds, Plenum Press, New York (1980, in press).
2. J. Melski, L. Tanenbaum, J.A. Parrish, T.B. Fitzpatrick, H.L. Bleich and 28 participating investigators: Oral Methoxsalen photochemotherapy for the treatment of psoriasis: A cooperative clinical Trial. J. Invest. Derm. 68:328-335 (1977).
3. E.I. Saltzer, Relief from uremic pruritus, A therapeutic approach, Cutis 16:298-299 (1975).
4. B.A. Gilchrest, J.W. Rowe, R.S. Brown, T.I. Steinman, K.A. Arndt, Relief of uremic pruritus with ultraviolet phototherapy, N. Engl. J. Med. 297:136-138 (1977).
5. B.C. Schultz, H.H. Roenigk, Jr., Uremic pruritus treated with ultraviolet light, J. Am. Med. Assoc. 243:1836-1837 (1980).
6. B.A. Gilchrest, J.W. Rowe, R.S. Brown, T.I. Steinman, K.A. Arndt, Ultraviolet phototherapy of uremic pruritus, Ann. Int. Med. 91:17-21 (1979).
7. H.E. Alderson, Heliotherapy in psoriasis, Arch. Dermatol. 8:78-80 (1923).
8. R.E. Bowers, D. Dalton, D. Fursdon, J. Knoweldon, The treatment of psoriasis with U.V.R., dithranol paste and tar baths. Brit. J. Dermatol. 78:273-281 (1966).
9. J.A. Parrish, H.A.D. White, T. Kingsbury, M. Zahar, T.B. Fitzpatrick, Photochemotherapy of psoriasis using methoxsalen and sunlight: A controlled study, Arch. Dermatol. 113:1529-1532 (1977).
10. J.A. Parrish, T.B. Fitzpatrick, L. Tanenbaum, M.A. Pathak, Photochemotherapy of psoriasis with oral methoxsalen and longwave ultraviolet light, New Engl. J. Med. 291:1207-1212 (1974)

11. K. Wolff, T.B. Fitzpatrick, J.A. Parrish, F. Gschnait, B. Gilchrest, H. Hönigsmann, M.A. Pathak, L. Tanenbaum, Photochemotherapy for psoriasis with orally administered methoxsalen, Arch. Dermatol. 112:943-950 (1976).
12. J.A. Parrish, Action spectrum of phototherapy of psoriasis, J. Invest. Dermatol. 74:251 (1980).
13. T. Fischer, UV-light treatment of psoriasis. Acta Dermatovener. (Stockh.) 56:473-479 (1976).
14. E. Young, J.C. van der Leun, Treatment of psoriasis with long-wave ultraviolet light. Dermatologica 150:352-354 (1975).
15. J.A. Parrish, The treatment of psoriasis with longwave ultraviolet light (UV-A). Arch. Dermatol. 113:1525-1528, (1977).
16. H. Tronnier, H. Heidbüchel, Zur therapie der psoriasis vulgaris mit ultravioletten Strahlen. Z. Hautkr. 51:405-424 (1976).
17. H. Tronnier, Derzeitiger stand der photochemotherapie für die dermatologische. Praxis Dt. Derm. 25:265-276 (1977).
18. P. Mischer, Erste Erfahrungen mit der "selektivan ultravioletten Phototherapie" (SUP) der psoriasis vulgaris. Österreichische Dermatologische Gesellschaft, Jahressitzung 18:6 (1977).
19. H. Pullmann, A.C. Wichmann, G.K. Steigleder, Praktische erfahrungen mit verschiedenen phototherapieformen der psoriasis - PUVA, SUP-, Teer-UV-Therapie. Z. Hautkr 53:641-647 (1978).
20. F. Schröpl, Zum heutigen Stand der technischen Entwicklung der selektiven Fototherapie. Dt. Derm. 25:499-504 (1977).
21. H. Hönigsmann, P. Fritsch, E. Jaschke, UV-therapie der psoriasis: Halbseitenvergleich zwischen oraler photo-chemotherapie (PUVA) und selektiver UV-phototherapie (SUP). Z. Hautkr. 21:1078-1082 (1977).
22. N. Sapeika, Treatment of psoriasis at the Dead Sea, S.A. Med. J. 50:2021 (1976).
23. I.L. Schamberg, Treatment of psoriasis at the Dead Sea. Int. J. Derm. 17:524-525 (1978).
24. L.H. Goldberg, A. Kushelevsky, Ultraviolet light measurements at the Dead Sea, in: "Psoriasis: Proceedings of the Second International Symposium," E.M. Farber, A.J. Cox, P.H. Jacobs, N.L. Nall, eds, Yorke Medical Books, New York (1976).
25. W.H. Goeckerman, The treatment of psoriasis. Northwest Med. 24:229-231 (1925).
26. W.H. Goeckerman, Treatment of psoriasis: continued observations on the use of crude coal tar and ultraviolet light. Arch. Dermatol. Syphilol. 24:446-450 (1931).
27. H.O. Perry, C.W. Soderstrom, R.W. Schulze, The Goeckerman treatment of psoriasis. Arch. Dermatol. 98:178-182 (1968).
28. W.M. Sams, Jr., Phototherapy of psoriasis, in: "Sunlight and Man," M.A. Pathak, L.C. Harber, M. Seiji, A. Kukita, eds, T.B. Fitzpatrick, consulting ed, University of Tokyo Press, Tokyo (1974).

29. E. Young, Ultraviolet therapy of psoriasis: A critical study, Brit. J. Dermatol. 87:379-382 (1972).
30. J.E. Rasmussen, The crudeness of coal tar. Prog. Dermatol. 12:23-29 (1978).
31. J.W. Petrozzi, J.O. Barton, K. Kaidbey, A.M. Kligman, Updating the Goeckerman regimen for psoriasis. Brit. J. Dermatol. 98:437-444 (1978).
32. M.J. LeVine, H.A.D. White, J.A. Parrish, Components of the Goeckerman regimen, J. Invest. Dermatol. 73:170-173 (1979).
33. A.R. Marsico, W.H. Eaglstein, G.D. Weinstein, Ultraviolet light and tar in the Goeckerman treatment of psoriasis. Arch. Dermatol. 112:1249-1250 (1976).
34. L. Tanenbaum, J.A. Parrish, M.A. Pathak, R.R. Anderson, T.B. Fitzpatrick, Tar phototoxicity and phototherapy for psoriasis, Arch. Dermatol., 111:467-470, (1975).
35. A.R. Marsico, W.H. Eaglstein, Role of longwave ultraviolet light in Goeckerman treatment, Arch. Dermatol. 108:48-49 (1973).
36. J.A. Parrish, W.L. Morison, E. Gonzalez, T. Krop, H.A.D. White, R. Rosario, Therapy of psoriasis by tar photosensitization, J. Invest. Dermatol. 70:111-112 (1978).
37. M.J. LeVine, J. Hu, R.R. Anderson, J.A. Parrish, Reflectance of psoriatic plaques, Abstract presented at the 7th Annual Meeting of the American Society for Photobiology, June 24-28, p. 175, Pacific Grove, California, Programs and Abstracts (1979).
38. N.R. Schleider, R.S. Moskowitz, D.H. Cort, S.N. Horwitz, P. Frost, Effects of emollients on ultraviolet-radiation-induced erythema of the skin, Arch. Dermatol. 115:1188-1191 (1979).
39. R.M. Adrian, M.J. LeVine, J.A. Parrish, Treatment frequency for outpatient phototherapy of psoriasis. A comparative study, J. Invest. Dermatol. 74:251 (1980).
40. M.J. LeVine, J.A. Parrish, Outpatient phototherapy of psoriasis. Arch. Dermatol. 116:552-554 (1980).
41. P. Frost, S.N. Horwitz, R.V. Caputo, S.M. Berger, Tar gel-phototherapy for psoriasis, Arch. Dermatol. 115:840-846 (1979).
42. C.C. Ellis, The treatment of psoriasis with liquor carbonis detergens, J. Invest. Dermatol. 10:455 (1948).
43. M.A. Everett, E. Daffer, C.M. Coffey, Coal tar and ultraviolet light, Arch. Dermatol. 84:473-476 (1961).
44. M.A. Everett, J.V. Miller, Coal tar and UV light. II. Cumulative effects, Arch. Dermatol. 84:937-940 (1961).
45. J.T. Ingram, The approach to psoriasis, Brit. Med. J. Sept. 12, 591-594 (1953).
46. S. Comaish, Ingram method of treating psoriasis, Arch. Dermatol. 92:56-58 (1965).

47. H. Beierdorffer, A. Wiskemann, Kombinierte therapie der psoriasis mit einem aromatischen retinoid (RO 10-9359) und UVB-Bestrahlungen, Akt Dermatol 4:183-187, (1978).
48. G.K. Steigleder, C.E. Orfanos, H. Pullmann, Retinoid-SUP-therapie der psoriasis, Z. Hautkr 54:19-23 (1978).
49. J.F. Walter, R.B. Stoughton, P.R. DeQuoy, Suppression of epidermal and proliferation by ultraviolet light, coal tar and anthralin, Brit. J. Dermatol 99:89-96 (1978).
50. M.A. Pathak, R.K. Biswas, Skin photosensitization and DNA crosslinking ability of photochemotherapeutic agents. In the Society of Investigative Dermatology, Inc. National Meeting, Washington, D.C., p. 21 (1977).
51. R.S. Stern, S. Zierler, J.A. Parrish, Skin carcinoma in patients with psoriasis treated with topical tar and artificial ultraviolet radiation, Lancet, April 5, pp. 732-735 (1980).

ADVANCES IN PHOTOTHERAPY OF NEONATAL

HYPERBILIRUBINEMIA

Thomas R. C. Sisson

Department of Pediatrics, Rutgers Medical School
Piscataway, N.J. and Perth Amboy General Hospital
Perth Amboy, N.J. 08861

The advent of phototherapy for the treatment of jaundice in the newborn, originally reported in the studies of Cremer et al.[1] in 1958, has stimulated further investigation of the transport, binding, and metabolism of bilirubin. It is not an exaggeration to say that this mode of therapy is considered by many to have been responsible for much of the recent broad medical interest in the photobiological effects of visible light. Because of this there has been a further interest in the study of varied light sources — fluorescent to laser.

The measurement of the bilirubin concentration in plasma has been — and still remains — the most widely used prognostic index of bilirubin encephalopathy in the neonate. It is, however, a woefully inexact and even deceptive guide to the management of hyperbilirubinemia and the prevention of brain toxicity from it. Although statistical wisdom indicates that severe brain damage occurs only infrequently at plasma bilirubin concentrations below 18-20 mg/dl, experience has shown that significant injury will occur at levels as low as 15 mg/dl in full-term infants[2], and at 9-10 mg/dl (or less) in sick, immature infants[3].

It is generally agreed that unconjugated bilirubin, not its conjugated form, is neurotoxic. If in some way released from its transport in plasma, it will become attached to phospholipids in the neurone, leading to the cell's damage. It is now thought that bilirubin acid molecules enter brain cells because of an attachment to membranes, especially mitochondrial[4], not because of lipophilic characteristics[5].

As plasma bilirubin concentrations cannot be reliable guides to treatment, great effort has been made to find alternative techniques. Measurement of reserve albumin-binding capacity[6] and the saturation index of albumin[7] are widely used.

Broderson[4] and Levine[8] have pointed out[4] that measure of "free" bilirubin in plasma has not been successfully accomplished by the Sephadex gel filtration or any other method. Though this may be disputed by adherents of these methodologies, it is difficult to give whole-hearted reliance to such measurements in any event.

In practical terms, the combined determinations of plasma concentration and binding capacity are far more trustworthy and allow for more certain assessment of safety.

A new instrument has been described by Lamola et al.[9] in which very small amounts of blood are studied by front-face fluorometry and the amount of bound bilirubin, the total bilirubin, and the bilirubin binding capacity of the blood are determined. This technique combines the best features of concentration and binding measurements in virtually one operation. The instrument is not yet in general use, but has intriguing capabilities for clinical application and as a research tool, for its use offers new information on the partition of bilirubin among blood components, and on the production of photoisomers of bilirubin in blood.

New insights have been gained into the fate of bilirubin in vivo under visible light irradiation. Ostrow et al.[10] demonstrated that the photoproduct in Gunn rat bile was a true isomer of bilirubin IXa. Shortly after, Stoll et al.[11] produced two photoisomer pairs anaerobically, and that they were excreted in rat bile after I. V. administration. Confirmation of these observations was soon made by Lightner et al.[12], and it was concluded that two pairs of photoisomers are formed from irradiation of unconjugated bilirubin IXa, and that the photoisomerization is rapidly reversible, although one pair readily converts to UCB-IXa in the dark, and the other is stable in the dark. The detailed biochemistry of the photoisomers of bilirubin has been reported by McDonough et al.[13] and by Lightner et al.[12,14] as well as by Cohen and Ostrow[15].

Jori, Rossi and Rubaltelli[16] have shown that visible light irradiation of bilirubin removed from albumin by gel filtration induces a covalent binding of the pigment or its photoadducts. This binding appears to be dependent upon the amount of light energy involved. These observations have relevance in consideration of the apparently increased efficiency of phototherapy as the energy output of the light source is increased[17].

Potential hazards of intense visible light irradiation exist. Detrimental effects upon platelet number and function have been demonstrated by Maurer et al.[18]. Likewise, such irradiation has been shown to adversely affect erythrocyte structure and enzymes[19]. Haemolysis under phototherapy was found to be induced by the inhibition of erythrocyte G-6-PD and glutathione reductase indirectly by the reduction of riboflavine in red cells[20].

Visible light irradiation can produce DNA modifications in pro- and eukaryotes[21], and will cause an increase of DNA cross-linkage and chromatid breakage in cultured embryonic mouse tissues[22].

New studies indicate that phototherapy causes jejunal lactase deficiency in the newborn. This observation was made originally by Bakken[23] in infants fed human breast milk.

Recently Sisson[24] in a controlled study of jaundiced infants fed diets containing lactose who received phototherapy, demonstrated significantly abnormal stools, frequent, discolored, watery, with low pH and containing reducing substances, in those subjects. Lactose tolerance tests were also abnormal. In contrast infants under phototherapy fed lactose-free formula did not exhibit these abnormalities. Infants who were jaundiced but who did not receive phototherapy did not show abnormal stools or lactose tolerance tests, indicating that the presence of hyperbilirubinemia alone does not induce lactase deficiency, but rather that phototherapy causes these unwanted effects — either directly on the cells of the intestinal lumen, or, more plausibly, as an inhibitory influence of unconjugated bilirubin and/or its photoisomers on lactase production (Tables 1, 2).

The use of phototherapy for the purpose of reducing plasma bilirubin concentrations in the infant obliterates the normal biorhythms of human growth hormone[25,26]. Plasma HGH, under cycled light has not only regular ultradian rhythms of about 4-6 hours duration, but also a circadian rhythm whose peak, as in the adult, occurs near midnight. These rhythms are erased in infants under constant light irradiation (Figure 1). Similarly blood calcium and glucose levels observed under cycled light are distorted under constant phototherapy[26]. The normal pattern of REM and non-REM sleep was shown to be distorted, as well, by phototherapy[27].

In these studies it was demonstrated, however, that within a few hours of discontinuance of phototherapy these abnormalities disappeared.

Intriguing evidence has been developed to indicate that visible light (especially 420-470 nm) will induce liver enzyme activity[28]. Gunn rat livers, homozygous and heterozygous, were perfused with bilirubin added to perfusate in concentration of 20 mg/dl. Half of the livers were exposed to fluorescent light (420-470 nm).

Table 1. Number of Infants with Abnormal Stools and Reducing Substances

Formula	Reducing Substances −	Reducing Substances ±	Reducing Substances +	Stools Normal	Stools Abnormal*
I Control					
Isomil	8	2	0	10	0
Similac With Iron	8	3	1	10	2
II Jaundice					
Isomil	8	2	0	9	1
Similac With Iron	7	1	1	8	1
III Jaundice & Phototherapy					
Isomil	10	2	0	12	0
Similac With Iron	3	2	15	2	17

*Abnormal stools = green, loose, watery.

Table 2. Lactose Tolerance Test (LTT)

Formula Group	Normal LTT*	Abnormal LTT
	— No. of infants —	
I Control		
Isomil	9	1
Similac With Iron	11	1
II Jaundiced		
Isomil	10	0
Similac With Iron	9	1
III Jaundiced & Phototherapy†		
Isomil	11	2
Similac With Iron	4	16

*Normal LTT defined as a rise in blood glucose of greater than 20 mg/100 ml.
†Tests were performed 20-36 hours after phototherapy started.

Cytochrome P450, cytochrome b5, p-nitrophenol glucuronidation, benzo-a-pyrene and aniline hydroxylation, and aminopyrine demethylation were significantly increased — almost two-fold — at the end of one hour. These observations, if they can be confirmed, have implications in the enhancement of drug metabolism by hepatic cell enzyme induction from light (Table 3).

Studies of the efficacy of various fluorescent light sources[17,20] have been supplemented by experimental evidence that quartz-halide lamps of high intensity are also effective in producing a decline in plasma bilirubin in jaundiced infants[29].

It has been suggested that infants with hyperbilirubinemia might have impaired responses to infection. Rubaltelli et al.[30] and Granati et al.[31] have shown that the lymphoproliferative response to phytohemagglutinin (PHA-M) was inhibited in jaundiced infants before and after light treatment. The effect of bilirubin and photobilirubin on T-cell function was equal, suggesting that

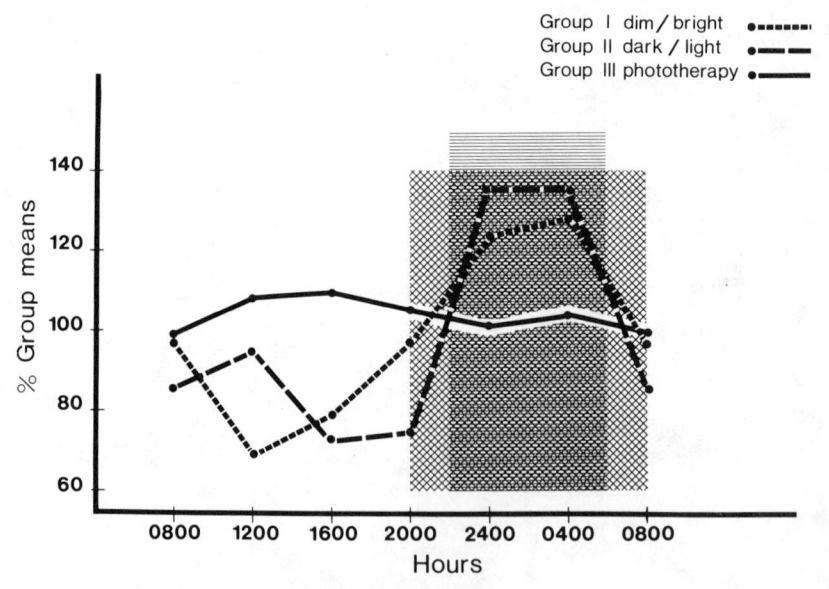

Fig. 1. Level of neonatal plasma HGH under two light/dark cycles; I = dim (5 ftcd)/light (100 ftcd), 2: dark (0 ftcd)/light (100 ftcd); and under constant phototherapy (420-470 nm, 24 μw/cm^2.nm). The observed circadian rhythm under cycled light is erased under constant light.

Table 3. Enzyme Activity of Perfused Gunn Rat Livers — μmole/gm

ENZYME	DARK jj	DARK jj	DARK Jj	LIGHT jj	LIGHT jj
Aminopyrene demethylation	3.028	3.77	1.835	5.175	5.089
Aniline dehydroxylation	0.151	0.187	0.147	0.301	0.335
Cytochrome P450	0.435	0.442	0.387	0.608	0.762
Cytochrome b5	0.792	0.551	0.554	0.808	0.794
Benzo (a) pyrene hydroxylation	0.569	0.772	0.351	0.924	1.743
p-Nitrophenol glucuronidation	2.592	1.350	2.082	4.607	4.639 / 5.020

the photoadducts of bilirubin have similar inhibitory effect upon immune responses.

These considerations of neonatal phototherapy are but the latest of an accumulated body of phenomena, mostly isolated, found to be involved in this clinical procedure, and are not intended to form a compendium of studies on bilirubin metabolism. The breadth of investigations stimulated by phototherapy, encompassing photochemistry, photophysics, and photobiology in a wide sense proves the interest excited by such a simple therapeutic tool. The scatter of this variets of studies in so many disciplines suggests that it is necessary to refine our understanding of the effects of visible light at cellular and molecular levels. The effects of light may be directly photochemical, photosensitizing and may have indirect and direct influences upon essential enzyme activity with clinically observable results. The manipulation of total light exposure alters biorhythmic metabolic activity with perhaps profound biologic effects, especially on the developing human organism. The possibility that drug and poison detoxification may be enhanced by phototherapy is of interest to yet another discipline-pharmacology. The use of intermittent and pulsed light exposure has received scant attention but may prove experimentally fruitful.

It seems reasonable to say, in the end, that studies in the many aspects of phototherapy call for coherent research effort.

REFERENCES

1. Cremer, R. J., Perryman, P. W., and Richards, D. H.: Influence of light on the hyperbilirubinemia of infants, Lancet 1: 1994, 1958.
2. Boggs, T. R., Jr., Hardy, J. R., and Frazier, T. M.: Correlation of neonatal serum total bilirubin concentrations, and developmental status at age 6 months, J. Pediat. 71:553, 1967.
3. Gartner, L. M., Snyder, R. N., Chabron, R. S., and Bernstein, J.: Kernicterus: High incidence in premature infants with low serum bilirubin concentrations, Pediatrics 45:906, 1970.
4. Broderson, R.: Bilirubin: transport in the newborn infant, reviewed with relation to kernicterus, J. Pediat. 96:349, 1980.
8. Levine, R. L.: Bilirubin: worked out years ago? Pediatrics, 64:380, 1979.
5. Mustafa, G., and King, T. E.: Binding of bilirubin with lipid. A possible mechanism of its toxic reaction in mitochondria, J. Biol. Chem. 245:1084, 1970.
6. Porter, E. G., and Waters, W. J.: A rapid micromethod for measuring the reserve albumin binding capacity in serum from newborn infants with hyperbilirubinemia, J. Lab. Clin. Med. 67:660, 1966.

8. Odell, G. B., Storey, G. N. B., and Rosenberg, L. A.: Studies in kernicterus. III. The saturation of serum proteins with bilirubin during neonatal life and its relationship to brain damage at 5 years. J. Pediat. 76:12, 1970.
9. Lamola, A. A., Eisinger, J., Blumberg, W. E., Patel, S. C., and Flores, J.: Fluorometric study of the partition of bilirubin among blood components: basis for rapid microassays of bilirubin and bilirubin binding capacity in whole blood, in Analyt. Blodiem. 100:25-42, 1979.
10. Ostrow, J. D., Berry, C. S., Knodell, R. G., and Zarembo, J. E.: Effect of phototherapy on bilirubin in Man and Rat, in "Bilirubin Metabolism in the Newborn (II). Birth Defects: Orig. Art. Ser. 12:81-92, 1976. Excerpta Medica, N. Y. ed. D. Bergsma and S. H. Blondheim.
11. Stoll, M. S., Zenone, E. A., and Ostrow, J. D.: Preparation and properties of bilirubin photoisomers. Biochem. J. 183: 139, 1979.
12. Lightner, D. A., Wooldridge, T. A., and McDonough, A. F.: Configurational isomerization of bilirubin and the mechanism of jaundice phototherapy. Biochem. Biophys. Res. Commun. 86:235, 1979.
13. McDonough, A. F., Lightner, D. A., and Wooldridge, T. A.: Geometric Isomerization of bilirubin IX-a and its dimethyl ester., J. Chem. Soc. Chem. Commun. Jan. 1979, pp. 110-112.
14. Lightner, D. A., Wooldridge, T. A., and McDonough, A. F.: Photobilirubin: an early bilirubin photoproduct detected by absorbance difference spectroscopy., Proc. Nat'l Acad. Sci. USA 76:29, 1979.
15. Cohen, A. N., and Ostrow, J. D.: New concepts in phototherapy: Photoisomerization of Bilirubin IXa and potential toxic effects of light. Pediatrics 65:740, 1980.
16. Jori, G., Rossi, E., and Rubaltelli, F. F.: Evidence for visible light—induced covalent binding between bilirubin and serum albumin "in vitro" and "in vivo." Trans. Amer. Soc. Photobiol., 1979.
17. Sisson, T. R. C., Kendall, N., Shaw, E., and Kechavarz-Oliai, L.: Phototherapy of jaundice in the newborn. II. Effect of various light intensities. J. Pediat. 81:35, 1972.
18. Maurer, H. M., Fratkin, M., McWilliams, N.: Effects of phototherapy on platelet counts in low birth weight infants and on platelet production and life span in rabbits., Pediatrics 57:506, 1976.
19. Blackburn, M. G., Orzalessi, M. M., and Pigram, P.: Effect of light on fetal red blood cells in vivo. J. Pediat. 80:640, 1972.
20. Sisson, T. R. C., Slavin, B., and Hamilton, P. B.: The effects of broad and narrow spectrum fluorescent light on blood constituents, in "Bilirubin Metabolism in the Newborn (II)," D. Bergsma, S.H. Blondheim, eds., Excerpta Medica, N. Y. Birth Defects: Orig. Art. Ser. 12:122-133, 1976.

21. Speck, W. T., Rosencranz, H. S.: Intracellular DNA-modifying activity of phototherapy lights. Pediat. Res. 10:553, 1976.
22. Gantt, R., Parshad, R., Weig, R. A. G.: Fluorescent light-induced DNA cross-linkage and chromatid breaks in mouse cells in culture.
23. Bakken, A. F.: Intestinal lactase deficiency as a factor in the diarrhea of light-treated jaundiced infants., N. E. J. Med. 295:1483, 1976.
24. Sisson, T. R. C.: Advantages of lactose-free formula for jaundiced infants undergoing phototherapy. Ross Clin. Res. Conf.: LBW Infants Fed Isomil., 1979, p. 101.
25. Sisson, T. R. C., Katzman, G., Shahrivar, F., and Root, A. W.: Effect of uncycled light on plasma human growth hormone in neonates. Pediat. Res. 9:280, 1975.
26. Park, T. S., Padget, S., Fiorentino, T., Root, A. W., and Sisson, T. R. C.: Effect of phototherapy and nursery light on neonatal biorhythms. Pediat. Res. 10:429, 1976.
27. Sisson, T. R. C., Ruiz, M., Wu, T-K., and Afuape, O. S.: Sleep patterns of newborn infants under phototherapy. Pediat. Res. 11:411, 1977.
28. Sisson, T. R. C., Granati, B., Sonawane, R., Fiorentino, T.: Effect of light on the perfused Gunn rat liver. Pediat. Res. 12:399, 1978.
29. Sisson, T. R. C., Ruiz, M., Wu, K-T., and Afuape, O. S.: Comparison of "incandescent" and fluorescent light sources in phototherapy. Pediat. Res. 12:535, 1978.
30. Rubaltelli, F. F., Piovesan, A. L., Granati, B., Neidhart, C., and Semenzato, G.: The effects of phototherapy on immunologic competence: studies "in vivo" and "in vitro," Trans. Amer. Soc. Photobiol., 1978.
31. Granati, B., Colleselli, P., Felice, M., and Rubaltelli, F. F.: The effects of bilirubin, phototreated bilirubin and light on spontaneous motility and chemotaxis of leucocytes. Trans. Eur. Soc. Pediat. Res., 1980.

PHOTODYNAMIC THERAPY OF INFECTIONS

Caldas, L.R., Menezes, S. and R.M. Tyrrell

Biophysics Institute, Federal University of Rio de Janeiro, Rio de Janeiro, R.J., 21941, Brazil

1. INTRODUCTION

Photodynamic action involves the photo-oxidation of biological molecules in the presence of a sensitizer dye and oxygen (1). The phenomenon was first observed by Marcacci (2), but it is usual to credit the discovery to Raab (3) who in 1900 described the inactivation of Paramecia cultures with visible light in the presence of acridine orange. The same year Tappeiner (4) observed that the photodynamically active dyes were fluorescent. Two years later Ledoux-Lebard (5) reported that molecular oxygen was required for the reaction. Work during the next four decades mainly involved the determination of the nature of the substrates which could be damaged by photodynamic action and the chemical nature of the dyes. Along these lines Tappeiner (6) described the inactivation of the proteins diastase, invertase, papain and trypsin by the photodynamic action of eosin and noted that this was due to the destruction of aminoacids.

Harris (7) and Carter (8) showed that the aromatic aminoacids were much more photodynamically sensitive than the aliphatics, a finding confirmed more recently by Spikes and Straight (9). It is now known that all molecules of biological importance including proteins, DNA and RNA are susceptible to inactivation by appropriate photodynamic treatment. The ease with which such treatment also inactivated viruses, bacteria and animal cells led many researchers to consider the possibility of using photodynamic action in phototherapy. Table one summarizes some of the

TABLE I

A - PHOTODYNAMIC ACTION - POSITIVE CLINICAL RESULTS

Condition	Procedure	Results	Authors
Herpes simplex virus Types I and II	Proflavine plus white light (150 watt bulb)	Decrease of recurrence	Moore, C.D., 1972 (10).
Herpes simplex virus Types I and II	Neutral red and visible light	Disappearance of recurrence	Wallis and Melnick, 1973 - (11).
Vulval infection (Herpes genitalis)	Neutral red plus visible light. Repeated sessions	Increased healing. Decreased recurrence	Friedrich, 1973 (12).
Herpes simplex Types I and II	Proflavine plus light. Re-exposure at 8 hrs and 24-36 hours	More successful than neutral red. More successful with repetition	Jarrat and Knox,1974 (13). Melnick and Wallis, 1977. (14).
Herpes simplex Types I and II	Neutral red plus light. Re-exposure in 1 to 6 hours.	Improvement in healing time. Decrease in recurrence (some placebo effects)	Felber et al., 1973 (15).
Vulval infection (Herpes genitalis)	Proflavine plus light. Two treatments	Primary infection - symptoms relieved. Reduced recurrent infection	Kauffman et al.,1973 (16).
Herpes simplex Types I and II	Methylene blue (in phosphate buffer, pH7) plus white light	Increased healing. Some cases of decreased recurrence. Some cases of recurrence disappearing	Studies in progress (Rio de Janeiro)

B - PHOTODYNAMIC ACTION - NEGATIVE CLINICAL RESULTS

Genital herpes	Neutral red or phenol red (non photodynamic) plus light	No difference between the two treatments	Roome et al., 1975 (17).
Genital herpes	Proflavine plus light at 0,24 and 48 hours or iodoxuridine. Normal saline solution.	No difference between treatments	Taylor and Doherty, 1975. (18).
Oral, genital herpes	Neutral red plus light or placebo plus light. Repeated 4 to 6 hours	No difference	Myers et al., 1975 (19); 1976 (20).

clinical experience in this area.

This table illustrates the extreme variation in effectiveness of such treatments. We will return to this point later. Since a wide review of phototherapy based on photodynamic action is not practical in this session, we shall first summarize some of the molecular aspects of photodynamic action and then return to clinical aspects of the phototherapy of infections, including a description of our experience with methylene blue.

2. PHOTODYNAMIC DYES

Soon after the discovery of photodynamic action, it was thought that only fluorescent dyes could be photodynamically active. This was later found not to be true, and most attempts to find shared chemical, photochemical or structural properties among photodynamic compounds were unsuccessful. The agents belong to many different chemical groups including vitamins, antibiotics, polycyclic hydrocarbons, thiazines, oxazines, etc. Santamaria and Prino (21) listed more than 400 photodynamically active dyes. This list could now be expanded considerably.

Based on their physico-chemical properties, Bourdon and Schnuriger (22) proposed a classification of the most commonly used dyes into three classes : Class I - those dyes which have great affinity for O_2, good reducing agents ; Class II - dyes which react easily with hydrogen donors, good sensitizers of textile fibres ; Class III - dyes with chemical structure similar to those of class II, but without effect on textile fibres. However this classification did not encompass all the known dyes. In an extensive series of tests the efficiency of a large list of dyes in sensitizing phages to visible light was measured and it was found that the thiazines, oxazines and acridines are the most efficient (23). As expected the photodynamic effects depend not only on the dye used and the specific treatment conditions but also the localization of the dyes and their sites of action. Variation in experimental conditions could readily account for the striking differences reported by some authors with the same dye. For instance Ito (24) concluded that methylene blue is located outside of the cells in his experiments with the photodynamic inactivation of S. cerevisiae using this dye (Fig. 1). In our own laboratory we have shown (unpublished data) that the same dye localizes in the interior of E. coli.

SENSITIZERS AND DAMAGE SITES

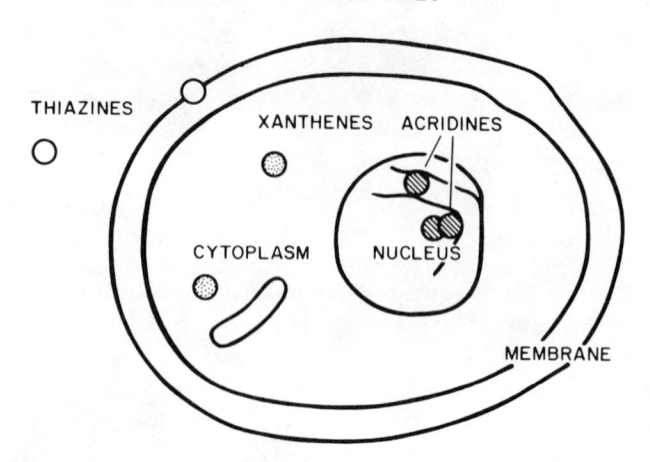

Fig. 1 : A schematic representation of localization of dye sensitizers and presumed sites of photodynamic damage. Redrawn from Ito (1978)

3. BIOLOGICAL LEVELS POTENTIALLY DAMAGED BY PHOTODYNAMIC ACTION

Both in vivo and in vitro experiments have provided strong evidence that all levels of biological organization are potentially damaged by photodynamic action under appropriate conditions. We shall mention here some of the molecules of biological importance which are damaged or destroyed by such treatment.

a) Aminoacids

It has been clearly demonstrated that aromatic aminoacids are more susceptible to photodynamic action than the aliphatic aminoacids (25, 26, 27). The reactive intermediates and final products are varied because the reactions may occur via Type I, Type II or both processes (see later) depending on the experimental conditions (28).

b) Proteins

The effect of photodynamic action on proteins was the subject of the earliest studies in this field. After Tappeiner's work (see introduction reports of studies on a wide variety of proteins such as enzymes (transferases, lyases, lipases, etc), hormones (insulin and angiotensinamid), casein, cytochrome C, snake venoms, globins and many others appeared. A detailed list can be found in references (25 and 26). The structure of the proteins plays an important role in determining their sensitivity to photodynamic action. Since peptide bonds are not sensitive to this treatment (29), protein sensitivity is determined by the sensitivity of the specific amino acid components and whether they are exposed or "buried" as a result of protein folding.

Damage induced in proteins by photodynamic action leads to changes in various physico-chemical properties including alterations in ultraviolet absorption spectra, solubility, viscosity, surface tension, sedimentation behavior, etc, as well as biological changes such as loss of catalytic or hormonal activities, loss of toxicity, etc. A detailed list of references can be found in (3).

c) Nucleic acids

Probably the most important damages induced in organisms by photodynamic action are those affecting nucleic acids. Again it is difficult to establish mechanisms of action, because they will depend on the experimental conditions. However, there is no doubt that the most effective dyes are those that are able to penetrate the cells and bind to nucleic acids. Chromosome aberrations, strand breaks and photodynamically induced mutations have been reported (31, 32). Yamamoto reported that damage to DNA led to bacteriophage inactivation by photodynamic action (33, 34). The T odd phages are more sensitive to photodynamic treatment than the T even phages because of differences in the permeability of their protein coats to dyes (33, 34, 35). Simon and Van Vunakis (36) clearly demonstrated that guanine bases are more sensitive to methylene blue and light treatment than other DNA bases. A similar result was found by Wacker et al. (37), with thiopyronine and light. Photosensitization of transforming DNA by lumichrome leads to the selective destruction of guanine (38). Knowles (39) has reviewed the changes in physical properties of DNA, loss of biological activity, relative sensitivity of bases and the appearance of modified bases that result from photodynamic action.

d) Organisms

Viruses, bacteria, fungi and algae all undergo a variety of effects such as membrane permeability changes (40), interference with respiratory functions (41), synthesis of macromolecules (42) and mobility (43), when exposed to appropriate photodynamic treatment. Mammalian cells have also been used in studies of photodynamic action and effects such as cessation of protein synthesis, inactivation of cellular enzymes (44) and cell-killing (45) have been reported. Multi-cellular plants and animals when submitted to photodynamic action often present damage to vital functions (see 30). After receiving sensitizing dyes orally or by subcutaneous injection, small mammals are killed under intense illumination (46). Disorders of porphyrin metabolism (47) and contact with or ingestion of sensitizing compounds used for medical purposes are common causes of photodynamic phenomena in man.

4. MOLECULAR MECHANISMS OF PHOTODYNAMIC ACTION

An important aspect of basic studies on photodynamic action involves the investigation of the dependence of the phenomenon on the dye, the substrate and the experimental conditions. The precise mechanism of photodynamic action will vary according to the combination of such factors involved. However, some general features relating to molecular mechanisms are evident. The earliest experiments on photodynamic action showed that oxygen was required (5) for the reaction to occur. The work of Kautsky and De Bruijn (48) demonstrating the quenching of excited dye molecules by molecular oxygen and the consequent production of singlet oxygen led to the proposal of several models, all involving singlet oxygen as the principal intermediate of the reaction. In 1939 Kautsky (49) produced strong evidence in support of this idea by photodynamically inactivating a biological substrate that was physically separated from the dye but under conditions that allowed diffusion of small molecules between dye and substrate. The singlet oxygen theory was further supported by work of Debey and Douzou in 1970 (50), Churakova et al. in 1973 (51) and Kornhauser et al. in 1973 (52) who obtained typical photodynamic effects with singlet oxygen either produced chemically or by microwave-discharge. Nilsson and Kearns in 1973 (53) showed effects of deuterated buffer and azide on the photodynamic action of methylene blue (Fig. 2) thus confirming the role of singlet oxygen in the reaction.

The existence of two general patterns of mechanisms for photodynamic reactions was recognized and described by Gollnick (54) and Foote (55) and these were designated Type I and Type II reactions. In Type I reactions the excited dye reacts directly with the substrate by hydrogen or electron abstraction. The subsequent reactions may be quite varied but involve free radicals and are

PHOTODYNAMIC THERAPY OF INFECTIONS 355

Fig. 2

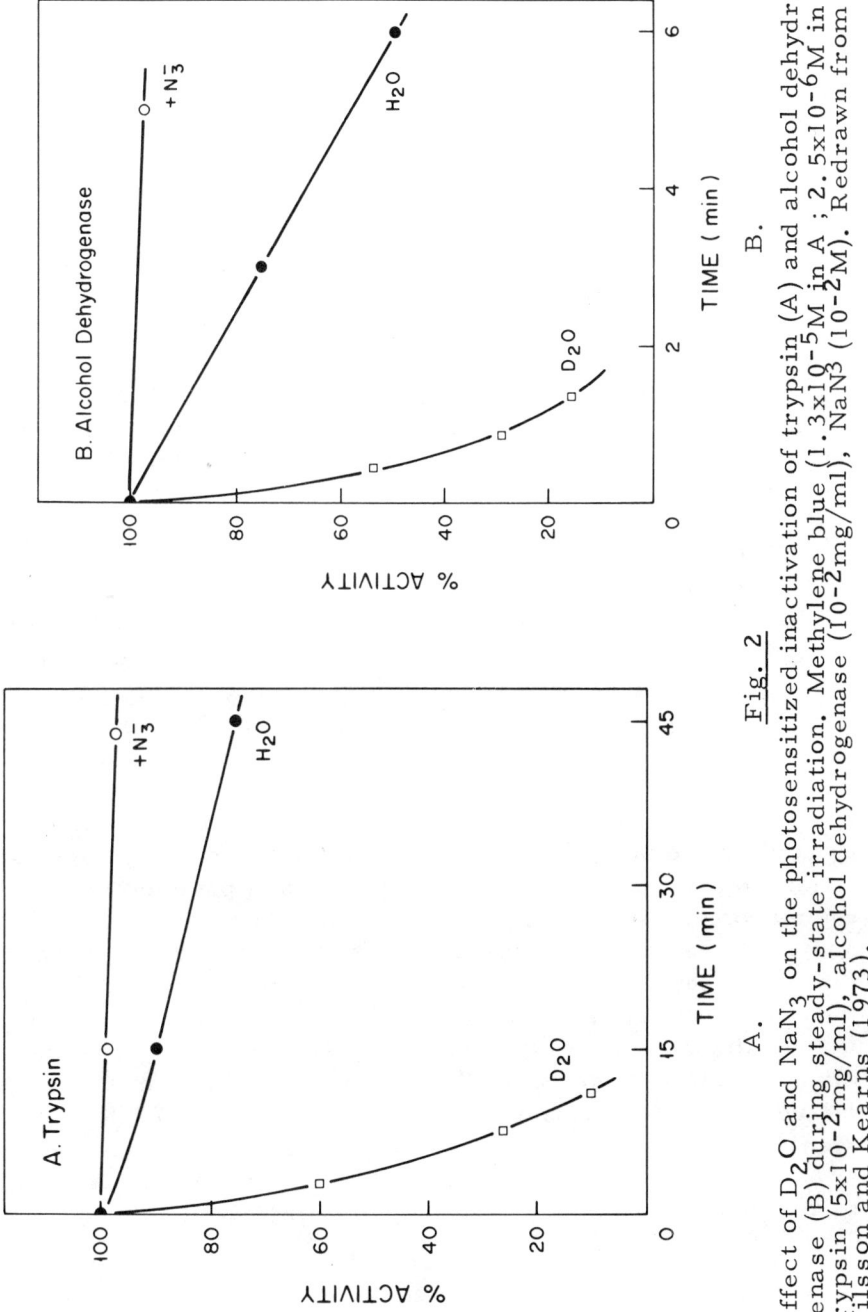

Effect of D_2O and NaN_3 on the photosensitized inactivation of trypsin (A) and alcohol dehydrogenase (B) during steady-state irradiation. Methylene blue (1.3×10^{-5}M in A ; 2.5×10^{-6}M in B, trypsin (5×10^{-2}mg/ml), alcohol dehydrogenase (10^{-2}mg/ml), NaN_3 (10^{-2}M). Redrawn from Nilsson and Kearns (1973).

difficult to follow because of the short lifetimes of the species produced. In Type II reactions, the excited dye is quenched by molecular oxygen via energy transfer and singlet oxygen production or via electron transfer and superoxide production. In most systems, both types probably occur, and the relative contribution of each depends on the nature of the substrate and the oxygen concentration.

Regarding the specific damage caused by photodynamic action, it has been demonstrated (36) that the guanine moieties in nucleic acids are the principal targets for the inactivation of bacteria by the photodynamic action of methylene blue. This finding was later confirmed by several authors, among them Rosenthal and Pitts (56) who demonstrated that guanine is more sensitive to photodynamic damage than are the other nucleic acid bases. This correlates well with the fact that the Highest Occupied Molecular Orbital energy for guanine is lower than for the other bases (57) and that the reaction of singlet oxygen with unsaturated compounds is electrophilic.

5. REPAIR OF DNA DAMAGE INDUCED BY PHOTODYNAMIC ACTION

The difference in the mechanism by which the photodynamic action of different dyes inactivates biological material is clearly demonstrated by studies of the repair of photodynamically induced damage to DNA. For example, it was shown (58) that rec (recombination deficient) but not uvr (excision deficient) strains of E. coli are more sensitive to photodynamic action sensitized by thiopyronine than are wild type strains. However, using acridine orange or acriflavine as the sensitizer, Harm (59) found a strict correlation between the sensitivity to far-UV and photodynamic action when he used strains with the four possible combinations of the rec and uvr genes. This result indicates that the excision repair system is at least as efficient in repairing photodynamic damage as the rec-dependent system. Jaffe-Brachet et al. (60) obtained similar results with proflavin, and calculated that the excision system alone is able to restore 67 % of the lesions induced in DNA, while the rec-dependent system is able to repair about 28 % of these lesions. Janovska et al. (61) found no correlation between far-UV or X-ray sensitivity of ten strains of E. coli and their sensitivity to acridine orange sensitized photodynamic action.

We have studied the repair of lesions induced in E. coli by photodynamic action of methylene blue. Fig. 3 shows the sensitivity of the strains carrying the four possible combinations of rec and uvr genes, indicating that both systems contribute to this repair.

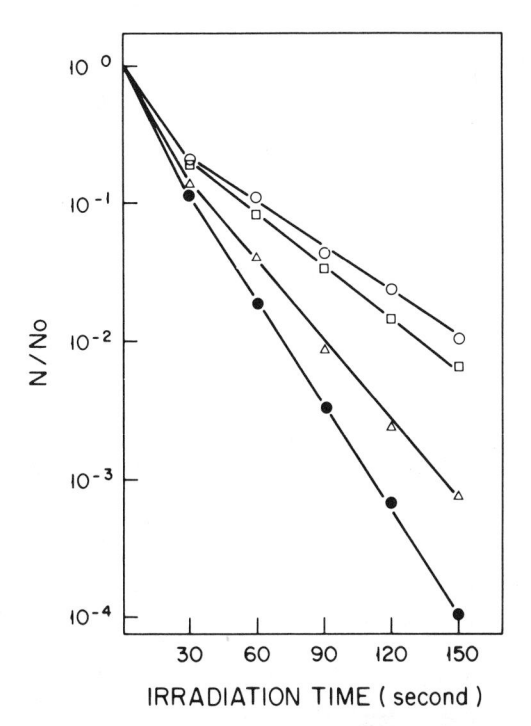

Fig. 3 : Photodynamic action of methylene blue on repair proficient and repair deficient strains of E. coli K12. Exponential phase cells were incubated with methylene blue (2µg/ml) for 60 min at 37°C and exposed to white light (GE PAR 38 "cool-beam", 150W, 5 cm distance) at ambient temperature : AB1157 (wildtype), O ; AB1886 (uvrA), ■ ; AB2463 (recA), ▲ ; AB2480 (uvrA recA), ● .

The rec-dependent system seems to be more effective. In another experiment we found that polymerase I also plays an important role in repairing these lesions since the polA1 strain is also more sensitive to treatment with light and methylene blue than the isogenic wild-type strain (62). It is interesting that bacteria grown in nutrient broth medium supplemented with glucose (0,4%) are more resistant to the photodynamic action of methylene blue than those grown in the same medium without glucose (62). Fig. 4 shows that this difference exists only for the strains possessing the rec-dependent system of repair (AB1157, wild type ; AB1886, uvrA6) and not for the other strains (AB2463 recA13 ; AB2480, uvrA6 recA13).

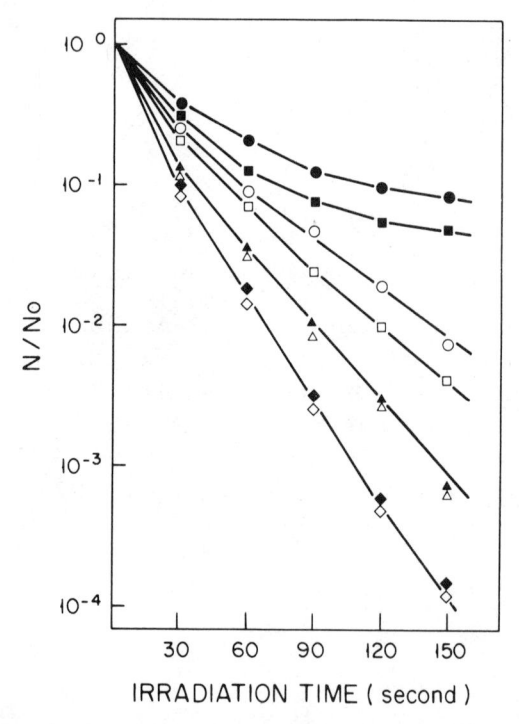

Fig. 4 : Photodynamic action of methylene blue in different strains of E. coli grown to exponential phase in tryptone broth with or without added glucose (4g/l). Experimental conditions as described in Fig. 3 : AB1157, with (●) or without (o) added glucose ; AB1886 (uvrA), with (■) or without (□) added glucose ; AB2463 (recA), with (▲) or without (△) added glucose ; AB2480 (uvrA recA), with (◆) or without (◇) added glucose.

A similar resistance is found in the strain deficient in polymerase I. In 1959 Hollaender et al. reported that E. coli strains grown in nutrient broth medium supplemented with glucose were more resistant to X-rays than when grown in the same medium without glucose (63). Stapleton and Engel (64) confirmed these results and suggested that the glucose-induced resistance to X-rays in E. coli could be related to the functioning of repair systems. Similar phenomena were found with gamma radiation (65) and it was suggested that bacteria grown in media supplemented with glucose could have their ability to repair lesions by the rec-dependent system enhanced. Additional work has supported the relationship between glucose and rec-system enhanced repair (66).

Makman and Sutherland (67) found that cells grown in the presence of glucose have low levels of cyclic AMP, due to 1) extrusion of intra-cellular cAMP (which is an immediate effect); and 2) decrease in the biosynthesis of cAMP (late effect). We have used two strains of E. coli deficient in the metabolism of cAMP (68, 69) in order to see if the enhanced resistance of E. coli strains to the photodynamic action of methylene blue is related to the decrease of intra-cellular levels of cAMP. The E. coli strains GY2615 (deficient in adenylcyclase and therefore unable to synthesise cAMP) and GY2606 (a strain that synthesises cAMP normally but is deficient in the protein receptor for cAMP and is therefore unable to use this nucleotide) were used. These strains were both found to be very resistant to the photodynamic action of methylene blue. Under the same experimental conditions these strains are 2 times as resistant to photodynamic action than the wild-type E.coli K12 AB1157 (62). When grown in medium supplemented with cAMP (5mM), GY2615 was almost as sensitive as AB1157, while the sensitivity of GY2606 was unaltered (70). We conclude that the internal level of cyclic AMP is an important factor in determining the sensitivity to the photodynamic action of methylene blue, but the mechanism is not yet clear.

6. VIRAL DESINFECTION

In 1966 (71) a report appeared concerning lysogenic induction by photodynamic action. More recently we attempted to induce a lysogenic strain of Staphylococcus albus P (ω) with methylene blue and visible light. Not only was this strain not induced by the treatment, but we observed that infectious centers (a constant percentage of cells in the culture that are "spontaneously" induced and are producing mature phage particles) are more sensitive than the cells whose viral genome remains integrated in the genome of the bacterium (72). Fig. 5 illustrates the contrast in sensitivity between cells and infectious centers subjected to the photodynamic action of methylene blue. The difference in sensitivity was further enhanced when the culture was induced with UV (254 nm) or X-rays prior to the photodynamic treatment. In order to see if this difference in sensitivity between cells and infectious centers existed

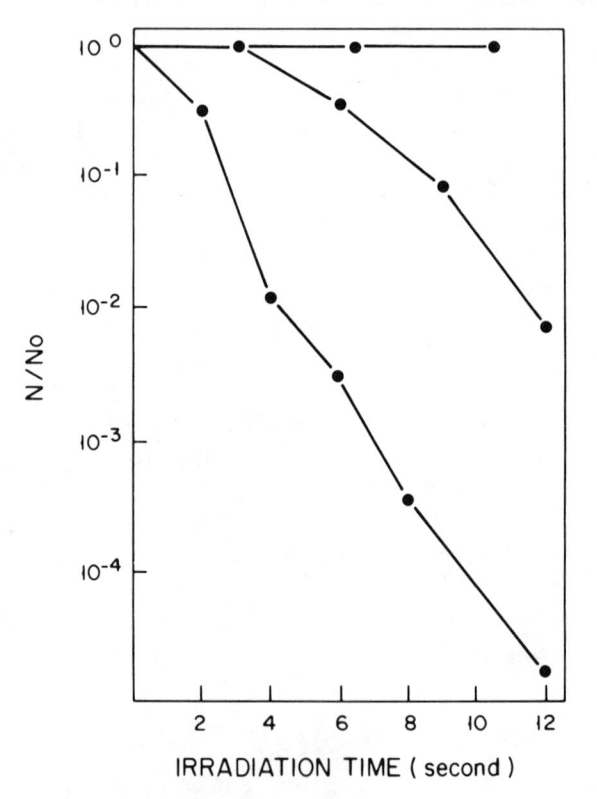

Fig. 5 : Viral desinfection of lysogenic Staphylococcus albus (ω) by methylene blue + light. Experimental conditions as described in Fig. 3 : Upper curve, bacteriophage ω alone ; middle curve, lysogenic population of Staphylococcus albus ; lower curve, induced lysogenic population of Staphylococcus albus. Curves redrawn from Caldas et al. (1974).

also for an infection complex, a non-lysogenic bacterial culture (K12S) was infected with bacteriophage T2 at a multiplicity of infection greater than one. Eight minutes later, the infection process was stopped by cooling to 0°C, and the culture filtered through a Millipore membrane to eliminate the free phages and ressuspended in phosphate buffer. The fraction which contained complexes and cells, was then submitted to the photodynamic action of methylene blue. Fig. 6 shows that the complexes are much

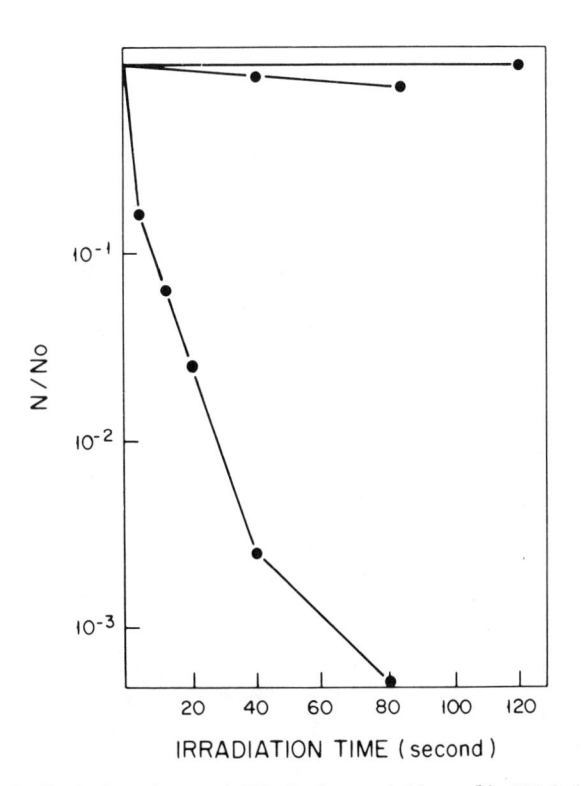

Fig. 6 : Viral disinfection of T2 infected E. coli K12 by methylene blue + light. Experimental conditions as described in Fig. 3 : Upper curve, bacteriophage T2 alone ; middle curve, E. coli K12 A15 non-infected ; lower curve, E. coli K12 A15 irradiated 8 min after infection with bacteriophage T2. Curves redrawn from Caldas et al. (1974).

more sensitive to this treatment than are the cells or phages treated separately. This may be important since a bacterial infection complex is a better model of a mammalian cell infected by a virus than is a bacterial induction complex.

7. CLINICAL APPLICATIONS OF PHOTODYNAMIC ACTION

A controversy exists concerning the benefits and risks involved in photodynamic therapy. The diverging viewpoints have been well reviewed by Bockstahler et al (73) and Little et al. (74). However, we would like to stress a few points that are sometimes overlooked.

Choice of the dye and experimental conditions

A lack of basic information makes the rational choice of a dye or dyes from the more than 400 currently available, extremely difficult. The questions as to whether viral or non-viral dyes should be used and the choice of cytoplasmic or nuclear dyes remain to be evaluated. The varied success of such treatments (Table 1) almost certainly results from the wide spectrum of treatment protocols employed. Among factors which could affect the final clinical result are the following :

a) Choice of dye : A dye that is selectively absorbed by infected cells is desirable. Methylene blue comes into this category and may be particularly effective for this reason.

b) Light sources and dose : The source must include the appropriate wavelength region absorbed by the dye. In addition, some negative studies may have used insufficient illumination.

c) The concentration of the dye and the specific dye solvent employed.

d) Repeated applications may be necessary. Variation in frequency and spacing adds to the diversity of the treatment protocols currently employed.

The risks

The widest use of photodynamic therapy has been in the treatment of Herpes simplex infections . Since this virus is clearly oncogenic, the few cases of cancer in humans that have been reported following therapy may have resulted from a transformation that occurred prior to the treatment. However, photodynamic action is clearly mutagenic in bacteria. This means that test systems currently available for screening for carcinogens at all levels of organization should be applied to photodynamic drugs in order that a clearer answer as to the carcinogenic risk may be obtained.

Other applications

Various dermatosis of mycotic and bacterial origin are a serious health problem, particularly in poorer communities. Photodynamic therapy may be particularly effective in the treatment of such conditions.

SUMMARY AND PERSPECTIVES

Photodynamic action is effective at all levels of biological organization. The mechanism of action and the possibility of atinging a specific cellular target will depend on the dye employed. Damage to nucleic acids has been measured and is known to be susceptible to various cellular repair systems. The clinical success of the treatment is varied. However, repeated success with particular dyes indicates that negative results may be due to inadequate protocols. A careful evaluation of the risks involved in therapy employing a particular dye system should be undertaken.

ACKNOWLEDGMENTS

This work was supported by the following Brazilian granting agencies : CNPq (National Research Council), CNEN (National Nuclear Energy Council), CEPG/UFRJ (University Council for Post-Graduate studies) and FINEP (B-76-79-074-0000-00).

REFERENCES

1. Blum, H. F. (1941) in Photodynamic action and diseases caused by light, Heinhold, New York.
2. Marcacci, A. (1888) Arch. Ital. Biol. 9:2
3. Raab, V. O. (1900) Z. Biology 39:524
4. Tappeiner, H. (1900) Munch. Med. Wschr. 47:5
5. Ledoux-Lebard (1902) Ann. Inst. Pasteur, 16:587
6. Tappeiner, H. (1903) Ber. Deut. Chem. Ges. 36:3035
7. Harris, D. T. (1926) Biochem. J. 20:288
8. Carter, C. W. (1928) Biochem. J. 22:575
9. Spikes, J. D. and Straight, R. (1967) Ann. Rev. Phys. Chem. 18:409
10. Moore, C., Wallis, C., Melnick, J. L. and Kuns, M. D. (1972) Infect. Immunol. 5:169
11. Wallis, C. and Melnick, J. L. (1973) Program and Abstracts, First Annual Meeting of the American Society for Photobiology, 133.

12. Friedrich, E.G., Jr. (1973) Obstet. and Gynecol. 41:74
13. Jarrat, M. and Knox, J.M. (1974) Prog. in Dermatol. 8:1
14. Melnick, J. and Wallis, C. (1977) Annals New York Acad. Sci. 284:171
15. Felber, T.D., Smith, E.B., Knox, J.M., Wallis, C. and Melnick, J.L. (1973) J. Am. Med. Assoc. 223:289
16. Kaufman, R.H., Gardner, H.L., Brown, D., Wallis, C., Rawls, W.E. and Melnick, J.L. (1973) Am. J. Obstet. Gynecol. 117:1144
17. Roome, A.P., Tinkler, A.E., Hilton, A.L., Montefiore, D.G. and Walter, D. (1975) Br. J. Vener. Dis. 51:130
18. Taylor, P.K. and Doherty, N.R. (1975) Br. J. Vener. Dis. 51:125
19. Myers, M.G., Oxman, M.N., Clark, J.E. and Arndt, K.A. (1975) N. England J. Med. 293:946
20. Myers, M.G., Oxman, M.N., Clark, J.E. and Arndt, K.A. (1976) J. Infect. Dis. 133:(Suppl.A) 145
21. Santamaria, L. and Prino, G. (1972) in Res. Prog. Org. Biol. Med. Chem. edited by U. Gallo and L. Santamaria, Vol. III, Part 1, p.13, North-Holland Publ. Co. Amsterdam
22. Bourdon, J. and Schnuriger, B. (1967) in Physics and Chemistry of the Organic Solid State edited by D. Fox, M. Labes and A. Weissberger, Vol. III, p.139, Interscience, New York
23. Yamamoto, N. (1972) in Res. Prog. Org. Biol. Med. Chem. edited by U. Gallo and L. Santamaria, Vol. III, part 1, p;297, North-Holland Pub. Co. Amsterdam
24. Ito, T. (1978) Photochem. Photobiol. 28:493
25. Vodrazka, Z. (1959) Chem. Lysty 53:829
26. Weil, L. (1965) Arch. Biochem. Biophys. 110:57
27. Spikes, J.D. and Livingston, R. (1969) Advanc. Rad. Biol. Vol.3, p.29, Edited by L.G. Augenstein, R. Mason and M. Zelle, Academic Press, New York
28. Spikes, J.D. and M.L. Macknight (1972) in Res. Prog. Org. Biol. Med. Chem. edited by U. Gallo and L. Santamaria, Vol. III, part 1, p.124, North-Holland Publ. Co.Amsterdam
29. Ghiron, C.A. and Spikes, J.D. (1965) Photochem. Photobiol. 4:13
30. Spikes, J.D. and Straight, R. (1967) Ann. Rev. Phys. Chem. 18:409
31. Kumar, J. and Natarajan, A.T. (1965) Mut. Res. 2:11
32. Naekai, S. and Saeki, T. (1964) Genet. Res. 5:158
33. Yamamoto, N. (1956) Virus 6:522
34. Yamamoto, N. (1958) J. Bacteriol. 75:443
35. Welsh, N.J. and Adams, M.H. (1954) J. Bacteriol. 68:122

36. Simon, M.I. and Van Vunakis, H. (1962) J. Mol. Biol. 4:488
37. Wacker, A., Dellweg, H., Trager, L., Kornhauser, A., Lodeman, E., Turck, G., Selzer, H., Chandra, P. and Ishimoto, H. (1964) Photochem. Photobiol. 3:369
38. Sussenbach, J.S. and Berends, W. (1965) Biochim. Biophys. Acta 103:360
39. Knowles, A. (1972) in Res. Prog. Org. Biol. Med. Chem. edited by U. Gallo and L. Santamaria, Vol. III, part 1, p. 183, North-Holland Publ. Co Amsterdam
40. Mathews, M.M. (1963) J. Bacteriol. 85:322
41. Tereza, G.W., Tereza, N.L. and Millius, P. (1955) Can. J. Microbiol. 11:1028
42. Garvin, R.T., Julian, G.R. and Rogers, S.J. (1969) Science 164:583
43. Duijn, C. (1962) Exp. Cell. Res. 26:373
44. Hill, R.B. (1960) Exp. Cell. Res. 21:106
45. Klein, S.W. and Goodgol, S.H. (1959) Science 130:629
46. Mathews, M.M. (1964) Nature 176:1212
47. Burnett, J.W. and Pathak, M.A. (1964) Arch. Dermatol. 89:257
48. Kautsky, H. and De Bruijn, H. (1931) Naturwissenschaften 19:1043
49. Kautsky, H. (1939) Trans. Faraday Soc. 35:216
50. Debey, P. and Douzou, P. (1970) Israel Journal Chem. 8:115
51. Churakova, N.I., Kravchenko, N.A., Serebryakov, E.B., Lavrov, I.A. and Kaversneva, E.D. (1973) Photochem. Photobiol. 18:201
52. Kornhauser, A., Krinsky, N.I., Huang, P.K.C. and Glagett D.C. (1973) Photochem. Photobiol. 18:63
53. Nilson, R. and Kearns, D.R. (1973) Photochem. Photobiol. 17:165
54. Gollnick, K. (1968) Adv. Photochem. 6:1
55. Foote, C.S. (1968) Science 162:963
56. Rosenthal, I. and Pitts, Jr., J.N. (1971) Biophys. J. 11:963
57. Pullman, B. and Pullman, A. (1963) in Quantum Biochemistry, Interscience Publishers Ltd., London
58. Bohme, H. and Gessler, E. (1968) Mol. Gen. Genet. 103:228
59. Harm, W. (1968) Biochem. Biophys. Res. Comm. 32:350
60. Jaffe-Brachet, A., Henry, N. and Errera, M. (1970) Mut. Res. 12:9
61. Janovska, E., Zhestjanikov, V. and Vizdalova, M. (1970) Int. J. Rad. Biol. 18:317
62. Menezes, S. (1975) PhD Thesis, Institute of Biophysics, Federal University of Rio de Janeiro, Brazil

63. Hollaender, A., Stapleton, G.E. and Martin, F.L. (1951) Nature 167:103
64. Stapleton, G.E. and Engle, M.S. (1960) J. Bacteriol 80:544
65. Friesen, B.S., Iyer, P.S., Baptist, J.E., Meyn, R. and Rodgers, J.M. (1970) Int. J. Rad. Biol. 18:159
66. Town, C.D., Smith, K.C. and Kaplan, H.S. (1971) J. Bacteriol. 105:127
67. Makman, R.S. and Sutherland, E.W. (1965) J. Biol. Chem. 240:209
68. Schwartz, D. and Beckwith, J. (1970) in The Lactose Operon, edited by D. Zipser and J. Beckwith, p. 417, Cold Spring Harbor Laboratory on Quant. Biol.
69. Zubay, G., Schwartz, D. and Beckwith, J. (1970) Proc. Natl. Acad. Sci. USA, 66:104
70. Coelho, A.M. (1976) Master of Sciences Thesis, Institute of Biophysics, Federal University of Rio de Janeiro, Brazil
71. Freifelder, D. (1966) Virology 30:567
72. Caldas, L.R., Menezes, S., Leitão, A. and Latarjet, R. (1974) C.R.Acad.Sc. Paris, 278D:2369
73. Bockstahler, L.E., Lytle, C.D. and Hellman, K.B. (1974) U.S.Depart. of Health, Educ. and Welfare, PHS, Food and Drug Adm. Bureau of Radiol. Health, DHEW Publication (FDA) 75-8013
74. Lytle, C.D., Bockstahler, L.E., Hellman, K.B., Goddard, J.G. and Brewer, P.P. (1978) in International Symposium on Current Topics in Radiobiology and Photobiology, p. 99, edited by R.M. Tyrrell, Academia Brasileira de Ciências, Rio de Janeiro, Brazil.

THE USE OF LASERS IN OPHTHALMOLOGY

Gabriel Coscas, Gabriel Quentel, and Michel Binaghi

Ophthalmology Clinic of Créteil University
University of Paris-Val de Marne (40, avenue de Verdun)
94010 Créteil, France

The appearance of the laser, and more precisely, the argon laser, constituted a considerable advance in ophthalmology. Retinal photocoagulation permitted a radical modification in the prognosis of a number of ocular diseases by allowing intraocular surgery of a very delicate type in the unopened eye.

I - REVIEW OF ANATOMY AND PHYSIOLOGY

The human retina has numberous layers, from the interior of the eye to the exterior, that can be distinguished:

- the internal limiting membrane which separates the retina from the vitreous cavity,
- the layer of nerve fibers
- the internal plexiform layer in which run the retinal vessels,
- the internal nuclear layer,
- the external plexiform layer,
- the external nuclear layer, containing the photoreceptors
- the pigment epithelium, very rich in melanin, which separates the external retina from the choroid

The light which penetrates through the eye must, therefore, after tranversing the transparent media of the eye (cornea, aqueous humor, crystalline lens, vitreous) as well go through the internal layers of the retina, normally transparent, before it can affect the photoreceptors. The light cannot, however, go through the highly adsorbent pigment epithelium layer immediately adjacent to the photoreceptors.

II - THE PRINCIPLES OF USE OF LASERS IN OPTHALMOLOGY

1. History

The first attempts at the utilization of the laser in retinal diseases were started by the photocoagulations by Meyer Schwickerath, who in 1946 used direct sunlight as a source, then made with the help of the Zeiss firm, the Xenon photocoagulator (1957).

In 1964 the first patients were treated with the ruby laser. L'Esperance in 1968, then Zweng in 1969 made the first argon laser photocoagulators. This is the type of laser which is, actually, for the most part, used in opthalmology.

2. Principles of the Argon Laser

*The argon laser consists schematically of:
- the tube, the active element, a gas mixture of argon and krypton in an ionized state (from excitation by electric current of a high voltage) which produces Photochromatic emissions.

A system of electron pumping and a resonant cavity which allows the radiation to be made more monochromatic. Circulating water to cool the tube all the time (7 liters/second minimum with three bars pressure).

- the observation system: a slit lamp
- a delivery system connects the laser to the slit lamp which consists of a articulated optical arm or an optical fiber bundle.
- a system to select the area of photocoagulation
- a panel to control the different parameters which characterize the laser exposure
- a security system

*The spectral composition of the emission is such that 95 percent of the energy is concentrated in two wavelengths.

- the first one, 4990 Å., of blue color
- the second 5145 Å, of green color

At low power essentially only the blue component is emitted. The cornea, aqueous humor, crystalline lens, vitreous, and internal layers of the retina <u>transmit</u> the wavelengths between 4000 and 14,000 Å: All these structures have a high transmission for the argon laser beam. The pigment granules contained in the cells of the pigment epithelium, on the contrary, adsorb strongly

USE OF LASERS IN OPHTALMOLOGIE

the wavelengths between 4500 Å and 8500 Å.

The hemoglobin in the red blood cells does adsorb all the radiation shorter than 6000 Å, therefore only the wavelengths between 4500 and 6000 Å will be selectively adsorbed by both melanin and hemoglobin. This is the case with an argon laser.

Light emitted by the argon laser will be adsorbed, in principle, at the level in the eye of each of the pigmented structures (essentially the pigment epithelium layer) and the red cells in the retinal vessels.

In fact, the photocoagulation of the vessels (through the hemoglobin) is very difficult even when using high intensities. From a practical point of view, it is essentially the melanin pigment which permits the adsorption of the argon laser light: the absorbed luminous energy is transformed into heat, which produces a burn resulting in tissue destruction. This necrosis initiates an inflammatory reaction which itself is followed by a scar formation and destroys structures surrounding the initial burn.

Thermal elevation is important at the moment of photocoagulation at the place where the adsorption is produced, but the thermal effect remains localized and can only go away by conduction. Photocoagulation of the pigment epithelium goes through the choroid, and through the immediately adjacent retina.

Thus:

the area
the diameter } of the exposure
the power

determine: the position
the extent } of the scar
the severity

In conclusion, we can say that for carefully chosen intensities, a scar essentially composed of glia, is produced which:

(1) creates a strong adherance between the retina and the choroid,

(2) initiates a secondary degeneration of the internal layers of the retina which is moderate, but will be more marked when the intensity is increased,

(3) stops at the internal limiting membrane in the retina, and, therefore, does not affect the vitreous body.

These are the properties of the laser that are utilized in ophthalmology.

We have to mention that the argon laser light is coherent. This permits precise focus without adsorption by the transparent media, and without scattering during its passage.

3. In Practice

Photocoagulation is made with patients sitting before the slit lamp. The pupil has been previously dilated with effective drops, and the eye has been anesthetised by anesthetic drops. A special contact lens (or a three mirror contact lens) is put on the eye which permits observation of the fundus with the aid of the slit lamp. This gives a binocular view, at high magnification, of the retina and the laser exposures.

Different parameters are shown (duration, size and intensity of the exposures on a control panel and an indication of the tetinal area to be exposed with the help of a small movable spot. The exposure is controlled by a foot switch. The photocoagulation of the argon laser is not painful, and may be done on an out-patient basis.

III - INDICATIONS FOR LASER TREATMENT IN OPHTHALMOLOGY

The main indications for photocoagulation in ophthalmology are:

- prevention of retinal detachment
- prevention or destruction of retinal neovascularization
- destruction of certain abnormal structures such as lesions of the pigment epithelium, angiomas, tumors, vascular anomalies which cause retinal edema.

It must be emphasized that before deciding on photocoagulation in ophthalmology it is first necessary to do a good examination of the eye with the help of fluorescein angiography. This procedure consists of photographing the ocular fundus after the injection in one vein of the arm of 5 cc. of fluorescein of 10 or 20 percent concentration. The stain which progressively runs through the different retinal and choroidal vessels from the largest to the smallest, then will be eliminated progressively by little. The fluorescein stain with the help of complimentary filters will expose the photographic film, giving thus a very precise image of the ocular fundus and its vessels at different stages of the passage of the fluorescein. This examination permits a very precise analysis of the morphology and physiology of the different structures of the ocular fundus. It seems to be indispensable before photocoagulation (especially showing evidence of occluded vessels and non-perfused retinal areas as you will see later).

1. The Prevention of Retinal Detachment

Retinal detachment is a serious disease. It consists of a separation between the sensory retina and the pigment epithelium. Without any doubt, this develops from a retinal tear.

This retinal tear can occur on its own, or can develop in an area of fragile retina. It is important to note that all of these lesions are situated in the periphery of the retina. (That means far away from the physiologically useful areas). It occurs very often in certain kinds of patients: high myopes, aphakes, fellow eyes to ones that have already had a retinal detachment. These subjects are called "high risk" as far as retinal detachments are concerned.

We have seen above that the laser principally acts on the eye to create a very adherent scar between the retina and choroid through an intermediate burning of the pigment epithelium. As separation between the retina and the choroid is not possible anymore in the treated areas, further retinal detachment is thus prevented. Thus, an excellent prevention of retinal detachment can be accomplished by a "barrage" of laser lesions killing the susceptible fragile retina which could cause retinal detachment in the high risk subjects.

The same can be done when a beginning retinal tear is diagnosed (before the retinal detachment itself begins). The tear can be surrounded by laser burns, preventing the detachment of the surrounding retina.

It should also be said that, on the contrary, when the retina is already detached, photocoagulation is absolutely not effective. The retina is too far away from the choroid and cannot stick to it anymore.

Photocoagulation permits the prevention of retinal detachment even when there is a retinal tear, especially when there is reason to suspect that there is a "flat" retinal tear, but not after any retinal detachment has occurred.

2. Abnormal Neovascularization in the Retina

Numerous diseases are complicated by occlusions of a certain number of retinal capillaries, producing an ischemia and hypoxia of the dependent retinal areas (vasculatization of the retina is a terminal type).

These diseases are numberous, and among the most important onces, we will mention diabetic retinopathy, branch vein occlusion, certain types of retinal vasculities, and the hemoglobinopathies.

In all these cases, the mechanism of the formation of neovascularization is identical: the retinal areas rendered hypoxic have the property of secreting a vasoproliferative factor (not yet identified with certainty), which diffuses through the eye and produces new vessel growth. One can say that proliferation of the new vessels is an attempt to supply the ischemic areas in a natural way. Unfortunately, this attempt is not only ineffective but also harmful. Indeed, these new vessels have a primitive wall which is abnormally fragile, and they bleed very easily. Furthermore, they have a tendency to grow not only in the retina but on its surface and into the vitreous cavity.

These things explain the major complications of neovascularization: the massive hemorrhages in the vitreous which leads to blindness.

The more extensive the ischemic retinal areas, the more numerous the neovascularizations, which increase the risk. The initial idea was first to destroy the new vessels directly by photocoagulation. We encounter two major obstacles which make this useless:

- the difficulty of occluding a vessel with a laser,
- major risk of bleeding at each laser exposure

It was noticed by all that the production of the secretion of the vasoproliferative factor from an area of retinal hypoxia could be diminished by total destruction of these areas, and that there were no more living cells in this area.

The hypoxic area is too much affected to be functional. Furthermore, the lesions are irreversible. As the retinal tissue is composed of nervous structures, its survival in hypoxia is very brief, and it cannot be regenerated when it is destroyed.

We can sum all this up in the following scheme:

```
                          nonfunctional area
                         ↗
area of retinal hypoxia ──→ secretion of vasoproliferative
                              factor
                                ↓
                            neovascularization
                                ↓
                              hemorrhages
```

```
                              nonfunctional area
                            ↗
area of total retinal necrosis
                            ↘
                              absence of secretion of vaso-
                                  proliferative factor
                                          ↓
                              absence of neovascularization
```

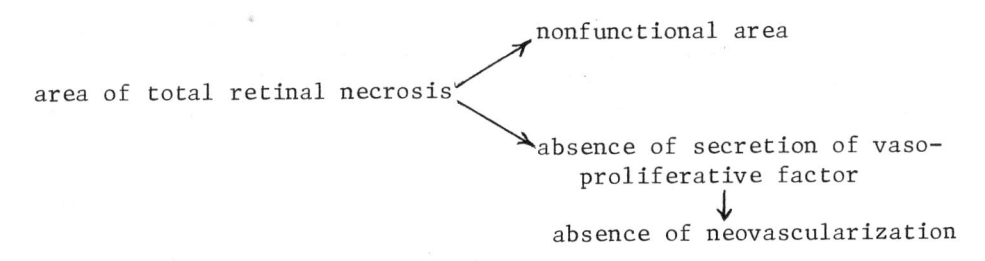

The clinic permits the daily verification of these facts. The total destruction by photocoagulation in these areas of retinal ischemia (already nonfunctional) prevents the appearance of neovascularization and may even produce regression of that already present.

In the current practice of ophthalmology, the photocoagulation of the areas of retinal ischemia, more or less extensive, (these areas have been shown up by fluoresein angiography), spares, of course, the central area of the retina which is responsible for visual acuity. This type of application of photocoagulation has permitted a radical transformation of the prognosis in diabetic retinopathy. Before the advent of the laser, this was the leading cause of blindness in our country.

3. Destruction of Certain Abnormal Structures

One utilizes in certain cases the possibility to destroy by direct photocoagulation, certain abnormal structures.*

   a. Certain anomalies in the pigment epithelium (central serous choroidoretinopathy). This type of disease is a localized disturbance in the permeability of the pigment epithelium, which doesn't carry on anymore its function as a physiological barrier. The focal coagulation of the zone of abnormal permeability restores its function.

   b. Certain tufts of neovascularization of choroidal origin, responsible for a number of cases of "macular degeneration", will stop having a negative effect when they are totally destroyed. Again, only argon laser photocoagulation (with the wavelengths be well adsorbed by the hemoglobin in the abnormal vessels) will accomplish this destruction. (It should be noted also that senile macular degeneration is one of the main causes of blindness).

   c. Certain microvascular abnormalities (particularly in diabetics) have an abnormal permeability and can create retinal edema around them. Photocoagulation of these abnormalities causes regression of the edema.

d. Certain angiomas can be totally destroyed by the laser, preventing a number of complications of retinal angiomatosis (in these particular cases, the adsorptive properties by hemoglobin of the laser are again especially important.)

e. Finally, certain choroidal melanotic tumors of small size.

*I must mention that laser photocoagulation of abnormal structures also causes in all these cases the destruction of the surrounding retinal areas. Thus, before thinking of laser treatment, one must be sure the lesion to be treated is sufficiently far from the central retinal area (called the fovea) to avoid the worst complication of the treatment, profound and irreversible lowering of the visual acuity.

IV - FUTURE

A huge field of research is apparent in the development of new types of lasers. The krypton laser can be used for treatment of the macular region, the CO_2 laser for surgery, and the dye laser will be tested in the treatment of glaucoma.

IN CONCLUSION

Let me emphasize how important it is with the help of an angiogram to examine properly the indications for laser photocoagulation. Although such a treatment has drastically transformed the prognosis of numerous ocular diseases, it can in other cases be ineffective or even dangerous.

Whatever it is, the laser is now an indispensable instrument to the ophthalmologist and we are going to present some figures as an example.

- There are in France: 1,200,000 diabetics of which 100,000 must be treated.
  500,000 high myopics of which 50,000 must be treated.
  There are 50,000 aphakics per year, which 10,000 must be treated.
- All of these diseases take about one-third of our time, which represents one to two million treatment sessions per year...

ULTRAVIOLET PROPHYLAXIS OF ADVERSE EFFECTS OF ENVIRONMENTAL CHEMICALS ON ORGANISMS

Jury I. Prokopenko

The A.N. Sysin Institute of General and Communal

Hygiene of the U.S.S.R. Academy of Medical Sciences

Moscow, U.S.S.R.

The use of UVB radiation as a treatment for different adverse effects caused by environmental factors is widely recognized in photobiology and medicine. Prophylactic UV irradiation of the poppulation is used most extensively in those areas where there is a deficit of UV radiation during most of the months of the year. Approximate calculations show that about 10 % of the entire population live under such conditions. These data clearly indicate the importance and the necessity of UV irradiation of the population as a prophylactic measure, particularly when one takes into consideration that a prolonged deficit of UV radiation produces a number of unfavorable effects in the body, including rickets in children and a depression of body resistance in adults. As a result of the latter state the probability that different diseases will develop is increased (1, 2, 3).

The global changes which have been occurring in the environment and which can be related to the influence of chemicals increase acutely the necessity for prophylactic irradiation of the poulation. There are two aspects of this problem. The first aspect concerns the decrease in the transparency of the atmosphere to UV radiation which takes place in industrial towns. Pollution of the atmosphere with different chemical substances such as sulfur dioxide, nitrogen dioxide, soot, dust, etc. is the cause of this decrease in transparency. Data presented by numerous investigators show that the total amount of UV radiation (both direct and diffused) is decreased by 30 to 50 % in the large industrial towns (see Graceva et al)(4). Taking this data into account, it may be assumed that a deficit of natural UV radiation also occurs for prolonged periods in those towns which are situated between 55° and 60° north latitude. On the basis of this assumption it may therefore be postulated that the proportion of the population subject to a deficit of UV radiation is also increased. The probability, therefore, of the development of diseases due to a lack of adequate UV radiation is also increasing.

The second aspect of this problem is as follows : Among the different unfavorable consequences of a prolonged deficit in UV radiation, as mentioned above, the most important is the decrease in the body's resistance to different pathogenic factors in the environment. The same environmental pollutants which decrease the transparency of the atmosphere in the UV radiation range, as well as other pollutants, induce numerous adverse effects. These chemicals affect the body because of their presence in air, food and drinking water. In large quantities over a short time they may cause acute poisoning. The more widespread and prolonged action of small doses of such chemicals induces different kinds of chronic disorders both functional and anatomic, such as tumors, allergic diseases, genetic damage, congenital malformation and degenerative processes.

In accordance with this evidence, two further questions arise. One, can the UV deficit change the susceptibility of the body to unfavorable environmental factors ? Two, is the use of UV irradiation useful as a prophylactic against the unfavorable effects of chemical pollutants ? In order to study these problems, an investigation has been carried out over the last decade. In summary, the objects of the investigators were as follows :

1. Establishment of the magnitude of the UV prophylactic effect on the most widespread harmful effects of environmental chemical pollutants.

2. Study of the possible mechanism of the favorable effects of UV radiation as far as the harmful action of the environmental chemical pollutants is concerned.

3. Establishment of the "dose-effect" relationship for UV radiation and definition of its optimum doses.

Among the various unfavorable effects of environmental chemicals the following four are widespread : the general toxic effect, the carcinogenic, allergenic and mutagenic effects.

On the basis of these observations, the possible prophylactic action of UVR in relation to chemicals which produce these kinds of harmful effects has been studied. Altogether the effects of more than 20 different chemicals have been observed. The collaborators in this study were the members of the Kiev Medical Institute under the guidance of Professor Galovich.

Investigations of the UV prophylaxis of pneumoconiosis and the toxic effect of carbon monoxide carried out earlier preceded the study of the prophylactic effect of UV radiation in relation to the action of the following chemical substances which possess a general toxic effect : salts of lead and its organic compound - tetraethyl lead, aniline, nitrate, chloroform and others. In these investigations the chemicals were used in concentrations of 10 to 1000 times of their threshold limits.

What did the prophylactic effect of UV radiation show ? First of all, a manifestation of the UV prophylactic effect was a decrease in the degree of direct toxic effect produced by the chemicals. For example, in the case of exposure to aniline nitrate which possesses an effect on the blood, there was a decrease in the level of methemoglobin. The red blood cell counts and hemoglobin levels were higher than among the animals which were not treated with UV radiation but were exposed to the chemicals. Higher levels of hemoglobin and red blood cells were also observed in animals which were treated with tetraethyl lead and UV radiation as compared to animals which were not treated with UV radiation and exposed only to chemicals. In this group of animals reticulocyte counts were, on the contrary, higher. In the case of the inorganic lead salt in combination with UV radiation, the level of the urinary coproporphyrin was lower than in animals which were exposed to lead only without the UV radiation treatment. The mercury content of the organs, in animals which had been exposed to this chemical, was also lower with UV radiation treatment. In the case of the organo-phosphorus compounds, the action of cholinesterase, the enzyme which is inactivated by these chemicals, was higher in combination with a suberythemal dose of UV radiation. Other examples of the positive effect of UV radiation in relation to chemicals with general toxic effects have also been obtained.

There is considerable contradictory data in the literature about the effects of a combination of UV radiation and chemicals on carcinogenesis. There is a possibility that the contradictory results can be related to the doses used and to the spectrum of the UV radiation used with different chemical carcinogens used in the different investigations. In order to obtain a clearer picture of this problem, our laboratory carried out investigations in two directions : These investigations were intended to study the influences of UV radiation upon the development of tumors under experimental conditions and to compare the carcinogenic indices of two different chemical carcinogens, benzpyrene and nitrazodimethylamine. The first compound was rubbed into the skin to induce the development of tumors. The second was given orally in drinking water to induce mainly tumors of the liver.

As a result of these investigations it was proven that in all cases, irrespective of the kind of tumor (transplanted Jensen sarcoma, skin cancer or hepatoma), suberythemal doses of UV radiation had a noticeable prophylactic effect. For example, the statistical decrease of the volume of transplanted tumors was approximately 50 %. A decrease in the number of tumors induced and a lengthening of the latent period before the appearance of tumors was also observed when the exposure to the carcinogenic chemicals was combined with suberythemal doses of UV radiation. It is important to note, moreover, that the number of malignant tumors of the skin in UV untreated animals were twice the number in UV treated animals.

The prophylactic effect of UV radiation was also shown in relation to chemicals which cause allergic reactions. Among the allergens studied were benzene, dinitrochlorobenzene, paraphenylenediamine, etc. It was considered necessary to differentiate between the indirect effect of UV radiation upon the intensity of the sensitivity reaction induced by the chemicals and the direct effect of the UV radiation on the molecular structure of the chemicals to provide them with the ability to induce a state of sensitivity in the organism. The latter phenomena are well known in photobiology as phototoxicity and photoallergy.

In our studies we did not use phototoxic or photoallergic chemicals. Our object was to study the influence of UV radiation upon the organism which had been sensitized by other agents not connected with the development of phototoxicity or photoallergy. In other words, a study was directed to investigate the possibility of correcting the altered reaction of the body with the use of UV radiation.

Earlier investigations were carried out concerning the influence of UV radiation on the development of immune reactions following vaccination and upon allergic diseases such as asthma. These investigations showed that UV radiation may normalize the changed reactivity of the body. In addition, the UV radiation may accelerate the production of the immune substances. Taking this data into consideration, it was decided to investigate the influence of UV radiation upon the allergic process in general. The experimental model of the allergic process used was the introduction of horse serum in guinea pigs and rats. It was shown that the intensity of the anaphylactic shock was less in those animals subjected to suberythemal doses of UV radiation than it was in nonirradiated animals.

The same results were obtained when a study was made of the influence of UV radiation upon the development of the allergic process induced by the introduction of chemicals with a known allergic action. In this investigation UV radiation reduced the manifestation of specific reactions to the antigen. The decrease in the symptoms of these specific reactions may be considered as being due to a decrease in the production of the protein-chemical complex which is the initial link in the chain of the allergic reaction. In addition it was also observed that the intensity of the skin reactions in the animals exposed to suberythemal doses following the introduction of the chemicals was less than in nonirradiated animals. At the same time it was observed that there was an increase in the activity of the nonspecific immunological reactivity in the body. All of the above data allowed the conclusion that suberythemal doses of UV radiation have a favorable effect on those systems exposed to chemical allergens.

In order to understand the influence of UV radiation upon the process of reproduction, the experimental model of spontaneous mutagenesis which occurs during close inbreeding was used. The reproductive process over five generations of mice (Balb/c strain) was observed. The following indices were used to characterize the reproductive cycle : The number of newborn mice, the number of stillborn mice, the number of mice which died during the lactation period, the indices of postnatal development and others.

Evidence of the favorable influence of UV radiation upon the animal's reproductive process was obtained in these investigations. It was found that the animals which were treated by UV radiation in suberythemal doses had better indices for the reproductive process. In particular the number of newborn mice in the UV irradiated groups was greater than in the nonirradiated groups. In addition the number of stillborn mice in the UV treated groups was only one-quarter the number in the untreated groups. The rate of development of the newborn animals was also faster when their parents were exposed to suberythemal doses of UV radiation. It would therefore appear possible to consider the possibility of correcting the frequency of spontaneous lethal mutations in close inbreeding and to improve the nature of postnatal development. The question about the possibility of correcting the mutagenic action of chemicals, however, is still open. There have been no literature reports concerning this possibility, and for this reason investigations are now taking place in our laboratory. Preliminary data have shown that UV radiation does have a positive action in correcting both mutagenic and embryotoxic effects of several chemicals.

The data obtained in these investigations testifies to the considerably wide range of the prophylactic action of UV radiation in relation to the various harmful effects of environmental chemicals. Different mechanisms are responsible for the toxic effects of these chemicals. Different organs and systems of the body are responsible for the manifestations which are observed in the body after exposure to the chemicals. UV radiation, however, appeared to be effective in all these reactions. Because of this, the possibility must be considered that there are some common nonspecific elements which are responsible for the state of the body's resistance and the degree to which the intoxication develops.

Summarizing the data obtained in our own experiments and that available in the literature (see Grabovic et al)(7), it is possible to enumerate the main changes which occur in the body as a result of UV treatment and which form part of the defensive mechanism against different chemicals. Earlier it was shown that the main defense action was due to an increase in the activity of its sympathetic adrenal system. In addition, the specific effects of UV radiation include correction of the metabolism of calcium and phosphorus (2, 5, 6).

The above data, however, cannot explain the ability of UV radiation to increase the resistance of the body to different chemicals which possess different harmful effects. We believe that several factors may be involved in these processes. An increase in the activity of some enzymes such as those of the mitochondria and microsomes was observed. These latter enzymes are known to be involved in the metabolims of drugs and environmental chemicals, resulting in their transformation into nontoxic and noncarcinogenic compounds. The increased activity of microsomal enzymes leads to a decrease in the amount of carcinogenic product in the system. As a result of this decrease the number of tumors are reduced and the latent period before their appearance following exposure is lengthened. These enzymes are responsible for the destruction of carcinogenic products by UV irradiation.

The ability of UV radiation to increase the activity of mitochondrial enzymes was also observed. UV radiation increases the activity of these enzymes in animals treated by chemicals which can reduce their activity. It is known that the reduction in the activity of these enzymes, which are involved in the utilization of oxygen by tissues, leads to the phenomenon of oxygen starvation, interruption of DNA synthesis and the development of degenerative processes which have been observed by various investigators following chronic exposure to certain chemicals. Taking these ideas into account, it may be considered that the increase in activity of both mitochondrial and microsomal enzymes following UV irradiation is one of the possible mechanisms by which such radiation acts as a prophylactic against certain chemicals.

In addition, an increase in the concentration of neuroaminic acid was also observed. This acid plays an important role in stabilizing the cell membrane and thus enhances the common resistance of the body. Neuroaminic acid, because of its complex reaction with cell membranes, limits the penetration of high molecular weight chemicals into the cells with a resulting decrease in the toxic action of such chemicals.

Other researchers have observed an acceleration in the removal of toxic chemicals from the body. These chemicals included fluorides, mercury, lead and manganese (7).

We believe that investigations into the mechanism of action of UV radiation _must be continued_ and that some of these mechanisms must be _studied in more detail._ The mechanisms of UV prophylaxis in relation to the mutagenic effect of chemicals _have not been studied._

Investigations have shown that the UV prophylactic effect in respect to certain chemicals bears a dose-relationship. Only suberythemal doses possess such prophylactic activity. The most favorable amelioration of the toxic effects was found to occur in animals which were treated with UV radiation representing 0.5 to 0.75 of an erythemal dose. An increase in the dose of UV radiation resulted in an increase in the _harmful_ effects of the chemicals. Studies also found that the number of transplanted tumors in animals irradiated with 3 erythemal doses was the same as in the untreated animals. Dose increases to the levels of 6 or 10 erythemal doses resulted in a considerable increase in the number of tumors. The "dose-effect" relationship was the same in the case of the activity of the mitochondrial and microsomal enzymes. The greatest activity occurred in animals exposed to erythemal doses. Both UV deficit and UV excess caused a reduction in enzyme activity.

The same results were obtained in respect to the allergic action of chemicals and the reproductive process in animals. It is thus possible to say that suberythemal doses cause a weakening of the toxic effect induced by environmental chemicals and large UV doses increase the toxic activity.

This data demands that there be a strict control over the dose of UV radiation when it is used for the purpose of prophylaxis on the one hand and on the other hand it also raises a question about the protection of the population against exposure to large doses of UV radiation when there is contact with harmful environmental chemicals.

In conclusion, the experimental data and accumulated experiences with the prophylactic use of UV radiation have made possible the establishment of a widely available system which can be used where the population is exposed to environmental chemicals under conditions where a UV deficit exists. As a result of the use of this system a weakening of the adverse effects of environmental pollutants and a decrease in morbidity associated with these agents should be expected.

REFERENCES

1. Dancing, N. M., The hygienic basis for preventing light starvation among ship's crews, in : The Biological Effects of Ultraviolet Radiation, 1975, Nauka, 168
2. Galanin, N. F. and Pivkin, V. M., 1971, The rational utilization of the sun's radiant energy with a view to preventing ultraviolet deficiency, in : Ultraviolet Radiation, Medicina, 209
3. Petrova, N. N., 1971, Study of ultraviolet deficiency in persons living at different latitudes in Western Siberia in springtime, in : Ultraviolet Radiation, Medicina, 244
4. Graceva, I. P., Bagrov, E. W. and Mandyc, L. F., 1975, Losses of incident UV radiation due to industrial pollution of the atmosphere, Gig. i. Sanit., 9:20
5. Koskin, M. L., The ultraviolet irradiation of premises as a problem of hygiene, in : Ultraviolet Radiation and Hygiene, Academy of Medical Sciences of the U.S.S.R., Moscow, 81

6. Dancig, N.M., Gorkin, Z.D. and Galanin, N.F., 1966, Prophylactic ultraviolet irradiation, in : Ultraviolet Radiation, Medicina, 208
7. Gabovic, R.P., Mihaljnk, I.A. and Motuzkov, I.N., 1975, The combined effect on the organism of chemical factors in the environment and various schedules of UV irradiation, in : The Biologic Effects of Ultraviolet Radiation, Nauka, 142.

ROUND TABLE SUMMARY : USAGE AND TESTING OF SUNSCREENS

Franz Greiter

Institute for Applied Physiology

Elisabethstr. 51-53, A-3400 Weidling, Austria

INTRODUCTION

In his introduction F. Greiter showed parameters with strong correlation to the evaluation of sun protection factors (SPF), which represent the only identification-possibility for the quality of a sunscreen (Fig.).

FIRST SPEAKER : F. URBACH

He mentioned that the shown parameters are of importance and emphasized some critical points of testing as :

- Quality of light source, whereby he compared XENON and OSRAM, the systems mostly used, regarding them as acceptable as long as a qualified dosimetry controls the emission data.

- Only lab-testing assures comparable test situation.

- Skin types evaluated for SPF-testing should represent I - III, but this has to be measured.

- Layer thickness of an applied sunscreen is important. The recommendation of 20 µm seems useful, but one should have in mind the different viscosities of the different sunscreen-types (milk, oil, etc..).

SECOND SPEAKER : H. IPPEN

He showed his method for UVB- and UVA-testing.

He believes that natural test-situations would be desirable but in praxis not possible. Himself he made good experience with lab-testing up to a limit of SPF 10.

He said that SPF-numbers higher than 10 should not be stated on the packings ; he found that mentioning "SPF > 10" is sufficient.

In future, he emphasized, UVA-protective screens would become more and more important.

Furthermore, not only "sunburn-protection" would be important in future, but "protection from too much light" ; sunscreens should reduce the complete amount of light acting on the skin.

THIRD SPEAKER : M. PATHAK

Presented impressive slides, which made clear what the difference between lab and field testing is. Products with SPF > 10 showed poor results under combined stress-testing (swimming and sweating).

He mentioned the waiting interval (time between application and exposure) and the use of standards (4 and 8) as specially important.

The amount of SPF should be limited to 12 and the "number business" has to be stopped.

He expressed with emphasis that the existing lab testing is nothing else than UVB-testing.

FOURTH SPEAKER : A. WISKEMANN

Recommended 3 different layer thicknesses for the 3 mostly used sunscreen types.

The best exposure interval (arithmetically or logarithmically used) should be changed to 1 minute.

FIFTH SPEAKER : G. GROVES

Reported that the difference between XENON and OSRAM is + 14 % for the Xenon. That means that the SPF evaluated under the Xenon-lamps is 14 % higher.

He was against the indication of numbers and recommended classifications as : maximum protection sunscreen (SPF15 & over)
high protection sunscreen (SPF8 - under 14)
moderate protection sunscreen (SPF4 - under 8)
minimum protection sunscreen (SPF2 - under 4)

SIXTH SPEAKER : T. FITZPATRICK

Mentioned that sun is one of the 4 most important things to men.

He believes in field testing and said -besides this meeting- that lab-testing is like testing a sailing boat in a swimming pool.

He cannot understand that industry did not take the challenge to recommend people to protect themselves against everydays' radiation, mainly during summer time, because of the fact that the daily direct or indirect radiation harms the skin.

He recommended to form a group of "academia" to formulate the conditions, which are asked for a serious SPF.

Sunscreens are the only possibility to protect people against elastosis and cancer.

The audience was in close discussion for nearly half the time of the whole session.

Concerning field and lab testing opinions turned out to be different. The numbering system was discussed. Some participants are against it and prefer classifications as they exist in Australia. Lip protection would have to be taken into consideration as well.

H.-Chr. Wulf reported that a trial on 150 hairless, pigmented mice showed clearly, that the development of tumors can be retarded by effective sunscreens.

CONCLUSION BY F. GREITER

The forming of a body of "academia" would be desirable.

The "number business" of more or less serious SPF's should be stopped.

The forcing up of higher and higher SPF's should be given up.

Standardized field testing would be desirable, since only these conditions represent natural exposure conditions.

Sunscreens have to meet all exposure conditions as swimming and sweating, cold, etc...

All sunscreen-producers should use only one range of SPF-numbers as for instance 2, 4, 6, 8, 10, 12, etc...

Numbers are the best information to the consumer, the more as he is already used to them in Europe and the USA.

The recommendation for the international formulation of the SPF should be finished in one year.

In regard to the question of recommending only one layer thickness for all screen-types, to which F. Greiter principally agrees, he reports that measurements of the layer thickness at 300 volunteers have shown that in praxis people never apply one and the same layer thickness.

THE TRENDS AND FUTURE OF PHOTOBIOLOGY : MEDICAL ASPECTS

T. B. Fitzpatrick

Chairman, Department of Dermatology
Harvard Medical School
Chief, Dermatology Service
Massachusetts General Hospital
Boston, Mass. 02114

This meeting has witnessed the emergence of a new collaboration of science and medicine--a society of scholars, studying the effect of light on biologic systems at all levels of organization from bacteria to humans. Photobiology has joined with photomedicine, both using the scientific method to plumb new basic truths and to apply them to the problems of human disease.

For the physician,

--Light is a probe into the mechanisms of disease. A good example is erythropoietic protoporphyria in which it was ascertained that the action spectrum was in the Soret band, which made it highly suspect that the chromophore was a porphyrin. Indeed, this porphyrin was ultimately found in the plasma and the red cells, but not in the urine.

--Light is an approach to the control of cell metabolism, ultimately expressible in terms of wavelength/intensity, as a drug is given in milligrams. It is even feasible, with new technology, that this can be expressed in specific wavelengths, for example, 330 nm/mJ/cm^2.

--Light plus chemicals now seems to be a possible pathway for the future control of disease, expressed as mg drug/mJ/cm^2.

Nonionizing radiation, especially ultraviolet, has the apparent advantage of lack of penetration and is therefore a "magic bullet" for skin-targeted therapy. This may be an oversimplification because ultraviolet can have a profound action on the immunologic system, at least in mice, with very important effects on tumorigenesis. This occurs through the mechanism of induction of suppressor lymphocytes. Also, blue light can induce, at least in mice, enzymes in the liver, and it can obliterate, in humans, the biorhythms of human growth hormone. Therefore, the action of light is more than skin deep, and there is much that needs to be learned about its systemic effects.

What are the trends in photomedicine expressed at this meeting?

If we compare the meeting at Bochum in 1972 with the meeting in Strasbourg in 1980 you will note that there was very little discussion of photomedicine in Bochum; most of the presentations were concerned with basic mechanisms using for the most part 254 nm. In Strasbourg, photomedicine has emerged as a partner.

What are the reasons for this new thrust of photomedicine in the past decade to a position of major importance?

1. Ozone depletion→increased UVB flux→skin cancer, which was postulated in the early '70s, spawned a whole field of research not only in skin cancer but in the effect of UVB on nonhuman biologic systems. The United States Government through its support of the National Academy of Sciences and the Department of Transportation made possible a huge study effort. Reams of paper were produced in several reports.[1] It was quickly discovered, however, that UVB was the stepchild of photobiology. In a sense, this past decade has seen a shift of attention by basic photobiologists to UVB, and more recently, in addition, to UVA.

The recent predictions on the effect of halocarbons point to a disastrous increase of UVB by the middle of the next century, even if controls are now put into effect in the United States. A 16% decrease in the average ozone concentration, which is now predicted (best guess), would produce a UVB increase of 44% at middle latitudes, leading possibly to several thousand more patients with malignant melanoma, a fatal tumor, and certainly to several hundred thousand more cases of nonmelanoma skin cancer. But this is only the tip of the iceberg--the nonhuman biologic effects of UVB could have a more catastrophic effect; this area of research is just now being explored and, based on the work to date, no statements are possible. There are some worrisome findings, however; for example, some species, such as young anchovies and certain crab and shrimp larvae, are now living near the limit of their sensitivity when at the surface of the water.

2. Skin cancer incidence and mortality is increasing at a rapid rate in the white population, and increased exposure to ultraviolet radiation is the most probable cause.[1] Based on current cohort data, for example, and without regard to changes in the ozone layer, there is a rise of melanoma mortality from a 1971-1975 rate of 26.3 per million for whites in the United States to 28.7 in 1976-1980, and a predicted 33.5 in 1981-1985. Proof for UVB as an etiologic agent for malignant melanoma is still circumstantial but more and more evidence is accumulating.[1] There is a latitude dependence, and a low incidence in pigmented peoples (a recent study of malignant melanoma incidence in 59 population-based worldwide cancer registries has shown a clear negative correlation between skin color and malignant melanoma incidence for exposed sites). There is a strong cohort effect (persons who are age 30 now have a higher incidence of melanoma than do persons of the same age born ten years earlier). This suggests a new causative agent, and the most likely cause is a change in behavioral patterns (increased leisure time, less concealment, more vacations in sunny areas). A recent case-control study of malignant melanoma queried for sun exposure habits revealed that painful or blistering sunburns during either childhood or adolescence were associated with subsequent increased risk of developing melanoma.

This new increase in malignant melanoma is occurring in the white, middle class, professional indoor worker who receives intermittent intense exposure--weekend sun worshippers!

We are not yet certain of the action spectrum for malignant melanoma because there is no animal model, and it is entirely possible that both UVB and UVA may be implicated or, less likely, UVA alone.

How has the medical community responded to these challenges?

1. It is difficult to change the personal habits of individuals, but we can control the absorption of ultraviolet radiation by the skin by the judicious use of the very effective topical sunscreens now available. These could modify the incidence of ultraviolet-induced nonmelanoma skin cancer and possibly modify the incidence of malignant melanoma. The need for standardization of efficacy by an international task force was discussed in a round table held at this meeting.

2. We now know who are the high-risk population for the development of skin cancer.[1] Based on their personal history of sunburning following a first 30-minute exposure of summer sun at northern latitude, susceptible white persons can be classified into four grades of reactivity. About one-third of the U.S. population is susceptible and, with a concerted public education program on

the use of sunscreens, ultraviolet damage can hopefully be prevented in this group.

3. <u>The use of light as a treatment for the diseases of humans has burgeoned in the past few years</u>. Phototherapy of neonatal hyperbilirubinemia, a serious problem, results in excretion of large amounts of unconjugated bilirubin in the bile and intestinal tract. The action of blue light causes an isomerization; these isomers can then be excreted. Large-scale prospective studies are now under way. Also, there are new methods recently developed for the detection of total bilirubin and its binding capacity by automated front-face fluorimetry.

Ultraviolet light therapy has been used since the early part of this century; Finsen, in fact, used ultraviolet light to treat skin tuberculosis and received the Nobel Prize for this effort. Ultraviolet light therapy (UVB) for other diseases, such as psoriasis, was introduced in 1925 but was really not very effective and therefore not widely used prior to the past three or four years when new high-intensity UVB light sources became available; dosimetry problems are being worked out, and good treatment protocols are being developed.

The combination of light and chemicals for the treatment of disease was never successfully accomplished. Considerable research has been done on the use of dyes and light, especially for the treatment of herpes simplex infections of the skin. Controlled prospective trials have produced conflicting results; this approach is still promising, but requires more research. There are over 400 available dyes, and new protocols are needed.

The first successful application of light and drug for a beneficial effect on disease occurred in 1974 with the use of orally administered furocoumarins combined with UVA irradiation.[2] Prior to 1974, all light/drug interactions were deleterious. In Bochum, in 1972, the furocoumarins had been discussed for the first time by Cole and by Musajo's group as interesting tools for the study of DNA interactions. They were shown to form mono- and bifunctional photoadducts with DNA; cross-links were also formed. 8-Methoxypsoralen was known to be mutagenic and to cause cancer in mice on topical application and exposure to long-wave ultraviolet light. Again, at the Bochum meeting evidence was presented to the effect that <u>topically</u> applied psoralens could cause a regression of a common <u>skin</u> disorder, psoriasis, presumably by inhibition of cell replication. This report at Bochum resulted in a backlash by the molecular biologists who warned that such drug/light therapy could be dangerous because of its action on DNA.

In 1974, the introduction of orally administered 8-methoxypsoralen in combination with newly developed high-intensity artificial UVA irradiators provided a highly effective modality for the control of severe psoriasis. This new procedure was called PUVA (psoralen + UVA), and the technique photochemotherapy.[2]

PUVA photochemotherapy came at a time when there were few options for treating generalized psoriasis. Controlled clinical trials were required to establish safety and efficacy. Soon, following publication of the effectiveness of PUVA in psoriasis, in 1974, there was widespread interest in the use of this new therapy--the drug (8-methoxypsoralen) was commercially available by prescription, and some electrical manufacturers had begun to make UVA boxes. This use of PUVA without protocols, however, was premature because the treatment regimen had not yet been evolved and there were initial problems in dosimetry of drug and of light that required solution by large controlled clinical trials. This was accomplished between 1975 and 1979 from three multicenter clinical trials, with the accumulation of a large data base. Patients (5,275) have been treated in a 28-center trial in the United States[3,4] and a 20-center trial in Europe.[5]

As a result of these clinical trials, a detailed outline of the technique and side effects of PUVA photochemotherapy was prepared and published in 1979.[6] The FDA Bureau of Radiation Health then published specifications for the UVA irradiator.

A large world patient population now receives PUVA. During the past five years PUVA centers have been established in all parts of the world, including Russia, Poland, China, several countries in South America, and in the Middle East. Approximately 80,000 patients are now being treated. PUVA therapy has been granted government approval in several countries--Australia, Austria, Belgium, Canada, Finland, Greece, Italy, Malaysia, The Netherlands, South Africa, Switzerland, and West Germany.

This comprehensive clinical experience from Europe and from large multicenter controlled trials in the United States and Europe has confirmed PUVA's high degree of efficacy for psoriasis and its short-term safety when administered by a standardized method using a suitable UVA irradiator.

From the beginning of these clinical trials the warnings pronounced at Bochum by the photobiologists that psoralens could be oncogenic were kept in mind, and careful follow-up of 1,300 patients in the United States suggests now that PUVA acts as a promoter of nonmelanoma skin cancer in a special subset of patients who have previously received treatment for their psoriasis with ionizing radiation[7,8] or with arsenic.[8] I say "suggest" because these protocols did not include a control population.

There is as yet no evidence of an increased incidence of malignant melanoma in PUVA photochemotherapy.

Now, newly developed high-intensity UVB light sources have been made available for the treatment of psoriasis and these are effective in clearing psoriasis.[9,10] Curiously, the discovery of PUVA has led to the rediscovery of UVB. The long-term effects of UVB, however, are probably no different from PUVA, as UVB is also photocarcinogenic.[11] The 50-year safety record of UVB is not valid because of the higher exposure doses used in these new UVB lighting sources.

How do we resolve the risk/benefit problems in these highly effective treatment modalities?

There are also similar risk/benefit problems in the phototherapy of hyperbilirubinemia and in dye/light therapy. This will be accomplished by more research and more interdisciplinary cooperation among teams of synthetic chemists, molecular biologists, pharmacologists, and physicians.

The French have proposed 3-carbethoxypsoralen, 3-methoxypsoralen, and 3-cyanopsoralen; these form only monoadducts, are not photocarcinogens, and have been reported to be effective in clearing psoriasis when applied topically.

The Italians have synthesized a series of nonlinear psoralens, such as 5-methylangelicin, that also form only monoadducts and are less mutagenic.

There has been a rippling effect from PUVA throughout all of photobiology: in 1978-1979 there were 900 published reports on PUVA and on basic aspects of psoralen photochemistry. We have heard here at this meeting that psoralen unwinds the helix in DNA, that it has been useful as a molecular probe into elucidating the structure of 5-S RNA of Drosophila melanogaster.

Finally, we can ask ourselves: Is ultraviolet hazardous to health?

The answer is yes and no!

Yes: it is one of the four deadly pleasures:

        1. alcohol,
        2. food,

3. tobacco,
4. sun,

all four of which used in excess can be very hazardous to health.

No: if ultraviolet is used wisely, we have heard today from Prof. Franz Greiter that it seems to improve your physical performance and increase your level of aspiration. And used intelligently, ultraviolet light alone or ultraviolet light in combination with chemicals can improve the quality of life for those unfortunate several million people with disfiguring and disabling diseases.

If we can improve the quality of life, we photobiologists can be deemed worthy to serve the suffering!

References

1. Committee on Impacts of Stratospheric Change, "Protection Against Depletion of Stratospheric Ozone by Chlorofluorocarbons," National Academy of Sciences, Washington, D. C. (1979).
2. J. A. Parrish et al, Photochemotherapy of psoriasis with oral methoxsalen and long wave ultraviolet light, N. Engl. J. Med. 291:1207 (1974).
3. J. W. Melski et al, Oral methoxsalen photochemotherapy for the treatment of psoriasis: a cooperative clinical trial, J. Invest. Dermatol. 68:328 (1977).
4. Clinical cooperative study of PUVA-48 and PUVA-64: photochemotherapy of psoriasis, Arch. Dermatol. 115:576 (1979).
5. T. Henseler et al, PUVA study: results of oral photochemotherapy for psoriasis conducted in 17 European centers, (in preparation).
6. J. H. Epstein et al, Current status of oral PUVA therapy, J. Am. Acad. Dermatol. 1:106 (1979).
7. R. S. Stern et al, Risk of cutaneous carcinoma in patients treated with oral methoxsalen photochemotherapy for psoriasis, N. Engl. J. Med. 300:809 (1979).
8. H. Honigsmann et al: Keratoses and non-melanoma skin tumours in long-term photochemotherapy (PUVA), J. Am. Acad. Dermatol. (submitted for publication).
9. M. J. LeVine et al, Components of the Goeckerman regimen, J. Invest. Dermatol. 73:170 (1979).
10. M. J. LeVine and J. A. Parrish, Outpatient phototherapy of psoriasis, Arch. Dermatol. 116:552 (1980).
11. R. S. Stern et al, Skin carcinoma in patients with psoriasis treated with topical tar and artificial ultraviolet radiation, Lancet 1:732 (1980).

IV - PHOTOPHYSIOLOGY

VISUAL RHODOPSIN AND PHOTOTRANSDUCTION IN THE VERTEBRATE RETINA

Marc Chabre

Laboratoire de Biologie Moléculaire et Cellulaire
(E.R. 199 du CNRS)
DRF - CENG - 85 X
38041 - Grenoble

INTRODUCTION

Progress in the elucidation of the structure and functional mechanism of visual rhodopsin has not been as fast as for its younger homonym bacteriorhodopsin. This is due essentially to the lack of crystalline ordering of rhodopsin in the visual receptor cell membrane, and to the complexity of the visual transduction and amplification process. The visual receptor cell membrane (i.e., the retinal rod outer segment disc membrane) and the purple membrane of Halobacterium are indeed the opposite extreme examples of membrane structure : respectively the most fluid and disordered membrane for rhodopsin, and the most rigid and ordered for bacteriorhodopsin. As for the functions, bacteriorhodopsin is a simple photo-energised proton pump, and rhodopsin in the retinal rod cell seems to be only the first component of a rather complex multi-enzymatic photosensitive chain. Simpler photosensitive systems may exist in more primitive organisms, but the choice of the vertebrate retina for investigations is dictated by its unique morphological advantages : the high specialization of the receptor cell and the segregation of the photopigments and the associated components of the phototransduction chain in the outer segments of these cells make them an ideal material for biophysical and biochemical studies which require large quantities and a high degree of purification.

The analogies between rhodopsin and bacteriorhodopsin might therefore not go far beyond their common chromophore retinal, bound in both cases by a protonated Schiff base to a lysine of the protein. Some structural analogy might be claimed for the proteins themselves, as it appears now that the hydrophobic part of rhodopsin is also constituted by a bundle of transmembrane alpha helices. But this might not be very significant, as it is probably a general feature of transmembrane proteins.

The functionnal analogies remain very hypothetical : the fact that the protonation state of the Schiff base nitrogen changes during the relaxation sequence following photo-excitation of visual rhodopsin has led to many speculations as to a direct role in visual transduction of this phenomenon and of the associated pH and potential variations across the membrane. But one must always keep in mind the fact that visual transduction is a very sensitive process, able to detect the absorption of a single photon in a cellular organelle of large size : the variation of transmembrane potential or pH for this single photon absorbed are much too small to constitute detectable signals above noise : the large number of rhodopsin molecules in the membrane is there to optimise the light catching efficiency, but these pigment molecules are not supposed to work in parallel and sum up their responses, as it is the case in photosynthetic membranes. The large amplification required to build a detectable response from a single photon event results from a cascade of steps, some of which being certainly enzymatic. There is plenty of time for that : visual receptors, if they are very sensitive, are also very slow at low level of illumination : for a single photon event, the rise time of the cellular response is of the order of a few hundred milliseconds. It has been amply proven in the past few years that photoexcited rhodopsin, in one of the long lasting Meta states, activates an enzymatic system which controls the level of guanine cyclic nucleotide (cGMP) in the cell. Whether this mediates or only modulates the receptor physiological response is still debated, but it is certain that this system provides a very large amplification factor : the photoexcitation of one rhodopsin molecule can lead to the hydrolysis of about 10^4 cGMP molecules within 100 msec (1). Cyclic GMP has largely outshadowed calcium as the putative transmitter modulated intracellularly by light in the retinal receptor cells (2). However, very recent results of two independant groups (3-4) seem finally, after 10 years of unsuccessful quest, to give indirect but rather convincing evidence for a role of calcium ions in the visual transduction process. If this is confirmed, it will lead to an even more complex scheme, where both cGMP and Calcium will interfere in the transduction mechanism.

In this presentation we shall concentrate on three points on which very significant advances have been made in recent years.

The primary photochemical event, a rather old problem supposedly settled twenty years ago by the isomerization hypothesis of Wald, has been the object of intense reinvestigation : picosecond kinetics measurements at low temperature had raised a great turmoil in this field with the provocative proposal of a proton tunnelling transition. This has instigated a new set of very elaborate spectroscopic studies, which have finally led to a refined version of the original isomerization hypothesis.

Our second point will concern the structure of rhodopsin. It has been approached by a large variety of biophysical and biochemical techniques which converge now toward a common picture of the location of the protein in the membrane, of its orientation and gross structure.

We shall discuss finally the conformational changes of the chromophore and of the protein upon photo-excitation, particularly in the long lasting Meta states : it is during those steps that rhodopsin becomes "active" and couples with other peripheral proteins present on the membrane and in the cytosol to initiate and/ or control the physiological response.

THE PRIMARY PHOTOCHEMICAL EVENT

Wald and his collaborators had proposed in the early sixties that the primary photochemical event was the isomerization of retinal from its original 11-cis conformation in native rhodopsin to an all trans conformation in the first detectable photoproduct bathorhodopsin. The most convincing argument was that bathorhodopsin is also obtained when illuminating rhodopsin in which 9-cis retinal had been substituted to the natural 11-cis chromophore : a common intermediate to a 11-cis and to a 9-cis pigment should have an all-trans conformation. Bathorhodopsin can be blocked indefinitely at liquid nitrogen or liquid helium temperature, but the chromophore cannot be extracted from this state, and all-trans retinal was only extracted from later intermediates such as Meta II rhodopsin, and at room temperature. Very few spectroscopic measurements, besides the absorption spectra, had been performed on bathorhodopsin, and the various explanations for the large bathochromic shift associated to this step were still largely speculative.

Kinetics of the formation of bathorhodopsin became measurable with the development of picosecond technique ; bathorhodopsin, at room temperature, was found to be formed within the 6 picoseconds resolving time of the technique (5). But more surprisingly, this rise time remained very short at low temperature ; it was still only of 36 picoseconds at 4° K (6), and the Arrhenius dependence on temperature of the rate of formation seemed non linear at these low temperatures. This, combined with a strong deuterium isotopic effect on the kinetics, led Rentzepis and his coworkers (6) to propose that bathorhodopsin was the result of a proton translocation along the retinal chain and that this state was reached through a tunneling mechanism (see fig. 1). This attractive proposal was very stimulating for the spectroscopists, which have made in the last three years remarkable progress on the knowledge of retinal conformation in bathorhodopsin but also in native rhodopsin, and on the nature of the transition. It seems now that the proton translocation hypothesis must be given up for a refined version of the original isomerization hypothesis, but the challenge has proven to be very fruitfull. I cannot review here all the experimental approaches, and I'll give only a few major arguments :

In favor of a geometrical change of the chromophore at this stage was first the observation that bathorhodopsin is formed at room temperature with the same fast kinetics ($\simeq$ 3 psec) when starting from isorhodopsin (with a 9 cis chromophore) than when starting from rhodopsin (7). One at least of these transitions requires an isomerization, which proves that isomerization can be very fast. That isomerization could occur at very low temperature was demonstrated by the observation that at 4° K isorhodopsin could be formed photochemically from bathorhodopsin (8), although the kinetics was not studied in this case.

Detailed information on the retinal conformation in bathorhodopsin has been obtained from Raman spectroscopy at low temperature, using pigments reconstituted with synthetic retinal molecules specifically deuterated at various positions (9). This allowed unambiguous line assignments. The fingerprint region of the spectra, between 1100 and 1400 cm^{-1} indicated that, in bathorhodopsin the conformation of the 11-12 double bond is trans, but assignments through specific deuteration and model molecules, of intense lines around 900 cm^{-1} indicated these to be due to out of plane vinyl hydrogen motion (9). The intensity of those lines is revelative of twists of about 20° around single bonds in the middle of the chain. The structure of the chromophore in bathorhodopsin is therefore "distorted all trans" (10). A very elegant demonstration that the structure is not of the "retro" type, as would be implied by a proton translocation, is the assignment, through specific deuteration of an out of the plane vibration line of the two hydrogens on C_{11} and C_{12} (11). This requires $C_{11} = C_{12}$ to be a double bond and excludes a retro conformation where the double bond would be transferred to $C_{10} = C_{11}$.

A seemingly simple and direct evidence for a geometric change of the retinal molecule upon the transition to bathorhodopsin is the observation of a change of orientation of the main absorbing dipole of retinal which is roughly parallel to the polyene chain. But the measurement is not straight forward, and only one group has so far published data on this point (12). Their experiment is based on the comparison of the absorbance changes observed when illuminating with unpolarised light isotropic solutions of rhodopsin in digitonin and the absorbance changes observed under similar

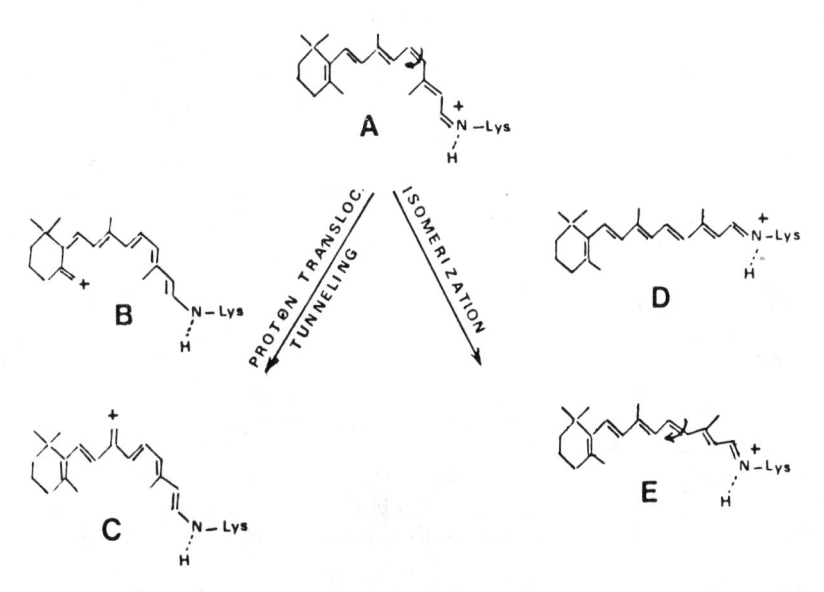

Figure 1 - The conformation of retinal in Rhodopsin (A) and the various models (B,C,D,E) for the primary photoexcited state bathorhodopsin. The "Distorted All Trans" conformation E is presently the most favored model.

conditions (that is at liquid nitrogen temperature) with an oriented retina. This has to be supplemented by photoselection experiments with polarised light. It has been estimated from these experiments that the absorbing dipole of retinal undergoes an orientational change of about 26° upon the rhodopsin to bathorhodopsin transition (12). This is a large geometrical change in the retinal molecule, such as only an isomerization could induce. More direct measurement may be based on linear dichroism measurements on retinal rods oriented transversally with respect to the polarised beam. We have recently prepared such samples by orienting isolated rod outer segments by a magnetic field and trapping them at low temperature in a glycerol-water mixture. Our preliminary results (M. Michel-Villaz and M. Chabre, unpublished) confirm the orientational change of the absorbing dipole, but the angular change seems to be much smaller than 26°. This is not against the isomerization hypothesis : in fact, when one considers the exact conformations of 11-cis and "distorted" trans retinal, the change of orientation of the polyene chain upon the transition is surprisingly small if one assumes that the β-ionone ring of retinal remains fixed in the protein. As it will be stressed below, the constraint of the protein on the retinal molecule is very tight, the movement of retinal is restricted and involves a lot of energy to distort the binding site.

A remarkable experimental result on this point is the recent direct measurement by photocalorimetry of the energy uptake in the first photochemical step (13). The ground state of bathorhodopsin has been found to lie 35 Kcal above that of rhodopsin. The energy difference between 11-cis retinal and all-trans retinal in solution is neglegible ($\simeq$ 0.5 Kcal). A considerable amount of energy must therefore be involved in the interaction between the retinal and its binding site : more than 60 % of the absorbed photon energy is stored in the interaction between the distorted all-trans retinal molecule and its strained binding site which hinders its displacement. This is a very high efficiency for a photoenergetic conversion. This system, however, is aimed at gathering information, that is detecting the event of one photon capture with the best possible dark noise rejection, it is not collecting energy. In the later steps of photoproduct decay the stored energy will be relaxed in the protein, inducing informative conformationnal changes and triggering enzymatic activation resulting in an overall negative energy balance. But the high energy step required by the primary event is a very efficient defence against thermal noise.

From the photocalorimetric measurement and the various reaction rates, potential energy profiles can be constructed for the ground and excited states in the primary step of photoexcitation (13). This appears to be in very good agreement with the detailed "external point charge model" proposed recently by Honig and his collaborators (14) (15) for the interaction of the retinal molecule with its environment in rhodopsin and bathorhodopsin. This model integrates the recent Raman results and a set of spectroscopic data obtained with locally deuterated retinal chromophore. It postulates the presence of a counterion near the protonated Schiff base nitrogen and a second negatively charged group situated around the middle of the polyene chain, about 3 Å from the carbon 12 and carbon 14. The model explains the bathochromic shifts in rhodopsin and bathorhodopsin and leads to a photoreaction scheme which seems to account for all the known reaction rates and spectroscopic data.

A controvery still exist however on the existence and the nature of an earlier state, hypsorhodopsin, which is observed by the group of Yoshizawa at liquid helium temperature (16). It has been speculated that it could be an early deprotonated state of the chromophore Schiff base, and could be a side pathway in the scheme of Honig (14).

The basic mechanism of the primary photochemical event being apparently well established, the next step now is to explain how this high energy state, where the "transoid" retinal is tightly strained in its proteic site, relaxes inducing significant conformational changes of the protein. Before discussing that, we need to review the knowledge on the structure of the protein.

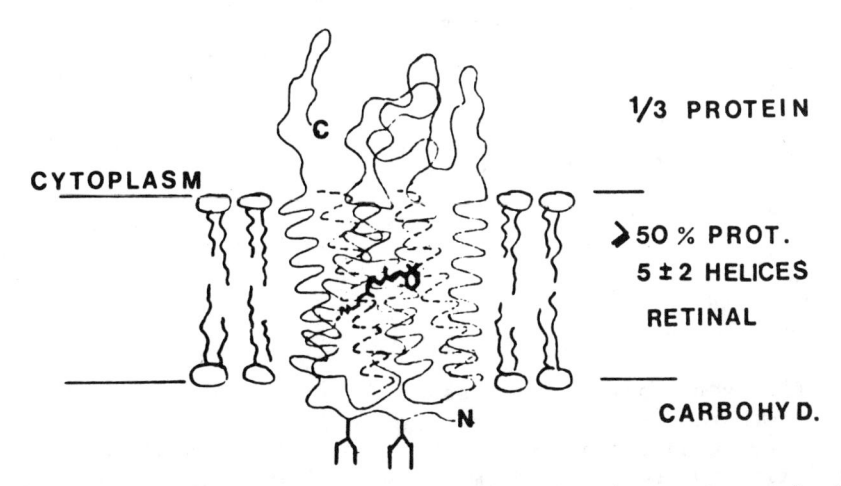

Figure 2 - Model for the structure of rhodopsin in the retinal rod disc membrane.

RHODOPSIN STRUCTURE

In the retinal rod outer segments, the disc membranes, which contain rhodopsin as the major and quasi unique intrinsic protein, are regularly stacked in a quasi crystalline manner. But, contrary to the situation in chloroplasts and in the purple membrane for that matter, there is no rigid organisation within the disc membrane plane. Individual rhodopsin molecules float freely in a two dimensional "oil" bilayer, with a very high degree of lateral and rotational mobility. Rotation is only constraint around an axis perpendicular to the membrane plane, so that the angle of the chromophore is kept fixed with respect to the rod axis. This is a very significant observation with respect to the mechanism of visual transduction, where protein mobility plays an important role, but a very undesirable one when one is interested in structure : there is absolutely no hint of crystallisation, which explains the slow rate of progress in structure determination. Small pieces of information have to be gathered from as many biophysical and biochemical techniques as possible. This has proven to be quite successfull in the past few years, and good agreement is reached among the biophysicists and the biochemists on the model presented on fig. 2. It is not a high resolution modele, only the main features are well defined :
1) - The protein is transmembranous. A large part of the proteic mass is buried in the hydrophobic layer of the membrane ; it has a

small hydrophilic contact on the intradiscal (or extracellular) side where two carbohydrate chains are located. A larger hydrophilic part of the protein (of the order of 1/3 of the total) protudes into the cytoplasm. This is the site of interaction of rhodopsin with peripheral membrane proteins and soluble proteins in the cytoplasm (18 - 22).

2) - The core of the protein is very hydrophobic. This was demonstrated by the observation that a large proportion of the peptide protons are not exchanged with deuterium when D_2O is substituted to H_2O in the medium (23). This would be hardly compatible with the concept of an ionic channel to be formed within this protein.

3) - This hydrophobic core is essentially constituted by a bundle of transmembrane α-helices, analogous, although less oriented, to that found in bacteriorhodopsin. This was first proposed as an explanation for the large diamagnetic anisotropy of the rod outer segments (24), which orient very easily in magnetic fields. It has been calculated, and demonstrated experimentally, that straight α helical segments have a significant diamagnetic anisotropy along their axis (25) and no other reasonable explanation can be found for the magnetic orientation of the rods. This overall transmembrane orientation of the α-helix segments has been later demonstrated by polarised infra-red studies of the $C = O$ and $N - H$ peptide groups in oriented samples (26).

4) - The location of the retinal molecule in the protein is still much debated. From fluorescence transfer data with probes attached to the protein, a location close to the intradiscal side of the membrane has been proposed (27) (28). But other approaches would be needed to confirm that, as there are uncertainties on the exact binding sites of the probes. On a recent partial sequence obtained by combining the results of two independant groups (22) (29), the lysine to which retinal is bound may be tentatively identified. This lysine 53' is in a hydrophobic segment not far from the C terminal hydrophilic peptide which is known to be on the cytoplasmic side.

5) - The biochemical data and sequence information are making quick progress. There is no more ambiguity on the molecular weight (~ 38000), the sequence has been published for more than 1/3 of the total lenght of the polypeptide chain and much more is already known by the groups which use the classical proteolytic methods. On the other hand, the mRNA coding for bovine opsin has been isolated and purified (30). Various groups are attempting the transcription of the DNA and the cloning. DNA sequencing being a much easier task than proteolytic sequencing of hydrophobic proteins, a fruitfull competition of the two techniques should result soon in the obtention of the total sequence. The example of bacteriorhodopsin has shown that for such type of proteins knowledge of the total

sequence is of great interest, in particular for the determination of the retinal site and interactions in the protein.

CONFORMATIONAL CHANGES UPON PHOTOEXCITATION AND "PHOTOACTIVATION" OF RHODOPSIN

It remains now to elucidate how the primary photochemical event, localised around the retinal, transforms the protein into an "active" species able to control the level of a diffusible transmitter into the cytoplasm of the cell. The simplest scheme (31), in which rhodopsin itself would be a light dependant channel controlling directly the release of an ion (Ca^{++} ?) has been much investigated but still lacks a solid experimental support. In order to be significant for visual transduction, such a channel should allow the release of the order of 10^3 calcium ions into the cytoplasm within about 100 milliseconds. This might only be performed by a gramicidine-like channel of high conductance. We have already noted that this seems not easily compatible with the high hydrophobicity of the rhodopsin core. This is not a very strong argument, but many attempts to detect such a release have never succeeded in demonstrating a light controlled release in excess of one calcium per bleached rhodopsin, and often much less, on the required time scale. Comparable permeability changes are observed upon strong illumination for other ions, and are probably unsignificant. These studies have mostly been made with purified and washed disc membranes or even reconstituted systems in which rhodopsin remains the unique protein component. It seems therefore unlikely that rhodopsin itself is a calcium channel. It has also been remarked by R. Cone (personal communication) that the fixed latency of the electrophysiological response at low level of illumination is hard to reconcile with such a simple scheme : if rhodopsin itself was the channel, the latency should depend on the distance between the particular rhodopsin that has absorbed the photon and the cell membrane. It would be very surprising that the delay could be due to the mechanism of opening the channel, since the major conformational changes in rhodopsin seem to appear within one millisecond at physiological temperature. But we have already mentionned that a light induced extracellular release of Ca^{++} ions by the rod outer segments has just been demonstrated on intact retinae preparations (3) (4). This might be a hint of a still undetected intracellular release. If this was confirmed, one might still keep the concept of ionic channels distinct from rhodopsin, and imagine models in which the photoexcited rhodopsin molecule triggers either directly after lateral diffusion, or indirectly through the enzymatic control of a first transmitter (cGMP), the opening of a few ionic channels spread on the disc membrane with a very low relative stoichiometry. The fact that these channels have not been biochemically identified is certainly not a valid objection : about 30 years after the existence of Na channels in the axonal membrane has been definitely proven by the electrophysiologists, and after a very intensive biochemical work, these channels are still very elusive entities for the biochemists ! The temporal characteristics of the electrophysiological response of the photoreceptor cell are in favor of a first step involving lateral diffusion of protein in the membrane to account for the latency.

But other schemes, without any calcium channel are also proposed : cGMP itself would be the transmitter. The light dependent multicomponent enzymatic chain which controls the hydrolysis of this nucleotide provides the required amplification, and the lateral diffusion times required for the assembly of the different components of the chain would explain the latency and the slow rise time of the electrophysiological response.

The biochemical analysis of the native rod outer segment has made decisive progress in the last two years and has revealed that rhodopsin is far from being the only protein involved in the phototransduction process at the level of the disc membrane (32-35). In particular an elegant experiment by H. Kuhn (36) has demonstrated that illumination results in the binding to the disc membrane of a fairly abundant GTP dependent protein which, in the dark, can be eluted from the membrane at low ionic strength. This protein is part of the enzymatic complex which controls the level of cGMP in the cell. The binding is reversible and not observed with opsin. This proves that photoexcited rhodopsin is specifically recognised by this enzyme. At this stage, the high energy primary state in which large constraints are localised around the "transoid" retinal, has decayed into a lower energy state in which the retinal is probably less constraint and the protein has developped on its cytoplasmic surface a high affinity site for the GTP dependent protein. The "active" state, identified as Meta II rhodopsin in the standard spectral classification, is reached in about a millisecond at physiological temperature, a very long time on the time scale of molecular motion, suggesting the occurence of large macromolecular rearrangments. This state is transient, and it must not be confused with the last state of the decay, in which retinal is released from the protein. In Meta II rhodopsin, retinal is still in its original binding site, as demonstrated by the orientation of its main absorbing dipole (37), but, not surprisingly, its spectral characteristics are very close to that of free retinal : this is related with the fact that the interaction of the chromophore with the protein is very relaxed. One has therefore an experimental problem : photobiologists tend to classify and to study photoexcited states through their spectral characteristics, and at this crucial step the spectral sensitivity in the visible and near UV range is lost. In fact Meta II rhodopsin, as defined spectrally, covers various states of the protein, with for example different degrees of protonation : the significant events are not anymore located exclusively around the chromophore and different probes must

be found to study them. Later stages in the final decay toward
opsin and free retinal are also difficult to differentiate on the
basis of their absorbance characteristics alone. Linear dichroism
measurements on oriented cells may be very helpful at this point
to follow the fate of the loosely bound retinal (37). But after
the Meta II state(s), one enters into the regeneration cycle of
rhodopsin, the coupling with the phototransduction process is switchef off.

In the dark evolution from the primary photoexcited state,
batho, to this functionally important Meta II state, the major
changes occur with the Meta I → Meta II transition. The earlier
intermediates Lumi and Meta I which can be blocked only at low
temperature, correspond probably to very local rearrangments of
the retinal site. Associated with the large spectral shift of the
Meta I → Meta II transition, related to the deprotonation of the
Schiff base, various electrical, structural and biochemical effects
have been detected :

The charge displacement and the changes of protonation states
in the protein, detected in vivo a long time ago by the electric
signal termed "early receptor potential" (ERP), has been thoroughly reinvestigated recently on various purified membrane preparations
and model systems, and its origin is understood (38-41). Its functional significance is very dubious. My personal feeling is that
it is an unspecific signature of a conformational change, which is
made measurable under the very special experimental conditions in
which a very large number of proteins molecules, all oriented in
a membrane, are excited simultaneously so that minor effects add
up. Similar comments may apply to the transmembrane pH gradient
that may be observed under similar circumstances : Recents results
indicate that they can be simply related to a proton uptake on the
cytoplasmic side of the membrane and need not to imply a transmembrane proton transport (42).

The structural modifications clearly demonstrated are only
minor and localised : a change of orientation of the retinal molecule (43-44) and a rotation of one tryptophane residue. But one
may just lack the appropriate conformational probes in the sensitive region. It is however demonstrated that no overall change
occurs, in the native membrane, on the secondary structure of rhodopsin. The circular dichroism spectrum in the far U.V. is not modified (44) and the infra red absorbance spectra (45) and linear
dichroism (26), as well as the diamagnetic anisotropy data (24),
confirm that the α-helical content is not modified and there is
no major change of orientation of these α-helical segments. One
can also exclude now an overall shift of the whole molecule across
the lipid bilayer of the membrane, such as it had been claimed
10 years ago, on the basis of early X ray measurements (47). The

very small changes observable upon illumination on X ray and neutron diffraction patterns of intact rods can be entirely accounted for by the light dependent binding of the peripheral membrane proteins (48) (18).

Large structural changes upon bleaching have often been observed with purified rhodopsin in various detergents or other artificial systems. This reflects only the fact, known since the early studies, in the fifties, that the Meta II state, and also opsin, are much less stable than rhodopsin and are irreversibly denaturated in such conditions although rhodopsin is well preserved. It has clearly been shown, in proton exchange kinetics measurements for example (46) that the so called "structural change" occurs only upon the addition of detergent if this addition is made after the illumination.

As an example of light induced biochemical modification at a site which seems remote from the retinal, one can quote the uncovering, at the Meta II stage, of several phosphorylation sites on the hydrophilic C terminal peptide, located in the cytoplasmic region (49, 22). The other light induced biochemical modifications seem also to concern exclusively the part of the protein accessible from the cytoplasm. There is therefore no reason to expect a major change in the hydrophobic core. But then, if rhodopsin has no vectorial transport property through the membrane, what would be the function of an inert hydrophobic core ? It seems too important to be just an anchorage for the protein on the membrane. The main limitation to further speculations on the extent of the conformational changes induced in the protein by the photoisomerization of retinal is the lack of precise date on the location of the retinal binding site.

CONCLUSIONS

The spectroscopists have done an excellent job in the analysis of the primary photochemical event, and the molecular models at this stage have attained a high degree of precision. More remains to be done on the early dark reactions, but this is progressing fast. The biophysicists have been less successful in the determination of the structure of the protein, which limits now the interpretation of the precise spectroscopic data. Progress in this direction seems now more in the hands of the biochemists, for the primary structure at least.

The main problems now are in the late dark reactions, when spectroscopy ends and enzymology begins. For a long time people in the visual transduction field hoped they would avoid enzymology and get along with nice and simple channels and ions mechanisms. But, whether we like it or not, visual rhodopsin upon photoexcitation activates other enzymes, this is certain. It is much less probable that it would also be a calcium channel or a simple proton transducer... In this photobiological problem the "biology" part appears even more complex than the "photo" part !

This contribution is only a very partial view of the recent progress in this field, with a strong personnal bias. The quotation of the literature is far from complete. A very complete review of the recent literature has recently appeared (50).

REFERENCES

1. P.A. Liebman and H.G. Pugh, Vis. Res. 19 375-380 (1979)
2. M.L. Woodruff and M.D. Bownds, J. Gen. Physiol. 73 629-653
3. S. Yoshikami, J.S. George and W.A. Hagins, Nature 286 395-398 (1980)
4. G. Gold and M. Montal, Proc. Natl. Acad. Sc. U.S. under the press
5. G.E. Busch, M.L. Applebury, A.A. Lamola and P.M. Rentzepis, Proc. Natl. Acad. Sc. U.S. 69 2802-2806 (1972)
6. K. Peters, M. Applebury and P. Rentzepis, Proc. Natl. Acad. Sc. U.S. 74 3119-3123 (1977)
7. B.H. Green, T.G. Monger, R.R. Alfano, B. Aton, R.H. Callender, Nature 269 179-180 (1977).
8. B. Aton, R. Callender and B. Honig, Nature 273 784-786 (1978)
9. G. Eyring and R. Mathies, Proc. Natl. Acad. Sc. U.S. 76 33-37 (1979)
10. G. Eyring, B. Curry, R. Mathies, R. Fransen, I. Paling and G. Lugtenburg, Biochemistry 19 2410-2418 (1980)
11. R. Mathies, private communication and J. Am. Chem. Soc. In press.
12. S. Kawamura, T. Tokunaga, T. Yoshizawa, A. Sarat, T. Takitan, Vis. Research 19 879-884 (1979)
13. A. Cooper, Nature 282 531-533 (1979)
14. B. Honig, T. Ebrey, R.H. Callender, U. Dinur and O.M. Ottolenghi J. Am. Chem. Soc. 101 7084-7086 (1979)
15. B. Honig, U. Dinur, K. Nakanishi, V. Balogh-nair, M.A. Gawinowics, N. Arnaboldi, M.G. Motto, Proc. Natl. Acad. Sc. U.S. 76 2503-2507 (1979)
16. Y. Shischida, F. Tokunaga and T. Yoshizawa, Photochem. Photobiol. 29 343-351 (1979)
17. M.T. Mas, J.K. Wang and P.A. Hargrave, Biochemistry. In press.
18. H. Saibil, M. Chabre and D.L. Worcester, Nature 262 266-270 (1976)
19. H.B. Osborne, C. Sardet, M. Michel-Villaz and M. Chabre, J. Mol. Biol. 123 177-206 (1978)
20. A.D. Albert and B.J. Litman, Biochemistry 17 3893-3900.

21. W.L. Hubbell and B.B.K. Fung, in "Membrane Transduction mechanisms", R.A. Cone and J. Dowling editors, Raven Press N.Y. (1979).
22. P.A. Hargrave, S.L. Fong, J.H. Mc Dowell, M.T. Mas, D.R. Curtis J.K. Wang, E.Juszczak and D.P. Smith. In "Neurochemistry of the retina", G. Bazan and R. Lolley Editors, Pergamon Press, London (1980).
23. H.B. Osborne and E. Nabedryk-Viala, F.E.B.S. Letters 84 217-280 (1977).
24. M. Chabre, Proc. Natl. Acad. Sc. U.S. 75 5471-5474 (1978)
25. D.L. Worcester, Proc. Natl. Acad. Sc. U.S. 75 5474-5476 (1978)
26. M. Michel-Villaz, H. Saibil and M. Chabre, Proc. Natl. Acad. Sc. U.S. 76 4405-4408 (1979).
27. C.W. Wu and L. Stryer, Proc. Natl. Acad. Sc. U.S. 69, 1104 (1972)
28. J.S. Pober, V. Iwanij, E. Reich and L. Stryer, Biochemistry 17 2163-2169 (1978)
29. C. Pellicone, P. Bouillon and N. Virmaux, Cpte Rend. Acad. Sc. Paris 290 D.567-569 (1980)
30. J. Schechter, Y. Burstein, R. Zemell, E. Ziv, F. Kantor and D. Papermaster, Proc. Natl. Acad. Sc. U.S. 76, 2654-2658 (1979)
31. W.A. Hagins, Ann. Rev. Biophys. Bioeng. 1 131 (1972)
32. H. Kuhn, Biochemistry 17 4389-4395 (1978)
33. W. Godchaux and W.F. Zimmerman, J. Biol. Chem. 254 7874-7884 (1979)
34. W. Baehr, M.J. Delvin and Applebury M.L., J. Biol. Chem. 254 1169-11677 (1979)
35. P. Schnetkamp, A.A. Klompmaker and F.J.M. Daemen, Biochim. Biophys. Acta 552 379-389
36. H. Kuhn, Nature 283 587-589 (1980)
37. M. Chabre and J. Breton, Vision Research 19 1005-1019 (1979)
38. D.S. Cafiso and W.L. Hubbel, Biophys. J. 30 243-264 (1980)
39. N. Bennett, M. Michel-Villaz and Y. Dupont, Eur. J. Biochemistry, In press (1980).
40. Y. Chapron, Cpte Rend. Acad. Sc. Paris 288 D 155-158 (1979).
41. V.I. Bolshakov, C.P. Kalamkarov and M.A. Ostrovski, Doklady Akademie, Nauk, S.S.S.R. 249 1485-1488 (1979)
42. N. Bennett, Eur. J. Biochemistry, In press (1980)
43. M. Chabre and J. Breton, Photochem. Photobiol. 30 295-299 (1979).
44. C.N. Rafferty, Photochem. Photobiol. 29 109-120 (1979)
45. H.B. Osborne and E. Nabedryk-Viala, Eur. J. Biochem. 89 81-88 (1978)
46. H. Shichi and E. Shelton, J. Supramol. Struct. 2 7-16 (1974).
47. J.K. Blasie, Biophys. J. 12 191-213 (1972)
48. M. Chabre and A. Cavaggioni, Biochim. Biophys. Acta 382 336-343 (1975)
49. H. Kuhn and W.J. Dreyer, FEBS Letters 20 1-6 (1972)
50. H. Shichi and C.N. Rafferty, Photochem. Photobiol. 31 631-639 (1980)

PHOTOREGULATION OF E.COLI GROWTH

A. Favre

Laboratoire de Photobiologie Moléculaire
Institut de Recherche en Biologie Moléculaire
UNIVERSITE PARIS VI - PARIS V ème (FRANCE)

INTRODUCTION

Fundamental to life's existence is the earth balance with respect to optical radiation. Sunlight not only provides a suitable physical environment it also produces important chemical effects in matter and in the organisms it has cradled. From the beginning life has been dependent upon the actinic effects of solar optical radiations. Light is absolutely required as an energy source in photosynthesis and can therefore be considered the driving force for life. On the other hand the deleterious effect of light on organisms it can effectively penetrate is well known. Amongst the innumerable number of biological effects triggered by light there are phenomena which are not immediately and obviously advantageous or deleterious to the cell and can be considered as an "adaptative" response to a change in the environment.

Probably the most important and characteristic of these effects is the cell inhibition of growth and division triggered by near-UV light (300-380 nm), the most energetic radiation that cells normally encounter in large amounts. This phenomenon is widespread in nature and has been observed in bacteria, fungi, protozoa, algae, higher plants and animal cells (see review in 1). It occurs at sublethal doses and can be considered the primary effect of near-UV on biological systems. Finally if one considers following F. JACOB (2) that "le rêve d'une cellule est de donner deux cellules" this phenomenon should reveal of considerable biological interest.

Our purpose here is to summarize the knowledge we have at present of the molecular mechanism of growth delay (GD) in the very simple but highly sophisticated organism E. coli.

I - GROWTH DELAY IN E.COLI

The inhibition of E. coli growth by near-UV light was discovered in 1943 by HOLLAENDER (see 1) and is illustrated in Fig.1 for bacteria illuminated at 366 nm in the stationnary phase and then resuspended at 37°C in glucose containing growth medium. In the illuminated cell, growth is delayed but then resumes at a normal rate. Since in these conditions killing is negligible the effect is due to an inhibition of growth and division (1). Hence near-UV converts the cell to a new transient state with reduced metabolism from which the cell is able to recover.

$$|\text{normal state}| \xrightarrow{h\nu} |\text{transient blocked state}| \xrightarrow{\text{"repair"}} |\text{normal state}|$$

Obviously the G.D. effect is mediated by light, by the growth conditions and by the cells genotype and history. As we will see, it depends of the cell genotype but also whether illumination is performed on growing or not growing cells. The measure of the effectiveness of light to induce G.D. has been described (1) and this allowed the determination of the action spectrum.
It exhibits a narrow band peaking at 340 nm. Of course, the lag time increases with the light dose, but although this is the only variable for nongrowing cells, the dose rate becomes critically important for growing cells as we will see later. For cell illuminated in the stationnary phase the fluence that yields the half effect is about 30 kJ/m^2 at 334 nm (1). Finally growth conditions plays an important role in the expression of the G.D. We have systematically observed that cells grown in amino-acid supplemented media experience longer G.D. than in minimal media (3).

The critical step in understanding a light-induced effect is the identification of the target. Previous studies (1) pointed to the oxydative respiratory system as the probable site of damage leading to G.D. It will be shown here that the main target is transfer RNA.

II - NEAR-UV PHOTOCHEMISTRY of tRNAs

Transfer RNA are molecules of approximately 25 000 daltons that play a pivotal role in the cell. Not only do they participate in protein biosynthesis but they are also implicated in an increasing number of regulatory roles in the cell.

In order to study the structure-function relationships in tRNA, I proposed in 1969 to my colleague M. YANIV to attempt to induce covalent links between non adjacent bases of this molecule. tRNA$_1^{val}$ the molecule we were interested in, was known to contain a rare residue 4-thiouridine, S^{4}U, (Fig.2) which strongly absorbs near-UV light (λ_{max} 337 nm) and can thus be selectively excited. Furthermore tRNA$_1^{val}$ contains a unique S^{4}U residue in position 8 (Fig.3).

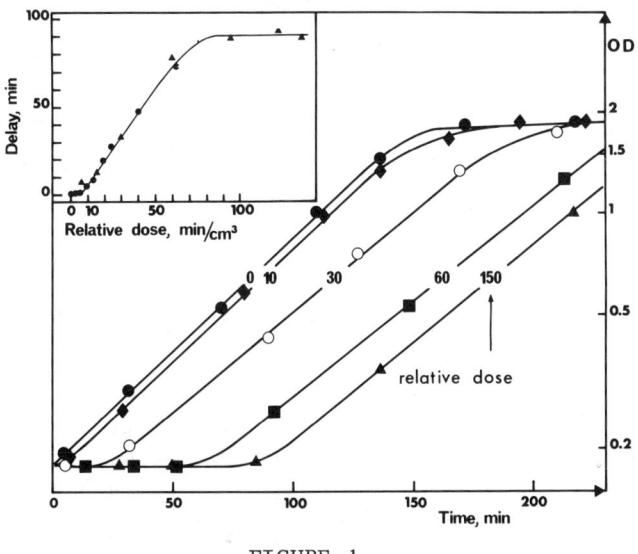

FIGURE 1

Effect of near-UV light (366 nm) on E.Coli K12 cells growth. Illumination was performed on the cell in the stationnary phase. The insert shows the extent of the lag phase as a function of the light fluence

FIGURE 2

Formation of the 8-13 link (5-6). The relative orientation of cytidine 13 and 4-thiouridine 8 is that found in yeast tRNA^{Phe} crystal.

335 nm illumination of native tRNA triggers a specific photoreaction of S^4U to a new product with a quantum yield of 7.10^{-3} and a final yield of 1. Sequence analysis revealed that in the illuminated tRNA$_1^{val}$ the 8th position is found covalently linked to cytosine 13 (4). The photoproduct was obtained from illumination of a mixture of cytosine and 4-thiouracil, or purified from bulk tRNA and its structure established (Fig. 2). In bulk tRNA, 75 % of the molecule have a unique S^4U residue in position 8, and approximately 60 % of the molecules can be cross-linked. At least 18 pure tRNA species all containing S^4U and C_{13} and corresponding to twelve distinct amino-acids belong to the cross-linking class (7).

To examine to which extent tRNA polymorphism is required for its functioning in protein biosynthesis, considerable efforts have been made to study the biological activity of cross-linked tRNA (see 7). In unspecific processes such as tRNA nucleotidyl-transferase repair or protein biosynthesis, all tRNA species behave similarly. Their activity is little or not affected. On the other hand the different cross-linked tRNA species behave quite differently in the highly specific acylation reaction. Some tRNA species are acylated with unchanged parameters while others such as tRNA^{Phe} and tRNA^{Pro} appear extremly sensitive. In an in vitro system, the apparent rate of acylation of cross-linked tRNA^{Phe} is decreased to 5 % of the control.

III - tRNAs ARE THE TARGETS FOR G.D.

Our in vitro data open the possibility that cross-linked tRNAs may have some in vivo effects. That is may trigger G.D. was substantiated by the similarity of the S^4U absorption spectrum and the action spectrum for G.D. This hypothesis was examined conclusively by THOMAS and FAVRE in Paris (8) and by RAMABHADRAN and JAGGER in Dallas (9).

Biochemical evidence was obtained from cells illuminated in the stationnaryphase and can be summarized as follows : i) the 8-13 link formation occurs in vivo. The amount of cross-linked tRNA is intimately correlated to G.D. The lag appears when 50 % of the tRNAs becomes cross-linked and is maximum when all tRNAs are cross-linked. Formation of the 8-13 link is accompanied by "inactivation" of tRNA^{Phe} as judged on cell extracts. ii) Resumption of growth occurs when 8-13 linked tRNA is no longer detectable in the cells. iii) the action spectrum for G.D. fits more closely the S^4U absorption spectrum than any other cellular chromophores.

Even more demonstrative is the fact that we have been able by a cyclic procedure to isolate mutants exhibiting a reduced lag. This was based on the fact that the mutant should grow immediately after illumination and thus be selected. After five selection cycles the level of S^4U in the population began to decrease and reached a stable figure of 20 % of the original amount after the 8th cycle. One of these mutants revealed particularly interesting. It

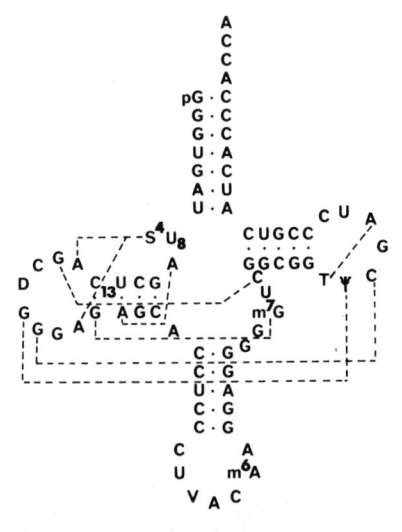

FIGURE 3

Clover-leaf representation of tRNA$_1^{Val}$. The dotted lines represent some of the tertiary interactions deduced from cristallographic studies

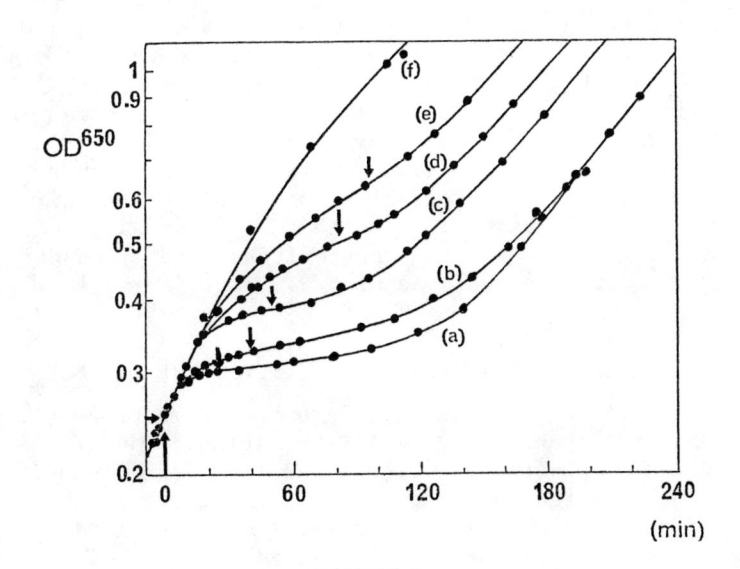

FIGURE 4

Growth of E.Coli K12 AB 1157 illuminated at 336 nm at various fluence rates (kJ. m^{-2}.min^{-1}) a) 5, b) 2.85, c) 2.3, d) 1.5, e) 1.3, f) 0.75. The arrows indicate the begining and the end of illumination

is completely devoid of S^4U and exhibits a five time reduced near-UV induced G.D. The S^4U deficiency is due to a single mutation which maps at 9.3 min on the E.coli genetic map (10) and does not alter the behaviour of the cell when grown in the dark.

The mechanism of G.D. can be pictured as follows. During the light phase near-UV photons absorbed by S^4U_8 trigger the following events :

$$S^4U \xrightarrow{h\nu} S^4U^* \longrightarrow S^4U^{**} \begin{cases} S^4U \\ S^4U + h\nu' \\ 8-13 \text{ link} \end{cases}$$

where S^4U^* and S^4U^{**} are respectively the S^4U first excited singlet and triplet states. S^4U^{**} deactivates mainly non-radiatively but also by emitting a weak phosphorescence ($\tau \simeq 5$ sec) and also by reacting on C_{13} to yield the 8-13 link through a thietane intermediate (5). In the dark phase which follows immediately the thietane formation the tRNA structure rearranges. The tertiary base pair $S^4U_8 - A_{14}$ is disrupted and residue 8th initially stacked upon C_{13} at a distance of 3.4 A° moves in order for the two residues to be coplanar. The angle of their glycosidic bonds increases from a value of approximatevely 8° to 60°. "Deconstruction" of the yeast tRNA^Phe model derived from tRNA crystal in order to introduce the cross-link shows that this can be achieved with minor alterations of the overall structure in agreement with physical studies (7). Only phosphates 7, 8, 9 need to be displaced (unpublished results of A.FAVRE, G. QUIGLEY and A. RICH). The next questions is why is tRNA^Phe acylation so sensitive to the 8-13 link ? Recently we showed that when bound to its cognate acylation enzyme tRNA^Phe is distorted in the region of tertiary structure (11). By reducing the number of tRNA allowed conformations the cross-link should reduce the affinity of tRNA^Phe for its synthetase. In vivo this should result in a lower acylation level of this tRNA and concomitantly to a reduction in the rate of protein synthesis leading ultimatly to a block in growth and division.

IV - GROWING CELLS : AMPLIFICATION OF G.D. BY THE STRINGENT RESPONSE

Illumination of exponentially growing cells triggers a block in the stable RNA synthesis as well as a stimulation of the ppGpp level (9). Hence growing cells respond to light exactly as if they were starved for a required amino-acid. ppGpp is synthetised on the ribosome in a reaction catalysed by the stringent (relA dependent) factor when uncharged tRNAs come to the A site.

$$pppG \xrightarrow[pppA \quad pA]{relA} pppGpp \longrightarrow ppGpp \xrightarrow{spoT} \text{degradation products}$$

In a relA⁻ strain no ppGpp synthesis occurs and as expected

stable RNA synthesis in unaffected by illumination. The extent of G.D., 90min, is approximately the same in K_{12} and B/r relA⁻ strains. The relA⁺ allele triggers the stringent response and amplifies the lag by a factor of 1.5 to 2 in K_{12} strains and a factor of 5 to 7 in B/r strains where the ppGpp synthesis is more efficient (3,9). It is striking that in relA⁺ cells, growth resumes only when the ppGpp level has decreased to its normal level. The same observation apply to spoT⁻ mutants. The gene spoT control ppGpp degradation and in a spoT⁻ strain the growth lag is increased by a factor of two.

It is possible now to describe accurately the behaviour of exponentially growing cells immediately after illumination. As shown in Fig.4, the fluence rate ψ plays a critical role in the cell response. Also the growth rate is not immediately affected. After a time interval t^* however the growth rate declines and reaches a reduced value. Both t^* and the growth rate are controlled by ψ. They exhibit decreased values when ψ increases. This can be easily understood if one assumes : i) that the cell growth rate k remains constant, $k = k_0$, until the ratio C/CT of intact to total tRNA is higher than a critical value α^*. When C/CT < α^*, k becomes proportional to C/CT. ii) the concentration of intact tRNA in the cell is given by $\frac{dc}{dt} = \beta c + \frac{Co}{No}\frac{dN}{dt}$, where $-\beta c$ accounts for the tRNA photochemical inactivation ($\beta \sim \psi$) and $\frac{Co}{No}\frac{dN}{dt}$ for tRNA neosynthesis. Expressions for t^* and N can be derived that accounts quantitatively for the observations with $\alpha^* = 0.6$ (12).

V - CONCLUSIONS - PERSPECTIVES

There is no doubt at present that S^4U is the main chromophore triggering G.D. The observation that a residual lag occurs in nuv⁻ cells point out to the existence of secondary targets. One of these targets has been recently identified (THOMAS, THIAM and FAVRE - manuscript in preparation). Unexpectedly near-UV illumination of nuv⁻ strain trigger a relA dependent stringent response. This is obtained selectively with light of wavelengths shorter than 360 nm. A likely chromophore to trigger this effect is mam⁵S²U a minor base present in the anticodon of $tRNA^{Lys}$ and $tRNA^{Glu}$. mam⁵S²U is photoreactive at wavelengths shorter than 360 nm and the illuminated tRNAs become poor substrates for their acylation enzyme.

Although the mechanism triggering G.D. becomes increasingly well understood, very little is known of the mechanism by which a light blocked cell resumes growth. Some information can be gained from the observations that the level of ppGpp decreases smoothly during the lag. This could be due i) to tRNA neosynthesis, an hypothesis made unlikely under the conditions of stringent response, ii) degradation of the cross-linked tRNA, iii) "repair" of the 8-13 linked tRNAs. At present no evidence has been obtained supporting i) and ii) although some experimental date in favor of iii) were obtained.

The final question is what is the use of S^4U ? A strain deficient in S^4U grows perfectly well in the dark and everything happens as if S^4U has been designed and built in tRNA in order to trigger G.D. It is not yet obvious which selective advantage G.D. confers to the cells. It has been proposed by JAGGER (1) that G.D. acts mainly by leaving more time to the cells to repair their damaged DNA. Photoprotection i.e. protection conferred by pre-illumination against killing by 254 nm radiation is related to G.D. since it is nearly suppressed in a nuv^- mutant. The role of photoprotection in nature is open to question however owing to the fact that this phenomenon requires extremly peculiar conditions to be observed in the Laboratory and requires the presence of additional mutations, such as recA in order to be observed in E.Coli K12 strains. An interesting possibility lies in the fact that S^4U may help adapt growing cells to peculiar illumination conditions.

REFERENCES

(1) JAGGER J. - in Gallo and Santamaria ed. - Res. Prog. in Organic Biological and Medical Chemistry, vol. 21, 383-401 (1972).
(2) JACOB F. - La logique du Vivant - Paris - Ed. Galimard (1970).
(3) THIAM K. - Thèse de 3ème Cycle - Université Paris VI (1980).
(4) FAVRE A., YANIV M. and MICHELSON A.M. - Biochem. Biophys. Res. Commun., 37, 266-271, (1969).
(5) BERGSTROM D.E. and LEONARD N.J. - Biochemistry, 11, 1-9 (1972).
(6) FAVRE A., ROQUES B. and FOURREY J.L. - FEBS Letters, 17, 236-240, (1972).
(7) FAVRE A. - In Frontiers of Matrix Biology. Karger Ed. - Basel, Vol. 4, 199-217, (1977).
(8) THOMAS G. and FAVRE A. - Biochem. Biophys. Res. Commun., 66, 1454-1461, (1975).
(9) RAMABHADRAN T.V. and JAGGER J. - Proc. Natl. Acad. Sci. USA, 79, 59-69, (1976).
(10) THOMAS G. and FAVRE A. - C.R. Acad Sci., Paris 284, Série D, 1485-1488, (1977).
(11) FAVRE A., BALLINI J.P. and HOLLER E. - Biochemistry, 13, 2887-2895, (1979).
(12) FAVRE A. - C.R. Acad. Sci., Paris 290, Série D, 1111-1114, (1980).

PHOTOMOVEMENTS OF MICROORGANISMS

Francesco Lenci

Biophysics Institute - CNR

Via S.Lorenzo 24-28, 56100 Pisa (Italy)

INTRODUCTION

Light not only means energy for living organisms (photocoupling → photosynthesis,f.e.), it can also be a signal, a stimulus (photosensing → vision, f.e.). Actually alterations in the illumination conditions (quite similarly to alterations in other environmental parameters, such as pressure, temperature, pH, concentration of some chemical species) can elicit in living organisms characteristic behavioral responses such as, for instance, sudden variations in their motion (photomotile reactions). Of course in complex living systems the structures involved and the chain of physico-chemical events between stimulus perception and behavioral response are in general quite complicated. Microorganisms, and especially unicells, besides being a rather interesting system in their own right, might also offer a suitable model for studying photosensory transduction processes at molecular level. In fact in these systems the connection between receptive and effector units should be through relatively simple molecular pathways, without integration and organization levels of cellular type (Feinleib, 1978; Haupt and Feinleib, 1979; Lenci and Colombetti, 1978; Diehn, 1979; Nultsch and Häder, 1979; and references therein). From a teleological point of view it can be easily understood why living beings, including aneural microorganisms, have developed the capability of searching for the best illumination conditions for their survival and growth. However the molecular reactions which initiate and control this photobehavior are still unknown in most cases.

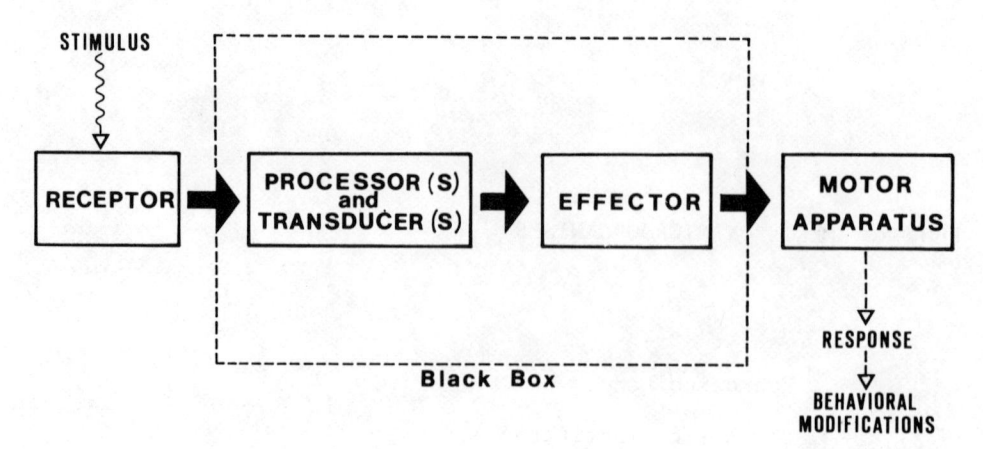

Fig.1 : Schematic Representation of Sensory Transduction Chains

In this lecture we will discuss the most important features of photosensing in some of the most significant microorganisms, trying to stress, whenever possible, the molecular aspects of the different problems when clarified and to discuss the open questions. Our approach and analysis of photosensory transduction chains (the multistage molecular processes through which photomotile reactions are exhibited) will follow the simplified scheme reported in Fig.1. According to this block-diagram the stimulus (in our case a photic one but, as already mentioned, it might also be of chemical, mechanical, thermal nature) is perceived by the receptor, in which the primary photophysical and photochemical reactions take place. The outputs of these molecular reactions will, in their turn, be the input signals for the transducers and processors controlling the effector activation, which will operate on the motor apparatus causing the final motile response.

PHENOMENOLOGY OF PHOTOMOVEMENTS

The most important types of photomotile reactions (Diehn et al., 1977) of aneural organisms are reported in Table I. Photokinesis (light intensity-dependent variation of the velocity modulus, $|\vec{v}|$) will not be discussed further because it costitutes a typical

TABLE I : MOST IMPORTANT TYPES OF LIGHT-INDUCED MOTILE REACTIONS

TYPE OF PHOTOREACTION	LIGHT CHARACTERISTIC DETECTED	RESULTING PHOTOBEHAVIOR
Photokinesis	Light intensity I	Increase (positive photokinesis) or decrease (negative photokinesis) of speed, referred to the state of movement in the dark. No adaptation.
Photophobic response	dI/dt	Stop-response, tumbling, transient alterations in the swimming direction. Step-down ($dI/dt < 0$) or step-up ($dI/dt > 0$) photophobic responses. Adaptation to the new light conditions.
Phototaxis	Direction of the light source	Movement toward (positive phototaxis) or away from (negative phototaxis) light source.
Phototropism	Direction of the light source	Bending toward the light source.

example of photocoupling process, through the photosynthetic system, in which no adaptation to the new constant "stimulation" occurs (Nultsch, 1980 and references therein).

Photophobic step-down and step-up responses are elicited by a sudden decrease or, respectively, increase over a threshold value of light intensity. The photophobic reactions consist in cessation of forward movement (stop-response), change in direction of movement independent of the direction of the photic stimulus, turning in place (tumbling). Photophobic responses cease either almost immediately upon removal of the stimulus (restoration of the previous light conditions) or because of adaption to the stimulus. Of course a spatial gradient of light intensity (dI/ds) can be perceived as a temporal change in light intensity (dI/dt) by a moving organism, which moves along or crosses this spatial gradient, so that step-down and step-up photophobic responses can result either in photoaccumulation in or in photodispersal from light traps (Feinleib, 1980; Nultsch, 1980; and references therein).

Oriented movement toward (positive phototaxis) or away from (negative phototaxis) the light source implies the capability of the microorganism of detecting the direction of the photic stimulus. Basically this may be achieved in two ways. If the organism has two spatially separated photoreceptors, the light direction can be detected because of the difference in the light intensity simultaneously impinging on the two photoreceptive units (one-instant-mechanism). Otherwise the organism can compare at different instants of time the light intensity impinging on a single photoreceptive unit, i.e. a temporal variation in light intensity is detected (two-instant-mechanism).

Phototropic organisms, finally, are not freely moving systems, and therefore asymmetric illumination will induce bending of their organ(s) toward the stimulus source (Hertel, 1980; and references therein). In any case a light-absorption gradient across the organism is to be established. This can be accomplished either by attenuation (screening organelles, shading pigments) or by refraction (cell acting as a converging lens) or by oriented photoreceptor molecules, absorbing different amounts of light depending on the orientation of the electric dipole transition moments with respect to the direction of light propagation (Feinleib, 1980).

PHOTORECEPTIVE STRUCTURES AND PHOTOPIGMENTS

Living organisms interact with the external world through specialized molecules, sometimes localized in specialized structures, which are able to undergo physico-chemical modifications in response to environmental stimuli, thus initiating the process of signalling to the organism that the external sorroundings are changing. Of extreme importance is, therefore, the question of which subcellular structure(s) and molecules are devoted to light perception and of the nature of the physicochemical alterations occurring in these photoreceptive units. The unambiguous characterization of the photoreceptor pigments is, actually, a prerequisite for formulating reasonable hypotheses on the primary molecular photoreactions occurring in the photoreceptor system, which in turn represent the very first steps of the photosensory transduction chain.

Concerning the photoreceptive structures, prokaryotes like, e. g., Halobacterium halobium and the blue-green alga Phormidium uncinatum, seem not to have specialized structures dedicated to light detection for photomotile responses. In Halobacterium the receptor for the step-down photophobic response would be located in the patches of purple membrane, within the surface membrane, which is used by the bacterium also for photoenergy conversion (Oesterhelt and Stoeckenius, 1970; Hildebrand and Dencher, 1975; Spudich and Stoeckenius, 1979), whereas the receptor for the step-up response would be embedded in the membrane areas adjacent to the purple membrane (Hildebrand and Dencher, 1975; Hildebrand, 1978). In Phormidium, like in other blue-green algae and in photosynthetic bacteria, photosensing processes originate from light absorption by the photosynthetic systems PSI and PSII (Nultsch and Häder, 1979; Häder, 1980; and references therein). It can be interesting to observe that this "multipurpose" characteristic of the photopigment apparatus of prokaryotic organisms could be related to their low level of evolution, and therefore specialization.

In eukaryotes localized and specialized structures dedicated to photosensing can exist, but in only very few cases they have been identified with a sufficient degree of confidence. In Euglena gracilis a swelling inside the membrane of the emergent flagellum, the paraflagellar body (PFB), a highly ordered quasi-crystalline structure (Kivic and Vesk, 1976; Piccinni and Mammi, 1978), is presently thought to be the site of actual photoreception. In Chlamydomonas reinhardtii (Walne and Arnott, 1967) and Volvox (Schletz, 1976) a thickening (approximately 120 Å, in Chlamydomonas) of the plasmalemma membrane has been suggested as the possible photoreceptor site. However no definite experimental support to these hypotheses is presently available.

The question as to the determination of the nature of the pigments involved in a photomotile response can be solved, at least in principle, by different approaches.

The isolation of the photoreceptor molecules can provide a great deal of information on their chemical, biochemical and spectroscopic properties. Unfortunately there are not very many examples of this: besides the bacteriorhodopsin containing purple membrane of Halobacterium and the case of those prokaryotes using the photosynthetic apparatus as photosensing receptor, only for the ciliate protozoan Stentor coeruleus this goal has been completely attained. The typical blue-green pigment of Stentor, localized in small vesicles in the ciliary rows region, has been isolated and characterized by Song's group (Walker et al., 1979; Song et al., 1980b) by means of both chromatographic and spectroscopic (absorption, fluorescence, fluorescence lifetimes and polarization, circular dichroism) techniques. The photoreceptor pigment has been identified as hypericin covalently bound to a peptide. This peptide-linked stentorin, constituting a new unique class of photoreceptor pigments, would significantly increase its acidity upon light excitation. As we will see in the following this property is of fundamental importance in the photosensing process. Recently Diehn and Doughty (in preparation) did succeed in isolation of Euglena flagella with the PFB.

The alternative route of "in vivo" microspectroscopy, which was successfully used also in the case of Stentor by Wood (1976), can be followed when the location of the presumed photoreceptor in the cell is known, provided its dimensions and its pigment content are not too small. In the case of Euglena gracilis, for example, the presence of flavins in the PFB has been shown by fluorescence microscopy (Benedetti and Checcucci, 1975), by means of a fast scanning microspectrofluorometric technique (Benedetti and Lenci, 1977) and using a tunable dye-laser microspectrofluorometer (Colombetti et al., 1980a)

The most widely used technique to identify photoreceptor pigments is based on action spectra determination. Action spectra can be defined as the ratio between the magnitude of the response and the light quanta flux, plotted as a function of the wavelength, provided that the response is a linear or a monotonous function of the number of photons falling on the photoreceptor per unit time (Colombetti and Lenci, 1980a). In the simplest case a correctly measured action spectrum will be proportional to the absorption spectrum of the pigments responsible for reaction. The occurrence of more than one "antenna" pigment or of screening organelle, which can mediate the photobehavioral response and whose absorbing properties therefore affect the action spectrum, can complicate the problem. Moreover, when action spectra are determined by means of population methods ("phototaxigraphs", e.g.), the relation between the "elementary" response(s) of the cells and their macroscopic behavior must be clearly understood, to avoid misleading conclusions. Of course the biochemical state of the organism being studied (age, environmental conditions, composition of suspension medium) is of primary importance (Diehn, 1980). As a final remark, it is to be stressed that photoreceptor isolation, "in vivo" microspectroscopy and action spectra determination are by no means alternative approaches, but rather complementary and mutually integrating techniques.

In the case of Halobacterium the action spectra determined for the step-down and the step-up photophobic responses (Hildebrand and Dencher, 1975; Dencher, 1978; Hildebrand, 1978) seemed to suggest bacteriorhodopsin and a retinylideneprotein, plus bacterioruberine as additive pigment, as photoreceptor pigments respectively for the step-down (PS 565) and for the step-up (PS 370). According to Spudich and Stoeckenius (1979), however, each PS could perceive and transduce both step-up and step-down photic stimuli. The action spectra reported by Hildebrand and Dencher (1975) might therefore actually be "difference" action spectra of both PS' and even the hypothesis that PS 370 is a precursor of PS 565 (Hildebrand, 1978 and references therein) seems to be under discussion (Spudich and Stoeckenius, 1979). These discrepancies clearly indicate that much more experimental results are necessary to settle the questions related to photosensing pigment systems in Halobacterium.

Using a double irradiation technique, Häder has clarified the role of PS II and PS I in eliciting photophobic responses of Phormidium uncinatum : step-down responses (mediated by PS II) and step-up responses (mediated by PS I) can be both exhibited under simultaneous PS II and PS I irradiation, depending on the relative number of quanta absorbed by the two photosystems (see, f.e., Häder, 1980 and references therein).

Action spectra for the step-up photophobic responses (Song et al., 1980b) and for ciliary reversal (Wood, 1976) in single cells of Stentor coeruleus unambiguously support stentorin as the photoreceptor pigment for both responses.

Of particular interest is the case of Euglena gracilis. In fact all the action spectra, published in the last years, are quite similar to each other and proportional to the optical absorption spectrum of flavin-type chromophores (Colombetti et al.,1980b), regardless of the type of cells examined (wilde type, dark-bleached cells, streptomycin-treated mutants) and of the experimental procedure, and therefore of the particular photomotile reaction being examined (photoaccumulation, photodispersal, photophobic reactions). Here we only want to mention that these results contribute to keep the discussion open on the mechanism (stigma-mediated vs. oriented photoreceptor pigments) and even on the actual occurence of an actively oriented movement of the cells toward the light source (positive phototaxis) (for a more complete discussion, see Colombetti et al., 1980b). Even though the question raised by Mikolajczyk and Diehn (1975 and 1978) whether the step-down and the step-up photophobic responses may employ physically distinct photoreceptive units and the same transduction chain or a unique photoreceptor and two transduction pathways is still open, the afore-mentioned close similarity of all action spectra seems to indicate that wherever they can be localized, flavins are the photoreceptor pigments for Euglena photobehavior.

The phototactic action spectrum for Chlamydomonas (Nultsch et al.,1971), determined using a population method, shows maxima at about 440 nm and 503 nm. The resolution is unfortunately not very good, but a peak at 503 nm could suggest a carotenoprotein as photoreceptor pigment. Whereas a great deal of results is available and research is in progress on the sensory transduction chain (see below) as well as on the mechanism of light direction detection (Feinleib, 1980), only relatively little knowledge exists about the photoreceptor structure(s) and the photoreceptor pigments of Chlamydomonas.

Action spectra of the photomotile responses of Cryptomonas, determined using a population method (Watanabe and Furuya, 1974), point to phycoerythrin as the candidate photoreceptor pigment. Interestingly phycoerythrin is a photosynthetic pigment (and actually O_2 photosynthethic evolution is very efficient around 560 nm, the maximum of the action spectrum and of phycoerythrin absorption as well), but photosynthesis seems not to be directly involved in photosensing process, as shown by the use of photosynthesis inhibitors.

In the case of Phycomyces phototropism, finally action spectra structure indicates flavins as photoreceptor pigments (see,e.g., Delbrück and Shropshire,1960).

PRIMARY MOLECULAR PHOTOREACTIONS

As shown in Table II, in very few cases experimentally supported hypotheses can be put forward about the primary molecular reactions occurring in the photoreceptor.

Bacteriorhodopsin (BR) photocycle has been carefully investigated and researches in progress continuously yield new and more refined data. One of the fundamental features of this reaction cycle is, for our purposes, the complete photoisomerization of BR to the trans form following moderate irradiation. Upon illumination, trans-BR (absorption max around 568 nm) undergoes a rapid (less than 10 ms at 35°C) reaction cycle during which proton are released and subsequently taken up, thus giving rise to a net translocation of protons across the bacterial cell membrane. As the actual stimulus eliciting a step-down photophobic response in Halobacterium is a decrease in light intensity, it is reasonable to assume that this step-down photic stimulus, interrupting the BR photocycle, causes a decrease in the proton translocation rate, which would constitute a first step in the sensory process finally resulting in the step-down photophobic response (Hildebrand, 1978 and 1980; Wagner, 1979; and references therein). We will not discuss the much more hypothetical picture for the step-up response of Halobacterium.

The mechanism proposed for Halobacterium step-down response is consistent with and in some way similar to that suggested for Phormidium photophobic responses. In fact (Häder, 1979 and 1980; and references therein) also for this blue-green gliding alga the step-down photophobic response is thought to be triggered by a sudden "dark-induced" decrease in the photosynthetic electron flow. More precisely, an electron pool, whose biochemical correlate would be plastoquinone, localized in the non-cyclic electron transport chain between PSII and PSI would measure the electron flow rate. This flow rate can be unbalanced either when the number of electrons entering the pool via PS II decreases or when the pool is drained via PS I. Any perturbation of this dynamic equilibrium signals that the environment light conditions are changing, and the sensory process is initiated to elicit the behavioral response.

Proton gradient photogeneration seems to be a primary molecular event also in the case of Stentor. As we already mentioned, stentorin, the photoreceptor pigment whose chromophore is hypericin (Walker et al., 1979), gets its acidity significantly enhanced upon illumination and functions as a proton source in a light-induced proton flux across the membrane (Song et al., 1980a). This hypothesis of a light-driven proton conduction network, with the photoreceptor as the initial proton source, being the primary transducing mechanism in Stentor is supported by recent findings of Song et al. (1980b) on the specific inhibition of the step-up photophobic response by protonophorus uncouplers (TPMP$^+$ and FCCP) but not by ionophores like A23187.

Not very much can be said about the primary photoreactions of the photoreceptor pigments of Chlamydomonas and Cryptomonas. In the case of Chlamydomonas (in whose sensory transduction Ca ions play a crucial role, as we will see in the following) one might imagine the photoreceptive pigments serving as light-activated gates for the passage of ions, mainly Ca^{++}, across the cellular membrane; but these are, at present, only speculations.

Flavins in their first excited triplet state as the primary reactive species in photoreception process has been suggested by Diehn and Kint (1970) and Mikolajczyk and Diehn (1975) for Euglena and by Delbrück et al. (1976) for Phycomyces. The first excited triplet state as the primary reactive state in Euglena photobehavior is a reasonable candidate, but the experimental support is, at present, rather poor. Moreover a much more short-lived state, the first excited singlet, could well be, at least theoretically, an even more efficient intermediary. Actually, in this context, the most important property of the triplet state is its rather long lifetime ($\sim 1.0 \mu s \div \sim 1.0 ms$) in comparison with that of the singlet state (~ 1.0 ns); this long life-time, however can actually be a prerequisite for the occurrence of a secondary reaction mainly in diffusion-controlled processes. In an apparently rigid system like Euglena PFB the diffusion of the reactant(s) should not play a crucial role and the short-lived more energetic singlet might initiate reactions which could not start from the less energetic triplet. Finally the population of the singlet is for sure much higher than that of the triplet (Colombetti et al., 1980b). A light-dependent dark-reversible conversion of photoreceptor pigments of Euglena from a photochemically active, P, to a photochemically quiescent form, Q, has been suggested (Creutz et al., 1978) to be responsible for a stoichiometric inward and outward delivery and transport of charges across the PFB membrane. The hypothesis that these charges might be protons, even though attractive (see the already discussed cases of Halobacterium and Stentor) seems to be ruled out by recent findings (Doughty et al., in preparation) excluding the intervenction of proton transport processes in Euglena photosensory transduction.

TABLE II : PHOTOPIGMENTS AND PRIMARY PHOTOREACTIONS INVOLVED IN SOME PHOTOBEHAVIORAL RESPONSES

ORGANISM	PHOTOPIGMENT	PRIMARY PHOTOREACTIONS
Halobacterium	Bacteriorhodopsin (PS 565) (Step-down) Rethinylidene protein (PS 370) (Step-up)	Bacteriorhodopsin photocycle - light driven proton pump.
Phormidium	PSII, PSI	Non-cyclic photosynthetic electron transport. Filling of an electron pool (plastoquinone) via PSII and draining of the pool via PSI.
Stentor	Stentorin - hypericin	Acidification of the photoreceptor molecule upon illumination. Light driven proton pump.
Euglena	Flavins/Flavoproteins	Triplet as primary reactive state?? Photodependent dark reversible conversion of pigments with stoichiometric charge delivery and transport.
Chlamydomonas	Carotenoids/Carotenoproteins	?
Cryptomonas	Phycoerothrin (No involvement of the photosynthetic system)	?
Phycomyces	Flavins	Triplet as primary reactive state?

SENSORY TRANSDUCTION

Because of its relative "distance" (in terms of the molecular processes represented in Fig.1) from the direct effect of light, this last part of molecular sensory physiology will be discussed rather concisely.

The decrease in light intensity at 565 nm, could cause, through the interruption of the H^+ translocation process, a membrane depolarization in Halobacterium halobium. Depolarizing signals spreading over the cell membrane would open electrically excitable Ca^{++} channels, which can be hypothesized to exist nearby the flagella. An influx of Ca^{++} triggered by these depolarizing signals would cause the flagella to reverse their sense of rotation, inducing a conformational change at their basal body (Hildebrand, 1980 and references therein). This role of Ca^{++} ions in eliciting the step-down response of Halobacterium is, at present, rather hypothetical but definitely in agreement with the general role of Ca^{++} ions in controlling sudden changes of the activity of motor organelle.

Also in the case of Phormidium light-dependent alterations in the dynamic equilibrium of the electron pool, affecting in their turn the vectorial proton transport across the thylakoid and plasma membrane, would cause membrane potential changes. These light-induced potential variations have been measured by Häder (1979 and 1980 and references therein), who found that the action spectra of the photo-dependent potential change closely resemble those of the photophobic responses, and that inhibition of phobic responses occurs under conditions affecting the light-induced potential changes (inhibitors, uncouplers, ionophores, external electric fields). Incidentally, in this case too a role of Ca^{++} ions seems not to be excluded.

The role of Ca^{++} ions seems to be well established in photosensory physiology of Chlamydomonas. Actually a series of experimental results by different authors using very different experimental approaches (Marbach and Mayer, 1971; Stavis and Hirschberg, 1973; Stavis 1974; Schmidt and Eckert, 1976; Nichols and Rikmenspoel, 1978; Hyams and Borisy, 1978; Nultsch, 1979) are all substantially consistent with the hypothesis that in Chlamydomonas flagellar response and photic stimuli are coupled through Ca^{++} ions and Ca^{++} - mediated membrane phenomena: photostimulation results in an influx of Ca^{++} through the cell membrane, causing a transient increase in the intracellular Ca^{++} concentration (Colombetti and Lenci, 1980b and references therein).

In the case of Euglena, finally, neither the photosynthetic process, in particular non-cyclic photophosphorylation and electron transport, nor oxydative phosphorylation seem to be directly connected with photosensory transduction phenomena (Barghigiani et al., 1979). The "suspected" ionic nature of photosensory transduction in Euglena has been shown by Doughty and Diehn (1979) and by Doughty et al.(1980): flagellar reorientation, and therefore photophobic responses, is triggered by a transient increase in Ca^{++} concentration in the intraflagellar space. Light, however, does not directly regulate Ca^{++} permeability of the flagellar PFB membrane. Modifications of Euglena photobehavior and photosensitivity are induced by monovalent and divalent cations, the fluxes of which might be controlled by a light-activated oubain-blocked Na^+/K^+ pump.

Of course very many problems, both solved and unsolved, have not been even touched in this lecture, and most of the mentioned questions would deserve a much deeper and more extensive discussion (e.g. adaptation phenomena, interaction of different types of stimuli). Hopefully, however, this lecture can have revealed some of the fascinating features of the wide field of photoreception and photosensory physiology in aneural organisms.

REFERENCES

Barghigiani,C.,Colombetti,G.,Lenci,F.,Banchetti,R.,and Bizzaro,M.P. (1979) Arch.Microbiol. 120,239-245
Benedetti,P.A.,and Checcucci,A. (1975) Plant Sci.Letters 4,47-51
Benedetti,P.A.,and Lenci,F. (1977) Photochem.Photobiol. 26,315-318
Colombetti,G.,Ghetti,F.,Lenci,F.,Polacco,E.,and Posudin,Y. (1980a) Proc. 2nd Congress Quantum   Electronics and Plasma Physics C.N.R. Group, in press
Colombetti,G.,and Lenci,F. (1980a) in "Photoreception and Sensory Transduction in Aneural Organisms" (F.Lenci and G.Colombetti,Eds) pp. 173-188,Plenum,New York
Colombetti,G.,and Lenci,F. (1980b) in "Photoreception and Sensory Transduction in Aneural Organisms" (F.Lenci and G.Colombetti,Eds) pp. 341-354,Plenum,New York
Colombetti,G.,Lenci,F.,and Diehn,B. (1980b) in "The Biology of Euglena" (D.E.Buetow,Ed) Vol. III,Acad.Press,New York,in press

Creutz,C.,Colombetti,G.,and Diehn,B. (1978) Photochem.Photobiol. 27,611-616

Delbrück,M.,Katzir,A.,and Presti,D. (1976) Proc.Natl.Acad.Sci.US 73,1969-1973

Delbrück,M.,and Shropshire,W. Jr. (1960) Plant Physiol. 35,194-204

Dencher,N. (1978) in "Energetics and Structure of Halophilic Microorganisms" (S.R.Caplan and M.Ginzburg,Eds) pp. 67-88,Elsevier, Amsterdam

Diehn,B. (1979) in "Handbook of Sensory Physiology" VII,6A (H.Autrum Ed) pp. 23-68,Springer Verlag,Berlin

Diehn,B. (1980) in "Photoreception and Sensory Transduction in Aneural Organisms" (F.Lenci and G.Colombetti,Eds)pp.107-125,Plenum,New York

Diehn,B.,Feinleib,M.E.,Haupt,W.,Hildebrand,E.,Lenci,F.,and Nultsch, W. (1977) Photochem. Photobiol. 26,559-560

Doughty,M.J.,Diehn,B. (1979) Biochim. Biophys. Acta 588,148-168

Doughty,M.J.,Grieser,R.,and Diehn,G. (1980) Biochim.Biophys.Acta, in press

Feinleib,M.E. (1978) Photochem.Photobiol. 27,849-854

Feinleib,M.E. (1980) in "Photoreception and Sensory Transduction in Aneural Organisms" (F.Lenci and G.Colombetti,Eds) pp. 45-68,Plenum,New York

Häder,D.P. (1979) in "Encyclopedia of Plant Physiology"NS7 (W.Haupt and M.E.Feinleib,Eds) pp.268-309,Springer Verlag,Berlin

Häder,D.P. (1980) in "Photoreception and Sensory Transduction in Aneural Organisms" (F.Lenci and G.Colombetti,Eds) pp.355-372,Plenum,New York

Haupt,W.,and Feinleib,M.E. (Editors) (1979) "Encyclopedia of Plant Physiology" NS7 "Physiology of Movements",Springer Verlag,Berlin

Hertel,R. (1980) in "Photoreception and Sensory Transduction in Aneural Organisms" (F.Lenci and G.Colombetti,Eds) pp.89-106,Plenum,New York

Hildebrand,E. (1978) in "Receptors and Recognitions - Taxis and Behavior" (G.L.Hazelbauer,Ed) pp.1-68

Hildebrand,E. (1980) in "Photoreception and Sensory Transduction in Aneural Organisms" (F.Lenci and G.Colombetti,Eds) pp.319-340, Plenum,New York

Hildebrand,E.,and Dencher,N.(1975) Nature 257,46-48

Hyams,J.S.,and Borisy,G.G. (1978) J.Cell Sci. 33,235-253

Kivic,P.A.,and Vesk,M. (1972) Planta 105,1-14

Lenci,F.,and Colombetti,G. (1978) Ann.Rev.Biophys.Bioeng.7,341-361

Marbach,J.,and Mayer,A.M. (1971) Isr.J.Bot. 20,96-100
Mikolaiczyk,E.,and Diehn,B. (1975) Photochem.Photobiol.22,268-271
Mikolajczyk,E.,Diehn,B. (1978) J.Protozool. 25,461-470
Nichols,K.M.,and Rikmenspoel,R. (1978) J.Cell Sci. 29,233-247
Nultsch,W. (1979) Arch.Microbiol. 123,93-99
Nultsch,W. (1980) in "Photoreception and Sensory Transduction in Aneural Organisms" (F.Lenci and G.Colombetti,Eds) pp. 69-87,Plenum,New York
Nultsch,W.,and Häder,D.P. (1979) Photochem.Photobiol. 29,423-437
Nultsch,W.,Throm,G.,and Von Rimscha,I. (1971) Arch.Mikrobiol. 80, 351-369
Oesterhelt,D.,and Stoeckenius,W. (1970) Proc.Natl.Acad.Sci.US 70, 2853-2857
Piccinni,E.,and Mammi,M (1978) Boll.Zool. 45,405-414
Schletz,K. (1976) Z.Pflanzenphysiol. 11,189-211
Schmidt,J.A.,and Eckert,R. (1976) Nature 262,713-715
Song,P.S.,Häder,D.P.,and Poff,K.L.(1980a) Arch.Microbiol. 126,181-186
Song,P.S.,Walker,E.B.,and Yoon,M.J. (1980b) in "Photoreception and Sensory Transduction in Aneural Organisms" (F.Lenci and G.Colombetti,Eds) pp.241-252,Plenum,New York
Spudich,J.L.,and Stoeckenius,W. (1979) Photobiochem.Photobiophys. 1,43-53
Stavis,R.L. (1974) Proc.Natl.Acad.Sci. US 71,1824-1827
Stavis,R.L.,and Hirschberg,J.Cell Biol. 59,367-377
Wagner,G. (1979) Biol.unserer Zeit 9,171-179
Walker,E.B.,Lee,T.J.,and Song,P.S. (1979) Biochim.Biophys.Acta 587, 129-144
Walne,P.L.,and Arnott,H.J. (1967) Planta 77,325-353
Watanabe,M.,and Furuya,M. (1974) Plant and Cell Physiol. 15,413-420
Wood,D.C. (1976) Photochem.Photobiol. 24,261-266

THE MECHANISM OF THE CIRCADIAN RHYTHM OF PHOTOSYNTHESIS

Hans-Georg Schweiger

Max-Planck-Institut für Zellbiologie

6802 Ladenburg bei Heidelberg, Germany

One of the characteristic features of living systems is the ability to adapt to varying environmental conditions. During evolution, features which favor adaptation to special conditions have been positively selected and genetically established. Those environmental phenomena which are due to the rotation of the earth deserve special attention. These phenomena are coupled to a 24 hour cycle and, to a high degree, are connected to the cyclic availability of light energy emitted by the sun. Keeping in mind the cyclic nature of irradiation, it is tempting to assume that plants, which convert light energy into chemical energy, would have a higher capability to utilize light energy during the day and a lower one during the night.

In fact, it has been known for a long time that plants possess a number of different mechanisms which enable them to increase their efficiency when light is available, i.e. during day (1). One of these mechanisms is to move the leaves into a favorable position such that they are more directly illuminated. Another way would be to control the efficiency of the chloroplasts and their photosynthesis machinery. In all these cases, the question has to be asked what is the mechanism which results in cycling efficiency of the photosynthetic capability of the plant.

An extremely interesting feature of the oscillating efficiency is that the cylic changes are sustained when the exogenous night-day regimen is suspended. In other words, the fluctuations are of an endogenous nature; they represent an endogenous rhythm. In addition, this rhythm fulfills a number of criteria of circadian rhythms (2); that is, the rhythm has a period of about 24 hours, an endogenous character, far-reaching independence of the frequency on temperature, an ability to shift the phase by pulses of physical or chemical nature resulting in characteristic phase response curves and the possibility to entrain the phase by an exogenous Zeitgeber (for references see 3). Such rhythms are well-known in the plant and animal kingdom. Therefore, the conclusion is justified that this is an ubiquitous phenomenon.

The parameters which are subjected to such oscillations are quite different i.e., body temperature, locomotor activity, photosynthesis, mineral excretion and heart-beat frequency. To date, there has been no unequivocal demonstration of circadian rhythms in prokaryotic organisms. All systems which exhibit such a rhythm are eukaryotic organisms. Given circadian rhythms are a special features of eukaryotic systems one may ask what the essential difference between prokaryotic and eukaryotic organisms might be. One answer is that the gene expression system in prokaryotic organisms uses 70S ribosomes while the characteristic translation in eukaryotic organisms is performed on 80S ribosomes. These, however, are not the only differences in the gene expression system.

A circadian rhythm of photosynthetic activity is found not only in higher, multicellular organisms but also in unicellular organisms. Among these unicellular, eukaryotic photosynthezising systems, special interest is deserved by those in which it is possible to monitor photosynthetic activity over long periods, that is over weeks or months, and, if possible, in an individual cell. If that is possible then not only sensitive biochemical methods but also a number of cell biology techniques can be used in a meaningful way.

Such an unicellular system is represented by the uninucleate marine alga Acetabularia. This unicellular organism grows up to a length of 200 mm and it is well suited for cell surgery (for references see 4). The nucleus is located in an extreme position in the basal part of the organism and the cell is capable of surviving without difficulties the amputation of the basal part of the cell, which contains the nucleus. Such an anucleate cell is capable of performing species-specific morphogenesis.

Photosynthetic activity of Acetabularia is subjected to a pronounced circadian rhythm (5, 6). This rhythm is endogenous since it persists for months under conditions of contant light and temperature (7, 8). The length of the period is more or less independent of temperature (8). Dark pulses given at different phases result in a characteristic phase response curve (8) and the rhythm can be entrained by an exogenous Zeitgeber.

The photosynthetic rhythm can be followed in a flow-through system in which the O_2 content of the medium is measured by means of a platinum electrode after it has passed the cell (9). A major advantage of the flow-through system is that the cell is always exposed to medium with constant composition. From the fact that an individual cell of Acetabularia exhibits a circadian rhythm over several months one may conclude that expression of a circadian rhythm does not depend on the cooperation of a large number of cells. Therefore, individual cells of Acetabularia represent a rather simple and favorable system to study the molecular mechanism of the circadian rhythm. Of course, such a question anticipates that one is ready to accept that there is indeed a molecular mechanism.

The features of Acetabularia cells opens the way to answer the question of what happens with a rhythm when the cell is fragmented by cutting. The answer is that if a cell is divided into two parts, both parts exhibit a circadian rhythm (10). This type of experiment has an additional meaning in that fragmentation of a cell into two parts necessarily means that one fragment is nucleate and the other one anucleate. By means of this simple experiment, it has been clearly shown that the cytoplasm, even in the absence of the nucleus, is capable of maintaining a circadian rhythm. This underlines the independence and autonomy of the cytoplasm from the nucleus. The result also discredits an earlier hypothesis that the cell nucleus might be the site of the circadian clock.

In other experiments on Acetabularia, it has been shown that the cell nucleus nevertheless might have a certain regulatory influence upon the circadian rhythm. This statement is based on the observation that if one takes two cells in which the phase of the endogenous photosynthetic rhythm is different by 180 degrees, isolates the nuclei from these cells and exchanges the nuclei the phase in these two cells is shifted (11). Since an anucleate fragment does not show such a phenomenon, it is obvious that the nucleus must have the capability to influence the phase. This result, together with the fact that circadian rhythms in general have been found in eukaryotic but not in prokaryotic systems, directs attention to the participation of gene expression in circadian rhythms.

The flow-through system enables very simple additions and removals of inhibitors to the medium. Experiments with a number of inhibitors of gene expression have shown that the oscillations are suspended in the presence of puromycin and of cycloheximide but not in the presence of chloramphenicol and of rifampicin (12). Together with the finding that anucleate cells are able to maintain the rhythm these experiments indicate that neither the transcription of the nuclear or organellar genomes nor the translation on 70S organellar ribosomes of the prokaryotic type, but rather the translation on 80S ribosomes of the eukaryotic type are involved in the expression of the circadian rhythm (13) Figure 1). A critical interpretation of these results indicates the participation of translation on 80S ribosomes in the clock, but does not prove it. It could be the case that cycloheximide and puromycin only affect the hands of the clock but not the clock itself.

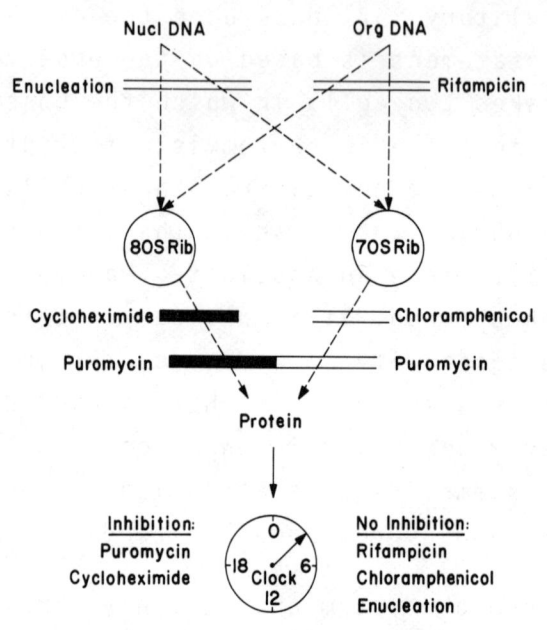

Figure 1. Schematic identification of the essential step of gene expression in the circadian photosynthesis rhythm in Acetabularia. The black bars indicate that the clock is affected while the open bars represent the absence of an effect (14).

The possibility that the changes which were observed in the presence of these inhibitors are not due to an effect on the clock itself may be excluded by the following experiments. The cells were exposed to 8 hour pulses of cycloheximide. The pulses were given at different times of the day. The experiments showed that the phase was shifted only when the pulse was given during the second half of the day. There was no effect if the pulses was applied during the first half of the day (14). Moreover, these experiments clearly showed that the cycloheximide indeed affects the clock itself and not its hands. Finally, these experiments demonstrated that the cell is sensitive to cycloheximide only during a distinct phase of the period. In molecular terms this means that it is not the translation of all proteins on 80S ribosomes which is essential for the circadian rhythm, but rather a specific part of these proteins.

In another set of experiments, cycloheximide pulses were given at 25° instead of 20°. These experiments demonstrated that a change in temperature by 5° resulted in a very pronounced effect on the phase in which the cell is sensitive to cycloheximide. In contrast to 20°, the experiments at 25° showed that at the higher temperature the cell was sensitive to cycloheximide in the first half of the day and insensitive during the second half of the day (16). A change in temperature obviously resulted in a pronounced change in the phase of sensitivity to cycloheximide.

Since the pattern of proteins which are synthesized during the first and the second half of the day are essentially identical and since there is no substantial change when the temperature is increased from 20° to 25°, one may assume that it is only a very few polypeptides or perhaps only one polypeptide which are synthesized on the 80S ribosomes and which are essential for the circadian rhythm (17). Such a polypeptide was looked for by means of incorporation experiments. One polypeptide which fulfills all these criteria has been found and purified. Its molecular weight is 39 000. It is a glycine rich polypeptide which occurs in the thylakoid membranes of the chloroplasts and carries glycine as its N-terminal group. The polypeptide is hydrophobic (18).

On the basis of the results discussed above a model was suggested which tries to explain the circadian rhythm in molecular terms (13)(Figure 2).

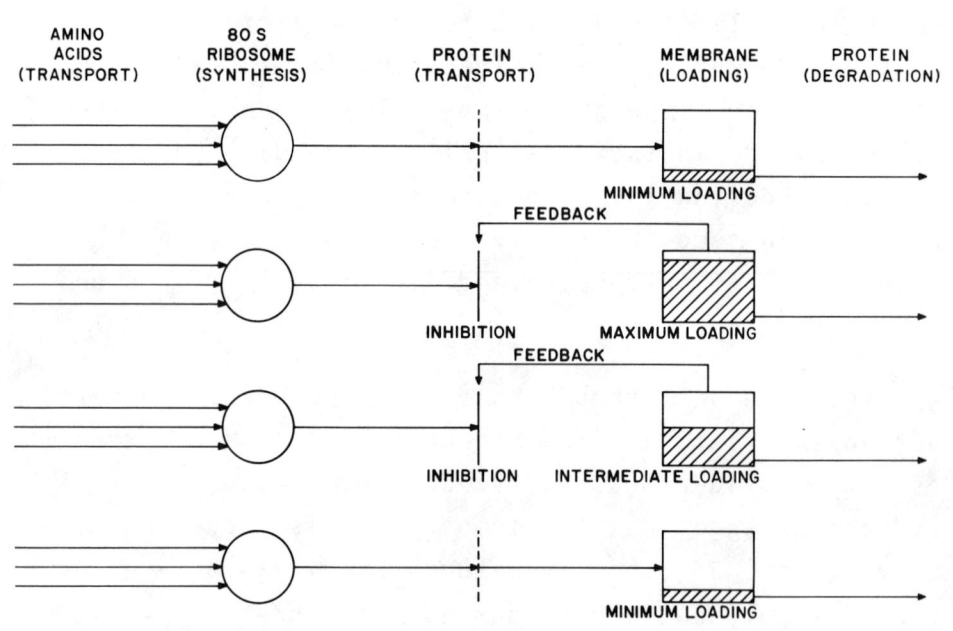

Figure 2. Schematic presentation of the coupled translation-membrane model (14).

It starts from the synthesis of an essential polypeptide on 80S ribosomes which is integrated into thylakoid membranes where it affects the rate of photosynthesis and by means of a feedback mechanism inhibits the synthesis and supply of itself. The synthesis of this essential polypeptide is switched on again when by turnover of the thylakoid membrane proteins, the concentration of the polypeptide in the active membrane decreases beyond a threshold value.

A special advantage of this coupled translation-membrane model is that it may help to explain the temperature compensation. Coupling two processes where the first, protein synthesis, has a temperature coefficient of two to three and the second, the integration of the polypeptide into the membrane, has a temperature coefficient of less than one results in a temperature coefficient of about one for the overall reaction. This, therefore, means that the frequency is to a large extent independent of temperature.

Since in living organisms, many other oscillating parameters exist besides the photosynthesis rhythm, one may ask how the different rhythms are interrelated and whether the mechanisms underlying these different rhythms are similar. If one starts from the results obtained with Acetabularia cells one may ask whether in this type of cell other parameters than that of photosynthesis occur which oscillate. Although there is quite a number of such rhythms there are only a very few which can be recorded in an individual cell over a long period.

Such a rhythm is the intracellular migration of chloroplasts. A cell belonging to one of the large species of Acetabularia at the end of its vegetative phase contains 10^7 or more chloroplasts (19). These, however, are not evenly distributed in the cell. During day time a significant percentage of the chloroplasts migrates into the apical part, while during night time they migrate back into the rhizoid and the basal part of the cell (20). This intracellular chloroplast migration persists under constant conditions, i.e. constant light and fulfills the characteristics of a circadian oscillation. The rhythm of the intracellular chloroplast migration can be easily measured and followed over a long period in an individual cell (21).

Experiments with inhibitors have shown that the rhythm of chloroplast migration depends essentially on the synthesis of polypeptides on 80S ribosomes and is thus similar to the photosynthetic rhythm (unpublished results). More recently, it has been possible to monitor both parameters over long periods in an individual cell (unpublished results).

An important result obtained with this method was that both rhythms have periods which are different from 24 hours, that the periods are not absolutely constant but rather change slowly and what seems to be very significant, that the changes in the periods of both rhythms run parallel. The experiment clearly shows that the two rhythms are closely related. It remains an open question whether this relation is due to a causal coupling or whether it is an indication that in one cell only one primary oscillator exists and that the different rhythms are coupled to this primary oscillator.

This is an interesting hypothesis since it is based on the assumption that the circadian rhythm of photosynthesis activity is not an isolated phenomenon but is rather closely connected to other rhythms. Such a generalization would mean that the features of the coupled translation-membrane model might be valid for all rhythms and that the mechanism includes the synthesis of a specific polypeptide, its integration into a membrane with a resulting feedback effect on the production of the polypeptide.

These features and, in particular, the assumption that there is an essential polypeptide translated on 80S ribosomes raises the question whether there is any additional evidence in favor of such a generalization. The generalization, of course, would mean that there is an essential polypeptide or may be a number of polypeptides which are closely related to each other, which are translated on 80S ribosomes and which by integration into membranes result in a change of the function of the membrane and via feedback inhibition regulate the translation of the essential polypeptide material. If there is more than one essential polypeptide, they should be closely related and they must be rather conservative, which means that during evolution there should have been no major changes in the amino acid sequences.

A number of experiments on different organisms have recently been performed. They included isolated eyes from Aplysia (22) and Gonyaulax (23). The parameters included were potentials and bioluminescence. All experiments revealed a characteristic effect of cycloheximide or other protein synthesis inhibitor pulses on the phase. They are, therefore, in good agreement with the assumption that a polypeptide plays an essential role in the different circadian rhythms.

A number of experiments indicate that membranes are involved in the circadian rhythm in Acetabularia (24). More recent studies have shown that spontaneous potentials which have been monitored over a long period in individual cells of Acetabularia are subjected to a diurnal rhythm which, apparently, is circadian by nature (unpublished results).

Summarizing, one may say that different rhythms and even those not coupled to chloroplasts have a common characteristic feature which underlines the similarity to the photosynthetic rhythm in Acetabularia. This is the sensitivity of the rhythm to cycloheximide and similar inhibitors and the production of phase response curves to pulses of cycloheximide. As it stands, the coupled translation-membrane model is a good basis to explain the circadian rhythm in molecular terms.

Acknowledgement

The author wishes to express his gratitude to Dr. Robert Shoeman for help in preparing the manuscript.

REFERENCES

1. Bünning, E. The Physiological Clock (1973)
   Springer-Verlag, New York.

2. Halberg, F. Halberg, E., Barnum, C.P. and
   Bittner, J.J.
   (1959) pp. 803-878.
   In: Photoperiodism and Related Phenomena in Plants
   and Animals. Withrow, R.B., Ed.
   Amer. Assoc. Adv. Sci. Washington, D.C.

3. Hastings, J.W., Schweiger, H.G.
   The Molecular Basis of Circadian Rhythms (1976)
   Dahlem Konferenzen Abakon Verlagsgesellschaft Berlin.

4. Schweiger, H.G., and Berger, S.
   Int. Rev. of Cytol. Suppl. 9 (1979) pp. 11-44,
   Academic Press, New York.

5. Sweeney, B.M. and Haxo, F.T.
   Science (1961) 134, 1361-1363.

6. Schweiger, E., Wallraff, H.G. and Schweiger, H.G.
   Z. Naturforschg. (1964) 19b, 499-505.

7. Mergenhagen, D. and Schweiger, H.G.
   Exp. Cell Res. (1973) 81, 360-364.

8. Karakashian, M.W. and Schweiger, H.G.
   Exp. Cell Res. (1976) 97, 366-377.

9. Mergenhagen, D. and Schweiger, H.G.
   First Europ. Biophys. Congr., (1971) Baden
   near Vienna, Austria, pp. 497-501.

10. Mergenhagen, D. and Schweiger, H.G.
    Exp. Cell Res. (1975) 92, 127-130.

11. Schweiger, E., Wallraff, H.G. and Schweiger, H.G.
    Science (1964) 146, 658-659.

12. Mergenhagen, D. and Schweiger, H.G.
    Exp. Cell Res. (1975) 94, 321-326.

13. Schweiger, H.G. and Schweiger, M.
    Int. Rev. Cytol. (1977) 51 pp. 315-342,
    Academic Press, New York.

14. Schweiger, H.G.
    Drug Research (1978) $\underline{28}$, 1814-1818.

15. Karakashian, M.W. and Schweiger, H.G.
    Exp. Cell Res. (1976) $\underline{98}$, 303-312.

16. Karakashian, M.W. and Schweiger, H.G.
    Proc. Natl. Acad. Sci. U.S.A. (1976) $\underline{73}$, 3216-3219.

17. Leong, T.Y., Woodward, D.O. and Schweiger, H.G.
    In Progress in Acetabularia Research Woodcock, C.L.F. Ed. (1977) pp. 153-157, Academic Press, New York.

18. Leong, T.Y. and Schweiger, H.G.
    Europ. J. Biochem. (1979) $\underline{98}$, 187-194.

19. Schweiger, H.G., Berger, S., Kloppstech, K., Apel, K. and Schweiger, M.
    Phycologia (1974) $\underline{13}$, 11-20.

20. Koop, H.-U., Schmid, R., Heunert, H.-H. and Milthaler, B.
    Protoplasma (1978) $\underline{97}$, 301-310.

21. Broda, H., Schweiger, G., Koop, H.-U., Schmid, R. and Schweiger, H.G.
    (1979) pp. 163-167
    In: Developmental Biology of Acetabularia
    Bonotto, S., Kefeli, O. and Puiseux-Dao, S. Eds.
    Elsevier, Amsterdam.

22. Jacklet, J.W.
    J. of exp. Biol. (1980) $\underline{85}$, 33-42.

23. Dunlap, J.C., Taylor, W. and Hastings, J.W.
    J. of comp. Physiol. (1980) $\underline{138}$, 1-8.

24. Vanden Driessche, T., Dujardin, E., Magnusson, A. and Sironval, C.
    Int. J. Chronobiol. (1976) $\underline{4}$, 111-124.

PHOTOREGULATION OF NEUROENDOCRINE RHYTHMS

I.Assenmacher, A. Szafarczyk, J.Boissin and
M.Jallageas
Lab. Neuroendocrinology, Dpt. Physiology, University
of Montpellier II - and CEBAS-CNRS, Chizé-Niort - France

It is generally believed that from the molecular bases of life up to the most integrated functions of the organism, most if not all biological functions display rhythmic variations within a wide range of frequencies, which might reflect a very basic temporal organization of living systems. In particular, for organisms living in the natural environment, most biological rhythms fluctuate with a 24 h (circadian) periodicity, while for many of them additional yearly (circannual) variations have also been observed.

Considering circadian and circannual biological rhythms in the light of Environmental Biology, it is obvious that their close temporal relationships with essential environmental cycles very efficiently contribute to an optimal adaptation of the animals' and/or species' vital functions to the unceasingly varying natural environment.

On the other hand, the mechanisms of biological rhythms, admittedly reside in a complex interplay between cycling environmental parameters (synchronizers) and internal (endogenous) factors. A great variety of environmental factors actually have been shown to be associated with the diurnal and annual timing of biological rhythms. Yet, non of them provide as regularly recurring signals as the 24 h daylight cycle (photoperiod), and its annual variations. Impressive experimental evidence has indeed accumulated indicating that photoperiodic environmental changes provide the most widespread and strong external cues for biological rhythms.

Two aspects of the photoregulation of neuroendocrine rhythms in representative mammals and birds will be considered here : the photoperiodic entrained circadian rhythm of the adrenocortical system in relation with the circadian activity rhythm, and the photoregulated annual reproductive cycle.

I - CIRCADIAN RHYTHM OF THE ADRENOCORTICAL SYSTEM

Although a variety of circadian neuroendocrine rhythms were recently described in mammals (ACTH, TSH, PRL, GH) (see Krieger, 1979), and in birds (ACTH, LH, PRL, GH, thyroxin, melatonin) (see Assenmacher and Jallageas, 1980 b), the circadian rhythm of the CRF-ACTH-corticosteroid system presently emerges as the best documented rhythm. In mammals most experimental data were collected in rats, in addition to enlightening observations in man, while quail and pigeons served as experimental tools of special interest in birds.

In the natural environment or under any experimental 24 h light/dark cycle, the adrenocortical system functions on a cyclic circadian basis in precise phase relationships with the sleep/wake cycle. In this conjunction, the hypothalamic concentrations in corticotropin-releasing factor (CRF), the driving hormone of the adrenocortical system, starts increasing shortly after the onset of sleep, i.e. in early morning hours for rats (Ixart et al., 1977) and at the beginning of night in pigeons (Sato and George, 1973), and this strong dependance of both biological responses on the change in the light/dark cycle appears as a fairly general pattern in mammals and birds. The initial rise of CRF is followed by an increment in the pituitary content in ACTH (rat, Ixart et al., 1977 - quail, Assenmacher and Boissin, 1973) and finally in plasma ACTH (rat, Szafarczyk et al., 1979) and corticosterone (rat, Ixart et al. 1977 - quail, Boissin and Assenmacher, 1970) leading to a maximal circulating load in this vital metabolic hormone at the daily transition from rest to activity.

That the daily photoperiodic variations in the environment are not a causal factor of the circadian adrenocortical and activity rhythms, but merely a potent synchronizer to an endogenous rhythmic system (pacemaker), clearly became evident from the demonstration of maintained "free-running" circadian rhythms in blind mammals or in intact animals living in non-fluctuating environments, e.g. constant darkness (DD) or light (LL). From a number of experimental data in this field, the period of the endogenous oscillator driving both the ACTH-corticosterone and the activity rhythms was estimated at 24.2 to 24.3 h in blind or DD-housed rats (see Szafarczyk et al., 1978, 1980a), 24.8 h in isolated blind humans (Aschoff, 1978a), and 24.1 h in quail kept under DD (Boissin et al, 1976). Under constant light (LL), the endogenous periodicity was shown to increase logarithmically, up to 26.2 h (600 lx) in rats, and to decrease down to 21.5 h (40 lx) in quail with increasing illuminance (Szafarczyk et al., 1978), although it is a common observation that, for still unknown reasons, most circadian rhythms are less clearly expressed in constant bright light.

Since the endogenous periodicity of circadian rhythms indeed ranges fairly close to be period of the normal day/night cycle, the easy entrainment of the former by the latter is quite understandable. In rats as in humans, entrainment of the circadian systems, actually was shown to be complete under experimental environments cycling between 23 and 27 periods. Beyond those limits two salient responses were observed : (1) The impaired entrainment of the endogenous pacemaker to the synchronizer resulted first in a "splitting" of the biological system's period into several components, - some of them being still entrained, and others displaying an endogenous circadian periodicity -, and finally in a free-running pattern of the whole circadian system. (2) On the other hand, once circadian rhythms started to free-run, several of them appeared to loose their normal mutual phase relationships, e.g. adrenocortical versus activity rhythm (internal desynchronization of the circadian system), (Aschoff, 1978a - Szafarczyk et al., 1978). Although the same general pattern holds for birds, the avian circadian system was shown to exhibit an amazing entrainability. In quail, entrainment actually appeared still complete under exogenous photoperiods of 13 or 35 h, whereas a free-running corticosterone rhythm appeared for photoperiods of 6 h, and a "split" activity rhythm only for a photoperiod of 3 h (Szafarczyk et al., 1978).

Another intriguing characteristic of the circadian system resides in the "asymmetry effect", following opposite phase-shifts of the synchronizer, (external desynchronization of the circadian system). Whereas in rats and humans undergoing transmeridian jet flights, re-entrainment of the adrenocortical and activity rhythms occurred faster after delay shifts (e.g. west-bound jet-flights) than after advance shifts (east-bound flights) (Aschoff, 1978 b) birds exhibited a reverse pattern with faster re-entrainment after advance than delay shifts (Boissin et al, 1976). That the difference presumably opposes organisms with noctural versus diurnal activity behavior, rather than mammals versus birds might be inferred from the fact that, if studied in standard bunker conditions rather than in conventional travel conditions, human subjects displayed an asymmetry effect similar to diurnal birds as opposed to noctural mammals (rats) (Aschoff - 1978b).

As for most circadian parameters hitherto studied, the structure of the circadian system generating the adrenocortical rhythm still awaits elucidation. However, from a series of recent neuroendocrinological studies, several brain areas clearly emerged as putative components of the system. Among them, a key role has been attributed to the suprachiasmatic nuclei (SCN) in the anterior hypothalamus, when Moore and Eichler (1972) first stated that bilateral destruction of the SCN in rat led to the suppression of the circadian rhythms in both plasma corticosterone and motor activity. Since several other behavioral and hormonal rhythms were also claimed to be impaired by SCN lesions, it was suggested that these nuclei might act as a biological "master clock" for a great variety, if not all CNS depending rhythms (Stephan and Nunez - 1977). Yet, such a simple theory does not fit adequately all observations in this field. With respect to the circadian rhythm in plasma ACTH, SCN lesions in rats indeed were shown to induce a dramatic depression (60 % to 80 %) in the mean level and amplitude of weak residual fluctuations which additionally were lacking any circadian periodicity. In contrast, however, free-running circadian motor activity rhythms still were detected in most animals, and, intriguingly, plasma corticosterone also displayed fluctuations with normal mean level and only slightly decreased (30 %) amplitude, whose periodicity sometimes evoked a circadian pattern (Szafarczyk et al., 1979).

One salient effect of SCN lesions, i.e. the loss of photic synchronization in residual circadian rhythms, clearly can be attributed to the impairment of the retino-hypothalamic tract terminating in the SCN (Moore and Lenn - 1972), which admittedly provides mammals with the major route for photoperiodic entrainment of circadian rhythms (Moore - 1979). Indeed, ocular enucleation by itself, actually resulted in free-running ACTH rhythms with 30 % depressed level and amplitude, while these parameters were almost normal for the corticosterone rhythm (Szafarczyk et al - 1980a).

On the other hand, the SCN are known to receive the most impressive serotoninergic (5-HT) innervation from the midbrain raphe nuclei in the hypothalamus (Moore et al., 1978), and this specific innervation clearly is involved in the control of neuroendocrine rhythms, since its suppression by either midbrain raphe lesions or local injections into the SCN of the specific 5-HT neurotoxin, 5,7-dihydroxytryptamine (5,7-DHT) decreased at least by 50 % both the mean level and the amplitude of the ACTH rhythm, without impairing the corticosterone rhythm (Szafarczyk et al., 1980b). Moreover, when rats were submitted to combined ocular enucleation and raphe lesion, the dramatic depression (60 % to 70 %) of mean level and amplitude in plasma ACTH were approaching the effects of SCN lesions, the more so than in several animals circadian fluctuations in ACTH were no longer detectable, while the free-running corticosterone and activity rhythms had normal parameters (Szafarczyk et al., 1980d). Finally, a pharmacological blockade of the whole central 5-HT system by a systemic administration of p-chlorophenylalanine (pCPA) led to a syndrome that resembled the SCN lesion syndrome even more, since a circadian rhythmic pattern was no longer discernible in the drastically depressed ACTH levels that were again contrasting with persisting corticosterone fluctuations. Interestingly enough, normal diurnal rises in ACTH could be restored in pCPA treated rats by a daily injection of 5-hydroxytryptophane (5-HTP) - a 5-HT precursor that bypasses the blocking effect of pCPA -, provided that 5-HTP was administred with a very precise timing with respect to the photoperiodic signal, which, incidently corresponded to a time of increased 5-HT content in the SCN (Szafarczyk et al., 1980c).

Taken all together these results might indicate that rather than a "master clock", the SCN may act as a major integrator for several facilitating inputs converging into the SCN in precise phase relationships with the photoperiodic entraining (and facilitting) afferences from the retina. In addition to the basic role of the serotoninergic midbrain raphe system other extra-hypothalamic areas, e.g. the limbic system (see Baylé, 1980) and/or neuroendocrine factors, e.g. pineal factors (see Binkley, 1980), as shown in birds, may be essential components of the complex pacemaker driving the adrenocortical and other neuroendocrine circadian rhythms.

II - PHOTOREGULATED ANNUAL REPRODUCTIVE CYCLES

The striking temporal correlations between the annual reproductive cycle of most wildlife birds and mammals, and annual variations in the external milieu is a well established feature. In this respect, most bird species of the temperate zone are spring breeders, whereas mammals include both spring breeders (e.g. vole, mink, mole, dormouse, hamster, hedgehog, badger, ferret, fox) and autumn breeders (e.g. chamois, red deer, goat, sheep).

The adaptive value of a well timed reproductive season obviously lies in an optimal chance for the offsprings to survive, depending on the season of birth. This include firstly the seasonal development of the gonadal function and of mating behavior, in addition to more sophisticated timing devices such as delayed ovulation in bats, or delayed ovoimplantation of the fertilized egg in mink, badger and roe deer.

Among the variety of environmental factors, whose seasonal fluctuations may account for the yearly recurrence of reproduction, the annual changes in photoperiod, actually were shown to provide the animals the main annual synchronizer for reproduction, whereas less dependable changes including temperature, rainfall and food availability may serve as secondary synchronizers. Experimental exposure in fall of various spring breeders to artificial "long days", i.e. photoperiods with longer light than dark phases, or imposing in summer "short days" (L/D < 1) to autumn breeders, inevitably will lead to off-season reproduction in a number of birds (see Assenmacher and Jallageas - 1980b) and mammals (see Herbert and Vincent, 1972 - Reiter, 1974). Most spectacularly, light regimens cycling on periods shorter than one year, were able to entrain breeding cycles up to 2 cycles/year in sheep (Ortavant et al., 1964), 3 cycles/year in ferrets (Herbert and Vincent - 1972), and even 5 cycles/year in European starlings (Gwinner, 1977).

Several hypotheses that may not be mutually exclusive, have been proposed to explain the close phase relationships between annual environmental and reproductive cycles. (1) According to a first theory, the basic concept regarding circadian clocks might be extended to circannual rhythms. The theory then postulates the occurrence of an endogenous circannual clock, that will be entrained by the annually cycling photoperiod. This concept is, indeed, strongly supported in several bird species which have been shown capable of generating persisting circannual sexual rhythms when maintained on either constant photoperiods or permanent darkness, e.g. weaver finch, junco, european starling, sylvia warbler, duck and indian spotted munia (rev. : Assenmacher and Jallageas, 1980b). Although the data are much scarcer regarding mammals, circannual reproductive cycles have been reported by Heller and Poulson (1970) in ground squirrels maintained for several years under a 12 h photoperiod. Similarly, blind hedgehogs displayed persisting circannual testicular cycles over two years (Saboureau, 1979), whereas in blind ferrets or in intact specimens kept under constant short days, persisting yet irregular estrous cycles were observed (Herbert and Klinowska, 1978). (2) According to the "external coincidence" theory, light admittedly has a dual function. Firstly, it will entrain an endogenous circadian rhythm in photosensitivity of the gonadotropic machinery ; and, secondly it will induce the photogonadal response when light is coincident with the "photoinducible phase", i.e. the phase of maximal photosensitivity of the entrained circadian rhythm. Very suggestive results conforming to this theory have been obtained in various birds, e.g. house finch, house sparrow, quail, green finch, white-crowned sparrow and indian weaver bird (rev. Assenmacher and Jallageas, 1980b) and in golden hamster (Elliott et al., 1972), and field vole (Grocock and Clarke, 1974). (3) Whereas the former hypotheses are considering the sole interplay between an external factor and the gonadal response, the "internal coincidence" model predicts that photoperiodic induction occurs when several circadian rhythms are in precise phase relationships to one another. Impressive data on opposite photoperiodic responses obtained in white-throated sparrows, depending on daily and annually varying plasma concentrations in prolactin versus cortisosterone, are fitting this theory (Meier, 1976).

Whatever the mechanism of photoperiodic control of annual reproductive rhythms, it must however, be emphasized that not all components of the reproductive cycle require direct driving by day length. In most spring breeders, whose sexual cycles start annually in close correlation with increasing day length, the breeding season, thus ends before the summer solstice, and will not be protracted by further experimental long day treatment. Although the physiological basis of "photorefractoriness" which impeds prolonged photoperiodic sexual stimulation is still speculative, multihormonal interactions may play an important role in this pecular state, and, more generally in seasonal sexual regression.

In this connection, thyroid-gonadal interactions appear of special interest, at least in species displaying an annual thyroid cycle whose climax coincides with sexual regression. Among birds, starling, blackbird, mallards, teal, Peking duck and several species of north american and indian passerines (rev. : Assenmacher and Jallageas, 1980a) conform this particular pattern, as do several mammals, e.g. badger, mink and fox (rev. : Boissin et al., 1980). Furthermore, in a few bird species (european starling, indian spotted munia and Peking duck), thyroxin administration was shown to depress both the natural breeding state and experimental photogonadal stimulation, whereas a prolonged (duck) or permanent (starling and spotted munia) breeding state was induced by thyroidectomy (rev. : Assenmacher and Jallageas, 1980a). These data strongly favour the thesis of seasonal hyperthyroidism as one potent factor in the determinism of seasonal regression.

On the other hand, a number of studies have been carried out to explore the possible role of the pineal in photoregulatory processes. The bird pineal, indeed, could a priori appear as a possible component in photogonadal control, for at least two good reasons (see Menaker and Oksche, 1974 - Oksche and Hartwig, 1979 - Binkley, 1980) : 1. the avian pineal may participate in extra-retinal photoreception, and 2. the pineal gland plays a major role in the control of the circadian organization in birds, which in turn is involved in photogonadal responses (see above). Most surprisingly, however, neither pinealectomy, nor administration of the putative pineal hormone melatonin were reported to have significant effects on the avian photosexual response (Turek, 1978 - Storey and Nicholls, 1978). This is in contrast to recent data on two well documented photosensitive mammals, i.e. ferrets (rev. : Herbert and Klinowska, 1978) and hamsters (rev. Reiter, 1974 - Hoffmann, 1978), in which the pineal was shown to interfer with photoperiodically dependent changes in the breeding season. Although both species are spring breeders and respond to artificial long photoperiods by off-season breeding, spontaneous, yet delayed sexual recrudescence may occur under constant short days or in blind animals. On the other hand, if maintained on long days, ferrets develop a state of photorefractoriness, that requires exposure to short days to be reversed. By contrast, prolonged long photoperiods induce permanent sexual development in hamsters, and spontaneous regression has never been observed thereafter in sighted animals. This basic difference must be kept in mind when considering the interference of the pineal with the photoperiodic control of both species. In ferrets pinealectomy actually prevents photogonadal stimulation, as does blinding, and the gland therefore appears essential for timing the onset of estrus to the annual cycle in day length. In hamsters, on the other hand, pinealectomy has no discernible effect if performed in animals kept under long photoperiods. However, the operation prevents gonadal regression, which is normally brought about by short photoperiods or blinding. In this case, the essential role of the pineal presumably is to inhibit temporally the gonadal function, in close association with short photoperiods and to play therefore a preeminent role in the seasonal arrest of breeding season. Regarding the possible role of melatonin, silastic implants of the pineal hormone, actually were shown to induce testicular atrophy in hamsters kept on long days (Turek, 1977), thus mimicking short day exposure. In the same line, it is of great interest that bilateral destruction of the suprachiasmatic nuclei, which obliterates the circadian rhythm in the activity of N-acetyltransferase, the rate-limiting enzyme of melatonin production in the pineal (Moore and Klein, 1974) was shown to suppress the effect of short photoperiods leading to sexual regression in hamster (Rusak and Morin, 1975). In ferrets melatonin appears to have a more complex role, but it clearly does not enable pinealectomized ferrets to respond to long days (Herbert and Klinowska, 1978).

III - PHOTORECEPTION AND NEUROENDOCRINE RHYTHMS

There is ample evidence suggesting that in mammals the photoperiodic stimuli entraining neuroendocrine rhythms are perceived by the retina as the exclusive photoreceptor. Bilateral enucleation or sectioning of the optic nerves always led to similar results as keeping the animals in continuous darkness, whether circadian (Szafarczyk et al., 1980a), or circannual (Hoffman and Reiter, 1965 - Herbert and Klinowska, 1978) neuroendocrine rhythms were considered. On the other hand, the optic pathway essential for the photoperiodic effects on neuroendocrine rhythms is the retino-hypothalamic tract terminating in the SCN (see above). Yet, it has already been stressed that, beyond its relaying role of photic information to the brain, the SCN might play a more important role in photoneuroendocrine regulations, since bilateral SCN destructions were repeatedly reported to induce more profound (e.g. circadian ACTH rhythm in rats), if not opposite effects (e.g. photogonadal responses in hamsters) as compared with blinding.

A similar retino-hypothalamic neural tract ending within the SCN has also been evidenced in birds (Bons and Assenmacher, 1970 - Hartwig, 1974). Its precise role in the photoperiodic regulation of neuroendocrine rhythms is, however, shadowed by the occurrence of very efficient extra-retinal photoreceptors (ERP) which have been located in the hypothalamus and the paraolfactive area of the forebrain on the basis of physiological studies (Benoît and Ott, 1944 - Benoît et al, 1950 - Baylé, 1980). The sensitivity of ERP is such that, at least in passerine birds, the involvement of the retino-hypothalamic route in neuroendocrine and behavioral regulations has been questioned (Menaker, 1971). From recent experiments in quail there is, however, evidence for a possible photogonadal response following the sole stimulation of the retina by the implantation of luminescent plates implanted into the anterior chamber of the eye (Baylé, 1980), and for a faster photogonadal response in intact versus blinded animals (Bons et al., 1975). These findings might point to a possible dual input of photoperiodic information into the CNS, although it will not be an easy task to get a deeper insight into when and how in the natural conditions either retina and/or ERP may be involved in photo-neuroendocrine or behavioral regulation.

In summary, the environmental light cycles appear to play a major role in the regulation of vital neroendocrine and behavioral rhythms. The site and mechanism of photoreception, especially in birds, and the complex components of the central neural machinery involved in the photoregulation of neuroendocrine rhythms, occasionally in association with endogenous pacemakers, still are far from being elucidated. Their current exploration belongs to one of the most fascinating fields in modern Photobiology.

REFERENCES

1. ASCHOFF J. (1980a) - Circadian rhythms within and outside their ranges of entrainment. In : Environmental Endocrinology (I. Assenmacher and D.S. Farner, edts). Springer Verlag, Berlin, pp. 172-181.

2. ASCHOFF J. (1980b) - Problems of re-entrainment of circadian rhythm : asymmetry effect, dissociation and partition. In : Environmental Endocrinology (I. Assenmacher and D.S. Farner edts). Springer Verlag, Berlin, pp. 185-195).

3. ASSENMACHER I. and BOISSIN J. (1973) - Aminergic mechanisms and control of circadian blood corticosterone cycle in birds. In : Brain Adrenal Interactions (P. Dell edt), Ed. INSERM, Paris, pp. 257-279.

4. ASSENMACHER I. and JALLAGEAS M. (1980a) - Adaptive aspects of endocrine regulations in birds. In : Hormones, Adaptation and Evolution (S. Ishii, T. Hirano and M. Wada edts). Springer Verlag, Berlin, pp. 91-100.

5. ASSENMACHER I. and JALLAGEAS M. (1980b) - Circadian and circannual hormonal rhythms. In : Avian Endocrinology, (A. Epple and M.H. Stetson edts). Acad. Press, New-York, in press.

6. BAYLE J.D. (1980) - The adenohypophysiotropic mechanisms. In : Avian Endocrinology (A. Epple and M.H. Stetson edts). Ac. Press, New-York, in press.

7. BENOIT J. and OTT L. (1944). External and internal factors in sexual activity. Effect of irradiation with different wave lengths on the mechanism of photostimulation of the hypophysis and testicular growth in the immature duck. Yale J. Biol. Med., 17, 22-46.

8. BENOIT J., WALTER F.X. and ASSENMACHER I. (1950) - Nouvelles recherches relatives à l'action de lumières de différentes longueurs d'ondes sur la gonado-stimulation du canard mâle impubère. C.R. Soc. Biol., Paris, 144, 1206-1211.

9. BINKLEY S. (1980) - The function of the pineal gland. In : Avian Endocrinology (A. Epple and M.H. Stetson edts). Acad. Press., New-York, in press.

10. BOISSIN J. and ASSENMACHER I. (1970) - Circadian rhythms in adrenal cortical activity in the quail. J. Interdisc. Cycle. Res., 1, 251-265.

11. BOISSIN J., JALLAGEAS M. and ASSENMACHER I. (1980) - Cycle annuel du fonctionnement testiculaire des oiseaux et des mammifères : régulation par les facteurs de l'environnement et influence des interrelations testo-thyroïdiennes. In : Rythmes et Reproduction (R. Ortavant and A. Reinberg, edts) Ed. Masson, Paris, pp. 141-145.

12. BOISSIN J., NOUGUIER-SOULE J. and ASSENMACHER I. (1976) - Free-running, entrained and resynchronized circadian rhythms of plasma corticosterone and locomotor activity in the quail. Int. J. Chronobiol., 3, 89-125.

13. BONS N. and ASSENMACHER I. (1970) - Nouvelles recherches sur les voies nerveuses retino-hypothalamiques chez les oiseaux. C.R. Acad. Sci., 287, 2529-2532.

14. BONS N., JALLAGEAS M. and ASSENMACHER I. (1975) - Influence des photorécepteurs rétiniens et extra-rétiniens dans la stimulation testiculaire de la caille par les "jours longs". J. Physiol., Paris, 71, 265A-266A.

15. ELLIOT J.A., STETSON M.H. and MENAKER M. (1972) - Regulation of testis function in golden Hamster : a circadian clock measures photoperiod-time. Science, 178, 771-773.

16. GROCOCK C.A. and CLARKE S.R. (1974) - Photoperiodic control of testis activity in the vole, microtus agrestis. J. Reprod. Fert., 39, 337-247.

17. GWINNER E. (1977) - Photoperiodic synchronization of circannual rhythms in the European Starling, Sturnus Vulgaris. Naturwissenschaften, 64, 44.

18. HARTWIG H.G. (1974) - Electron microscopic evidence for a retino-hypothalamic projection to the suprachiasmatic nucleus of Passer domesticus. Cell Tissue Res., 153, 89-99.

19. HELLER H.C. and POULSON T.L. (1970) - Circannian rhythms II. Endogenous and exogenous factors controlling reproduction and hibernation in chipmunks (entamices) and ground-squirrel (spermophilus). Comp. Biochem. Physiol., 33, 357-383.

20. HERBERT J. and KLINOWSKA M. (1978) - Daylength and the annual reproduction cycles in the ferret, mustella furro : the role of the pineal body. In : Environmental Endocrinology (I. Assenmacher and D.S. Farner edts), Springer Verlag, Berlin, pp. 87-93.

21. HERBERT J. and VINCENT D.S. (1972) - Light and the breeding season in mammals.Excerpta Medica, Int. Congr. Ser., 273, pp. 875-879.

22. HOFFMANN R.A. and REITER R.J. (1965) - Pineal gland, influence on gonads of male hamsters. Science, 148, 1609-1611.

23. HOFFMAN K. (1978) - Photoperiodic mechanism in hamsters : the participation of the pineal gland. In : Environmental Endocrinology (I. Assenmacher and D.S. Farner edts) Springer Verlag, Berlin, pp. 94-102.

24. IXART G., SZAFARCZYK A., BELUGOU J.L. and ASSENMACHER I. (1977) - Temporal relationships between the diurnal rhythm of hypothalamic corticotrophin releasing factor, pituitary corticotrophin and plasma corticosterone in the rat. J. Endocrinol., 72, 113-120.

25. KRIEGER D.T. (1979) - Endocrine rhythms. Raven Press, New-York, pp. 332.

26. MEIER A.H. (1976) - Chronoendocrinology of the white-throated sparrow. In : Proc. 16th Int. Ornithol. Congr. (H.J. Frith and H.J. Calaby edts). Austr. Acad. Sci., Camberra, p. 369-382.

27. MENAKER M. (1971) - Rhythms, Reproduction and Photoreception. Biol. Reprod., 4, 295-308.

28. MENAKER M. and OKSCHE A. (1974) - The avian pineal organ. In : Avian Biology. (D.S. Farner and J.R. King edts) Acad. Press., New-York, 4, 79-118.

29. MOORE R.Y. (1979) - The anatomy of central neural mechanisms regulating endocrine rhythms. In : Endocrine rhythms (D.T. Krieger edt), Raven Press, New-York, pp. 63-87.

30. MOORE R.Y. and EICHLER V.B. (1972) - Loss of a circadian adrenal corticosterone rhythm following suprachiasmatic lesions in the rat. Brain Res., 42, 201-206.

31. MOORE R.Y., HALARIS A.E. and JONES B.E. (1978) - Serotonin neurons of the midbrain raphe : ascending projections. J. Comp. Neurol., 180, 417-438.

32. MOORE R.Y. and KLEIN D.C. (1978) - Visual pathways and the external neural control of a circadian rhythm in pineal serotonin-N-acetyltransferase activity. Brain Res., 71, 17-33.

33. MOORE R.Y. and LENN N.J. (1972) - A retino-hypothalamic projection in the rat. J. Comp. Neurol., 146, 1-14.

34. OKSCHE A. and HARTWIG H.G. (1979) - Pineal sense organ. Components of photo-endocrine systems. In : The pineal gland in vertebrates including man. (J.A. Kappers and J. Pevet edts) Elsevier/North Holland, Amsterdam, pp. 113-129.

35. ORTAVANT R., MAULEON P. and THIBAULT C. (1964) - Photoperiodic control of gonadal and hypophysical activity in domestic mammals. Ann. N.Y. Ac. Sci., 117, 157-193.

36. REITER R.J. (1974) - Circannual reproductive rhythm in mammals related to photoperiod and pineal function : a review. Chronobiologia, 1, 365-395.

37. RUSAK B., MORIN L.P. (1976) - Testicular responses to photoperiod are blocked by lesions of the suprachiasmatic nuclei in golden hamsters. Biol. Reprod., 15, 366-374.

38. SABOUREAU M. (1979) - Cycle annuel du fonctionnement testiculaire du Hérisson, (erinaceus europaeus). Sa régulation par les facteurs externes et internes. Thèse Sci., Tours, 1979, pp. 198.

39. SATO T. and GEORGE C. (1973) - Diurnal rhythm of corticotropin releasing factor activity in the pigeon hypothalamus. Canad. J. Physiol. Pharmacol., 51, 743-747.

40. STEPHAN F.K. and NUNEZ A.A. (1977) - Elimination of circadian rhythms in drinking activity sleep and temperature by isolation of the suprachiasmatic nuclei. Behav. Biol., 20, 1-16.

41. STOREY C.R., NICHOLLS T.J. (1978) - Failure of exogenous melatonin to influence the maintenance or dissipation of photorefractoriness in the canary, serinus canarius. Gen. Comp. Endocr., 34, 468-470.

42. SZAFARCZYK A., ALONSO G., IXART G., MALAVAL F., NOUGUIER-SOULE J. and ASSENMACHER I. (1980b) - The serotoninergic system and the circadian rhythms of ACTH and corticosterone in rats. Am. J. Physiol. in press.

43. SZAFARCZYK A., ALONSO G., IXART G., MALAVAL F., NOUGUIER-SOULE J. and ASSENMACHER I. (1980d) - Effect of combined ocular enucleation and raphe lesions on ACTH, corticosterone and activity rhythms in rats. In preparation.

44. SZAFARCZYK A., BOISSIN J., NOUGUIER-SOULE J. and ASSENMACHER I. (1978) - Effect of ahemeral environmental periodicities on the rhythms of adrenocortical and locomotor functions in rats and japanese quail. In : Environmental Endocrinology (I. Assenmacher and D.S. Farner edts), Springer Verlag, Berlin, pp. 182-184.

45. SZAFARCZYK A., IXART G., MALAVAL F., NOUGUIER-SOULE J. and ASSENMACHER I. (1979) - Effects of lesions of the suprachiasmatic nuclei and of p-chlorophenylalanine on the circadian rhythms of adrenocorticotrophic hormone and corticosterone in the plasma, and on locomotor activity in rats. J. Endocrinol. 83, 1-16.

46. SZAFARCZYK A., IXART G., MALAVAL F., NOUGUIER-SOULE J. and ASSENMACHER I. (1980a) - Corrélation entre les rythmes circadiens de l'ACTH et de la corticostérone plasmatique, et de l'activité motrice, évoluant en "libre cours" après énucléation oculaire chez le Rat. C.R. Acad. Sci. Paris, 290, (D), 587-592.

47. SZAFARCZYK A., IXART G., MALAVAL F., NOUGUIER-SOULE J. and ASSENMACHER I. (1980c) - Influence de l'heure d'administration de 5-hydroxytryptophanne sur la restauration de la stimulation circadienne de la sécrétion de l'ACTH chez des rats traités à la p-chlorophénylalanine. C.R. Soc. Biol., Paris, 270, 53-58.

48. TUREK F.W. (1977) - Antigonadal effect of melatonin in pinealectomized and intact male hamsters. Proc. Soc. Exp. Biol. Med., 155, 31-34.

49. TUREK F.W. (1978) - Diurnal rhythms and seasonal reproductive cycles in birds. In : Environmental Endocrinology (I. Assenmacher and D.S. Farner edts) Springer Verlag, Berlin, pp. 144-152.

THE EFFECT OF ARTIFICIAL AND NATURAL SUNLIGHT UPON SOME
PSYCHOSOMATIC PARAMETERS OF THE HUMAN ORGANISM

F.Greiter, P.Bilek, N.Bachl, L.Prokop, R.Maderthaner,
H.Bauer, G.Kroyer, I.Steiner, P.Riederer, J.Washüttl and
G.Guttmann
Institut for Applied Physiology
3400 Weidling/Vienna, Austria

I - INTRODUCTION

The effect of sunlight upon the human organism is a controversial issue.

Western scientists tend to have a rather negative attitude (1-17), whereas research in Eastern countries, in particular in the Soviet Union, tends to be more positive (18-28). While the West in a rather simplistic way blames increased exposure to sunlight for the increasing incidence of malignant melanoma without taking into account any other environmental parameters, children in Russia are exposed to artificial sunlight during school-breaks in order to make up for the lack of natural sunlight.

The attitude of our Western society has to be regarded as a contradiction of "scientific findings" when we consider the increasing trend towards "sunshine holidays". More than 200 million people per year spend their holidays in extremely sunny regions.

An inquiry at the WHO on the effect of sunlight upon man resulted in the following survey (Table 1) which shows a certain balance of opinions prevailing in the East on the one hand and in the West on the other (29).

With regard to Table 1, J.A. PARRISH, however, takes issue with the spectulations made about the beneficial effects of UVA and visible light. He maintains that the beneficial effects listed under UVB might be variations of the vitamin D effect which can be artificially substituted (30).

Considerable food for thought is provided by the publications of the US Institute for Defense Analysis (31) and J. HIGGINSON (32) giving sound scientific reasons for the highly complex effect of environmental parameters on the increase of malignant melanoma, which are in line with the following summary that we compiled several years ago (Fig.1) (33-34).

TABLE 1

effect	UVB (285 - 320 nm)	UVA (320 - 400 nm)	VL (400 - 700 nm)
beneficial	mineral metabolism vitamin D synth. phosphor and calcium metabolism bone forming processes defensive power to diseases including dental caries	can augment biological effects of UVB	helps cells to repair UVR induced damages ?

harmful (primarily eyes and skin)	chronic
	pterygium and squamous cell cancer of the conjunctiva and cataracts
	solar elastosis (aging of the skin)
	premalignant and malignant skin tumors (actinic keratoses)
	non-melanoma and melanoma skin cancers
		doses which alone demonstrate no biological effects can - in presence of certain environmental chemicals - result:
		   phototoxicity
		   photoallergy
		   enhancement of photocarcinogenesis
			potentiates the detrimental effects of UVR ?

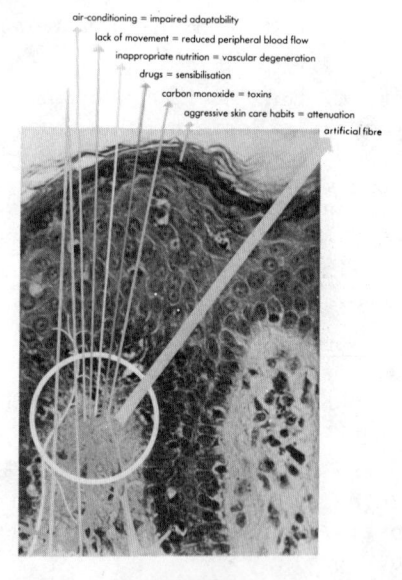

FIGURE 1

So far it is not known which correlations can be made between the selected parameters and how they potentiate or reduce each other.

What is known, however, is that each of the parameters is highly "suspicious" and that sunlight alone can probably no longer be blamed for skin carcinoma.

The controversal attitude adopted vis-à-vis the effect of sunlight upon the human organism has made sports medical researchers, physiologists, psychologists and biochemists all interested in this problem.

The result was the formation six years ago of a multidisciplinary research group which is trying to approach the subject matter from as many angles as possible (35-42).

II - EXPERIMENTAL SETUP

1. Study of human behavior regarding exposure to sunlight. The attitude towards sunlight and the motivations responsible for a certain form of behavior were determined by means of a questionnaire addressed to 20000 persons of both sexes, between 20 and 55 years of age, living in Australia, the Federal Republic of Germany, France and Australia. Multivariance analyses carried out in a randomized sample of 913 test subjects provided a clear two-factor structure as expression of the fact that behavior in exposure to sunlight is determined by two BIPOLAR ATTITUDES.

Factor I is a "pleasure dimension" ranging between the two extremes "pleasant" and "unpleasant". Persons very conscious of this dimension would regard sunbathing as extremely pleasant, restful and relaxing - or the opposite.

A second dimension of attitude, which has no bearing whatsoever on the first one, is the "health factor" which can be illustrated by the two extremes "beneficial to one's health" and "detrimental to one's health". Persons whose behavior is mainly determined by this factor are convinced that sunbathing is either good or - in the other extreme - bad for their health.

Since these two factors are independent of each other, any combination of attitudes can be observed - in particular combinations of marked attitudes in the same direction (pleasant and good or unpleasant and bad) - which should result in a marked tendency either in favour of (or against) sunbathing. Strong, but contradictory attitudes (pleasant and unhealthy) might result in the same weak motivation as the lack of any distinct attitudes.

The actual attitudes towards exposure to sunlight can be predicted with great accuracy when these are analysed by means of additive ties as is suggested by numerous theoretical data on teaching and behavior.

Since a number of theoretical studies on attitudes have shown that the weight of an attitude increases with the degree of its intensity, we introduced an exponent which can be determined empirically from our data. The result is the following expression for the actual attitude intensity in favour of (or against) sunbathing:

$$\text{BEHAVIOR VIS-A-VIS SUNLIGHT} = (\text{FACTOR I} + \text{FACTOR II})^n$$

$$= (\frac{\text{attitude intensity}}{\text{PLEASURE}} + \frac{\text{attitude intensity}}{\text{HEALTH}})^n$$

An analysis of the individual factor scores permits the placement of each individual on the attitude scale illustrated in Fig.2. The third dimension shows the actual intensity of attitude (upwards - positive, towards the sun).

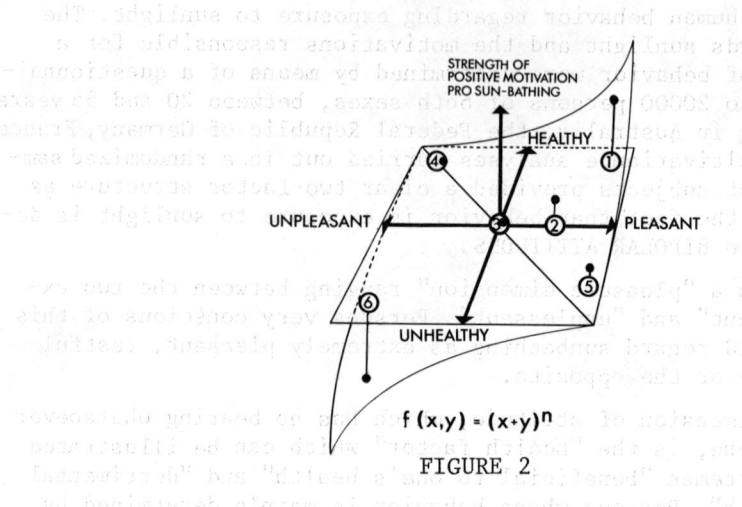

FIGURE 2

$f(x,y) = (x+y)^n$

Subject 1 thus is an enthusiastic "sun worshipper" whose positive attitude in both dimensions makes him a person at risk who tends to underestimate the harmful influences of an exaggerated exposure to sunshine.

Subject 6 on the other hand shows an equally exaggerated "escape behavior" with regard to sunshine. The empirical analysis of the data does, however, also confirm that neither intensive but opposing attitudes (subject 4) nor the actual absence of attitudes (subject 3) induce confidence - such an absence of attitude was found only in an extremely small number of subjects. A slight predominance of one of the two partial attitudes again results in a positive attitude (subject 5) or in an avoidance behavior. With only one of the two attitudes being intense, the actual readiness of behavior will also be markedly reduced (subject 2).

In our analyses the exponent n was always near 1.00 (highest value 1.10) which enabled us to use the simple linear additive model for the prediction of behavior. With regard to attitude intensity, direction, and attitude consistency we can distinguish the

following five groups :
- Group 1 : Weak intensity in the two dimensions. Neutral and indifferent attitude towards sunbathing.
- Group 2 : Positive attitude in both directions. Extreme "sun worshippers".
- Group 3 : Negative attitude in both directions - exposure to sun is avoided.
- Group 4 : Sunbathing is pleasant but harmful. "Sun worshippers" who are convinced that exposure to sunlight is detrimental to their health.
- Group 5 : Beneficial, but unpleasant. Persons belonging to this group take a sunbath as they would take a medicine or they deny themselves something that is good for their health.

There are clear-cut distinctions between these five groups. Group 1 was separated by limits parallel to the coordinate axis in a distance of 1 sigma. The remaining four groups are formed by the cases remaining in the respective quadrants. Because of the evident accumulations, a narrowing or extension of the separating lines hardly changes the individual groups.

The following table shows the distribution over the different classes of behavior both of the entire group and of the inhabitants of four selected countries with different sun experience due to difrent climatic conditions (Table 2).

TABLE 2

GROUP	ATTITUDE GROUP				
	1	2	3	4	5
TOTAL GROUP	5	20	19	36	20
F R G	5	23	18	36	18
AUSTRIA	5	16	17	35	27
FRANCE	3	21	30	30	15
AUSTRALIA	3	28	19	46	4

While the percentage of persons who are completely indifferent is very low, a nation-wide comparison of the distribution over the other four quadrants reveals remarkable differences.

While the values obtained in the Federal Republic of Germany approximate the total group distribution, in Austria a marked shift towards dissonant attitudes can be observed (4 and 5).

In France there is marked trend towards consistent avoidance behavior, while in Australia we see a striking trend towards the dimension "pleasant" : about 3/4 of the persons asked show this basic attitude, but at the same time 461 persons are convinced that sunbathing is bad for their health !

2. Material and Method

In 5 trials conducted at our Institute, the following parameters were considered :

- reflection of the entire solar spectrum from an artificial light source and natural sunlight.
- definition of the amount of reflected energy which, depending on the skin type, is sufficient to produce a minimal erythema reaction (MED) in all subjects.
- side effects were avoided by the selection of a homogeneous group, all subjects were familiar with the experimental setup and they were exposed only to average-submaximal ergospirometric stress.

a) Trial 1 - 4

. Volunteers

Sports students of the University of Vienna between 19 and 25 years of age ($\bar{x} = 21.63 \pm 2.08$). Healthy and not under medication. Not exposed to UV light prior to the test, thus not pigmented. Average training condition. Not under stress as regards sports achievements of pending examinations.

The individual susceptibility to sunlight was determined by an assessment of the intrinsic protection time.

Prior to exposure to the two forms of irradiation, each subject had to undergo a certain exercise program. On the basis of this initial performance it was not only possible to obtain the required reference point, but also to register any daily fluctuations. The exercises were repeated 2, 4 and 24 hours after randomized irradiation.

48 hours were allowed to pass between the first and the second irradiation test so that an overlapping effect could be excluded.

During irradiation the eyes of the subjects were covered, so that only the skin was exposed to sunlight.

The tests were repeated with different radiation intensities. While in trial 1 one minimal erythema dosis (+ MED) was used, in trial 2 $\pm$ MED and in trial 4 ++ MED were used. In trial 3 sports activities were performed under exposure to natural sunlight.

. Light

* Erythemogenic artificial sunlight
  (λ = 290 - 1300 nm)

4 Osram Vitalux lamps
of 300 W each, distance between the lamps and between the bottom side of the lamp and the irradiated skin 40 cm each.

* *Non-erythemogenic artificial sunlight*

    (same exposure time as in erythemogenic artificial sunlight)
    (λ = 320 - 1300 nm)
    Astralux Solarium 3000 consisting of :
    3 UV radiators of 150 W each
    18 IR dark light radiators
    with a total of 2100 W
    2 IR bright light radiators
    of 150 W each
    Distance between the irradiated skin and the bottom side of the lamps : 1.50 m.

* *Average incidence of sunlight during 24 hours from March to June* (Table 3)

TABLE 3

CLOUDLESS SKY	SLIGHTLY CLOUDY
297 nm $0,19 \times 10^{-4}$ J/s.cm^2	$0,06 \times 10^{-4}$ J/s.cm^2
313 nm $1,7 \times 10^{-4}$ J/s.cm^2	$1,38 \times 10^{-4}$ J/s.cm^2
365 nm $5,0 \times 10^{-4}$ J/s.cm^2	$3,78 \times 10^{-4}$ J/s.cm^2

Total exposure time 24 hours. No more than 1 MED was observed.

. Physical performance test

Ergospirometric measurements, Ergopneumotest according to Jäger.

An open system connected with a data processing system for on-line collection of performance physiological parameter.

On the basis of the data obtained from the subject, the reference values were determined by a method according to Bruce (43).

The following parameters were evaluated :

- ventilation   (VE, l/min)
- oxygen uptake (VO_2, l/min)
- oxygen pulse   (VO_2P, ml/min)
- heart rate   (HR, min^{-1})

The following exercises had to be performed : walking on the treadmill ergometer at a speed of 5 km/h with increases of speed by 2.5 % each. The individual exercise periods lasted for 2 minutes, the maximum increase was by 20 %.

This system was also used in trial 3 insofar as at the beginning of the semester (March) as well as at the end (June) the ergospirometric data were computerized.

In the meantime all kinds of open air exercises were performed under natural sunlight for a total of 24 hours during the whole period of time.

. Blood tests

The goal was to find out if GREITER's personal experience with the anti-coagulation effect of natural sunlight could be verified.

For this reason the volunteers of trial 3 had to undergo platelet and fibrinolytic coagulation tests before and at the end of the trial.

. Psychological test

To discover if in some psycho-physiological parameters such as activation, concentration, emotional situation and performance stability a light-induced change occurs.

b) <u>Trial 5</u>

. Volunteers

Psychology students of the Vienna University. Aged between 19 and 22, healthy and not under medication. Not tanned, skin type II. Before the trial control data (psychological, ergometrical, and biochemical) were computerized.

After this the students were selected at random and exposed alternately to erythemogenic- and non-erythemogenic artificial irradiation. Under both conditions the amount of 1 MED was emitted.

The trial lasted all day long, for a period of 4 weeks.

. Light

* *Erythemogenic*

   (λ = 290 - 1300 nm)
   6 Osram Vitalux (300 W each)
   3 Osram Siccatherm incand. Light (92 % heat, 1 % light)
   Total exposure time : 14 x 1 MED

* *Non-erythemogenic*

   (λ = 400 - 1300 nm)
   6 Philips Comptalux (150 W each)
   3 Osram Siccatherm incand. Light (92 % heat, 1 % light)
   Total exposure time : 14 x the time that was necessary to get 1 MED under erythemogenic light.

* *Control*

   8 Tungsram TL (40 W each)
   Total exposure time : the time that was necessary to get 1 MED

under erythemogenic light.

Physical performance test

The test arrangement of Physical performance test was used. Instead of the treadmill a bicycle with a work load of 2 W/kg was taken.

Psycho and Psycho-Physiological test

* *Cognitive selfassessment adjective list (Dubus & Janke) with scaling of the aspiration level.*
* *Behavioral-sensisorimotor coordination was measured by the pursuit rotor system.*
* *The spontaneous electro-encephalogram (EEC) was written in the range of 2-25 Hertz for a time of 160 sec.*
* *The contingent negative (CNV) was recorded by test with a press - and - push button.*

Biochemical analyses

The biochemical analyses of blood and urine were carried out to study the following parameters for the reasons which are indicated under the heading "expected effect after irradiation"(Table 4)

TABLE 4

SUBSTANCES	EXPECTED EFFECT AFTER IRRADIATION
NAD, NADH	Influence on respiration chain
Uric acid	Indication for purine metabolism
Total cholesterol	Possible change in the lipid metabolism
Tyrosine	Eventual change of the amino acid metabolism - Influence of ADR, thyroxine, melanin and turnover of phenylalanine
17-hydroxycorticoide	Production index for cortisol and cortison
17-ketosteroide	Function of adrenal cortex
Na, K, Ca, Mg	Change in mineral metabolism and membrane permeability
Serotonine, ADR, NADR, dopamine	influence on neurotransmitter, blood pressure, heart rate, ventilation, oxygen uptake, etc.
VAA, HMPG, DOPA, 5 HIAA, HVA, VA	Possible amount of metabolism of these neurotransmitters
ATP, ADP, AMP	Indicators for chemical energy (substrate chain phosphorylation and respiration chain phosphorylation)

III - RESULTS

1. Trial 1 (Table 5)

Non-erythemogenic light (320 - 1300 nm) resulted in an increased ventilation (VE) and decreased heart rate (HR) 24 hours after irradiation. Erythemogenic radiation (290 - 1300 nm) resulted in higher activation over the total period of time, mainly after 2 hours, which means increased ventilation and heart rate.

TRIAL 1 (time of + MED)

TABLE 5

Hours after irradiat	λ = 320 - 1300 nm				λ = 290 - 1300 nm			
	VE	VO$_2$	HR	VO$_2$P	VE	VO$_2$	HR	VO$_2$P
2	-0,9	-0,052	+3	-0,53	+4,1	+0,15	+5	+0,68
4	-0,8	-0,035	+3	-0,065	+0,4	+0,065	+4	+0,48
24	+1,2	-0,022	-3	+0,15	+0,7	+0,035	+3	+0,34

2. Trial 2 (Table 6)

showed interesting data under both light conditions : 4 hours after non-erythemogenic irradiation VE and HR have decreased remarkably. This indicates better performance economy.

TRIAL 2 (time of ± MED)

TABLE 6

Hours after irradiat	λ = 320 - 1300 nm				λ = 290 - 1300 nm			
	VE	VO$_2$	HR	VO$_2$P	VE	VO$_2$	HR	VO$_2$P
2	-1,0	-0,026	+2	-0,44	-0,9	-0,005	+1	-0,35
4	-4,1	-0,116	-3	-0,52	-3,2	-0,056	-6	-0,15
24	-0,2	+0,036	0	+0,69	-1,0	+0,040	+1	+0,57

3. Trial 3

- A comparison of the physical performance changes from March to June showed similar data under natural and artificial irradiation conditions.
  The most remarkable figures were found in ventilation (VE) and heart rate (HR), which decreases both to 4,73 l respectively 5 beats per minute.

- The psycho-physiological results showed a statistically significant to very significant correlation to the blood - and performance data. The following parameters were found in those volunteers who showed light-induced changes.
  - Assertiveness, extraversion, adaptability
  - Performance control, conscientiousness, emotional stability
  - Intelluctual ability (Fluid intelligence, which is a congenital intelligence).

4. Trial 4 (Table 7)

In the non-erythemogenic range a certain performance economy could be observed insofar as VE and mainly HR showed decreased data.

Under erythemogenic conditions less VE had to be compensated by a higher HR.

TRIAL 4 (time of ++ MED)

TABLE 7

Hours after irradiat.	λ = 320 - 1300 nm				λ = 290 - 1300 nm			
	VE	VO_2	HR	VO_2P	VE	VO_2	HR	VO_2P
2	-1,0	-0,003	-6	-0,41	-4,7	-0,031	+1	-0,30
4	-2,3	+0,021	-4	-0,14	-1,9	+0,012	+4	+0,09
24	+0,7	+0,021	-4	+0,12	-3,6	-0,056	+3	-0,24

5. Trial 5

   a) *The physical performance data*

did not show any remarkable change as it was obtained in the trials (1 - 4).

Anyway, the lower heart rate (- 4.min^{-1}) indicated a certain light-induced influence.

Of chief interest were the data of the individual cases, which the longitudinal study showed very clearly insofar as the heart rate ranged from 8 - 2 beats per minute, depending on the individual condition.

   b) *The most interesting light-induced change (290 - 1300 nm)*

could be seen in the psychological and psycho-physiological tests. More than 200 individual and combination variables were registered and evaluated in a blind test. Variance analyses (ANOVA) and multivariate variance analyses (MANOVA) supplemented by discriminance

and factor analyses resulted in the following principal findings :
the direct scaling of the subjective condition yielded statistically proved differences between the erythemogenic (EL) and the
non-erythemogenic (NEL) group in the dimensions :

> . EXCITATION (experienced more strongly after erythemogenic irradiation ; Figure 3) and

> . DREAMINESS (lower after erythemogenic irradiation ; Figure 4.

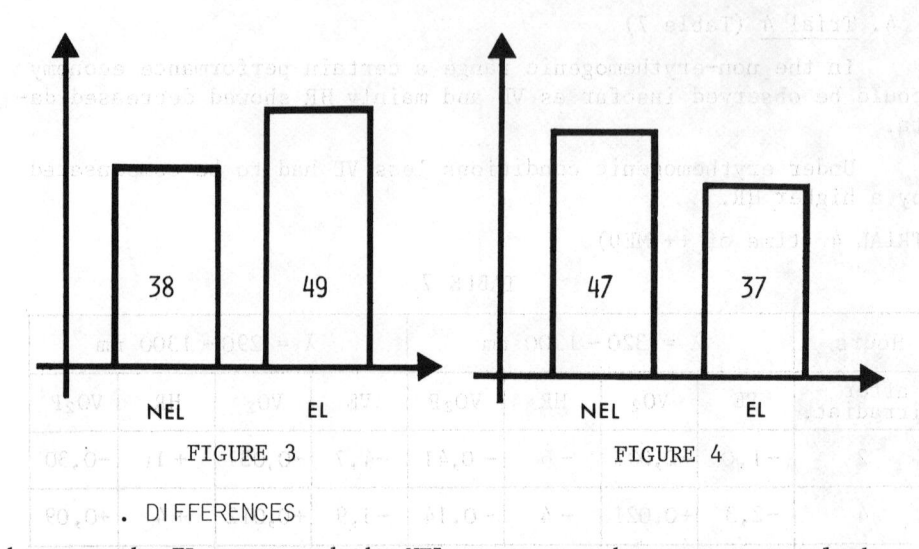

```
        38    49              47    37
        NEL   EL              NEL   EL
         FIGURE 3              FIGURE 4
```

. DIFFERENCES

between the EL group and the NEL group can, however, not only be
observed in the subjective condition, but even in certain areas
of performance. In a rotor pursuit test, where a point of light
has to be followed along an irregular path (sensori-motor coordination) performance under EL was significantly higher - Figure 5).

. STATISTICALLY PROVED

differences were also found with regard to the flicker fusion frequency, another good physiological indicator for activation, which
shows a drastic shift towards increased activation after exposure
to EL light (Figure 6).

. THE EXPECTATION WAVE,

the encephalo-electric correlate of waiting for an imminent event
and the expression of the cortical DC level, in its later course
shows statistically proved differences between EL and NEL which
can be interpreted as an expression of an increased reaction of
the cortical stimulation after EL.

If learning material is supplied in relation to the DC level
of the subject, and the subject is exposed to EL light, triggering
of the learning material by electropositive phases will increase

SUNLIGHT AND THE HUMAN ORGANISM

the learning performance (Figure 7).

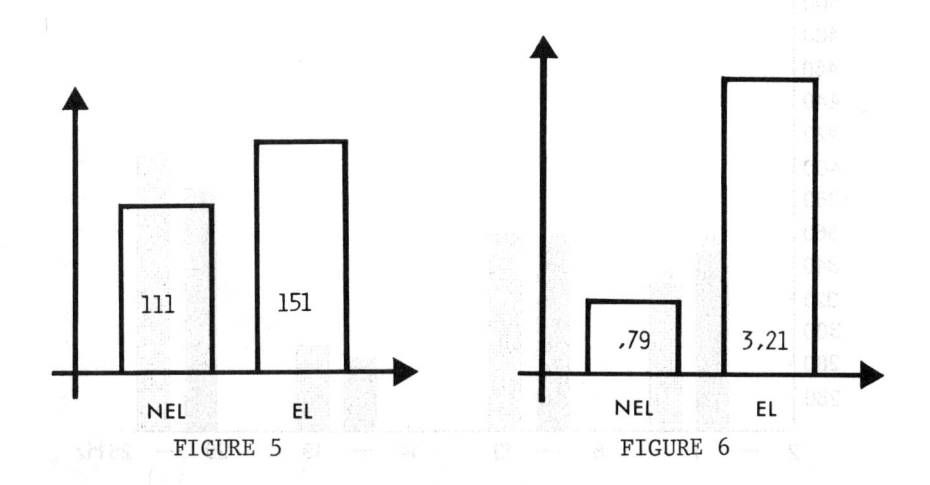

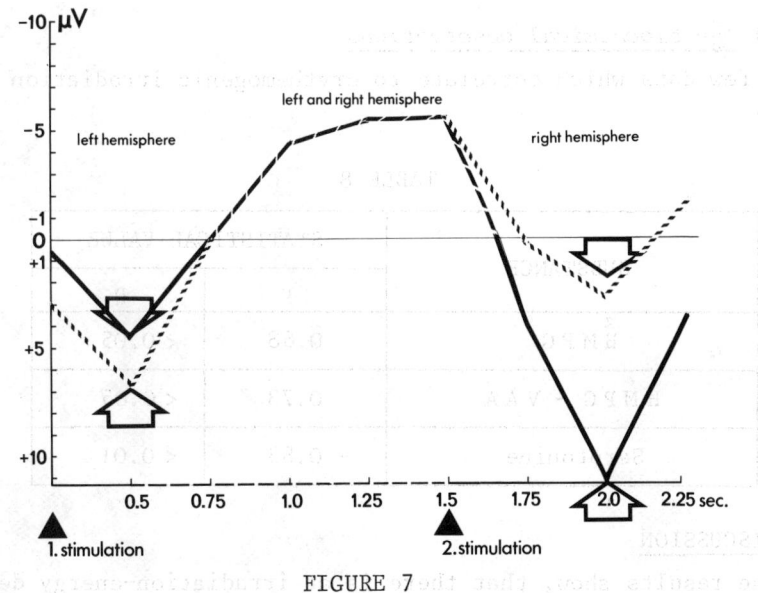

FIGURE 5

FIGURE 6

FIGURE 7

. AFTER EL
the frequency band between 4 and 7 oscillations per second (delta and theta range) is significantly less occupied, which indicates a reduction of the disactivated EEC phases (Figure 8).

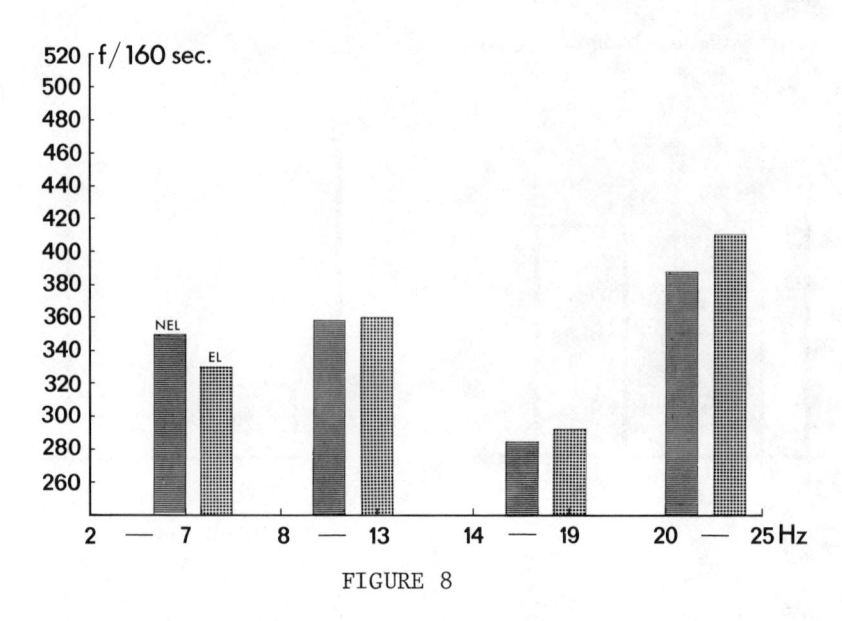

```
520 ┌ f/160 sec.
500
480
460
440
420
400
380
360    NEL
340        EL
320
300
280
260
      2 — 7   8 — 13   14 — 19   20 — 25 Hz
```

FIGURE 8

c) *The biochemical observations*

showed few data which correlate to erythemogenic irradiation (Table 8).

TABLE 8

SUBSTANCE	STATISTICAL VALUE	
	r	p
H M P G	0.68	< 0.05
H M P G - V A A	0.73	< 0.05
Serotonine	- 0.83	< 0.01

IV - DISCUSSION

The results show, that there is an irradiation-energy dependent effect upon the human organism.

The time in a range from ± MED and + MED seems to represent that amount of energy which confirms the individual experience of many people after exposure to sunshine. They feel more activated, more energetic and healthier.

The very surprising findings in trial 1, 2 and 4 were the facts that the non-erythemogenic irradiation also produces good performance data and that the ± MED reaction probably will represent the adequate light condition which can be described as the most beneficial one for the human organism.

The results, mainly those of trial 5, show that, from the psycho-physiological point of view, erythemogenic irradiation leads to a higher activation level. We learned also from this trial that the individual differences are important.

If we carefully analyse the attitude towards sunlight as shown in 2.1, we come to the conclusion that skin types I-III believe so intensively in the benefits of sunirradiation that they ignore the possible harm that could also result.

For this reason a program of public education about skin cancer seems more imperative than ever.

With all due respect to modern statistical methods, the following questions are still under discussion :

1. What importance

should be attributed to deviations of no more than 5 % from the mean value ? For instance a heart rate variation of ± 3 to 5 beats per minute may become statiscally highly significant ; from the physiological or even clinical aspect it is surely not noteworthy.

2. If non-erythemogenic irradiation

results in better performance conditions, is there any serious possibility of cutting out the erythemogenic part of the irradiation ?

3. What significance and correlations

may exist between the environmental parameters as mentioned in Figure 2 ? Could it be that less air-conditioning, more fitness activities, better and more natural food, less medication, less carbon monoxide, less agressive skin treatment and more natural fibres used for clothing would lead to a reduction of irradiation problems ?

4. Why should phototherapy

not be applied in fields other than dermatology ? The psycho-physiological data are encouraging.

V - CONCLUSION

The influence of irradiation on the human organism was studied in 5 trials. Psychological-, psycho-physiological-, ergometric-, and biochemical parameters showed irradiation-induced changes which demonstrate not only the influence of erythemogenic but also of the non-erythemogenic irradiation (expect trial 5).

The most interesting data are summarized in Table 9 which also raises the question of wheter light-therapy will not undergo a revival.

TABLE 9

CORRELATION COEFFICIENTS	
Variables with erythemogenic irradiat.	
perceptual : critical flicker frequency (amount of variation)	+ 0.48 (df = 23, p < 0.01)
behavioral : pursuit rotor (correlation time)	− 0.54 (df = 23, p < 0.01)
cognitive : self-assessment (adjective list "excited")	+ 0.41 (df = 23, p < 0.05)

** This paper is supported by the Forschungsförderungsfonds der Gewerblichen Wirtschaft.
** We would like to thank Mrs. H. GOTTWALD, Mrs. E. HEMEL, Mrs. A. DAVID, Mrs. S. DOSKOCZIL, Mrs. F. HASS and Mrs. B. FITZPATRICK for their valuable contributions.

REFERENCES

1. LATARJET R., Quantitative mutagenesis by chemicals and by U.V. and ionizing radiations : prerequisites for the establishment of rad-equivalences − VII International Congress on Photobiology, Rome, 1976, p. 17.
2. MOUSTACHI E., CHANET R. and HEUDE M., Ionizing and ultraviolet radiations : comparisons of their genetic effects and repair in yeast − VII International Congress on Photobiology, Rome, 1976, p. 19.
3. JUNG H., SONTAG W., LÜCKE-HUHLE C., WEIBEZAHN K.F. and DERTINGER H., Effects on vacuum-UV and excited gases on DNA ; VII International Congress on Photobiology, Rome, 1976, p. 39.
4. IPPEN H., Photophysiological research-preventive medicine − VII International Congress on Photobiology, Rome, 1976, p. 39.
5. WISKEMANN A., Sunlight and Melanomas − VII International Congress on Photobiology, Rome, 1976, p. 47.

6. MONTAGNA W., KLIGMAN A.M., Aging changes in human skin - VII International Congress on Photobiology, Rome, 1976, p. 55.

7. MAGNUS K., Epidemiology of malignant melanoma of the skin in Norway with special reference to the effect of solar radiation - VII International Congress on Photobiology, Rome, 1976, p. 59.

8. FITZPATRICK T.B., SOBER A., PEARSON B., LEW R., - Sunlight in the etiology of primary melanoma of the skin - VII International Congress on Photobiology, Rome, 1976, S 48.

9. Medical News, Oct. 18, 1976, Vol. 236, n° 16, S. 1817-1818.

10. REDDY A.L. et al, Ultraviolet Radiation - Induced chromosomal abnormalities in fetal fibroblaste from New Zealand blacke mice Science, Vol. 201, 1978, 8-9.

11. ERKLUND G. et al, Sunlight and incidence of cutaneous malignant melanoma. Effect of latitude and domicile - in Juedan. Scand. 3 Plast Reonstr. Surg. 12 (3), 231-241, 1978.

12. LINDQUIIJ C. et al, Epidemiological evaluation of sunlight as a risk factor of lip cancer. - Br. J. Cancer, 37, 583, 1978, 987-989.

13. SCOTT E.L., Estimating the increase in skin cancer caused by increases in ultraviolet radiation ; VII International Congress on Photobiology, Rome, 1976, S 60.

14. National Academy of Science - National Academy of Engineering, Washington, D.C. - Biological impacts of increased intensities of solar ultraviolet radiation, 1973, S. 8-19.

15. FREEMAN R.G., Action spectrum for ultraviolet carcinogenesis - Natl. Cancer Institute, 50, 27-29, 12/1979.

16. WHO, Effect of radiation on human heredity, 1959, S. 3-45.

17. VIOLA M.V. et al, Solar cycles and malignant melanoma - Medical hypotheses, 5, 1979, S. 153-160.

18. ISAEVA L., SARAFOVA N., RAIKOV K., Der Einfluss der allgemeinen UV-Bestrahlung auf einige thermoregulatorische Reaktionen bei Kindern in frühem Alter. - "Problemy Pediatri" (Volksrepublik Bulgarien), 1973, S. 7-15.

19. ALIVERDJEV A.A., BUKAROV N.G., Gasaustausch bei Ratten unter dem Einfluss von UV-Bestrahlung. - "Fiziol. Zurn. SSR", 53, n° 9, 1972, S. 1640-1642.

20. ALIVERDJEV A.A., ISRAFILOV G.I., LUKMANOV T.L., ALIVERDJEVA I.A., Der Einfluss des UV-Lichtes auf den Gas-Energieaustausch bei Kaninchen. - In : Wissenschaftliche Sammlung - "Ultrafioletovoe isluchenie i ego primenenie v biologii", Puschtchino-na-Oke, 1973, S. 89.

21. DESPRES S., Effects biologiques des infrarouges et des ultra-violets - Radioprotection, 13, n° 1, 1978, S. 11-21.

22. LITYNSKA A., Circadian activity rhythm of some liso-somatic hydrolases in supernatant homogenates of liver of mice parabiotically connected and influence of UV radiation of this rhythm. Monografic podreozinki schrypty AWF Poznanin, n° 106, 1978, S. 99-108.

23. DURDYEV B. et al, Dynamics of the densimetrical blood indices under influence of the concentrated solar rays. Biol. ylyml. ser., Izv. An Turkm. SSR. - Ser. Biol. n., n° 6, 1978, S. 81-83.

24. UDINCOV E. et al, Influence of sun activity on health status of human beings. Sposobnosti i organizaeii truda invalidov, n° 7, 1978, S. 43-46.

25. DANEŠVAR A.S. et al, Seasonal characteristics of functional state of the cortical layer of adronglands in practically healthy subjects in Turkmonie. - Fizjologija truda v uslovijakh zarkogo klimata. Askhabad, 1978, S. 63-69.

26. TARASOVA E.A., About the influence of UV-radiation on dynamics of respiratory function, gas exchange, oxygen saturation of blood of track and fiels athlets during muscular activity.

27. PROKHOROV V.G., Biophysics and geology-Krasnojarsk, 34 s, il. bibliograf 51 Nr. 2193-99, 1979.

28. DERJAPA N.R. et al, Programm of the global synchronic experiment for investigation of the medical - biological geographical and meteorological factor. - Vlijanije geofiz. i meteorol. faktorov na zizu dejatel nost organizma, 1978, 6-15.

29. WHO, Environmental Health Criteria for Ultraviolet Radiation (UVR) EHE/EHC/WP/77.15, 1978, p. 3-29.

30. PARRISH J., Hardward Medical School, Boston, Persönliche Mitteilung vom 2.1.1980.

31. Institute for Defense Analysis : On the linkage of solar ultraviolet radiation to skin cancer. Final Report, 1978, S. 1-7, 85, 92, 121-135.

32. MAUGH II T.H., Cancer and Environment : Higginson speaks out. Science, 205, 28.9.1979.

33. GUTTMANN G., GREITER F., Man and Sun. An analysis of attitudes. 6th annual meeting American Society for Photobiology, Burlington, Vermont, June 11-15.1978.

34. GUTTMANN G., GREITER F., Man and Sun. An empirical analysis. 7th annual meeting American Society for Photobiology Asilomar, California, June 24-28.1979.

35. GREITER F., BACHL N., PROKOP L., Die Wirkung künstlichen Sonnenlichtes auf die Leistungsfähigkeit des menschlichen Organismus. - VII International Congress on Photobiology, Rome, 1976, Osterreichisches J. für Sportmedizin, 6.Jahrg., nr.4,1976, S. 3-9.

36. GREITER F., GUTTMANN G., BACHL N., The influence of Artificial and Natural Sunlight Upon Various Psychological and Physical Parameters of the Human Organism. - 7th annual Meeting American Society for Photobiology, Asilomar, June 24-28.1979.

37. GREITER F., MADERTHANER R., BAUER N., PROKOP L., GUTTMANN G., Die Wirklung künstlichen und natürlichen Sonnenlichtes auf einige psychosomatische Parameter des menschlichen Organismus. Vortrag bei Union des Sociétés Pharmaceutiques de Yougoslavie, Portorose, Oct. 1976.

38. GREITER F., Die Einwirkung des Sonnenlichtes auf die Leistungsfähigkeit des menschlichen Organismus. - Diss. Dr. phil., 1977, S. 21-46.

39. GUTTMANN G., GREITER F., Man and Sun. An empirical Analysis of attitudes and behaviour. - 10th I.F.S.C.C. Congress Australia, Vol. 3, 1978, S. 721-726.

40. GREITER F., GUTTMANN G., OESER E., Die Rolle der Klassifikation bei der Entwicklung und Bewertung neuer Produkte. - 3. Fachtagung der Gesellschaft für Klassifikation e.V. Königstein/Ts. 5-6.4.1979, S. 165-183.

41. GREITER F., Vortrag Internationaler Kongress für Sportmedizin St. Christoph am Arlberg, 3/1978.

42. GREITER F., BACHL N., WESTPHAL G., Man and Sun. XXI World Congress in Sports Medicine, Brasilia, 9/1978.

43. GREITER F., Der Einfluss künstlichen Sonnenliches auf AMV, VO_2, VO_2P und HF des menschlichen Organismus. Internationales Meeting der Pharmazeuten und Kosmetologen in Portorose, Oktober 1979.

44. DEHN M., BRUCE R.A., Longitudinal variations in maximal oxygen intake with age and activity.
Journal of Applied Physiology, Vol. 3316, 1972.

INTRACELLULAR LOCATION OF PHYTOCHROME

Peter H. Quail

Department of Botany
University of Wisconsin
Madison, WI 53706

I. INTRODUCTION

Attempts to elucidate the molecular mechanism of phytochrome action must reconcile two superficially contradictory sets of data: (a) Physiological studies where rapid, phytochrome-mediated changes in membrane properties have led to the hypothesis that the pigment directly modifies cellular membrane function as its primary action upon photoconversion to Pfr;[1,2,3,4] and (b) Biochemical and immunocytochemical studies where most of the phytochrome is, respectively, extractable as a water soluble protein under conventional conditions and apparently uniformly distributed in the cytosol in situ - at least in unirradiated, etiolated tissue.[5]

Conceivable explanations for this disparity include the existence of a "second messenger" between soluble phytochrome and the membrane-localized response or multiple primary actions of the pigment,[6] some membrane-localized and others not. These remain viable hypotheses but will not be considered further here because of a lack of direct data permitting their evaluation. Instead the discussion will be restricted to examining the very explicit proposal: that the first molecular change induced by phytochrome photoconversion (the primary reaction) is localized in one or more cellular membranes and is the result of direct molecular interaction between phytochrome and the components of those membranes. Within this framework two fundamentally contrasting propositions have been explored experimentally:

(a) That phytochrome is a <u>permanent</u> component of cellular membranes exerting its <u>regulatory</u> influence there on location; and

(b) That phytochrome is <u>not</u> a permanent membrane component but is rather a soluble protein ligand that controls membrane function by interacting reversibly with membrane-bound receptors or reaction partners.

Cell fractionation and immunocytochemical data relevant to each of these propositions will be evaluated within the context of the following questions: (a) Is there direct evidence that phytochrome is associated at all with cellular membranes? (b) If so, with which membrane(s)? (c) Is the association biologically meaningful?

II. MEMBRANE STRUCTURE

Any approach to the question of putatively membrane-localized phytochrome requires an explicit understanding of current concepts of biological membrane organization. There is strong and broadly based evidence supporting the fluid mosaic model of membrane structure.[7] Two different types of association of bona fide membrane proteins with the bilayer are discernable: (a) Those involving <u>integral</u> membrane proteins where the protein interacts directly with the hydrophobic core of the phospholipid bilayer, in some cases traversing the bilayer entirely. (b) Those involving <u>peripheral</u> membrane proteins where the molecule interacts ionically with the phospholipid polar headgroups or exposed hydrophylic surfaces of integral membrane proteins. The operational criteria for distinguishing between the two classes of protein are well established.[8] Briefly, peripheral proteins are readily dissociated from the membrane by high ionic strength and/or chelators, are free of lipids when solubilized and are soluble in aqueous buffers. Integral proteins require hydrophobic bond-breaking agents such as detergents for solubilization, are usually associated with lipids when solubilized and are usually insoluble or aggregated in aqueous buffers.

Data pertaining to the postulated association of phytochrome with cellular membranes should be assessed against this background. For example, several of the models of the mechanism of phytochrome action proposed by various authors imply, intentionally or otherwise, that the protein spans the bilayer and functions by having access to the aqueous environment on both sides of the membrane.[2,9,10] The model of Brownlee et al.[10] proposes that soluble phytochrome becomes an integral membrane protein upon conversion to Pfr with the resultant formation of a hydrophylic channel that spans the membrane. Such models clearly carry specific and experimentally testable predictions.

It should also be emphasized in relation to cell fractionation experiments: (a) that pelletability upon differential centrifugation does not alone establish a membrane or organelle location; and (b) that even a demonstrated association with organelles does not establish a membrane localization - a lumenal location remains a possibility in the absence of direct evidence to the contrary.

III. PHYTOCHROME A PERMANENT MEMBRANE COMPONENT?

A. Evidence of Membrane-Associated Phytochrome?

In unirradiated tissue the bulk of the antigenically detectable phytochrome is uniformaly distributed throughout the cytoplasm[5] and >90% of the pigment is soluble upon extraction in aqueous buffers.[11] How can these observations be accounted for if phytochrome is indeed a permanent membrane component? Two opposing suggestions have been made:

1. "Bulk" versus "active" phytochrome. It has been suggested that the minor phytochrome fraction detected in particulate preparations is a bona fide membrane component and performs the molecule's regulatory function, whereas the major soluble and cytosolically distributed fraction has no such regulatory role and is therefore irrelevant to the molecule's function. The notion of two separate pools of phytochrome, one small and "active" and the other large and inactive, was initially proposed to rationalize certain physiological experiments (the so-called "paradoxes") where a lack of stoichiometry between spectrally detectable Pfr levels and degree of biological response was recorded.[12]

This notion has been misleadingly generalized. Other equally plausible alternative interpretations of the data have been advanced (e.g. S. Hendricks quoted by Vanderhoef et al.[13]). More importantly, however, there is direct evidence against the existence of a physiologically inactive "bulk" pool in many developmental responses. While some responses such as photoropic curvature sensitivity[14] and inhibition of lipoxygenase accumulation[15] saturate at very low Pfr levels, there are many other responses such as Avena mesocotyl inhibition[16] (Mandoli & Briggs, unpublished), inhibition of pea segment extension[17] and promotion of Avena coleoptile tip growth[18] (Mandoli & Briggs, unpublished) that continue to respond to increasing Pfr up to the maximum photoequilibrium level.

Such data indicate that in the cascade of biochemical events induced by phytochrome there is a spectrum ranging from (a) those for which a cellular parameter other than Pfr becomes rate limiting at very low Pfr levels; to (b) those where Pfr remains the rate limiting factor for the response up to the maximum Pfr level possible. This is evidence that while certain measurable responses are maximized at low Pfr levels the integrated development of the seedling, which presumably depends on the overall network of biochemical reactions, is subject to phytochrome regulation up to the maximum level of Pfr that can be formed. The proposition that the bulk of the phytochrome detected in etiolated seedlings is functionally irrelevant should be regarded with considerable circumspection until convincing evidence is presented in its favor. As Hillman[12] points out "... one of the greatest obstacles (in phytochrome research) is the belief that most of the phytochrome has no role ..."

2. <u>Phytochrome redistribution during fixation and fractionation</u>. The alternative suggested possibility is that <u>all</u> the phytochrome is permanently membrane-localized <u>in situ</u> but that most is dissociated from the bilayer and redistributed to the soluble phase during fixation for immunocytochemical studies and during or after homogenization in cell fractionation studies. This possibility might at first appear remote for the <u>in situ</u> studies. It cannot be dismissed out of hand, however, as <u>there is</u> direct evidence for substantial, fixative-induced cellular reorganization during the preparation of samples for microscopy.[19]

Dissociation of resident proteins from membranes during cell fractionation is a well-documented phenomenon for both integral and peripheral proteins.[8,20] The primarily electrostatic interactions that bind peripheral proteins may be readily disrupted by changes in ionic environment or pH during extraction. However, the observation that neither low ionic strength nor high divalent cation concentrations in the medium enhance the level of particulate phytochrome from unirradiated tissue[21,22] suggests that phytochrome is probably not a permanent peripheral membrane protein in the cell.

The release of the hydrophylic domain of amphipathic integral membrane proteins by discrete proteolytic cleavage during extraction is also well documented.[8,20] A prominent example is cytochrome b_5. This molecule is a single peptide consisting of a hydrophobic "tail" of 48 amino acid residues embedded in the hydrophobic core of the membrane and a hydrophylic "head" of 104 amino acids protruding from the bilayer into the aqueous phase. During extraction or deliberate protease treatment the heme-containing hydrophylic head is selectively cleaved from the tail and released as a water-soluble molecule.

We have attempted to determine whether phytochrome might behave similarly by extracting freeze-dried tissue in boiling detergent, immunoprecipitating the molecule and comparing its mobility on SDS gels with that of conventionally purified phytochrome. The apparent mobility of the detergent extracted molecule is slightly lower (~3 kilodalton equivalents) than that of conventionally purified phytochrome. These data indicate that one of the two preparative procedures causes modification of the molecule.

Since the mobility of the detergent-extracted molecule appears to be the same as the phytochrome peptide synthesized in vitro from Poly A-mRNA (G.W. Bolton and P.H. Quail, unpublished) it might be speculated that the modification involves cleavage of a small hydrophobic tail responsible for anchoring phytochrome to membranes in situ. Other alternative covalent modifications unrelated to whether or not the molecule is membrane-associated are equally plausible at present, however. These include chromophore and/or amino acid residue modifications such as phosphorylation, methylation, acetylation etc.[23] The present data supplement the original observations of Boeshore & Pratt[24] that the mobility of the phytochrome peptide changes as a function of extraction and irradiation conditions.

B. Membrane Identification.

Several attempts have been made to identify membranes or organelles with which the minor particulate fraction of phytochrome from unirradiated tissue might be associated (see Marmé,[25,26] and Pratt,[27] for reviews). Both preparative and analytical fractionation strategies[28] have been employed as well as both direct and indirect methods of assaying for the presence of phytochrome in the various fractions. Such studies have led to reports that phytochrome can be detected by direct spectral assay in fractions putatively enriched for mitochondria;[29,30] etioplasts,[31] etioplast envelopes,[32] and plasma membrane.[33] Recent immunoelectron microscope studies report the presence of antigenically detectable phytochrome associated with etioplasts, microbodies and outer mitochondrial membranes in organelle fractions from Avena.[34,35]

C. Is the Association Biologically Meaningful?

The obvious a priori limitation to all such studies involving only direct assay methods is that the simple demonstration of physical presence does not establish biological relevance. This problem is compounded in the case of fractions from unirradiated tissue since it can be readily argued that the phytochrome levels are such a small fraction of the total as to be attributed to contamination from the soluble phase. The indirect approach to phytochrome detection is designed to circumvent this limitation. In this appraoch, some molecular function intrinsic to the isolated fraction is tested for reversible modulation by red and far red light in vitro. Such a modulation is then interpreted to indicate that, not only is phytochrome present, but it is in addition exerting a potentially biologically meaningful influence on the isolated fraction.

Many reports claiming to demonstrate in vitro phytochrome-mediated changes have appeared but most either have proven to be irreproducible or do not withstand critical examination. These include: enhanced rate of NADP reduction in mitochondria-rich fractions[36,37] (Furuya, pers. comm.); enhanced NADH dehydrogenase activity in mitochondria;[30,38] modulation of peroxidase activity in crude particulate fractions[27,39] (Quail, unpublished); NAD kinase activity;[40,41] regulation of ultrastructural development of etioplasts;[26,42] enhanced succinate uptake in mitochondria and plastids;[26,43] and a change in choline acetyl transferase activity.[44]

An apparent exception to this catalog of false-alarms and fade-outs is the R/FR reversible alteration in vitro of the properties of GA-like substances extractable from etioplast-rich fractions.[31,45,46] This alteration is first detectable 5 min from the onset of R irradiation and involves the apparent conversion of non-acidic GAs to acidic GAs.[45,47] These observations suggest that the response involves enzymatic interconversion of the GA forms. Hilton & Smith[46] in a careful and thorough study have provided convincing evidence that the phenomenon is localized in intact etioplasts. There is no definitive evidence available, however, (a) that the phytochrome controlling the response is membrane-localized as opposed to residing in the lumen of the organelle nor (b) that the observed change in the partitioning properties of extractable GA-activity involves any alteration in the functional properties of the organellar membranes. A claim by Cooke and Kendrick[47] based on sub-fractionation studies that the phytochrome-GA response system is located in the plastid envelope has been strongly challenged. Graebe and Ropers[48] point out that the purported differences in GA-like activity reported are too small to be considered significant in the bioassay system used.

A recent study presents evidence for R/FR modulation in vitro of the rate of substrate-dependent Ca^{2+} accumulation by isolated mitochondria.[34] The data indicate that Pfr reversibly enhances the rate of passive Ca^{2+} efflux thereby reducing net active uptake. However, while the results suggest the direct modulation of a membrane property by phytochrome in vitro, a puzzling feature is evident: whereas the rate of Ca^{2+} flux is controlled by the inner mitochondrial membrane the phytochrome responsible for the effect is reported to reside on the outer membrane.[34,35] As a further cautionary note it should be added that red-light inhibition of mitochondrial Ca^{2+} uptake has been reported previously and found to result artifactually from the destructive photodynamic action of contaminating protochlorphyll(ide).[49] No far red reversal of this effect could be detected (D. Marmé, pers. comm.).

Jose and colleagues have described a complex series of experiments in which the effects of both in vivo and in vitro irradiations in various combinations have been tested for their effects on certain enzymatic activities in an operationally defined but as yet uncharacterized particulate fraction from mung bean.[50,51] All responses to in vitro irradiation were observed only when 90 sec or more red light had been given to the tissue before homogenization, a treatment that did not itself alter the amount of spectrally detectable phytochrome (<2% of total extractable) in the fraction. Two types of response to in vitro irradiation - termed Type I and Type II - have been reported.

Type I refers to a R/FR reversible change in the level of detectable enzyme activity that results when the particulate fraction is irradiated prior to assay. ATPase and adenylate kinase exhibit such changes interpreted to indicate in vitro phytochrome regulation.[51] Neither the mechanism of this regulation nor the requirement for prior in vivo R irradiations is understood. No balance sheets for the distribution of these enzymes have been provided so that it cannot be decided from the data whether the activities are bona fide membrane proteins or a minor proportion of a soluble activity non-specifically adsorbed to the particulate fraction. The latter might be expected for adenylate kinase since it appears to be a predominantly soluble enzyme in plant tissue.[52] Likewise, the existence of highly active soluble phosphatases in many plant tissues[53] renders equivocal the premise that the ATP-hydrolyzing activity measured is a native membrane protein.

The Type II responses refer to an observed promotion of enzyme activity by either R or FR light given during the assay. R given during the assay is only effective if FR is given prior to the assay and vice versa. The implied participation of phytochrome in these type II responses is highly questionable, however, firstly because the effect of the intra-assay irradiation whether R or FR always is to promote enzyme activity; and secondly because the action spectra of the response are not consistent with normal phytochrome absorption properties.

The proposition that the changes in enzyme activity detected in either response type represent a meaningful modulation of membrane properties in vitro is yet to be supported by definitive data. Even the purported involvement of phytochrome remains in question given the observed artifactual photodynamic action of protochlorophyll(ide) on enzymatic activities that include adenylate kinase.[49]

IV. PHYTOCHROME A SOLUBLE PROTEIN LIGAND?

The alternative considered here is that phytochrome is not a permanent membrane protein but rather a soluble effector capable of reversible interaction with cellular membranes. In this view the inactive Pr form of the pigment is soluble and uniformly distributed throughout the cytosol of unirradiated tissue but upon conversion to the active Pfr form rapidly binds to membrane-localized receptors inducing a change in the functional properties of those membranes (see Pratt,[5] for review). All the phytochrome is considered to be potentially physiologically active. Data consistent with, but, it should be stressed, by no means proving this hypothesis include the so-called "pelletability" and "sequestering" phenomena.

A. Pelletability Induced by Irradiation In Vivo.

Prior conversion of phytochrome to Pfr in the intact cell dramatically increases (10-fold, from $\sim$ 5% to $\sim$ 60%) the proportion of the pigment that is subsequently observed to be associated with crude pelletable material in cell free extracts, whether or not the molecule is reconverted to the Pr form again before homogenization.[5,22,25,27,54,55] Kinetic analysis in oats and corn has shown that the enhanced pelletability is the end product of a rapid intracellular reaction that is initiated upon Pfr formation, proceeds in the dark with a $t_{\frac{1}{2}}$ of 2 sec at 25°C whilever Pfr remains in the cell, but is terminated abruptly upon reconversion to Pr before the reaction is completed.[56] Reconversion to Pr after the reaction has been completed leads to a relatively slow ($t_{\frac{1}{2}}$ of 20-50 min, 25°C) in vivo decline in pelletability in the dark again to the preirradiation level.[22,54] In terms of the 'membrane hypothesis' of phytochrome action then, the most attractive interpretation is that conversion to the active Pfr form leads to the binding of the previously soluble molecule to membrane-localized receptors in the cell with subsequent preservation of this association upon homogenization. In this view the initial binding is rapid and Pfr-dependent, being prevented by reconversion to Pr before the interaction has taken place, but once bound reconversion to Pr results in a relatively slow dissociation of the molecule from the binding site.

1. Evidence of induced association with membranes? The pelletable phytochrome fraction bands isopycnically on sucrose gradients in the presence of 10mM Mg^{2+} in the density range 1.13 - 1.18 g. cm^{-3} with a distribution profile coincident with several membrane markers.[57] This result indicates that the pigment is in some way, directly or indirectly, associated with membranous material in these extracts. The association is readily disrupted, however, and the phytochrome released as a soluble molecule either by withdrawal of the Mg^{2+} or by addition of monovalent cations at moderate to high ionic strengths.[57] These data indicate that if phytochrome does indeed interact directly with membrane components the interaction is that of a peripheral protein.

2. *Is the association biologically meaningful?* The intracellular process that is ultimately expressed as enhanced phytochrome pelletability in oats and corn is clearly biologically meaningful since it occurs only in the intact tissue.[22,57] The more specific question of whether this intracellular process is the binding of Pfr to membrane-localized sites remains unsettled. Although some accumulated data tend to favor the notion that the association observed in the centrifuge tube was preformed inside the cell,[58,59] the trivial possibility that in vivo-modified phytochrome binds spuriously to particulate material during homogenization has not been rigorously excluded. Yamamoto et al.[60] have recently reported that purified pea phytochrome will aggregate upon Pfr formation in the presence of 10mM Ca^{2+} and that the aggregated material will then bind to a crude microsomal fraction in vitro. Although it is highly unlikely that intracellular Ca^{2+} levels even approach 10mM and despite the fact that Mg^{2+} does not substitute well for Ca^{2+} in this system as it does in the pelletability response,[22] these observations raise the possibility of Pfr-induced intracellular aggregation of phytochrome followed by artifactual binding of the aggregates to membranous material upon homogenization.

At a yet more fundamental level the irradiation-enhanced association is subject to the previously noted limitation common to all localization studies using direct phytochrome detection methods: that physical presence does not establish biological relevance. No attempts to demonstrate in vitro modulation of a membrane-associated function by pelletable phytochrome have been reported. Despite these limitations and potential trivial explanations the rate of the intracellular reaction leading to enhanced pelletability is rapid enough to make it a candidate for involvement in the primary action of phytochrome and therefore worthy of further investigation.

3. *Identity of the phytochrome bearing particles.* The question of the biological relevance of pelletable phytochrome would be greatly aided if the fraction with which the molecule is associated could be identified. Attempts to approach this question using conventional fractionation procedures, have been thwarted because the high levels of divalent cations required to maintain the association cause cross-aggregation of all the membranes in the homogenate precluding their separation.[57]

As a new approach to this problem we are currently investigating a technique using solid-phase antibodies for the identification and isolation of phytochrome-bearing particles (Quail, Bolton, Newcomb & Tokuhisa, unpublished). The procedure involves coating small polyacrylamide beads (4-8 μm) or Staphylococcus aureus cells (1 μm) with monospecific antiphytochrome or preimmune IgG and incubating them with phytochrome-containing pellets from R/FR irradiated Avena. The solid supports are then recovered by a brief centrifugation that does not pellet significant amounts of the particulate phytochrome in the absence of the added supports. The amount of spectrally detectable phytochrome associated with the solid phase is determined and an aliquot is processed for electron microscopy. The latter samples are then monitored for the presence of morphologically identifiable subcellular structures associated with the surface.

About 40% of the phytochrome in the particulate fraction is cosedimented with the antiphytochrome coated supports compared to only 3% for preimmune IgG-coated supports (Table 1). Preliminary

Table 1. Distribution of phytochrome among various fractions following incubation of a phytochrome-containing particulate fraction from R/FR irradiated Avena with or without polyacrylamide beads precoated with either monospecific antiphytochrome IgG (MAP) or the IgG fraction from non-immune serum (NIS).

	Phytochrome in fraction (%)		
	NIS-beads	MAP-beads	No beads
Phytochrome-containing particulate fraction	100	100	100
First bead supernatant	79	38	87
Second (wash) bead supernatant	18	19	12
Bead pellet	3	39	3
Recovery	100	96	102

survey electron micrographs show only erratic association of membrane vesicles with the bead surfaces. In one experiment, however, 58% of the antiphytochrome-coated beads had one or more membrane profiles associated with their surfaces compared to 22% of the preimmune IgG-coated beads (Quail & Newcomb, unpublished). The identity of the vesicles has not yet been determined. It is hoped that this procedure coupled with the use of membrane marker enzymes will permit us to determine whether phytochrome is preferentially associated with one or more identifiable cellular membranes or perhaps is present as an aggregate non-specifically adsorbed to a spectrum of such membranes.

We have also used this solid-phase antibody procedure in an attempt to isolate the putative molecular binding partner(s) with which phytochrome is associated in the particulate fraction (Quail & Bolton, unpublished). Antiphytochrome-coated S. aureus cells were treated with Triton X100 after incubation with a phytochrome-containing pellet from irradiated Avena and subjected to SDS polyacrylamide gel analysis. No preferential retention of any Coomasie blue-positive band or bands in parallel with the phytochrome was observed in response to the detergent treatment.

The phytochrome itself, however, was present as a doublet with or without the detergent treatment. The slower band had a mobility comparable to those of the in vitro-translated and detergent-extracted phytochrome peptides described earlier, whereas the faster band approached the mobility of the conventionally-extracted molecule. Boeshore and Pratt[24] have also previously observed differences in the apparent mobilities of pelleted and control phytochrome extracted under different conditions. One conceivable interpretation of our present observations is that phytochrome extracted in the pelleted condition may be partially protected from the modification that the conventionally purified molecule is subjected to upon extraction. This raises the possibility of a "mirror image" interpretation of the irradiation-induced pelletability phenomenon. It is possible that phytochrome is indeed a resident membrane protein that is redistributed to the cytosol and rendered water soluble by fixation and homogenization respectively of unirradiated tissue as previously discussed; but that Pfr formation prior to extraction somehow preserves or stabilizes the existing in vivo state rather than inducing a new one.

B. In Vitro-induced Association.

There are numerous reports in this category. Two basic types of experiment have been reported. More commonly crude phytochrome-containing homogenates from non-irradiated tissue have been irradiated to convert Pr to Pfr and any enhanced association with particulate material recorded by direct assay[33,61,62,63,64,65] (see Pratt,[27] for review). Alternatively, purified phytochrome has been added back to particulate fractions and any association monitored by direct assay.[66]

Such associations should be viewed with extreme caution until it can be convincingly demonstrated that they are not artifactual. One such artifact, the binding of phytochrome to partially degraded ribosomal material is already well documented.[67,68] This association was initially claimed to be a phytochrome-membrane association.[61,69] Reports that pea phytochrome will self-aggregate and then bind spuriously to a microsomal fraction upon Pfr formation in vitro [60] and that the isoelectric point of dicot phytochrome is about pH 7.0[70] increases the probability that much of the data assembled in this category of investigation is biologically meaningless. Even the use of the criteria for putative binding specificity proposed for mammalian hormone-receptor interactions[71] can be hazardous.[66] Reports of the binding of biological molecules to inanimate materials such as talc, cellulose nitrate filters and polystyrene tubes in a manner that satisfies these criteria of "specificity" are not uncommon.[71,72,73]

C. Irradiation Induced Phytochrome Sequestering In Vivo.

Immunocytochemically detectable phytochrome is distributed diffusely throughout the cytoplasm of unirradiated plants.[74] Conversion of the pigment to the Pfr form leads within seconds to its apparent relocation or sequestering into small (1 μm) discrete regions randomly scattered throughout the cytoplasm.[5,75,76,77,78] Upon reconversion to the Pr form the sequestered phytochrome slowly resumes its original diffuse distribution in the dark.

1. Evidence of induced association with membranes? The discrete areas into which the phytochrome is sequestered remain unidentified, but there is little indication thus far that the pigment is bound to any discernable cellular membrane. The regions do not appear to be mitochondria, plastids, nuclei, plasma membrane or tonoplast.[75,76,78] Thus, although the possibility has not been definitively precluded, there is presently no evidence that sequestering involves the binding of Pfr to membrane-localized receptors.

2. Nature of the phytochrome-containing regions. If the sequestered phytochrome is not associated with cellular membranes or membrane-bounded organelles what other possibilities exist? The possible intracellular formation of multimolecular aggregates upon Pfr formation, originally suggested by Mackenzie et al.[75] has been given new credance by Yamamoto et al's.[60] report of induced aggregation of purified phytochrome. Other subcellular structures such as cortical microtubule organizing centers might also be potential nucleating zones or target sites.[79]

3. Biological significance. The marked similarity in the kinetics of sequestering and in vivo-induced pelletability suggests that these two phenomena may reflect the same cellular process.[5,56,77] The rapidity of both responses indicates their potential involvement in the primary molecular action of phytochrome, regardless of whether or not that action ultimately proves to result from the direct interaction of the pigment with cellular membranes. On the other hand the kinetics of the phenomena are also compatible with spontaneous, intracellular self-aggregation. Such aggregates are formed in bacterial cells by abnormal proteins programed for degradation.[80] The aggregates are visible in electron

micrographs as dense, amorphous granules that are pelletable upon cell disruption.[81] This raises the possibility, originally noted by Mackenzie et al.,[75] that sequestering and pelletability are related to the phytochrome destruction mechanism. Supporting this view is the observation that limited Pr destruction occurs in vivo at the same rate as Pfr if the molecule has first been cycled through the Pfr form.[82,83] Moreover, in light-grown plants at least, the molecule must remain as Pfr for several seconds in the cell for the induction of this effect,[84] kinetics reminiscent of pelletability and sequestering induction.[56,77]

It is clear from the above discussion that, despite the attractiveness of the hypothesis, there is to date no definitive evidence that either in vivo-induced phytochrome pelletability or sequestering results from the intracellular binding of the pigment to membrane-localized receptors.

V. CONCLUSIONS

Critical examination of immunocytochemical and cell fractionation studies indicates that at present there is no unequivocal evidence that the pigment is or can become physically associated with cellular membranes in a biologically meaningful way. It remains premature, therefore, to conclude that phytochrome exerts its regulatory function by directly modifying cellular membrane properties.

REFERENCES

1. S. B. Hendricks and H. A. Borthwick, Proc. Nat. Acad. Sci. 58:2125 (1967).
2. H. Smith, Nature, 227:665 (1970).
3. R. L. Satter and A. W. Galston, in: "Chemistry and Biochemistry of Plant Pigments," T. W. Goodwin, ed., Academic Press, New York (1976).
4. W. Haupt and M. H. Weisenseel, in: "Light and Plant Development," H. Smith, ed., Butterworths, London (1976).
5. L. H. Pratt, in: "Photochemical and Photobiological Reviews," vol. 4, K. C. Smith, ed., Plenum Publishing Corp., New York (1979).
6. H. Mohr, Photochem. Photobiol. 20:542 (1974).
7. S. J. Singer, Ann. Rev. Biochem., 43:805 (1974).
8. S. J. Singer, J. Colloid and Interface Sci., 58:452 (1977).
9. C. B. Johnson and R. Tasker, Plant Cell Envir., 2:259 (1979).
10. C. Brownlee, N. Roth-Bejerano and R. E. Kendrick, Sci. Prog. Oxf., 66:217 (1979).
11. B. Rubinstein, K. S. Drury, and R. B. Park, Plant Physiol.,

44:105 (1969).
12. H. S. Hillman, in: "Phytochrome," K. Mitrakos and W. Shropshire, Jr., eds., Academic Press, New York (1972).
13. L. H. Vanderhoef, W. R. Briggs and P. H. Quail, Plant Physiol., 63:1062 (1979).
14. W. R. Briggs and H. P. Chon, Plant Physiol., 41:1159 (1966).
15. H. Oelze-Karow and H. Mohr, Photochem. Photobiol., 18:319 (1973).
16. L. Loercher, Plant Physiol., 41:932 (1966).
17. H. S. Hillman, Physiol. Plant., 18:346 (1965).
18. W. G. Hopkins and H. S. Hillman, Plant Physiol., 41:593 (1966).
19. B. Mersey and M. E. McCully, J. Microscopy, 114:49 (1978).
20. C. Tanford and J. A. Reynolds, Biochem. Biophys. Acta, 457:133 (1976).
21. P. H. Quail and E. Schäfer, J. Membrane Biol., 15:393 (1973).
22. L. H. Pratt and D. Marmé, Plant Physiol., 58:686 (1976).
23. R. Uy and F. Wold, Science, 198:890 (1977).
24. M. L. Boeshore and L. H. Pratt, Plant Physiol., in press.
25. D. Marmé, Ann. Rev. Plant Physiol., 28:173 (1977).
26. D. Marmé, in: "Plant Growth and Light Perception," B. Deutch, B. I. Deutch and A. O. Gyldenholm, eds., Aarhus (1978).
27. L. H. Pratt, Photochem. Photobiol., 27:81 (1978).
28. P. H. Quail, Ann. Rev. Plant Physiol., 30:425 (1979).
29. K. Manabe and M. Furuya, Plant Physiol., 53:343 (1974).
30. E. E. Billet and H. Smith, Ann. Europ. Sympos. Photomorph. Abstracts, p. 5 (1978).
31. A. Evans and H. Smith, Proc. Nat. Acad. Sci., 73:138 (1976).
32. A. Evans and H. Smith, Nature, 259:323 (1976).
33. D. Marmé, J. Bianco and J. Gross, in: "Light and Plant Development," H. Smith, ed., Butterworths, London (1976).
34. S. J. Roux and C. C. Hale, 8th Internat. Photobiol. Congress Abstr., p. 102 (1980).
35. R. D. Slocum and S. J. Roux, Plant Physiol. Supp., 65:101 (1980).
36. K. Manabe and M. Furuya, Plant Physiol., 51:982 (1973).
37. T. E. Cedel and S. J. Roux, Plant Physiol. Supp., 59:101 (1977).
38. T. E. Cedel and S. J. Roux, Plant Physiol. Supp., 61:52 (1978).
39. C. Penel, H. Greppin and J. Boisard, Plant Sci. Lett. 6:117 (1976).
40. T. Tezuka and Y. Yamamoto, Plant Physiol., 53:717 (1974).
41. D. W. Hopkins and W. R. Briggs, Plant Physiol. Supp., 51:52 (1973).
42. F. A. M. Welburn and A. R. Wellburn, New Phytol., 72:55 (1973).
43. H. W. Schmidt and R. Hampp, Z. Pflanzenphysiol., 82:428 (1977).
44. M. J. Jaffe, in: "Light and Plant Development," H. Smith, ed., Butterworths, London (1976).
45. R. J. Cooke, P. F. Saunders and R. E. Kendrick, Planta, 124:319 (1975).
46. J. R. Hilton and H. Smith, Planta, 148:312 (1980).
47. R. J. Cooke and R. E. Kendrick, Planta, 131:303 (1976).

48. J. E. Graebe and H. J. Ropers, in: "Phytohormones and Related Compounds: A Comprehensive Treatise," Vol. 1, S. L. Letham, P. B. Goodwin and T. J. Higgins, eds., Elsevier, Amsterdam (1978).
49. J. Gross, A. Ayadi and D. Marmé, Photochem. Photobiol., 30:615 (1979).
50. A. M. Jose, Planta, 137:203 (1977).
51. A. M. Jose and E. Schäfer, Planta, 146:75 (1979).
52. S. Frosch, E. Wagner and B. G. Cumming, Can. J. Bot., 51:1355 (1973).
53. R. T. Leonard, D. Hansen and T. K. Hodges, Plant Physiol., 51:749 (1973).
54. P. H. Quail, D. Marmé and E. Schäfer, Nature New Biol., 245:189 (1973).
55. P. H. Quail, Photochem. Photobiol., 22:299 (1975).
56. P. H. Quail and W. R. Briggs, Plant Physiol., 62:773 (1978).
57. P. H. Quail, Photochem. Photobiol., 27:759 (1978).
58. L. H. Pratt, Plant Physiol., in press.
59. P. H. Quail and W. R. Briggs, Plant Physiol., in press.
60. K. T. Yamamota, W. O. Smith and M. Furuya, Photochem. Photobiol., in press.
61. D. Marmé, J. Boisard and W. R. Briggs, Proc. Nat. Acad. Sci., 70:3861 (1973).
62. P. H. Quail, Planta, 118:357 (1974).
63. K. Yamamoto and M. Furuya, Planta, 127:177 (1975).
64. H. Smith, A. Evans and J. R. Hilton, Planta, 141:71 (1978).
65. N. Roth-Bejerano and R. E. Kendrick, Plant Physiol., 63:503 (1979).
66. G. Georgevich, T. E. Cedel and S. J. Roux, Proc. Nat. Acad. Sci., 74:4439 (1977).
67. P. H. Quail, Planta, 123:223 (1975).
68. P. H. Quail, Planta, 123:235 (1975).
69. D. Marmé, J. Supramol. Struc., 2:751 (1974).
70. M. M. Cordonnier and L. H. Pratt, in: "Photoreceptors and Plant Development," J. DeGreef ed., Antwerp, in press.
71. P. Cuatrecasas and M. D. Hollenberg, Biochem. Biophys. Res. Comm. 62:31 (1975).
72. J. L. Phillips, Biochem. Biophys. Res. Comm., 71:726 (1976).
73. M. R. Sussman and H. Kende, Planta, 140:251 (1978).
74. R. A. Coleman and L. H. Pratt, J. Histochem. Cytochem. 22:1039 (1974).
75. J. M. Mackenzie, R. A. Coleman, W. R. Briggs and L. H. Pratt, Proc. Nat. Acad. Sci., 72:799 (1975).
76. J. M. Mackenzie, W. R. Briggs and L. H. Pratt, Amer. J. Bot., 65:671 (1978).
77. L. B. Kass and L. H. Pratt, Plant Physiol. Supp., 61:13 (1978).
78. B. L. Epel, W. L. Butler, L. H. Pratt and K. T. Tokuyasu, in: "Photoreceptors and Plant Development," J. DeGreef, ed., Antwerp, in press.
79. B. E. S. Gunning, Eur. J. Cell Biol., in press.

80. A. L. Goldberg and A. C. St. John, Ann. Rev. Biochem., 45:747 (1976).
81. W. F. Prouty, M. J. Karnovsky and A. L. Goldberg, J. Biol. Chem., 250:1112 (1975).
82. J. M. Mackenzie, W. R. Briggs and L. H. Pratt, Planta, 141:129 (1978).
83. H. J. Stone and L. H. Pratt, Plant Physiol., 63:680 (1979).
84. M. Jabben, Planta, 149:91 (1980).

THE ROLE OF PHYTOCHROME IN THE NATURAL ENVIRONMENT

Harry Smith

Botany Department
Leicester University
Leicester LE1 7RH, U.K.

INTRODUCTION

   Almost everything we know about the structure and properties of phytochrome has been derived from studies of dark-grown plants. Inevitably, therefore, all serious attempts to achieve an understanding of the molecular mechanism of phytochrome action have also been directed towards the photoresponses of dark-grown plants. This work has, in general, been very productive, but it has not led to an agreed view, either in mechanistic, or formalistic, terms of the mechanism of action. It is generally assumed that "Pfr is the active form of phytochrome", but this can only be asserted with any degree of confidence for the very restricted situation of dark-grown plants given single pulses of light within the boundary conditions of reciprocity. Even under these conditions, the relationship between [Pfr] and response is not simple and uniform, exhibiting, for example, a biphasic form for the regulation of anthocyanin synthesis in mustard[1], whilst the control of lipoxygenase levels, also in mustard, displays an all-or-none threshold relationship[2]. It has long been known[3] that even simple pre-irradiation treatments which alter the total amount of phytochrome in subsequent darkness, can destroy the relationship between [Pfr] and response for subsequent light treatments[4]. When prolonged, or complex, light treatments are given, even to dark-grown plants, all relationships between Pfr and response break down, and the "Pfr as active form" view can only be maintained by considerable complication and elaboration of the original simple hypothesis. Such attempts have reached their apotheosis in the multi-component model of phytochrome action developed by Schäfer [5].

   Since concentration on the dark-grown plant has led

predominantly to confusion and complexity, is it possible that a
clearer picture may emerge from a study of the role of phytochrome
in the light-grown plant? Certainly, it is undeniably true that
phytochrome does not only operate in etiolated plants; its control
over vegetative growth[6], pigment formation[7], flowering[8], dormancy[9]
and many other processes in the fully-mature, light-grown, green
plant has been generally accepted for decades. Indeed, the very
first action spectrum for a phytochrome-mediated response was for
the inhibition of flowering by a night-break in soybean and
cocklebur[8]. It is, of course, difficult to study phytochrome in
green plants; firstly, the phytochrome content is normally only
ca. 1-5% of that of etiolated tissues, and secondly, the presence
of large amounts of the photosynthetic pigments renders in vivo
spectrophotometry of phytochrome impossible. Recent developments
using, on the one hand, immunochemistry[10] and, on the other,
bleaching herbicides[11], promise some alleviation of these difficulties, although both techniques have their limitations. In this
article, I wish to summarise briefly the data and conclusions which
have been derived from a different approach to the study of the role
of phytochrome in the light-grown plant - namely, the establishment
of quantitative relationships between the spectral characteristics
of the incident radiation, and the physiological responses of the
plant. Space precludes a comprehensive review, and the reader is
referred to more complete articles currently in preparation covering
the ecological[12], and physiological[13] aspects and, more specifically,
the properties of phytochrome in light-grown plants[14].

THE "FUNCTION" OF A PHOTORECEPTOR

The correlative exercise described here is based upon the
assumption, discussed more fully elsewhere[15], that each biological
photoreceptor has an identifiable "function". The "function" of a
regulatory photoreceptor, such as phytochrome - as opposed to that
of an energy-transducing photoreceptor, such as chlorophyll - is to
acquire specific information relating to the light environment to
which the organism is exposed, and to allow appropriate modulation
of development and metabolism to the temporal and spatial changes
which occur in that environment. An extension of this assumption
is that identification of the function of a photoreceptor should
lead to clues regarding the mechanism of action of the photoreceptor.
In principle, the identification of the function of a photoreceptor
should be possible through a comparison of the physical characteristics of the natural light environment to which plants are exposed,
and the physical properties of the photoreceptor itself.

The most conspicuous property of phytochrome is its photo-
chromicity, with the molecule existing, both in vivo and in vitro,
in two stable, photointerconvertible forms, Pr and Pfr, having
broad, widely separated absorption bands (Pr, λ_{max} = 660 nm;

Pfr, λ_{max} = 730 nm[16]). Although all biological photoreceptors may be photochromic[17], the stability of the two forms of phytochrome and the wide separation of their absorption maxima are exceptional. Thus, in broad-band radiation - e.g. in nature - a photoequilibrium between Pr and Pfr (usually expressed as Pfr/Ptotal, which is equivalent to the frequently-used symbol ϕ), is rapidly established, the proportions of the two forms being determined by the spectral distribution of the radiation, and thus, because of the high extinction of phytochrome in the red and far-red wavelength regions, largely by the red and far-red components of the incident radiation. It would seem logical, therefore, to search for a function for phytochrome which reflects the molecule's unique degree of photo-chromicity, and its high extinction in the red and far-red. A comparison with the physical characteristics of natural radiation shows that these properties could well be of major adaptive value to the green plant.

THE NATURAL RADIATION ENVIRONMENT

The natural radiation environment is highly complex and variable. Although the solar radiation reaching the earth's atmosphere is relatively constant as far as irradiance and spectral distribution are concerned, these parameters as experienced at the surface of the earth are subject to wide-ranging modifications and fluctuations. Radiation is selectively absorbed by atmospheric constituents, scattered by atmospheric molecules and aerosols, and attenuated and reflected by clouds. Consequently, weather and pollution significantly affect the radiation reaching a growing plant. In addition, the extent to which the above factors modify the global radiation varies regularly with the daily march of the sun, and with latitude and time of year, as a result of the variation in path length of the direct beam through the atmosphere, and the consequent changes in the relative proportions of direct and diffuse radiation incident upon a particular surface. Of greater ecological significance, within plant communities themselves, absorption, reflection and transmission of radiation by vegetation result in profound variations in the radiation micro-climate.

In temperate latitudes, the spectral photon distribution of the global radiation during the day (i.e. solar angle >10°C) is relatively constant (Fig. 1a); although clouds have large effects on the quantity of light reaching the earth's surface, their effects on light quality are small. When the solar angle is less than c. 10° (i.e. at sunrise and sunset or at high latitudes) substantial changes occur, with a relative increase in the long-wavelength contribution from the direct beam, as a result of refraction by the atmosphere, and a relative enrichment of the blue, because of the larger contribution of scattered radiation to the global radiation (Fig. 1b). However, the most striking changes in light quality occur

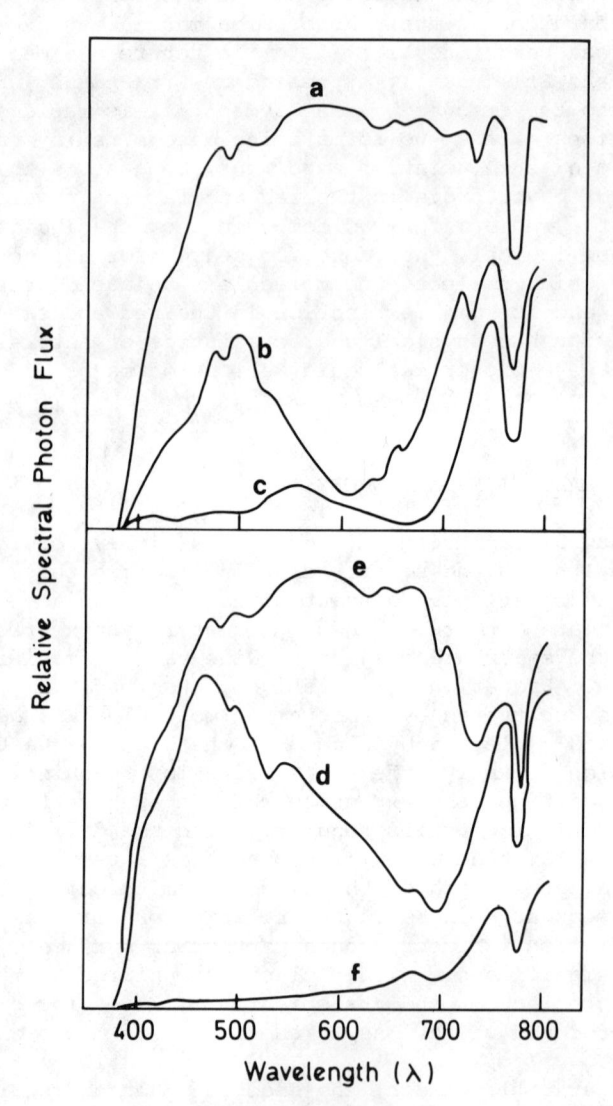

Fig. 1. Typical spectral photon fluence rate distributions of natural global radiation under various terrestrial conditions; (a) mid-day, clear skies; (b) sunset, clear skies; (c) under a vegetation canopy; (d) on the shaded side of a hedgerow; (e) as (a) but filtered through 30 cm of pure water (f) moonlight (x10). (All spectra obtained with a Gamma Scientific spectroradiometer.

upon attenuation of the radiation by vegetation; the red and blue wavelength regions are strongly absorbed by the photosynthetic pigments whereas the far-red is largely transmitted through vegetation (Fig. 1c). A further interesting situation occurs on the shaded side of a hedgerow or stand of vegetation, where the direct beam is filtered through the vegetation, giving a substantial drop in the red:far-red ratio, but where relative enrichment in the blue occurs as a result of the large amount of unfiltered skylight (Fig. 1d). Under water, the ratio of red:far-red radiation is increased because of selective attenuation of the far-red wavelengths (Fig. 1e). Finally, moonlight, which is principally reflected sunlight, may conceivably be of physiological and ecological importance (Fig. 1f).

LIGHT QUALITY AND PHOTOEQUILIBRIUM

Thus, conditions which are very important for the growth of plants in nature, particularly canopy-shade but also perhaps dusk/dawn and depth of immersion in water, substantially alter the relative proportions of red and far-red light in the radiation reaching the plant and thus should alter the photoequilibrium between the two forms (i.e. Pfr/Ptotal). Although it is impossible to determine directly the exact quantitative relationship between Pfr/Ptotal in green plants, and the R:FR ratio of the incident radiation, it is physically inevitable that Pfr/Ptotal, at any point, should bear a systematic relationship to the spectral photon distribution of the radiation incident upon the plant surfaces[18]. As an alternative to direct measurements, it is possible to relate the R:FR ratio (i.e. ζ, or the ratio of the photon fluence rates in 10 nm bandwidths centred on 660 nm and 730 nm) to Pfr/Ptotal in etiolated test material[19] (Fig. 2). In unfiltered daylight, ζ is remarkably constant at 1.15 ± 0.02, and canopy shade reduces this ratio to very low values (e.g. <0.1 under dense broad-leaved canopies) (see ref. 12 for survey). Thus, from the shape of the Pfr/Ptotal v. ζ curve, it can be seen that a small drop in ζ (e.g. due to slight shading) will result in a relatively large change in phytochrome equilibrium. Consequently, if phytochrome should act to perceive the R:FR ratio of the incident radiation, via the establishment of photoequilibria, then it would be a very sensitive mechanism of shade perception, well adapted to natural conditions.

PHOTOEQUILIBRIUM AND GROWTH

The above view is clearly based solely on correlative evidence, but more direct evidence has been derived from studies of the growth of plants under artificial light sources in which the total photosynthetically active radiation (PAR ≡ 400-700 nm) is held uniform and constant whilst the far-red (> 700 nm) content is varied. When

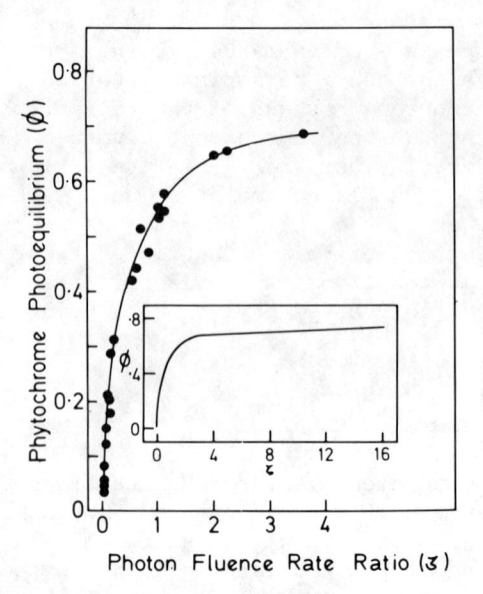

Fig. 2. Relationship between R:FR ratio (ζ) and Pfr/Ptotal (from reference 19)

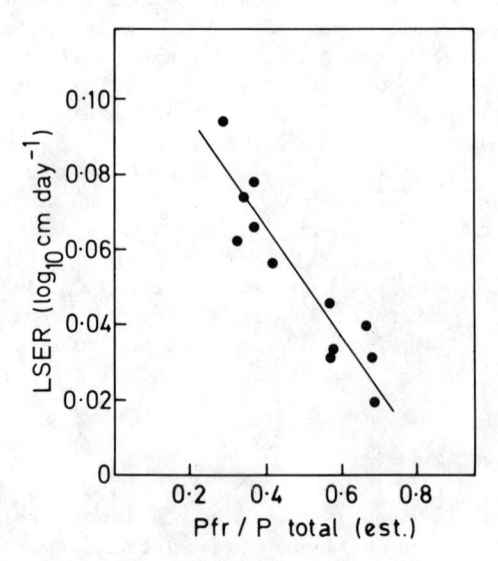

Fig. 3. Relationship between logarithmic stem extension rate (LSER) and Pfr/Ptotal (estimated from ζ) for _Chenopodium album_ L. (from reference 22).

aggressive ruderal herbs were grown in such cabinets, extensive modulation of growth and development was observed. Although a number of different developmental responses have been obtained[23-25] the most striking is an increase in extension growth with decrease in ζ. The relationship between ζ and growth rate, expressed as logarithmic stem extension rate, is a curve, and although this shows that growth rate is related to ζ, the shape of the curve provides little information on the mechanism. Transformation of ζ to Pfr/Ptotal using the calibration curve in Fig. 2, on the other hand, reveals a perfect linear relationship, as shown in Fig. 3 for Chenopodium album L. The linear relationship of Fig. 3 is strong, although still circumstantial, evidence that phytochrome functions to perceive the R:FR ratio of the incident radiation via the establishment of photoequilibria.

ECOLOGICAL RELEVANCE

A range of different weed species have been grown in the light quality cabinets and the results show that the response of extension growth rate (and a number of other responses) is in all cases linearly related to photoequilibrium, whilst the degree of response is systematically related to the normal ecological niche of the particular species[26]. Typical shade-avoiders, such as Chenopodium album L., Sinapis alba L. and Chamaenerion angustifolium (L.) Scop., show very marked increases in extension growth rate with decreasing Pfr/Ptotal. Normal woodland plants, which are shade-tolerators (e.g. Mercurialis perennis L., Circaea lutetiana L. and Teucrium scorodonia L.), show much less increase in extension rate with decreasing Pfr/Ptotal. These responses would, in each case, be of obvious value for growth and survival in the particular ecological niches of the different species.

PHYSIOLOGY OF THE GROWTH RESPONSE

The evidence described above is consistent with the view that the function of phytochrome is to obtain specific information from the environment on the R:FR ratio. This information is transduced biologically to provide appropriate responses to shade, and also, perhaps, to other natural situations where the R:FR ratio is different from daylight (i.e. dawn/dusk, underwater, etc.). The circumstantial nature of the above evidence, however, has stimulated attempts to obtain more direct quantitative relationships in the hope that a better understanding of the mode of action would result.

We were impressed, initially, by the rapidity of the extension growth responses to added FR. In Chenopodium and Sinapis, FR-mediated increases in height could be detected, by ruler, within 12 hours. When growth was measured continuously via a linear

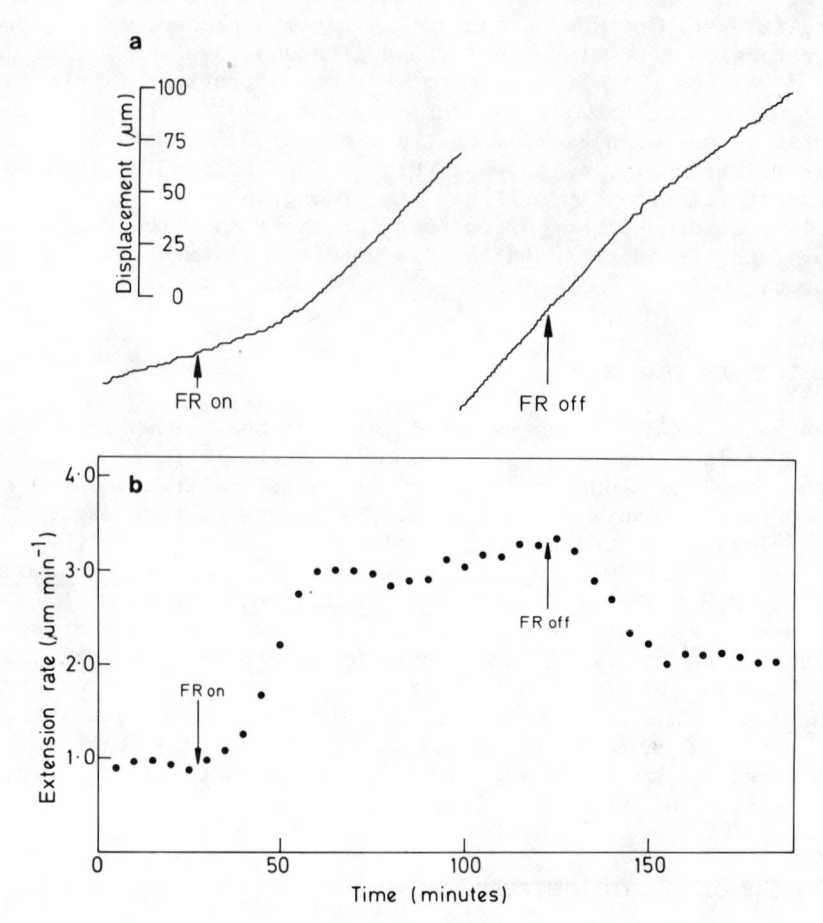

Fig. 4. Effect on extension growth rate of irradiating a whole
Sinapis alba seedling with supplementary FR for 90 min;
(a) copy of original chart trace; (b) transformation of
data to rate values, applying a three-point running mean.
Calculated Pfr/Ptotal in white light, 0.65; in white
plus FR, 0.46. (From reference 28).

displacement transducer, the extreme rapidity of these affects was revealed. In Chenopodium[27], extension growth rate increased within 7 minutes of the addition of FR to background white light; in Sinapis, the time lag was nearer 15 minutes[28] (Fig. 4). Using fibre-optic probes to direct the supplementary FR to specific regions of mustard plants, we were able to show that irradiation of the growing internode results in a sharp increase in growth rate within 15 minutes, whereas irradiation of the leaf has no such short-term effect[28] (Fig. 5). Leaf treatment, however, causes a significant increase in growth rate after a lag of ca. 3 hours, indicating the possible transmission of an elongation factor from the perceptive region to the growing zone[28]. Direct demonstration of the involvement of phytochrome is seen in Figure 6, in which the extension growth rate of a mustard seedling is shown to be continuously modulated by the R:FR balance of the incident radiation[28].

With such experiments, it is possible to construct quantitative relationships between FR fluence rate and extension rate. This has now been done for three wavelengths of supplementary radiation - 719 nm[28], 739 nm[28], and 700 nm[29]. The first two wavelengths, when added to background white light, establish identical photoequilibria at equal fluence rates, and the relationships between extension growth rate and fluence rate are similarly identical for these two wavelengths. On the other hand, 700 nm establishes much higher photoequilibria than 719 or 739 nm at comparable fluence rates, and correspondingly causes much smaller increases in growth rate. These data indicate that, at least for the short term responses, FR fluence rate per se is not important; the growth responses are related solely to photoequilibrium and not to phytochrome "cycling", i.e. the rate of $Pr \rightleftarrows Pfr$ photoconversions.

MECHANISM OF ACTION

The evidence summarised here and presented more fully in the cited papers strongly suggests that extension growth rate - and, a number of other developmental phenomena - in light-grown green plants, is controlled directly and continuously by the photoequilibrium of phytochrome. It is incumbent, consequently, to consider how such control might be achieved. The first question, and one which cannot currently be properly answered, is whether the apparent relationship of growth with Pfr/Ptotal is merely a manifestation of a real relationship with Pfr concentration - i.e. [Pfr]. If, under all the circumstances described here, Ptotal remained constant and uniform, then Pfr/Ptotal would be formally equivalent to [Pfr]. Indeed, for the short-term responses detected by the transducers, we can assume that significant changes in Ptotal would not occur within the observed short lag periods. Thus, in these cases, changing Pfr/Ptotal by the addition of FR light, would automatically result in an equivalent change in [Pfr].

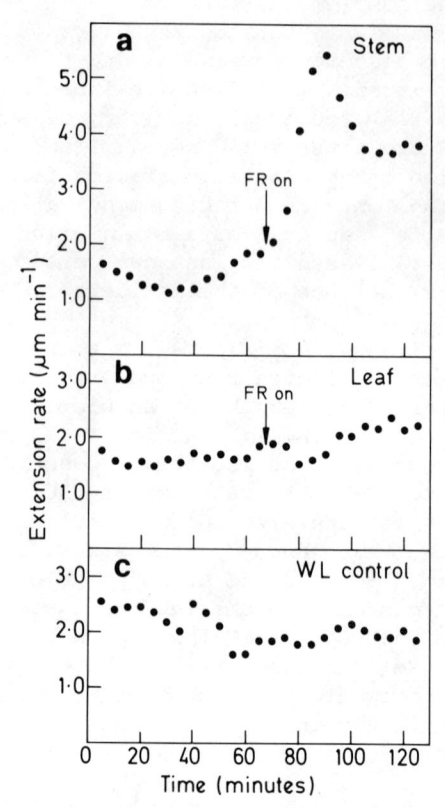

Fig. 5. Specimen recordings showing the effects of extension growth rate of Sinapis alba seedlings, growing in background white fluorescent light, of irradiating either the growing first internode (a) or the two primary leaves (b) with supplementary FR via fibre-optic probes. (c), white light control. Growth rate measured continuously with a transducer assembly. Calculated photoequilibria were: white light, 0.62; white plus FR (a) 0.23, (b) 0.31. (From reference 28).

In the longer term experiments, however, and if phytochrome behaves as it does in dark-grown plants, substantial changes in Ptotal would be expected under the different radiation regimes. Current phytochrome dogma would suggest that higher levels of Ptotal would be found under sources with added FR, than under the background white fluorescent light. Indeed, elegant calculations of the kinetics of phytochrome transformations under continuous irradiation in mustard seedlings have shown that, theoretically at least, [Pfr] should become independent of both wavelength and fluence rate[30].

Thus, until a way is found to measure Ptotal and Pfr/Ptotal in light-grown green plants, we can only conclude that the evidence is slightly in favour of photoequilibrium, rather than [Pfr], as the determinant of extension growth. If this is true, it would raise a major conceptual difficulty. How could the ratio of two inter-convertible components, independent of the total amount or concentration of those components, be the determinant of a biological process? Biological transduction mechanisms normally operate on the basis of changes in the concentration of effector molecules, often with an amplification step built in, conferring sensitivity. For a transduction mechanism to be based on a ratio, and compensated for concentration changes, would be most unusual, if not unique. One imagines that such a mechanism could only operate via the opposition of two vectoral actions, thus almost requiring the site of action to be a membrane, and implying the heresy of assuming that Pr, as well as Pfr, is actively involved in the overall mechanism.

A plausible model to account for the apparent relationship of response with photoequilibrium may be based on a proposal I put forward, admittedly to answer a rather different question, ten years ago[31]. In this scheme (Fig. 7) phytochrome is considered to be a membrane-associated transport factor, having anisotropic binding sites in the Pr and Pfr forms, for an unknown but critical metabolite, X. Thus, Pr would bind X only on one side of the membrane and, upon photoconversion, transport X across the membrane; conversely Pfr would bind X only on the other side of the membrane, and upon photoconversion, transport X across the membrane in the reverse direction. The required, two opposing vectoral actions are thus provided. In the original concept[31], it was suggested that the biological action was simply a function of the concentration of X on the Pfr side of the membrane, a view which does not now hold up well. An alternative, and more attractive, idea is that the biological action is determined by the concentration gradient of X across the membrane. For example, if X were an ion, e.g. H^+, K^+ or Ca^{++}, the charge separation across the membrane could be determined by photoequilibrium through control of the ion concentrations immediately adjacent to the two sides of the membrane. At higher photoequilibrium, each molecule would spend proportionately more time as Pfr than as Pr, favouring the binding of X to Pfr and leading to net flux of X from right to left in Fig. 7. At low photoequilibrium, the opposite would occur, leading

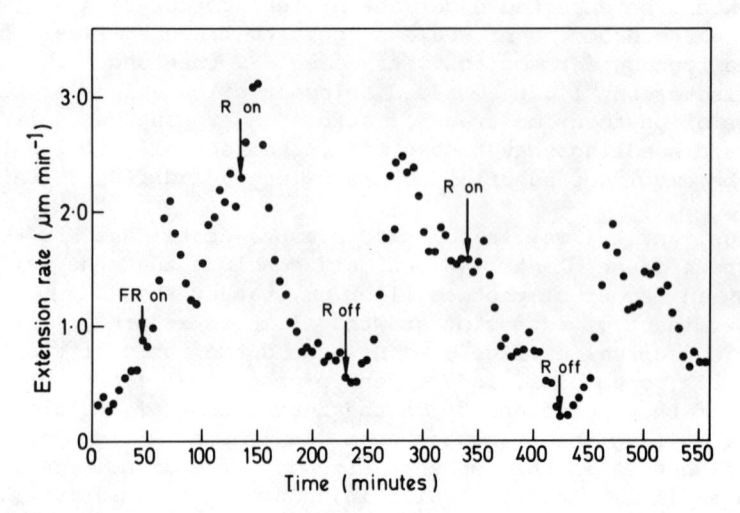

Fig. 6. Specimen recording showing the effect on extension growth rate of a <u>Sinapis alba</u> seedling of irradiating the stem with supplementary red light, in the presence of white light and FR. Calculated photoequilibria: white light, 0.65; white + FR, 0.52; white + FR + R, 0.62. (From reference 28).

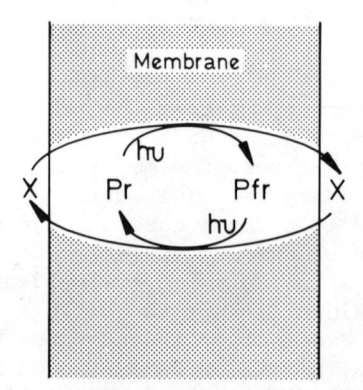

Fig. 7. A scheme which may account for the apparent relationship of response to photoequilibrium (modified from Smith [31])

to a net flux of X from left to right. If the availability of X
at the membrane surfaces was limited, then photoequilibrium would
determine the magnitude and direction of the gradient of X across
the membrane.

On this basis, phytochrome would operate as an electrogenic ion
pump, the direction and magnitude of the electrochemical potential
gradient thus established being determined by photoequilibrium, and
therefore by light quality. This model, at least has the merit of
being susceptible to experiment and to formalistic analysis,
particularly in those situations where phytochrome is known to be
associated with membranes, and to control measureable responses,
in vitro. Whether this model can also account for the action of
phytochrome in dark-grown plants, as originally proposed[31], has
yet to be considered.

CONCLUSIONS

1. The function of a photoreceptor is to obtain information from
   the environment, and should theoretically be deducible from a
   comparison of the physical characteristics of the photoreceptor
   and the radiation environment to which the organism is naturally
   exposed.

2. The radiation environment to which higher green plants are
   exposed displays marked fluctuations in light quality in the
   red and far-red regions; information on R:FR ratio (ζ) would
   therefore be of value to the plant.

3. R:FR ratio and photoequilibrium (Pfr/Ptotal) are systematically
   related.

4. Elongation growth rate is linearly related to "estimated"
   Pfr/Ptotal.

5. The function of phytochrome is to obtain information on the
   R:FR ratio and thus allow the plant to react to vegetational
   shading (and possibly to dawn/dusk and depth of immersion).

6. The growth responses are rapid, and continuously modulatable by
   R:FR ratio.

7. Evidence is slightly in favour of the view that Pfr/Ptotal,
   rather than [Pfr], is the determinant of growth.

8. A model of phytochrome action to account for the possible role
   of photoequilibrium is proposed; this envisages phytochrome as a
   bi-directional electrogenic ion pump establishing a gradient
   across the membrane, the direction and magnitude of which is
   determined by photoequilibrium.

REFERENCES

1. H. Drumm and H. Mohr. Photochem.Photobiol. 20:151 (1974)
2. H. Mohr and H. Oelze-Karow. in "Light and Plant Development" H. Smith, ed. Butterworths, London. p. 257 (1976)
3. W.S. Hillman, Ann.Rev.Pl.Physiol. 18: 301 (1967)
4. L.R. Fox and W.S. Hillman, Pl.Physiol. 43: 1799 (1968)
5. E. Schäfer. Jour.Math.Biol. 2: 41 (1975)
6. R.J. Downs, S.B. Hendricks and H.A. Borthwick. Botan.Gaz. 118: 199 (1957)
7. R.J. Downs, H.W. Siegelman, W.L. Butler and S.B. Hendricks. Nature 205: 909 (1965)
8. M.W. Parker, S.B. Hendricks, H.A. Borthwick and N.J. Scully Botan.Gaz. 108: 1 (1946)
9. P.F. Wareing. Ann.Rev.Pl.Physiol. 7: 191 (1956)
10. R.E. Hunt and L.H. Pratt. Plant, Cell & Environment 3: 91 (1980)
11. M. Jabben and G.F. Deitzer. Photochem.Photobiol. 27: 799 (1978)
12. D. Morgan and H. Smith. in Encycl.Pl.Physiol (New Series) "Physiological Plant Ecology" Vol. A. ed. P.S. Nobel. Springer, Berlin. Chapter 5.3 (in press)
13. D.C. Morgan and H. Smith in Encycl.Pl.Physiol (New Series) "Photomorphogenesis". eds. W. Shropshire Jr. and H. Mohr. Springer, Berlin. Chapter 18 (in press)
14. M. Jabben and M.G. Holmes. in Encycl.Pl.Physiol (New Series) "Photomorphogenesis". eds. W. Shropshire Jr. and H. Mohr. Springer, Berlin. Chapter 25 (in press)
15. H. Smith Nature (1980) (in press)
16. H.V. Rice and W.R. Briggs. Pl.Physiol. 51: 927 (1973)
17. B.F. Erlanger (this volume)
18. M.G. Holmes and L. Fukshansky. Plant Cell & Environment 2: 59 (1979)
19. H. Smith and M.G. Holmes Photochem.Photobiol. 25: 547 (1977)
20. M.G. Holmes and H. Smith Nature 254: 512 (1975)
21. M.G. Holmes and H. Smith Photochem.Photobiol 25: 551 (1977)
22. D.C. Morgan and H. Smith Nature 262: 210 (1976)
23. D.C. Morgan and H. Smith Planta 142: 187 (1978)
24. J.S. McLaren and H. Smith Plant, Cell & Environment 1: 61 (1978)
25. G.C. Whitelam, C.B. Johnson and H. Smith. Photochem.Photobiol. 30: 589 (1979)
26. D.C. Morgan and H. Smith Planta 145: 253 (1979)
27. D.C. Morgan and H. Smith Nature 273: 534 (1978)
28. D.C. Morgan, T. O'Brien and H. Smith Planta (in press)
29. D.C. Morgan and H. Smith (unpublished data)
30. E. Schäfer and H. Mohr J. Math.Biol. 1: 9 (1974)
31. H. Smith Nature 227: 655 (1970)

PHYTOCHROME AND GENE EXPRESSION

Hans Mohr

Biological Institute II
University of Freiburg
D-7800 Freiburg, W. Germany

Photomorphogenesis vs. Skotomorphogenesis

Photoautotropic plants have adapted to light during evolution.

The term photomorphogenesis has been used to designate the phenomenon that light controls development (that is, growth, differentiation, and morphogenesis) of a plant independently of photosynthesis (1). In fact, normal development in higher plants is photomorphogenesis. If light is excluded, development follows a conspicuously different route, which can be named (instead of the traditional 'etiolement') 'skotomorphogenesis' - from the Greek skotos = darkness. Clearly, skotomorphogenesis is an adaptation of growth to darkness, while photomorphogenesis is an adaptation of growth to the presence of light (1).

From the point of view of developmental genetics skotomorphogenesis vs. photomorphogenesis represent two different strategies of development on the basis of the same genome (or, genetic information). Naturally, light does not carry any morphogenetic information, but it is an 'elective factor' which deeply influences gene expression in the course of development (2).

Photomorphogenesis vs. skotomorphogenesis can be distinguished not only on the level of the organism but also in the case of intracellular morphogenesis of organelles: As an example, in the presence of light a proplastid develops to a green mature chloroplast while development in darkness follows a different strategy leading to an etioplast (Fig. 1).

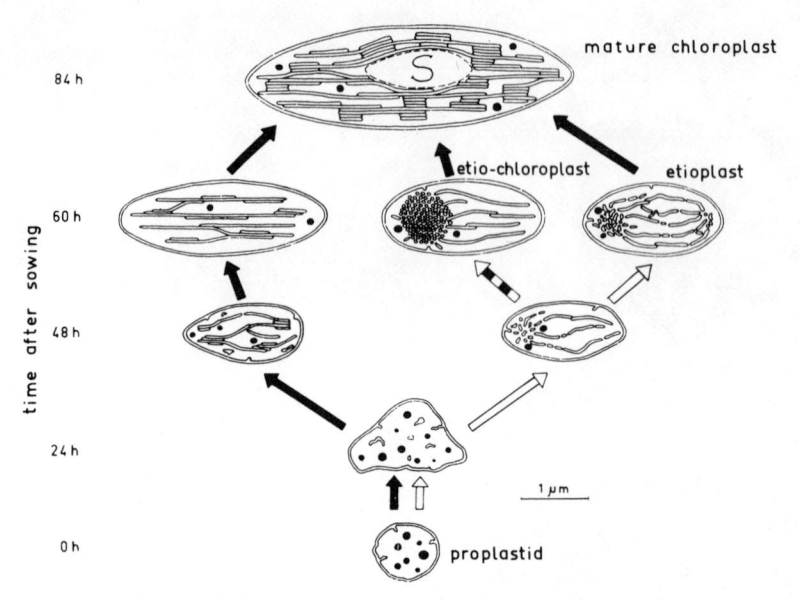

Fig. 1. A scheme showing the time course of plastid development in mustard cotyledons. White arrows: development in complete darkness. Dark arrows: development in continuous white light (7.000 lx). White/dark arrow: treatment of the seedling with 4 red light pulses given 36, 40, 44, and 48 h after sowing to form P_{fr} (and thus to activate the phytochrome system). It is indicated that light activation of phytochrome does not suffice to transform the ultrastructure of an etioplast into that of a chloroplast. From (11), after (14).

Differential phenotypic gene expression is the basis for photomorphogenesis vs. skotomorphogenesis (2). This can easily be demonstrated in thoses cases where a conspicuous trait of a light grown seedling, such as anthocyanin or chlorophyll synthesis, is completely lacking in a dark grown seedling. Clearly, all genes required for anthocyanin or chlorophyll syntheses are present in a dark grown seedling but phenotypic gene expression, i.e. the appearance of the trait, is totally blocked. The question is, why? More precisely: what is the biophysical and/or molecular 'mechanism' through which light affects phenotypic gene expression?

Sensor Pigments

There are at least two sensor pigments in higher plants for the detection of photosignals from the environment and for making use of these signals to regulate gene expression: the phytochrome system and the blue/UV photoreceptor (nickname 'cryptochrome'). While phytochrome is known to be a bluish chromoprotein with photochromic properties, the molecular identity of the blue/UV photoreceptor is not established unambiguously. While evidence points to its being a yellow flavoprotein, it is still defined operationally, in particular by its action spectrum which exhibits three peaks (or shoulders) in the blue spectral range (400 - 500 nm) and another peak at approximately 370 nm in the UV-A range.

Regarding the effector element of the phytochrome system ($P_r \rightleftharpoons P_{fr}$) it can be stated that there is no phytochrome action without P_{fr}. In some thoroughly investigated photoresponses, such as anthocyanin formation, it was found in light-pulse experiments that within the range of the validity of the reciprocity law, the extent of the response is determined by the amount of P_{fr} made by the light pulse (3). Even though in continued light (e.g. in continuous far-red light) cycling of the phytochrome system is believed to contribute to the extent of phytochrome action, this action depends on the presence of P_{fr} as well. Thus, for our present discussion P_{fr} can be considered as 'effector' of the phytochrome system in pulse as well as in long-term light experiments.

Scope of this Lecture

This lecture will consider the 'mechanism' of light-mediated anthocyanin and chlorophyll formation in seedlings of angiosperms. The red and green colors are conspicuous traits which appear during the transition from skoto- to photomorphogenesis and thus represent cases of light-mediated gene expression during this transition.

Anthocyanin Formation in Mustard (Sinapis alba L.)

Seedlings

Multiple action of phytochrome (4). As far as we know at present all photoresponses of the light-grown mustard seedling are consequences of the formation of the effector molecule P_{fr} in the cells of the seedling.

Conspicuous photoresponses include inhibition of hypocotyl lenghtening, enlargement of cotyledons, synthesis of anthocyanin in hypocotyl and cotyledons, and hair formation along the hypocotyl.

As far as we know at present, phytochrome, as a molecule, is the same in all cells of the seedling in which it occurs. But as we observe, the different organs and tissues of the seedling respond differently to the formation of P_{fr}. We have been calling this phenomenon the 'multiple action' of P_{fr}. This is to say that P_{fr} releases a great number of photoresponses in a plant whereby the specificity of the response does not depend on P_{fr} but on the particular competence of a cell or tissue at the moment when P_{fr} is formed (specific 'competence' for P_{fr}).

This crucial aspect of our argument is illustrated in Fig. 2. This figure shows segments of cross-sections through the hypocotyl of mustard seedlings on the left from a dark-grown seedling, on the right from a seedling of the same age which has been kept in continuous far-red light for some time. One sees that under the influence of phytochrome (P_{fr}) certain cells of the epidermis (trichoblasts) have formed long hairs and that all cells of the subepidermal layer - but no other cells - have formed anthocyanin. It is obvious from this simple drawing that phytochrome (P_{fr}) functions only as a 'releasing signal'; the specificity of the response - e.g., hair formation vs. anthocyanin synthesis - depends

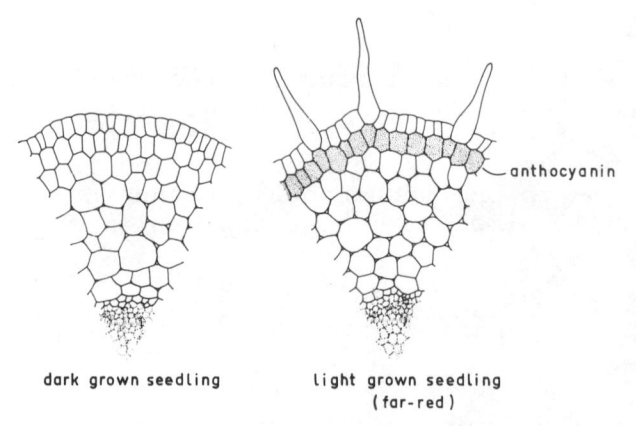

Fig. 2. Segments of cross-sections through the hypocotyl of mustard seedlings (Sinapis alba L.) grown in the dark or in standard far-red light. From (5).

on the specific competence of the epidermal and subepidermal cells at the moment when P_{fr} is formed in the seedling.

Specification and Realization of a Spatial Pattern. Figure 2 suggests that the hypocotylar cells were strictly 'determined' to respond to P_{fr} in a specific manner before the formation of P_{fr}. In view of the fact that the different states of determination of the cells are invisible before the formation of P_{fr} and become visible only after the formation of the releasing signal, we cannot but conceive anthocyanin (or hair) formation as a two-step process: 'pattern specification' (involving all processes which determine a particular cell to respond to the formation of P_{fr} with anthocyanin synthesis or hair formation) and 'pattern realization' (including those processes which lead to actual anthocyanin synthesis or hair formation after the formation of the releasing signal, P_{fr}). The term "pattern" is clearly justified: the cells that are able to form anthocyanin in the light are never distributed randomly within the organism but always form a rigid spatial pattern. The epidermal trichoblasts show a non-random distribution as well which can be described as the spatial pattern of trichoblasts within a cellular matrix of atrichoblasts.

Specification and Realization of a temporal Pattern. Anthocyanin synthesis in the mustard cotyledons takes place only in the epidermal cells (5). There is no anthocyanin synthesis without P_{fr} (6). However, even if P_{fr} is present in the cotyledons from the time of sowing onwards, anthocyanin synthesis only starts at 27 h after sowing (at 25°C) (7). This "starting point" (i.e., the point of time when the epidermal cells become capable of forming anthocyanin as a response to P_{fr} formation) cannot be shifted by light treatments or by the application of nutrients. It seems to be, at a given temperature, a system's constant.

Before 27 h after sowing the epidermal cells of the mustard cotyledons are not competent for P_{fr} with respect to anthocyanin synthesis. This means that before 27 h after sowing the phytochrome system (which functions immediately after sowing of the seed) cannot couple to that cell function which leads to anthocyanin synthesis, even though coupling of phytochrome to other cell functions, e.g. to that which controls growth of the cells, occurs many hours earlier (8).

Induction of Phenylalanine Ammonia-Lyase (PAL)

Phytochrome (P_{fr}) controls the appearance of many enzymes in the mustard cotyledons (1, 2). Among those which are inducible by light via phytochrome is PAL, a major enzyme of phenylpropanoid biogenesis which catalyzes the formation of the secondary phenylpropanoid compound, trans-cinnamic acid, from the primary phenylpropanoid compound, phenylalanin. The question has been whether phytochrome-mediated PAL induction is related to phytochrome-mediated induction of anthocyanin. The answer to this question is very probably, yes, for the following reasons:

1. Starting points and points of loss of full reversibility (coupling points) coincide as one might expect if PAL induction is related to anthocyanin induction (Table 1).

2. Initial and secondary lag-phases are compatible with the concept that PAL induction is a prerequisite of anthocyanin appearance (9).

3. Anthocyanin appearance depends on intact protein and DNA-synthesis (inhibitory effects of Actinomycin D, Puromycin, Cycloheximide) (1).

Mechanism of Phytochrome-mediated PAL Induction.

It must be expected that a light-mediated rise of the level of an active enzyme can be caused by different mechanisms, e.g. by a rise in the level of the mRNA (leading to an increase in the rate of synthesis of the enzyme protein) or by a decrease in the rate of degradation of the enzyme protein.

In the case of PAL from mustard cotyledons density labelling evidence indicates that the half-life of PAL is of the order of 4 h under all circumstances. This means that the time-course of phytochrome-mediated PAL activity must be attributed to changes in the rate of PAL synthesis rather than to changes in the rate constant of enzyme degradation (10).

Hahlbrock, Schröder and associates have measured the time course of PAL mRNA in a cell suspension culture of parsley which had been induced by light to synthesize PAL (Fig. 3). After onset of light the mRNA activity (as measured in a cell-free heterologuous in vitro system, rabbit reticulocyte lysate) strongly increases, reaches a maximum after 7 h, and then declines.

Table 1. Three enzymes from mustard seedling cotyledons are compared with regard to starting points and points of loss of full reversibility. From (4).

Enzyme[1)	Starting point[2)	Point of loss of full reversibility[3) (h)
Anthocyanin	27	26
PAL	27	26
Carboxylase	42	15
GR	48	18

[1) Phenylalanine ammonia-lyase (PAL), very low dark level; ribulosebisphosphate carboxylase (Carboxylase), low dark level; glutathione reductase (GR), considerable dark level.

[2) The point on the time scale[4) where a phytochrome mediated increase of the enzyme level is detectable.

[3) The point on the time scale[4) up to which the inductive effect of continuous red light (given from the time of sowing) remains fully reversible by a saturating 756 nm-light pulse.

[4) The time scale starts with the sowing of the seeds (time after sowing).

The solid line represents the calculated rate of PAL synthesis. Since this curve coincides with the values obtained in the PAL mRNA activity assay it has been concluded that the rate of PAL synthesis is limited by the availability of the PAL mRNA. The available evidence is only consistent with the concept that phytochrome control of the PAL level is exerted exclusively through a control of enzyme synthesis. There is no evidence in favor of alternative concepts, e.g. that phytochrome controls steps such as "activation" of a proenzyme or degradation of PAL.

Figure 4 shows that the time course of PAL levels is not changed markedly if the inductive effect of phytochrome is reduced by a reduction of total phytochrome ($P_{tot} = P_r + P_{fr}$). The time courses are to be explained exclusively by changes in the rate of PAL synthesis. There are no indications that the rate constant of PAL degradation is light (P_{fr}) dependent. The decrease of the PAL levels beyond 48 h after sowing is due to a strong decrease in the inducing effectiveness of P_{fr} This conspicuous reduction must be attributed to the

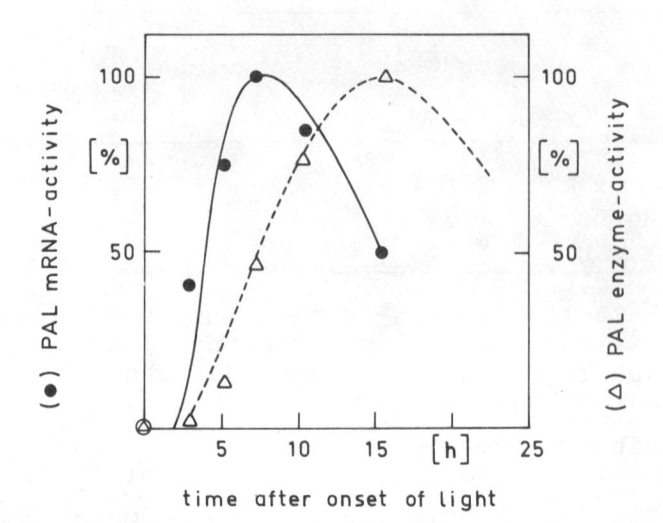

Fig. 3. ●, light-induced changes of the activity of PAL mRNA in a cell suspension culture of parsley (Petroselinum hortense). △, light induced changes of the PAL activity in the same cell suspension culture. The solid line was calculated. It describes the rate at which active PAL is being synthesized. From (17).

operation of "endogenous factors" which determine the temporal changes of responsiveness (= extent of competence).

Chlorophyll Formation in Mustard (Sinapis alba L.) Cotyledons

Most angiosperms cannot synthesize chlorophyll (Chl) a and b in the dark even though a significant pool of PChl is formed in the prothylakoids of the etioplast (see Fig. 1). Obviously, gene expression is arrested at the stage of PChl.

Significant accumulation of photoconvertible protochlorophyll(ide) (= PChl) in the cotyledons of the

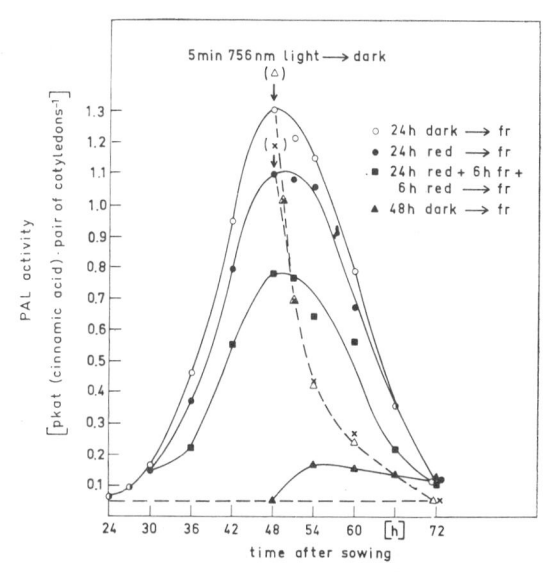

Fig. 4. Time courses (kinetics) of PAL levels in the mustard cotyledons under different light regimes. The light pretreatments before the onset of continuous far-red light (→ fr) serve to lower the level of total phytochrome (P$_{tot}$) and thus the level of [P$_{fr}$] established by the continuous far-red light. In addition, two far-red → dark kinetics are indicated (x, Δ). At the vertical arrows the standard far-red light was turned off and followed by a 5 min 756 nm light pulse. The postirradiation with 756 nm light returns almost all P$_{fr}$ back to P$_r$. Thus, the cotyledons are almost free from P$_{fr}$ at the beginning of the dark period. From (18).

mustard seedling takes place from 24 h after sowing onwards (25°C). The rate of accumulation in darkness is greatly increased by a pretreatment with red light (Fig. 5). The strong effect of continuous red light, given from the time of sowing, remains fully reversible by a 756 nm-light pulse up to slightly less than 18 h after sowing (Fig. 6). This means that up to this point the phytochrome system (which functions immediately after sowing of the seed) cannot couple to the cell function which produces PChl. (We recall at this point that in the case of anthocyanin induction the point of loss of full reversibility was found at 27 h rather than at 18 h, Table 1).

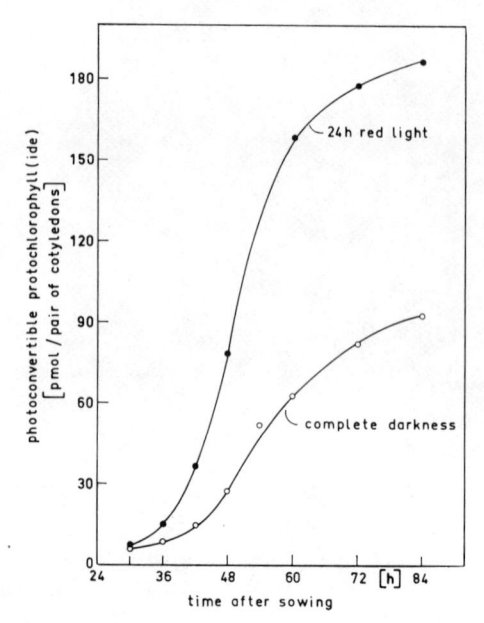

Fig. 5. Accumulation of photoconvertible protochlorophyll(ide) in the mustard seedling cotyledons in complete darkness (o) and in darkness following a red light treatment (●) during the first 24 h after sowing. From (4).

The process of chlorophyll biogenesis is light-dependent in two respects. (1) Photoreduction of protochlorophyllide to chlorophyllide a is an integral step of the biosynthetic sequence leading to chlorophyll a and b. (2) Phytochrome controls the potential rate ("capacity") of the metabolic sequence leading to protochlorophyllide through the control of 5-aminolevulinate (ALA) synthesis, and it controls specifically the flow of chlorophyllide a towards chlorophyll a and b (for references see 11).

It is well known that Chl a or b do not accumulate in the plastid. Rather, the pigment molecules form 'complexes' with particular proteins and become integral constituents of the thylakoid membrane. The light-harvesting Chl a/b protein complex, LHCP, is an example. The question has been how the syntheses of the pigments and of the apoprotein are coordinated.

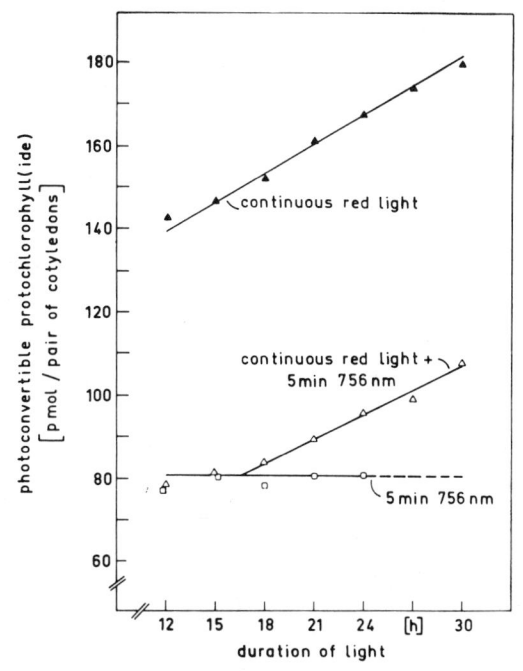

Fig. 6. Action of continuous red light of different durations and of 756 nm-light pulses on the amount of photoconvertible protochlorophyll(ide) 72 h after sowing. Onset of red light at time of sowing. ▲ , continuous red light only; △ , continuous red light, followed by 5 min 756 nm-light; the 756 nm light pulse returns almost all P_{fr} back to P_r. □ , 5 min 756 nm-light only. These pulses have no detectable effects as compared to the dark control (82 pmol photoconvertible protochlorophyll(ide)/pair of cotyledons). From (4).

Apel (12) showed that light (via phytochrome) mediates specifically the appearance in the cytoplasm of a prominent mRNA species which codes for the apoprotein of the light-harvesting chlorophyll a/b protein, LHCP, located within the thylakoid membrane. The mRNA activity is rapidly taken up into the polysomal fraction and translated in the dark. There is evidence that the apoprotein is coded for by nuclear DNA and that a precursor of the apoprotein (molecular weight 29,500) is being synthesized outside the plastid on 80 S ribosomes. The precursor has an apparent molecular weight larger than the authentic apoprotein from the thylakoids by approximately 4,000. Assembly of the LHCP within the plastid and its massive incorporation into the thylakoid membrane take place only under continuous illumination, which allows chlorophyll synthesis. In the absence of Chl, no incorporation of the newly formed LHCP into the thylakoid can be detected. It has been concluded that in the absence of Chl the apoprotein of LHCP is not stable and subjected to rapid degradation (13).

Figure 7 illustrates, as a convenient example, the interconnection of several light-dependent regulatory processes which are involved in the expression of a cell function, in this particular case accumulation of LHCP. Moreover, it is seen that the light-mediated appearance of the trait 'green colour' requires the subtle interplay of several distinct light dependent reactions, in addition to the availability of the pertinent genes.
Moreover, Fig. 1 indicates that the formation of the 'normal' plastidal fine structure (thylakoids, grana) requires phytochrome (P_{fr}) as well as chlorophyll (14).

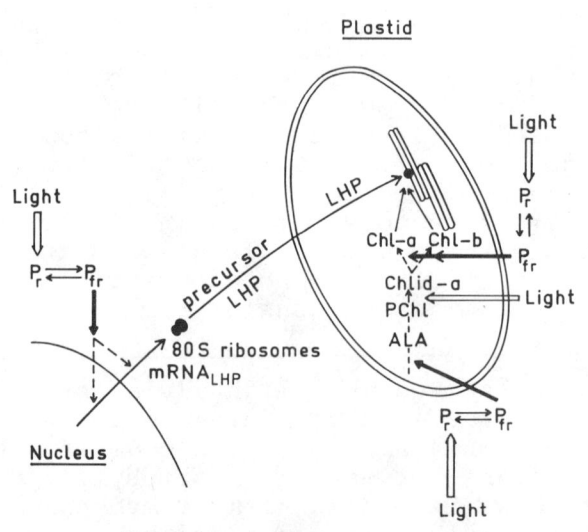

Fig. 7. An illustrating model for the effect of light on the formation of chlorophyll a and b (Chl-a, Chl-b) and on the synthesis of the apoprotein (precursor LHP) of the "Light-Harvesting-Chlorophyll a/b-Protein" (LHCP); (supplemented after (19). Chlid-a, chlorophyllide a; PChl, protochlorophyllide; ALA, 5-aminolevulinate.

Carotenoid Formation in Mustard (Sinapis alba L.) Cotyledons

Expression of the trait 'chlorophyll' (or, 'green color') in natural or fluorescent white light depends on the presence of carotenoids. If carotenoid synthesis is blocked genetically or inhibited specifically by the application of compounds such as Norflurazon (SAN 9789), gene expression with regard to greening is not possible. Why? The following picture has arisen from recent studies:

1. Chlorophyll <u>synthesis</u> in the mustard seedling cotyledons is <u>independent</u> of carotenoid synthesis insofar as Chl synthesis proceeds undisturbed even though carotenoid synthesis is totally inhibited by the application of the herbicide Norflurazon (SAN 9789 (15). While carotenoids are in fact required to protect Chl against self-photooxidation and are thus indispensable for Chl <u>accumulation</u>, Chl synthesis per se can proceed irrespective of concomitant carotenoid synthesis, (15).

2. Some carotenoid synthesis is taking place in an etiolated angiosperm seedling which grows in complete darkness while no Chl is being formed under these circumstances (Fig. 8). Carotenoid synthesis in the dark does not seem to be closely linked to protochlorophyll synthesis since several barley mutants have been identified which cause defects in protochlorophyll synthesis without affecting carotenoid synthesis (for references see 16).

3. Accumulation of carotenoids is promoted to some extent by phytochrome (for references see 16).

4. Accumulation of larger amounts of carotenoids is limited by the availability of Chl while small amounts are being accumulated even in the complete absence of Chl. This strategy implies first that carotenoids, as protectors against photooxidation, are already available when Chl appears in the thylakoids. Secondly, this strategy implies that carotenoids are being made in large amounts only when Chl is there and requires protection by carotenoids in order to remain functional (Fig. 8). From the fluence rate de-

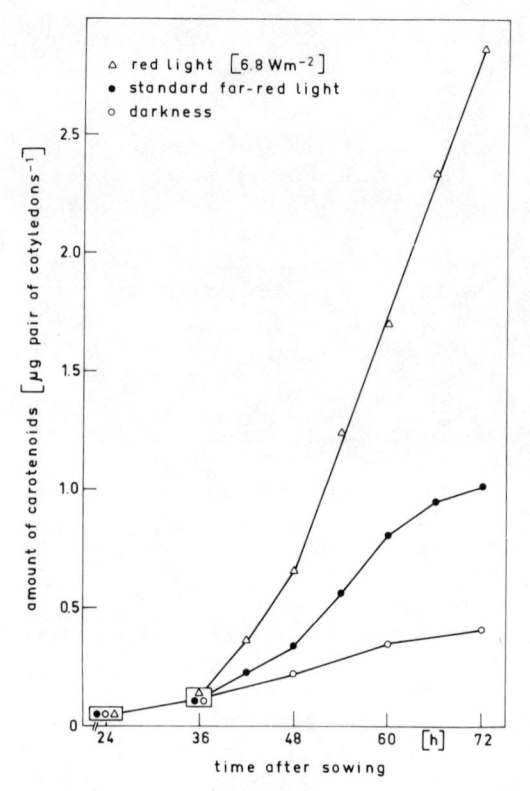

Fig. 8. Time courses of carotenoid accumulation in the mustard seedling cotyledons. The seedlings were kept in continuous light or in complete darkness from the time of sowing. There is no detectable light effect on carotenogenesis before 36 h after sowing. This means that the 'starting point' (see Table 1) is late. From (16).

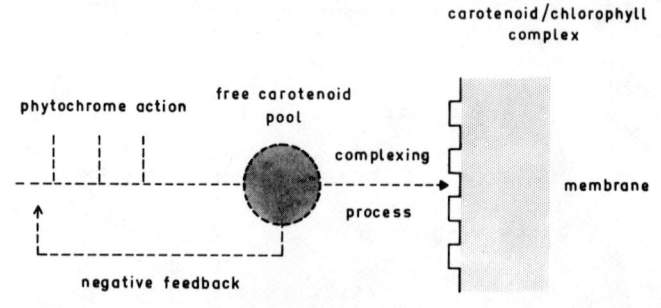

Fig. 9. A scheme to illustrate the proposed regulatory mechanisms involved in light-mediated carotenoid accumulation. Evidence for negative feedback control was discussed by Davies (20) with regard to fungal carotenogenesis.

pendency it can be concluded that carotenoids are always being synthesized somewhat ahead of Chl. Thus, a considerable 'free carotenoid pool' seems to serve as a safety measure (16).

The final scheme (Fig. 9) summarizes the proposed interplay of 'push and pull' regulations in controlling the accumulation of carotenoids. According to this scheme, the push regulation is due to phytochrome-mediated regulation of enzyme levels before a postulated pool of free carotenoids while the pull regulation is due to the complex-forming ability of Chl which drains off the pool of free carotenoids. As a result, light-dependent gene expressions in chlorophyll and carotenoid accumulation are interdependent processes.

CONCLUDING REMARK

Our goal is to understand the 'mechanism' of the transition from skoto- to photomorphogenesis, preferably in terms of differential gene expression.

We believe that anthocyanin synthesis and the appearance of the plastidal mass pigments during the transition from skoto- to photomorphogenesis are most useful biochemical model systems in our effort to understand - in principle at least - the morphogenetic power of light.

REFERENCES

1. H. Mohr, "Lectures on Photomorphogenesis", Springer, Heidelberg-New York (1972).
2. H. Mohr, Phytochrome and gene expression, in: "Cell Compartmentation and Metabolic Channelling", L. Nover, F. Lynen, and K. Mothes, eds., Elsevier, Amsterdam (1980).
3. B. Steinitz, E. Schäfer, E. Drumm, and H. Mohr, Correlation between far-red absorbing phytochrome and response in phytochrome-mediated anthocyanin synthesis, Plant, Cell and Environment 2:159 (1979).
4. H. Mohr, Pattern specification and realization in photomorphogenesis, Bot. Mag. Tokyo Special Issue 1:199 (1978).
5. E. Wagner and H. Mohr, Primäre und sekundäre Differenzierung im Zusammenhang mit der Photomorphogenese von Keimpflanzen (Sinapis alba L.), Planta 71:204 (1966).

6. B. Bühler, Untersuchungen zur Wechselwirkung von Phytochrom und Äthylen bei der Anthocyansynthese des Senfkeimlings (Sinapis alba L.), Doctoral Thesis, University of Freiburg (1977).
7. B. Steinitz, H. Drumm, and H. Mohr, The appearance of competence for phytochrome-mediated anthocyanin synthesis in the cotyledons of Sinapis alba L., Planta 130=23 (1976).
8. U. Wunsch, Der Effekt einer frühen Phytochrom-Prägung auf das spätere Hypokotylwachstum von Sinapis alba L., Diploma Thesis, University of Freiburg (1980).
9. R. Schmidt, personal communication (1980).
10. W.-F. Tong and P. Schopfer, Phytochrome-mediated de novo synthesis of phenylalanine ammonia-lyase: an approach using pre-induced mustard seedlings, Proc. Natl. Acad. Sci. U.S.A. 73:4017 (1976).
11. H. Mohr, Phytochrome and chloroplast development, Endeavour, New Series Volume 1:107 (1977).
12. K. Apel, Phytochrome-induced appearance of mRNA activity for the apoprotein of the light-harvesting chlorophyll a/b protein of barley (Hordeum vulgare), Eur. J. Biochem. 97:183 (1979).
13. K. Apel and K. Kloppstech, personal communication (1980).
14. C. Girnth, R. Bergfeld, and H. Kasemir, Phytochrome-mediated control of grana and stroma thylakoid formation in plastids of mustard cotyledons, Planta 141:191 (1978).
15. S. Frosch, M. Jabben, R. Bergfeld, H. Kleinig, and H. Mohr, Inhibition of carotenoid biosynthesis by the herbicide SAN 9789 and its consequences for the action of phytochrome on plastogenesis, Planta 145:497 (1979).
16. S. Frosch and H. Mohr, Analysis of light-controlled accumulation of carotenoids in mustard (Sinapis alba L.) seedlings, Planta 148:279 (1980).
17. J. Schroeder, Licht-induzierte messenger RNA in Pflanzen, Biologie in unserer Zeit 8:147 (1978).
18. H. Mohr, H. Drumm, R. Schmidt, and B. Steinitz, The effect of light pretreatments on phytochrome-mediated induction of anthocyanin and of phenylalanine ammonia-lyase, Planta 146:369 (1979).
19. K. Apel and K. Kloppstech, Light-induced appearance of mRNA coding for the apoprotein of the light-harvesting chlorophyll a/b protein, in: "Chloroplast Development", G. Akoyunoglou and J.H. Argyroudi-Akoyunoglou, eds., Elsevier, Amsterdam (1978).
20. B.H. Davies, Carotenoids - aspects of biosynthesis and enzymology, Ber. Dtsch. Bot. Ges. 88:7 (1975).

THE TRENDS AND FUTURE OF PHOTOBIOLOGY : PHYSIOLOGICAL ASPECTS

Daphne Vince Prue

Glasshouse Crops Research Institute
Littlehampton,
West Sussex, England

Attempting to sum up physiological aspects at a Photobiology Congress is a daunting task. Physiological responses of organisms are the final outcome of a series of events beginning with the absorption of light quanta:-

light stimulus → photoreceptor → transduction → effectors → physiological response.

They are the farthest point from the primary photochemical events and, even where these are similar, a multiplicity of effector chains may be involved in bringing about the observed response. Organisms are often characterised by diversity of response at the physiological level even where the underlying biochemistry and biophysics are essentially the same; to impose a common theme can thus be misleading. However this summary, which is only illustrative of the material presented at the Congress, will point to similarities between organisms rather than emphasise their differences.

Apart from its potential damaging effect, light is used by organisms as a source of energy for biological processes or as a signal giving information about their environment. M. Chabre emphasised that these two types of response have different characteristics. Where light is used as a signal, there is high sensitivity but low efficiency, information is stored (often as conformational change) and low thermal noise is important implying a high energy barrier. Where light is a source of energy, efficiency is high, energy is collected, transferred and summed, and there is no thermal noise problem. Thus one may expect many differences in the two systems. The main biological energy collecting process is, of course, photosynthesis and some physiological aspects were considered in the symposium on biosolar conversion; these will be discussed later.

Light as an informational signal for physiological processes can be considered in relation to the following questions:-

what kinds of processes are photoregulated?

what kinds of light signals are responded to?

how are the responses regulated at the cellular level?

how are they regulated at the whole organism level?

WHAT KINDS OF PROCESSES ARE PHOTOREGULATED?

Photosensitive systems are ubiquitous in living organisms and the range of physiological processes regulated by light is enormous. This range was amply illustrated both in the symposia and poster presentations. Among those referred to were photomovements in single celled organisms, both directional and non-directional; vision; growth and pigment responses of plants to light quality, and organogenesis in plant callus cultures; responses of both plants and animals to photoperiod, including a photoperiod-like response in the control of fruiting body development in the basidiomycete fungus, *Coprinus congregatus*; rhythmic responses in which light acts as a synchronising signal including reference to many circadian systems and also to less well documented circannual systems such as reproduction in the sub-tropical bird, *Lonchura punctulata* which shows a periodicity of about 10 months irrespective of photoperiod. Physiological and psychological responses of man to low doses of erythemogenic radiation were also reported.

Among such a variety of physiological responses to light it is useful to distinguish between developmental 'once and for all' events (e.g. light-induced seed germination or seasonal changes in reproductive cycles in animals and plants in response to photoperiod) and modulating effects of light (e.g. step down/step up photophobic responses of phototaxis or, in higher plants, leaflet movements and stem elongation rate). The effector chains are likely to be very different since the former involve cellular differentiation.

WHAT KINDS OF LIGHT SIGNALS ARE RESPONDED TO?

In the posters and symposia many examples of responses to irradiance, wavelength, duration, and direction of light and its plane of polarisation were given. In his review of photomovements, F. Lenci pointed out that in simple microorganisms such as the blue-green algae and photosynthetic bacteria, photomotile responses to changes in intensity are sensed through the photosynthetic system and this multipurpose characteristic and lack of specialisation of the photopigment apparatus may be related to the low level of evolution of prokaryotic organisms. Orientated movement towards or away from a light source implies the capability of detecting the direction of the light stimulus. Motile microorganisms may do this by comparing light simultaneously impinging on two separated photoreceptors or by comparing a temporal variation in light falling on a single photoreceptor unit. In plants, direction appears to be sensed by attenuation of light across the tissue, by refraction through a lens-like structure or by orientated photoreceptor molecules.

Intensity sensing in plants was discussed at a 'phytochrome' round table. Phytochrome, a photochromic pigment found in all higher plants, appears to detect the red/far-red ratio of environmental light. Under natural conditions (reviewed by H. Smith), plants are shaded by the foliage of other plants: the preferential absorption by chlorophyll of the red and blue wavelengths of sunlight means that light penetrating a plant canopy is enriched in far-red wavelengths (700-750 nm). This reduction in the ratio of red to far-red light establishes a phytochrome photoequilibrium with a low ratio of Pfr (the far-red absorbing form of the pigment) to Pr (the red absorbing form of the pigment). The extension growth rate of stems in many species is inversely proportional to the ratio of Pfr:Pr established in the tissue. Thus plants appear to utilise a wavelength sensing mechanism to detect canopy shade; there was no agreement, however, as to whether this is detected through the ratio of Pfr to Pr (which implies two vectorial actions) or through the amount of Pfr present in the tissue.

HOW ARE THE RESPONSES REGULATED AT THE CELLULAR LEVEL?

B. F. Erlanger pointed out that a common attribute of photosensitive systems is the regulation of a biologically active macromolecule by a low molecular weight compound capable of assuming at least two states, the relative concentration of each being a function of wavelength (as described above for the phytochrome system in plants). He postulated that the high molecular weight compound could be either an enzyme or a membrane protein which affects membrane properties, and described model systems for both.

It is possible to devise model systems in which enzyme activity is regulated by light and such a type of regulation may be the cellular basis for some physiological responses. An *in vitro* system in which a purified GTPase incubated with purified rhodopsin showed marked activation of the enzyme by light was described by H. Shichi and R. L. Somers. GTPase activity was manifested as a consequence of the formation of a protein-rhodopsin complex and this did not require a membrane environment. Nevertheless, *in vivo*, rhodopsin occurs in membranes and functions at membrane sites. M. Chabre reviewed current ideas of the *in vivo* mechanism of rhodopsin action and concluded that the weight of present evidence indicates that rhodopsin floats freely in a fluid membrane and, on photoexcitation, interacts with an enzyme chain controlling cyclic GMP levels. In this view, the primary action of rhodopsin is on enzyme activity. The role of the membrane is to provide a favourable environment for molecular interactions and amplification of the signal, and its properties are not directly influenced by light. An alternative hypothesis is that rhodopsin is a transmembrane protein which acts as a calcium ionophore: however, although the transmembrane disposition of rhodopsin is supported by various types of experiment, no sizeable release of calcium from the disks follows irradiation. Thus the mechanism of action of rhodopsin may be purely enzymatic leading on irradiation to altered levels of cyclic GMP which controls the sodium permeability of the outer segment membrane. If this hypothesis is correct, it provides a biological example of the first model of photoreceptor action proposed by Erlanger.

The second model system described by Erlanger illustrated the possibility that light acts *via* the regulation of membrane properties. Using the electrogenic membrane of the electric eel, acetylcholine was replaced by a synthetic photoisomerizable compound, Bis Q; the *trans* isomer acted like acetylcholine as an agonist which opens an ion gate and leads to depolarization, while the *cis* isomer had no effect. The system thus offers a model for the direct photomodulation of ion channels. Several *in vivo* examples of such a cellular mode of light action were considered during the Congress and, at present, models involving the photoregulation of membrane properties, particularly with respect to Ca^{2+} transport, are receiving much attention and considerable support. For example, F. Lenci concluded that there is good evidence that Ca^{2+} ions play a crucial role in the phototactic movement of *Chlamydomonas* and speculated that the photoreceptive pigment may serve as a light-activated gate for the passage of Ca^{2+} across the cellular membrane. In *Euglena*, on the other hand, although photophobic responses are triggered by a transient increase in Ca^{2+} concentration in the intraflagellar space, light does not appear directly to regulate the Ca^{2+} permeability of the paraflagellar body membrane; the possibility of a light-activated Na^+/K^+ pump was suggested.

Another example, with new data, was presented by S. J. Roux and C. C. Hale in the poster session. A mitochondrial preparation from oats showed a phytochrome modulated uptake of Ca^{2+} into the mitochondria with the evidence supporting a Pfr-induced efflux. The authors presented evidence that a plasmamembrane Ca ATPase in oats is modulated both by Pfr and calmodulin, and suggested that a light-dependent (= Pfr dependent) increase in cytosolic calcium could increase plasmamembrane ATPase activity *via* calmodulin. This hypothesis, therefore, is close to Erlanger's second model and assumes that the photoisomerization of phytochrome to Pfr leads to changes in membrane permeability to Ca^{2+} ions; enzyme activation is thought to be a secondary event. It is interesting to make comparisons with the calcium hypothesis for rhodopsin action in the vertebrate rod which now seems not to be supported by more recent evidence. However, as reviewed by P. H. Quail, a particular problem with phytochrome is that, unlike rhodopsin, its precise cellular location is still not known. Although it appears to be a largely cytosolic protein at least in dark-grown plants, Quail presented new evidence consistent with the concept that phytochrome is an integral membrane protein which is dissociated during normal extraction procedures. If this proves to be correct, the mode of phytochrome action may be easier to understand. P-S. Song suggested that the photoformation of Pfr may expose a photophobic site which can interact with particular components of the membrane. Quail reviewed the evidence that light causes an increased association of phytochrome with pelletable material which is largely membrane but concluded that the nature of the association and its functional meaning in the cell remains obscure. The possibility that phytochrome is always associated with membranes and that light renders it less easily dissociated during extracts is an intriguing one.

HOW ARE RESPONSES REGULATED AT THE WHOLE ORGANISM LEVEL?

Photoregulation at the whole organism level involves interactions between cells and organs. A complex interaction of this kind in plants was described by S. Obrenovic, where the expression of red/far-red reversibility in the stimulation of betacyanin formation in *Amaranthus* seedlings was shown to be dependent on the presence of the basal part even though photoperception occurred in the apical part of the seedling.

The problems of photocontrol at the whole organism level were considered by H. Mohr, who made particular reference to the complexities of the integration of the various light regulated steps in chloroplast development in plants. He emphasised that an important but often neglected aspect which can complicate the analysis of the chain of events from light absorption to observed response is that the latter is a function of the cell in which the photoevent occurs. Thus, at different stages of development of the organism, the overt responses can vary even though the primary photochemical processes remain the same. An example of this can be given from avian physiology where, after prolonged exposures there is loss of the sexual activity response to long photoperiods. I. Assenbacher *et al.* indicated that such photorefractoriness may be related to seasonal hyperthyroidism.

One of the commonest features of multicellular organisms is to exhibit a circadian rhythmicity in behaviour and/or function. Such rhythms also occur in single celled organisms, as in the *Acetabularia* rhythm of photosynthesis discussed by H. G. Schweiger. He presented evidence that, in this organism, the clock is not located in the nucleus and proposed that there is a rhythm of translation on 80s ribosomes leading to the formation of a particular polypeptide which is incorporated into membranes and affects their properties. Prokaryotes, with 70s ribosomes, do not have circadian rhythms. The problem for multicellular organisms is whether there is a master clock or only cellular clocks of the kind proposed for *Acetabularia*, which are kept synchronized by daily light signals. This question was considered by Assenbacher et al. who discussed the theory that a master clock which controls central nervous system rhythms is located in the suprachiasmatic nucleus (SCN) of the anterior hypothalamus but concluded that the weight of evidence does not support this concept. The role of the SCN appears to be to integrate signals converging into it. They pointed out that variations in daylength act to synchronize rhythms to particular phase relationship but if organisms are exposed to cycles which deviate too much from 24 hours, entrainment breaks down and the rhythms of the organism lose their mutual relationships. This internal desynchronization suggests that there is no 'master clock' but rather that the clock is the property of individual cells. From results with *Acetabularia*, the present indications are that the clock is not located in the nucleus, but what the clock is and how light interacts with it appear to remain as far from solution as ever.

LIGHT AS AN ENERGY SOURCE

The symposium on biosolar conversion illustrated well the effect of various hierarchies of complexity on physiological response. Despite the enormous amount of effort devoted to understanding the process of photosynthesis, J-L. Prioul reminded the meeting that the main limiting factors for productivity and yield in higher plants are not located in the chloroplast itself (which has a maximum yield of about 20%) but occur at canopy level where the overall annual yield is not usually more than 1-2%. In particular he drew attention on the one hand to the question of light interception and the generation of leaf surface and, on the other, to the partitioning of assimilates into harvestable yield which is often limited by developmental processes such as the transition to reproduction or storage organ formation. He concluded that, at present, improvements in light interception and in the partitioning of assimilates are more likely to lead to increases in productivity than are improvements in the efficiency of the photosynthetic process itself. Some of the components of light interception and assimilate partitioning are themselves regulated by light and are likely to repay further study; for example, photocontrol through the phytochrome system is important in leaf expansion, the transition to reproduction and the initiation of storage organs. J. F. Talling, in discussing aquatic ecosystems, also demonstrated that factors such as low temperature which limits population growth are most important in determining the productivity of phytoplankton.

V - PHOTOSYNTHESIS AND BIOCONVERSION OF SOLAR ENERGY

ORGANIZATION OF THE PHOTOSYNTHETIC PIGMENTS AND TRANSFER OF THE EXCITATION ENERGY

Guy Paillotin, André Verméglio and Jacques Breton

Service de Biophysique, Département de Biologie
Centre d'Etudes Nucléaires de Saclay, B.P. n° 2
91190. Gif-sur-Yvette. France.

INTRODUCTION

In the photosynthetic organisms four different functions are involved in the transduction of light energy into chemical free energy :
  A. Collection of energy by a light harvesting antenna which constitutes the large majority of the pigments.
  B. Transformation of the singlet excitation energy into electronic energy at the level of specialized sites.
  C. Formation of electrochemical potentials required for ATP synthesis.
  D. Separation of oxidized and reduced chemical products.

These functions imply different orders of organization of the photosynthetic apparatus. For instance the primary charge separation (function (B) also called primary reaction, PR) is achieved in a pigment-protein complex (the reaction center, RC) where the photoactive molecules have well defined positions and orientations (1). But in the same time a more "global" order is defined by the positioning of RC's within the membrane: they are orientated in such a way that the primary charge separation occur in the same direction across the membrane. This long range ordering permits the essential compartimentalization for functions (C) and (D) (2). These notions of local and long range ordering apply also to the function (A): the collection of light energy. The organization of the photosynthetic membrane required for this photophysical process is possibly, the most interesting one to study for the following reasons:
  1) it involves a rather large part of the membrane where the antennae are embedded
  2) singlet excitations (excitons) may migrate over large

distances. Thus their propagation depends on both local and global parameters.
3) different kinds of experimental information can be obtained on the antenna: some are "functional" and concern the propagation of excitons itself (for instance fluorescence measurements). Others are more "structural" and provide direct characterizations of the organization of the antenna (spectroscopy, electron microscopy)
4) finally this function involves regulation processes which imply long range effects occuring at the membrane level. The most striking one depends on the cation concentration, especially the Mg^{2+} concentration at the photosynthetic membrane surface (3).

According to the large body of experimental information obtained so far, different models of organization have been proposed (4): some are functional and some are structural. However several contradictions appeared between these two kinds of models. The aim of the present paper - essentially concerned with green plants - is to reconcile these different approaches.

BASIC PROPERTIES OF THE ANTENNA

In green plants there are two types of RC's (RC_I and RC_{II}) which work in series to drive electrons from water to $NADP^+$ (5) :

O_2^- evolving complex $\xrightarrow{e^-} RC_{II} \xrightarrow{e^-} RC_I \xrightarrow{e^-} NADP^+$ (CO_2 reduction).

500 antenna chlorophylls (75% Chl a and 25% Chl b) can be statistically associated to each electron transport chain. In this antenna, Chl a (and Chl b) are present as several spectroscopic forms.

Action spectra for the two primary reactions (PR_I and PR_{II}) are different: Chl b contributes essentially to the PR_{II} action spectrum while the long wavelength absorbing Chl a molecules feed the PR_I.

The basic properties of the antenna can be summarized in the framework of the connected units model first suggested by Joliot and Joliot (6). In this model the antenna is organized in different units (the photosynthetic units of PSU) containing about 250 Chl per RC. There are two kinds of units (PSU_I and PSU_{II}) connected respectively to PR_I and PR_{II}. Their pigment composition differ: PSU_{II} contains the Chl b molecules and PSU_I the Chl a. At the scale of one unit the fate of an exciton is essentially determined by the quenching property of the RC. The RC_{II}'s can have two quenching states: high (the open state) and low (the closed state) (for more details see (5)).

Concerning the whole light-harvesting apparatus there are two different kinds of excitation transfers (for details see the review of Williams (7)):
1) the intrasystem transfer: excitation energy can be exchanged between different PSU_{II}'s (and very likely between PSU_I's). This connection depends on the ion

concentration: it increases with the Mg^{2+} concentration.

2) the intersystem transfer: excitation energy transfer takes place from PSU_{II} to PSU_{I}. This directional transfer is called spillover. It also depends on Mg^{2+} concentration. The spillover increases when the Mg^{2+} concentration decreases.

This functional model of organization of the antenna must be taken with some care regarding the topological information it contains : the inner structure of the PSU's is not specified and the existence of morphological units is questionable. As a matter of fact the path of an exciton in the antenna can be very sinous even if it goes very fast from one point to another one. To clarify this point let us analyse the structural constraints imposed by the general properties of excitation migration.

EXCITATION ENERGY TRANSFER : GENERAL PRINCIPLES

The general principles that govern singlet excitation energy transfer in the antenna are now well-established (see the review of Knox (8)). After a very short time (less than 10^{-13} s) a singlet excitation becomes localized and its motion is then incoherent : it diffuses in the antenna by successive jumps. From a local point of view the Förster formula gives a good approximation of the pairwise transfer rate L. L decreases as R^{-6} with the distance R between the donor and acceptor molecules, depends on the mutual orientation of these molecules and on the energy gap which may exist between their first singlet excited states (9). A typical value of L is 10^{12} s^{-1} (R equal 15 Å, angular factor near unity). In comparison the rate of deactivation of the first singlet excited state of a free Chl a is about 2.10^{8} s^{-1}.

Concerning the overall functional organization of the antenna, these general principles have the following consequences:
i) <u>Fast equilibration</u> : if within a small part ( ) of the antenna the equilibration of the excitation amongst the different possible excited states is achieved before any deactivation occurs (including the excitation transfer to the outside), ( ) may be considered as a single site for the excitation propagation. For instance, in the antenna, Chl b is known to be closely associated with the same amount of Chl a within a Chl-protein complex (LHCP). It was shown (10) that the equilibration process within this complex is very fast. Thus since Chl b feeds essentially the RC_{II}'s, the Chl a of the LHCP must also feed the RC_{II}'s.
ii) <u>Localization and incoherence</u> imply two consequences :
   a) the cross section of a RC is a statistical notion and the same Chl molecule can feed two different RC's.
   b) an excitation does not know from where it is coming and where it is going. It follows that an excitation must visit a RC to "know" if it is open or closed. Thus the inter-PSU_{II} connection must be viewed as a transfer by successive jumps of an excitation from a closed RC to an open one (5).

iii) Uphill and downhill transfers. These mechanisms have been extensively discussed by many authors (see the review of Knox (8)). Let us recall that if the propagation of an excitation from a Chl b to a Chl a molecule is "allowed" the reverse propagation is practically forbidden because of the energy gap between their respective first singlet state. Moreover the rate of uphill transfer is strongly temperature dependent (9).

iv) Distance and orientation : If one considers two spatially distinct sets of Chl molecules, an efficient energy transfer between them occurs only if their separation distance is small enough. Typically the separation distance must be smaller than 45 Å if the rate of energy transfer is larger than 10^9 s^{-1}. Conversely the transfer cannot be prevented (except in the case of an uphill transfer) if this distance is small enough.

Very likely the orientation of the different Chl is not an important parameter for the major part of the antenna (8). This orientation may play a role in some regulation processes if the sequence of jumps performed by the excitation is constrained to go through a small and re-orientable part of the antenna (3).

FUNCTIONAL ORGANISATION OF THE ANTENNA

Taking into account these general principles and the most recent experimental informations obtained on energy transfer (see the reviews (5), (7) and (11) one can analyse in more detail the intra and inter-system transfers introduced in a previous section:

i) Intrasystem transfer. Owing to the relative lack of information on the PSU_I we will restrict our discussion to the inter-PSU_{II} transfers : at least four PSU_{II}'s are connected. This connection reflects a center to center propagation of the exciton. In the presence of Mg^{2+} a typical value of the rate of this transfer is 10^9 s^{-1}. It implies that if the PSU_{II}'s are morpholigically well-separated units, their side to side separation distance must be less than 45 Å.

ii) Intersystem transfer. In most cases the experiments which show the spillover actually reveal the possibility of an exciton transfer by successive jumps from a RC_{II} to a PSU_I (see principle concerning localization and incoherence). In the absence of Mg^{2+} a typical value of this transfer rate is 2.10^8 s^{-1}. It implies that the separation distance vetween PSU_{II} and PSU_I is less than 60 Å.

The reverse process (transfer from a RC_I to a PSU_{II}) is "forbidden" (uphill transfer). But there are some lines of evidence in favour of a reversible excitation exchange between the respective antennae of the two photosystems (see the review (5)). It is a PSU_{II} - PSU_I connection and not a RC_I - PSU_{II} or RC_{II} - PSU_I connection.

The following scheme (fig.1) summarizes these different results.

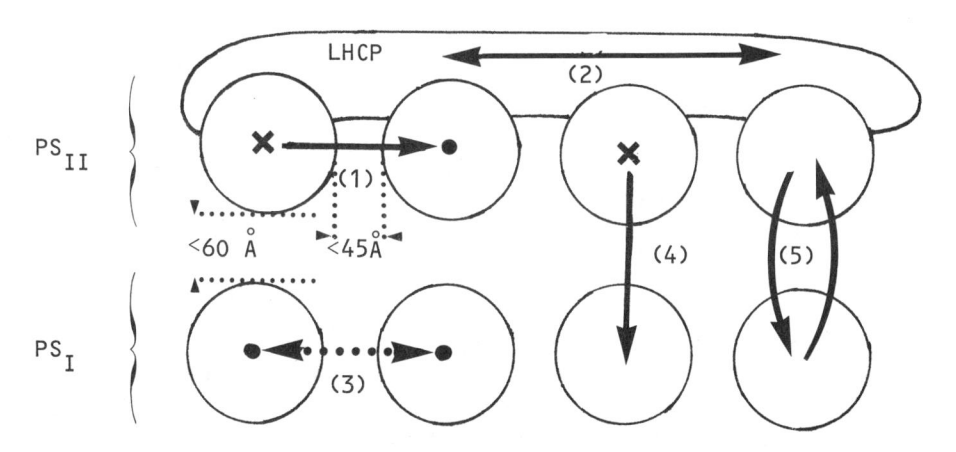

Fig. (1) Up to date functional model (see refs 6, 9 and 11).

(1) Four PS_{II} units connected; (2) LHCP feeds RC_{II}; (3) Inter-PSU_I connection; (4) Spillover $RC_{II} \rightarrow PS_I$ transfer; (5) Reversible exchange between PS_{II} and PS_I antennae.

STRUCTURAL ORGANIZATION OF THE PHOTOSYNTHETIC ANTENNA

The notions of short and long range order are commonly used in Physics. For example in the case of crystals, the short range order (or local order) is the elementary cell (or building block); the long range order results from the propagation of the local order on a three dimensional periodic lattice. For plastic or liquid crystals, only one symmetry element is preserved : regular distribution of centers of gravity, or of one axis of orientation (for details see (13)).

Analogous degrees of order can be inferred in the case of photosynthetic membranes :

i) Local order. Presumably all the chlorophyll molecules are bound to proteins. These building blocks (or elementary pigment protein complexes) have been isolated for both green plants and photosynthetic bacteria. Each building block contains less than 10 Chl. Several results (resonance Raman (14), photoselection and X ray diffraction (15) in some cases) suggest that both positions and orientations of the Chl in these complexes are well defined. This local order is probably due to Chl - protein and/or Chl - Chl interactions (1).

ii) Intermediate order. Aggregate of building blocks seems to occur in vivo. Two types of experiments suggest this intermediary order.

Different types of aggregates can be isolated by the action of mild detergents on the photosynthetic membrane. Besides the light-harvesting complex (LHCP, 3 Chl a + 3 Chl b) which can be obtained in monomeric form, relatively large particles containing either 150 or 40 Chl a in association with the PS_I reaction center (P_{700}) have been isolated. Finally particles of about 100 Chl are probably related to PS_{II} activity (16).

The second line of evidence comes from electron microscopy (freeze fracture photographs). Two kinds of particles can be observed. On the EF (extraplasmic face) there are large particles ($\sim$ 160 Å) associated with PS_{II} activity and LHCP, while on the PF (protoplasmic face) small ($\sim$ 85 Å) particles, very likely related to PS_I photoreactions (17-18), are observed.

Although these observations are still only qualitative, we can put forward that the antennae are organized in four super building blocks: the core of PS_I (CP_I), the core of PS_{II}, PS_I antenna and PS_{II} antenna (LHPC). This type of organization is summarized in the model proposed by Boardman (19).

iii) <u>Long range order.</u> EM pictures give no evidence of a long range conservation of a local ordering parameter (although some quasi crystalline regions are sometimes observed in the EF). There is however a long range ordering for the building blocks: it concerns their orientation with respect to the membrane plane. Linear dichroism measurements on oriented membranes (21) have clearly shown that the antenna Chl molecules are well oriented with respect to the normal of that plane. This preservation of the orientation of a local axis confers to the photosynthetic membrane some resemblance with a smectic liquid crystal.

A super organization of the photosynthetic membranes has been revealed by microscopy. The membranes appear to lie preferentially parallel to each other. To a large extent they are stacked in certain regions (grana) (20). This stacking phenomenon depends on both the ionic and the LHCP concentrations. It is worth mentioning that the PS_{II} activity is confined to these stacked regions.

CONCLUSIONS

Our analysis of the organization of the antenna discussed in the previous sections can be summarized by the following assertions.
1) The antenna consists of different building blocks (Chl - protein complexes) in which a definite local order exists: the Chl molecules are positionally and orientationally ordered
2) These building blocks are oriented with respect to the membrane plane. This orientational order is preserved at long distances.
3) Chloroplast fractionation and electron microscopy studies suggest that the centers of gravity of the building blocks are not distributed at random within the

photosynthetic membrane: they are condensed into aggregates. The most typical of these aggregates are the EF particles associated with the PS_{II} activity.
Two points must be emphasized :
- In these aggregates the building blocks are probably orientationally ordered: the EF particles for instance do not have a spherical symmetry.
- The separation distance between these aggregates must be small enough to allow energy transfer.

This suggests that the distribution of building blocks, although more important in the vicinity of special structural centers, is more continuous than generally assumed (see fig. 2). Moreover the resulting concentration profile of the building blocks depends strongly on the electrical properties of the membrane (ion concentration effects).

4) The composition of the different building blocks and the law of fast equilibration are in favour of the model propounded by W. Butler (11) since the LHCP plus the core of the PSU_{II} contain about 70% of the Chl molecules.

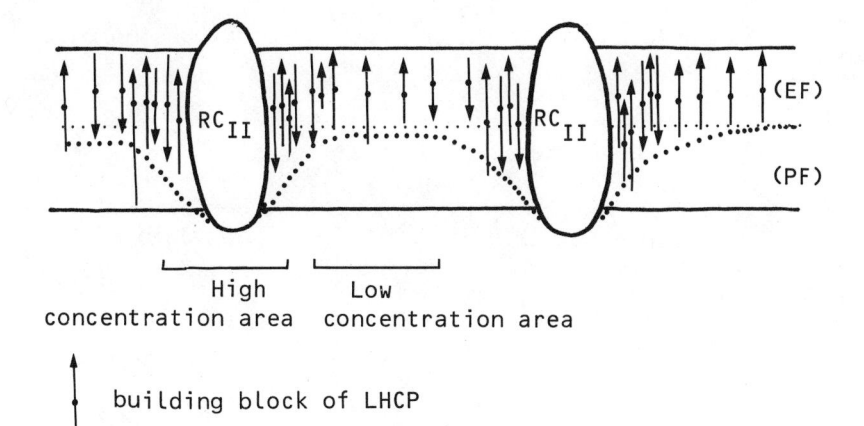

Fig. (2). Schematic organization of the PSU_{II} building blocks.

More experimental information is needed to built an accurate model of the organization of the antenna which has to take into account several unsolved problems (for instance : the balance of energy between the two photosystems and the cation concentration effects). In this respect the study of the exciton migration will probably still be our most fruitfull tool provided the right time scales are used :

. Steady state experiments which mostly average out the structural contribution over the rather large path of exciton diffusion will be sensitive to the feed-back and regulation mechanism involved in the partitioning of the energy among very large pools of pigments.

. Slow relaxation processes concerning the exchange of excitons among the various super-building blocks or even the various PSU's can be followed conveniently by phase fluorimetry (22).

. Within a given super-building blocks, picosecond time scale measurements can provide informations on the exchange of energy among the individual building blocks (23). If polarized light is used the local order between them can be investigated.

. Finally the very fast relaxation processes occuring within the individuals building-blocks might be observed by subpicosecond spectroscopy. However absorption spectroscopy, which is characterized by an otherwise experimentally inachievable time resolution (10^{-15} s) is still our most promising tool.

REFERENCES

1. Breton, J. and Verméglio, A. (1980) in "Integrated Approach to Plant and Bacterial Photosynthesis" (Govindjee ed.) Academic Press, New-York (to be published)
2. Witt, H.T. (1971) Q. Rev. Biophys. $\underline{4}$, 365-477
3. Barber, J. (1979) in Ciba Foundation Symposium 61, pp.283-304 Excerpta Medica, Amsterdam.
4. Thornber, J.P. and Barber J. (1979) in "Topics of Photosynthesis" vol. 3 :"Photosynthesis in relation to model system"(J. Barber, ed.), pp.27-70, Elsevier, Amsterdam.
5. Mathis, P. and Paillotin, G. (1980) in " The Biochemistry of Plants" (P.K. Stumpf and E.E. Conn eds) vol. 3 : "Photosynthesis"(M.D. Hatch and B.K. Boardman,eds), in press,Academic Press, New-York
6. Joliot, A. and Joliot, P. (1964) C.R. Acad. Sci. Paris, $\underline{258\ D}$, 4622-4625
7. Williams, W.P. ibidem ref(3), pp. 100-147
8. Knox, R.S. ibidem ref (3), pp. 55-97
9. Paillotin, G. (1978) in Proceed. 4th Internatl. Congr. on Photosynthesis (D.O. Hall, J. Coombs, T.W. Goodwin,eds) pp. 33-44, The Biochemical Society, London
10. Knox, R.S. and Van Metter, R.L. (1979) ibidem ref(3), pp. 177-190

11. Butler, W.L. (1978) Ann. Rev. Plant Physiol. 29, 345-378
12. Campillo, A.J., Shapiro, S.L., Geacintov, N.E. and Swenberg, C.E. (1977) FEBS Lett. 83, 316-320
13. De Gennes, P.G. (1974) in "The Physics of Liquid Crystals", Clarendon Press, Oxford
14. Lutz, M. ibidem ref (3), pp. 105-125
15. Fenna, R.E. and Matthews, B.W. (1977) Brookhaven Symp. in Biology 28, 170-182
16. Thornber, J.P., Markwell, J.P. and Reinman, s. (1979) Photochem. and Photobiol., 29, 1205-1216
17. Arntzen, C.J. (1978) in "Current Topics in Bioenergetics", vol. 3 (D.R. Sanadi and L.P. Vernon eds) pp.111-160, Academic Press, New-York
18. Staehelin and Arntzen, L.A. ibidem ref(3), pp. 147-175
19. Boardman, N.K., Anderson, J.M. and Goodchild, D.J. ibidem ref (17), vol. 7, pp. 35-109
20. Coombs, J. and Greenwood, A.D. (1976) in "Topics in Photosynthesis" vol. 1 : "The intact chloroplasts" (J. Barber, ed) pp. 1-52, Elsevier, Amsterdam
21. Breton, J., Michel-Villaz, M. and Paillotin, G. (1978) Biochim. Biophys. Acta 314, 42-56
22. Moya, I. (1979) Thesis, Paris-Sud, Orsay.
23. Breton, J. and Geacintov, N.E. (1980) Biochim. Biophys. Acta, Rev. on Bioenerg. (In press).

LASER STUDIES OF PRIMARY PROCESSES IN PHOTOSYNTHESIS

Nicholas E. Geacintov and Jacques Breton[*]

Chemistry Department, New York University,
New York, N.Y. 10003 U.S.A. and
*Département de Biologie, Centre d'Etudes Nucléaires
de Saclay, 91190 Gif-sur-Yvette, France

INTRODUCTION

In recent years the availability of picosecond lasers and of intense lasers of longer pulse duration has provided a new impetus to the study of primary processes in photosynthesis. A number of photophysical phenomena, which can be usually neglected when more conventional light sources are utilized for excitation, must be considered when intense laser beams are incident on a photosynthetic system. We will briefly discuss some of the most important non-linear effects which may occur under these conditions of excitation.

CHARACTERISTICS OF LASERS

Several different types of laser excitation have been used in photosynthesis research. These include single picosecond pulse, mode-locked picosecond pulse train, nanosecond and microsecond duration excitation. In each of these cases, the photophysical processes which may occur depend on the peak power, the total energy, and the duration of the laser pulse.

PHOTOPHYSICAL PROCESSES

Before undertaking an experimental study of laser-induced phenomena in biological systems, the characteristics of the laser excitation in relation to the properties of the biological system to be investigated, should be carefully considered. The phenomenon under investigation can be easily obscured by non-linear photophysical processes which may also be occurring. Some of these possible effects include :

    1.   Two-photon absorption

2. Ground state depletion
3. Excited state absorption
4. Stimulated emission and lasing
5. Irreversible photochemical effects
6. Bimolecular interaction between excited states (Exciton-exciton annihilation).

We now provide a brief discussion of these effects and the possibility of their occurrence in photosynthetic systems, or relevant model systems.

Typical values of the two photon absorption crossections σ_{02} for polycyclic aromatic molecules lie in the range of 10^{-49} - 10^{-50} cm^{4}s (1), while the single photon crossections σ_{01} lie in the range of 10^{-16} - 10^{-18} cm^2. The relative rates of two-photon and one-photon absorption processes are thus $R = (\sigma_{02} I(t))/\sigma_{01}$, where $I(t)$ is the instantaneous photon flux. In typical picosecond pulse experiments (2) utilizing a frequency-doubled Nd:Yag laser (530 nm, 20 ps fwhm), the maximum flux is $I(t) \approx 10^{27}$ photons cm^{-2}s^{-1}, thus $R = 10^{-7}$ - 10^{-4}. Two-photon absorption is thus expected to be important only when σ_{01} is small. Experimentally, it is sometimes difficult to distinguish two-photon absorption processes from the consecutive absorption of two photons in which the second photon is absorbed by an excited state, which itself was created by the absorption of the first photon (1).

Excited state absorption can be observed whenever there is a sufficient population of excited singlet or triplet states. These excited state absorption effects become particularly important when the depopulation of the ground state becomes significant. Since the ground state and excited state (both singlet and triplet) absorption coefficients of chlorophyll a are well known (3), the effects of these processes in photosynthetic systems can be estimated if the absolute intensity and the duration of the laser beam are known. Stimulated emission and lasing of chlorophyll a and bacteriochlorophyll solutions have been reported by Hindman et al. (4). These effects occur in solutions at relatively high laser intensities (instantaneous N$_2$-laser photon flux $I(t) \approx 4 \times 10^{25}$ photons cm^{-2}s^{-1}, 9 ns wide pulse at fwhm), but have not been observed for chlorophyll in photosynthetic systems (5). For chlorophyll in vivo the pigment concentration is about 0.1 M, and lasing effects are difficult to observe because the fluorescence lifetime in vivo is considerably lower than in solution. In addition, at the high intensities required to build up a significant population of excited states, bimolecular exciton-exciton annihilation processes take place which decrease the lifetime still further (5), and thus render the observation of lasing even less likely.

Bimolecular exciton interactions can occur when the average distance between molecules is of the order of ∼ 50-100 Å, or less,

LASER STUDIES OF PRIMARY PROCESSES IN PHOTOSYNTHESIS

i.e. when resonance energy transfer can occur between neighboring molecules. Therefore, exciton annihilation effects can manifest themselves whenever the pigment concentration is of the order of 10^{-2} M or higher. At relatively moderate intensities of laser pulse excitation ($10^{13} - 10^{15}$ photons cm^{-2} pulse^{-1} these exciton-exciton annihilation effects appear to be the most important non-linear phenomena in photosynthetic membranes (2,5-9).

Finally, destruction of chlorophyll in chloroplasts can be observed at high laser pulse intensities ($\sim 10^{18}$ photons cm^{-2}) particularly when microsecond duration pulse, or mode-locked picosecond pulse trains are used (8). The occurrence of these effects can be experimentally checked by remeasuring the fluorescence yield at low excitation levels after exposure of the sample to the high laser pulse intensity, since the fluorescence is proportional to the total number of chlorophyll molecules present.

LASER PULSE PARAMETERS

It is useful to consider the duration of the laser pulse, t_p, in relation to the excited state lifetime τ of the irradiated pigment molecules. In photosynthetic membranes, the fluorescence decay times are of the order of one nanosecond, or less, depending on the system. Thus, considering the commonly available lasers, three possible cases can be distinguished : (1) $t_p \ll \tau$, (2) $t_p \geqslant \tau$, and (3) $t_p \gg \tau$.

SINGLE PICOSECOND PULSE EXCITATION ($t_p \ll \tau$)

The pulse width is typically 10-30 ps and the peak power can be very high. This type of excitation has been utilized by Alfano and his co-workers in New York (10), Campillo and Shapiro at Los Alamos (2), by the present authors at Saclay (9), by Porter and his group in London (11), and by the Rubins (12) in Moscow, to study the fluorescence of photosynthetic systems.

In utilizing picosecond pulse excitation, the intensity of the pulse relative to the absorption crossection σ_{01} at the excitation wavelength must be considered. It is important to note the integrated pulse intensity I at which bleaching of the ground state becomes significant. At these intensities excited state absorption effects can also become important. Neglecting any deactivation processes during the excitation pulse itself, the fraction of molecules in the singlet state $[S_1]/[S_T]$ at the end of the pulse can be estimated according to the simple expression

$$\frac{[S_1]}{[S_T]} = 1 - \exp(-\sigma_{01} I) \quad (1)$$

where S_T is the total concentration of molecules. Using typical absorption spectra of chloroplast samples used in our experiments, and the values of the molar absorption coefficients of chlorophyll in chloroplasts given at several wavelengths (13), we have estimated a value of $\sigma_{01} = 1.6 \times 10^{-17}$ cm^2 for chloroplasts at 530 nm. Using this value, a plot of eq. (1) is shown in fig. 1.

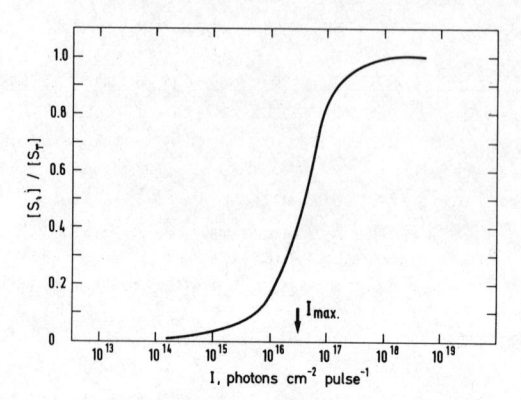

Fig. 1. Fraction of chlorophyll molecules in the excited state in chloroplasts (from eq. 1) as a function of picosecond pulse energy (530 nm) ; see text.

The maximum pulse intensities used by most research groups in their picosecond fluorescence studies are of the order of 3×10^{16} photons cm^{-2} pulse^{-1}, as shown in fig. 1. At these pulse energies the bleaching of the ground state is about 40 %. It is shown in fig. 2, under the same conditions of excitation, that there is a strong decrease in the fluorescence yield (fluorescence intensity/incident pulse intensity) which sets in at about 10^{14} photons cm^{-2} pulse^{-1}. This fluorescence quenching effect, which is accompanied by a sharp decrease in the lifetime (6), is attributed to singlet-singlet annihilation effects rather than to ground state depletion, since the latter becomes important only at much higher pulse intensities. For chlorophyll in solution, under the same conditions of excitation, no fluorescence quenching is observed (fig. 2) ; σ_{01} at 530 nm in solution is slightly lower (1.0×10^{-17} cm^2) than in chloroplasts.

The singlet-singlet annihilation model is summarized by the following processes :

$$S_0 + h\nu \longrightarrow S_1 \qquad (2)$$

$$S_1 + S_1 \xrightarrow{\nu} S_0 + S_1 \qquad (3)$$

Swenberg et al (14) have shown that the fluorescence yield in this model should follow the simple expression :

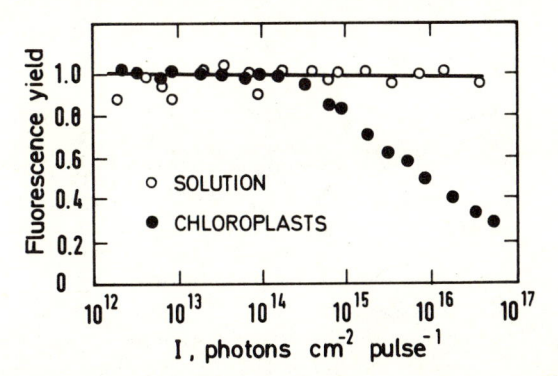

LASER STUDIES OF PRIMARY PROCESSES IN PHOTOSYNTHESIS

Fig. 2. Fluorescence yield as a function of picosecond pulse energy (530 nm) in chloroplasts and for chlorophyll *a* in solution (from reference 5) at room temperature.

$$\phi = \frac{1}{X} \log(1 + X) \tag{4}$$

where $X = I\nu\tau/2$, where ν is the bimolecular annihilation coefficient, and τ is the (low-intensity value) lifetime of the singlet exciton. A fit of this expression to experimental data obtained by Geacintov et al. (9) is shown in fig. 3. In this case the excitation was at 610 nm ($\sigma_{01} = 2.5 \times 10^{-17}$ cm^2), the fluorescence quenching sets in at about 10^{13} photons cm^{-2} pulse^{-1} (only about one photon hit per 2000-3000 chlorophyll molecules), and the extent of quenching is ~ 92 % at the maximum intensity of 2×10^{16} photons cm^{-2} pulse^{-1}. At the highest intensities used, contributions from bleaching and excited state absorption effects cannot be excluded.

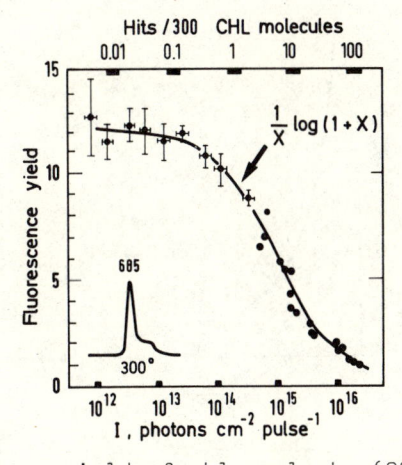

Fig. 3. Fluorescence yield of chloroplasts (300°K) as a function of picosecond pulse energy (610 nm) and theoretical fit according to eq. 4 (from reference 9).

A more complete theory describing exciton annihilation effects has been developed by Paillotin et al. (15).

NANOSECOND PULSE EXCITATION ($t_p \geqslant \tau$)

Typical applications of this type of excitation utilizing N_2 or ruby lasers in the study of photosynthetic systems include those of Mauzerall (16), Monger and Parson (17), Mathis et al. (18) and Hindman et al. (4). Intensity effects, particularly fluorescence quenching, have been considered in these studies. It is nevertheless useful to describe some of the important factors which should be considered with this type of laser excitation. The first difference between this case and the picosecond pulse excitation case is that the formation of triplet excited states can become important. Typical rate constants for intersystem crossing ($S_1 \to T_1$) are 10^9 - 10^8 s^{-1} (1). Thus, in contrast to the short time domain (10^{-11} - 10^{-9} s) following a picosecond pulse, a significant population of triplets may build up (19). Thus, triplet-triplet excited state absorption (3) and singlet-triplet annihilation (19) can become important. Using the known and estimated exciton parameters in chloroplasts, Breton et al. (19) have calculated the relative rates of annihilation by the singlet-singlet and by the singlet-triplet mechanisms as a function of the duration of a square-wave laser pulse (fig. 4) ; it is estimated that the singlet-triplet quenching mechanism begins to dominate the quenching of singlets at $t \geqslant 4$ ns, when $I(t) = 10^{24}$ photons cm^{-2} s^{-1}.

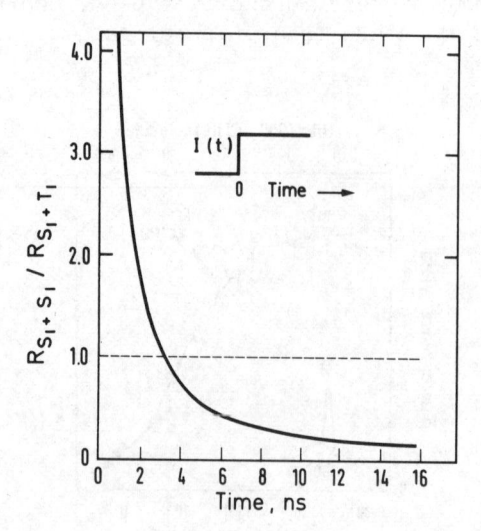

Fig. 4. Calculated relative rates of singlet-singlet to singlet-triplet annihilation rates in chloroplasts as a function of pulse duration of constant intensity $I(t) = 10^{24}$ photons cm^{-2} s^{-1} (from reference 19).

Using a simple model system consisting of solutions of Rhodamine 6G in ethanol, Husiak (20) has made a detailed study of the intensity dependence of the fluorescence yield decrease, transmittance changes during the laser pulse excitation, lasing effects and exciton annihilation. This system was chosen because of its simplicity, since the quantum yield of triplet formation is less than 3 %. Thus, triplet formation can be neglected to a first approximation. For dilute solutions, when exciton-exciton annihilations do not occur, the apparent fluorescence yield decrease due to ground state depletion occurs roughly when $I(t) \leqslant \sigma_{01} \tau$; for 337 nm N_2 pulsed laser excitation the fluorescence yield is well described in terms of a three-level model ($S_0 \to S_1$ absorption, $S_1 \to S_2$ excited state absorption, and $S_1 \to S_0$ fluorescence) ; the fluorescence yield $F(I)$ is given by (the steady-state approximation is assumed, which is a good approximation as long as t_p is at least 3-4 times larger than τ (20)) :

$$F(I) = \frac{1}{1 + \sigma_{01}\tau I(t)} \tag{5}$$

Fits of eq. (5) to experimental data yield values for $\sigma_{01}\tau$ of 2.2×10^{25} cm^2 s, for rhodamine 6G, whereas the independently known parameters σ_{01} and τ give a value of 1×10^{25} cm^2 s which is a reasonably good agreement. In order to avoid ground state depletion and the concomittant excited state absorption effects, it is thus necessary to maintain $I(t) \ll \sigma_{01}\tau$ in laser excitation experiments for which $t_p \gtrsim \tau$. The decrease in $F(I)$ as the laser intensity is increased is accompanied by changes in the transmittance of the rhodamine solutions ; however, this effect is dependent on the wavelength of the excitation. At 525 nm, the transmittance <u>increases</u> as the laser intensity is increased (bleaching). At 480 nm it <u>is independent</u> of excitation, while at 337 nm it is <u>decreased</u> at the <u>high intensities</u>. These effects may be understood in terms of the relative $S_0 \to S_1$ crossection (σ_{01}) and the $S_1 \to S_2$ crossection (σ_{12}). At 525 nm $\sigma_{12} < \sigma_{01}$, thus bleaching occurs as more and more molecules are promoted to S_1 ; at 480 nm $\sigma_{12} = \sigma_{01}$, while at 337 nm $\sigma_{12} > \sigma_{01}$. Fluorescence quenching described approximately by eq. (5) is observed at all of these wavelengths (20). At rhodamine 6G concentrations in excess of 5×10^{-3} M, singlet exciton-exciton annihilations take place. The onset of the decrease in $F(I)$ sets in at lower intensities than predicted by eq. (5), while the transmittance changes set in at higher intensities. These effects are a consequence of the fact that only a small fraction of molecules need to be excited in order to observe exciton-exciton annihilation, while an appreciable fraction must be excited in order to observe fluorescence yield decreases due to bleaching effects. Thus, the lack of a change in the transmittance with a sharp decrease in the fluorescence yield with increasing laser pulse intensity is a good criterion for establishing quenching by exciton annihilation, rather than by a ground state depletion effect. Such a comparison is shown for chloroplasts in fig. 5.

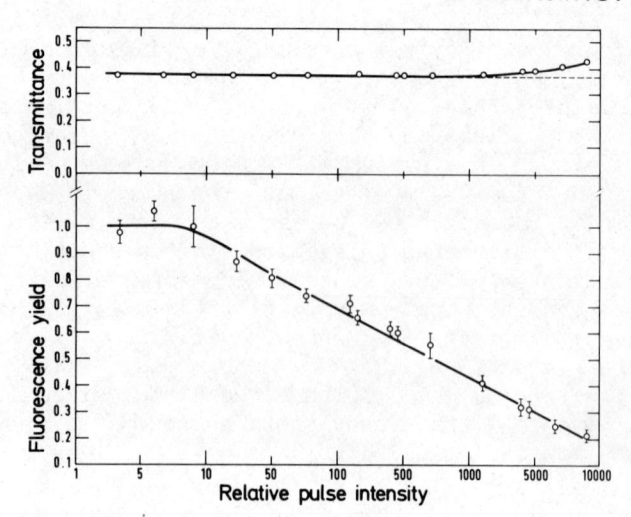

Fig. 5. Comparison of fluorescence yield and transmittance of chloroplasts as a function laser pulse energy (t_p = 1.5 μs, 600 nm) at room temperature.

While the fluorescence yield decreases with increasing pulse intensity, the transmittance remains constant, except at the highest pulse intensities.

MICROSECOND PULSE EXCITATION ($t_p \gg \tau$)

The most pronounced effect of this type of excitation in chloroplasts is that there is a significant buildup of triplet excited states. Thus, using non-mode locked or mode-locked picosecond pulse trains, the most pronounced non-linear effect in chloroplasts is the fluorescence quenching due to singlet-triplet annihilation. The fluorescence yield may be quenched by factors as large as 25 at the maximum pulse intensities (8). It is noteworthy that the buildup of triplets occurs preferentially in photosystem I rather than in the light harvesting pigment bed (7,8). This causes a distortion in the low temperature emission spectrum of chloroplasts since the light harvesting pigments display a fluorescence band at 685 nm, while the emission of pigments in photosystem I occurs at 735 nm ; as the duration, or the intensity of the excitation are increased, the emission at 735 nm decreases relative to the emission at 685 nm. This effect is demonstrated in fig. 6. Emission spectra, recorded with a spectrograph gated image intensifier-optical multichannel analyzer system, are recorded both at the onset and termination of a 100 ns dye laser pulse. During the pulse, triplets build up preferentially in photosystem I causing the relative intensity of the 735 nm emission band to be lower at the end than at the beginning of the 100 ns pulse.

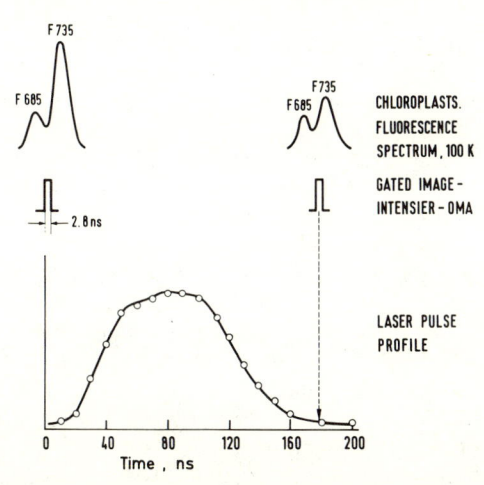

Fig. 6. Fluorescence spectra of spinach chloroplasts (100 K). Detection via a gated image intensifier (2.8 ns window) optical multichannel analyzer spectrograph system.

FLUORESCENCE LIFETIME STUDIES IN CHLOROPLASTS

During the years 1973-1975 the fluorescence decay times for chlorophyll <u>in vivo</u> were consistently lower than those determined by low-level excitation using conventional techniques (phase fluorimetry, single photon counting). In 1976 the effects of intensity on the fluorescence yield were demonstrated (6,16) and it was realized that pulse intensities (6,9) of the order of 10^{13} photons cm^{-2} (single picosecond pulse excitation 530 or 610 nm), or less, should be used to avoid exciton annihilation effects, and the concomittant lowered fluorescence yields and lifetimes. Since then, there are no fundamental disagreements between lifetime measurements obtained by picosecond and other, more conventional techniques (7).

BACTERIAL REACTION CENTERS-ABSORPTION STUDIES

In the past 5 years, the use of picosecond absorption techniques has brought about an increased understanding of the primary processes which occur in reaction center (RC) preparations isolated from photosynthetic bacteria (21). The basic technique involves the monitoring of the kinetics of absorbance changes induced by a picosecond laser pulse. However, most of the absorbance studies have been conducted under conditions of very high energies of picosecond pulse excitation. A reaction center particle contains two bacteriopheophytin, one quinone and four bacteriochlorophyll molecules, two of which constitute a dimeric species (P) which is oxidized within 10 ps or less by the picosecond laser flash. Multiple excitations (photon hits) of such RC particles are possible at high pulse intensities. In fact, it has been demonstrated that under these conditions of excitation (13 photons per RC) the quantum yield of oxidized P is less than under low-level steady state excitation conditions (22) ; this effect was interpreted in terms of a stimulated emission effect. Mauzerall (23), however, proposed that photo-reversal or multitrap effects were involved in this process. It appears that when more than one photon are absorbed by a single RC particle, exciton-exciton annihilation, significant ground state depletion, and the associated excited state absorption effects can occur. All, or some of these effects may be of relevance in explaining some of the fast absorption kinetic phases (∼ 35 ps) which have been observed at 790 nm in reaction centers of R. sphaeroides (R-26), and at 810 nm in reaction centers of R. viridis (21). Such a fast phase (∼ 30 ps) observed at ∼ 800 nm by Shuvalov et al. (24) has been attributed to electron transfer to a bacteriochlorophyll, rather than to a bacteriopheophytin molecule (21). This interpretation, however, may be questioned in view of the recent results of Akhmanov et al. (25). These workers have shown that fast kinetic phases in the picosecond laser-induced absorption occur under conditions of high excitation energies (multiple photon absorption by a single RC) but not at low energies of excitation.

ACKNOWLEDGEMENTS

This work was in part supported by a National Science Foundation Grant PCM-8006109, and in part by the Department of Energy (E(11-1) 2386) at New York University.

REFERENCES

1. J.B. Birks, Photophysics of Aromatic Molecules, Wiley-Interscience, London (1970).
2. A.J. Campillo and S.L. Shapiro, Photochem. Photobiol. 28:975 (1978).
3. J. Baugher, J.C. Hindman and J.J. Katz, Chem. Phys. Lett. 63:159 (1979).
4. J.C. Hindman, P. Kugel, A. Svirmickas and J.J. Katz, Chem. Phys. Lett. 53:197 (1978).
5. N.E. Geacintov, D. Husiak, T. Kolubaev, J. Breton, A.J. Campillo, S.L. Shapiro, K.R. Winn and P.K. Woodbridge, Chem. Phys. Lett. 66:154 (1979).
6. A.J. Campillo, V.H. Kollman and S.L. Shapiro, Science 193:227 (1976).
7. J. Breton and N.E. Geacintov, Biochim. Biophys. Acta Reviews on Bioenergetics, in press.
8. N.E. Geacintov and J. Breton, Biophys. J. 17:1 (1977).
9. N.E. Geacintov, J. Breton, C.E. Swenberg and G. Paillotin, Photochem. Photobiol. 26:619 (1977).
10. F. Pellegrino, W. Yu and R.R. Alfano, Photochem. Photobiol. 28:1007 (1978).

11. G. Porter, C.J. Tredwell, G.F.W. Searle and J. Barber, Biochim. Biophys. Acta 501:232 (1978).
12. L.B. Rubin and A.B. Rubin, Biophys. J. 24:84 (1978).
13. M. Schwartz, Meth. Enzymol. 24:139 (1972).
14. C.E. Swenberg, N.E. Geacintov and M. Pope, Biophys. J. 16:1447 (1976).
15. G. Paillotin, C.E. Swenberg, J. Breton and N.E. Geacintov, Biophys. J. 25:513 (1979).
16. D. Mauzerall, Biophys. J. 16:87 (1976).
17. T.G. Monger and W.W. Parson, Biochim. Biophys. Acta 460:393 (1977).
18. P. Mathis, W.L. Butler and K. Satoh, Photochem. Photobiol. 30:603 (1979).
19. J. Breton, N.E. Geacintov and C.E. Swenberg, Biochim. Biophys. Acta 548:616 (1979).
20. D. Husiak, Dissertation, New York University (1980).
21. D. Holten and M.W. Windsor, Ann. Rev. Biophys. Bioeng. 7:189 (1978).
22. E. Moskowitz and M.M. Malley, Photochem. Photobiol. 27:55 (1978).
23. D. Mauzerall, Photochem. Photobiol. 29:169 (1979).
24. V.A. Shuvalov, A.V. Klevanik, A.V. Sharkov, Yu.A. Maatveetz and P.G. Krukov, FEBS Lett. 91:135 (1978).
25. S.A. Akhmanov, A.Yu. Borisov, R.V. Danielius, R.A. Gadonas, V.S. Kozlowski, A.S. Piskarskas and A.P. Razjivin, Stud. Biophys. Berlin 77:1 (1979).

SYSTEMATIC MODIFICATION OF ELECTRON TRANSFER KINETICS IN A BIOLOGICAL PROTEIN: REPLACEMENT OF THE PRIMARY UBIQUINONE OF PHOTOCHEMICAL REACTION CENTERS WITH OTHER QUINONES

P. Leslie Dutton, M.R. Gunner, and Roger C. Prince

Department of Biochemistry and Biophysics
Medical School
University of Pennsylvania
Philadelphia, PA 19104

INTRODUCTION

The conversion of light energy to electrochemical energy in bacterial photosynthesis (we shall confine our discussion to Rhodopseudomonas sphaeroides) takes place within the photochemical reaction center (RC). Light excitation causes a charge separation which eventually is delocalized across the RC and the biological membrane, and generates an oxidation-reduction (redox) potential difference within the RC. These processes involve a series of electron transfer steps between two molecules of bacteriochlorophyll in the form of a dimer (BChl)$_2$, bacteriopheophytin (BPh), and ubiquinone-10 (Q). Monomeric BChl and possibly another BPh may also be involved, but this remains to be proven. In vivo, cytochrome c_2, an iron porphyrin containing protein, is associated with the reaction center, and the RC: cytochrome c_2 combination is associated with another membrane protein, the ubiquinone-cytochrome b-c_2 oxidoreductase. Reference 1 provides an extensive background.

Figure 1, although not comprehensive, shows the principal events that occur in the RC in solution, when isolated from the photosynthetic membrane and free from cytochrome c_2 and the Q-b/c_2 oxidoreductase. In this isolated state, both the reactions driven by light and the subsequent reactions which return the system to the ground state, occur within the RC protein.

Figure 1 shows the redox states of the (BChl)$_2$ BPh Q components presented on a potential energy scale, and the principal routes for the return of the system to the ground state [(BChl)$_2$ BPh Q]. The rates of electron transfer (shown in solid lines) are virtually

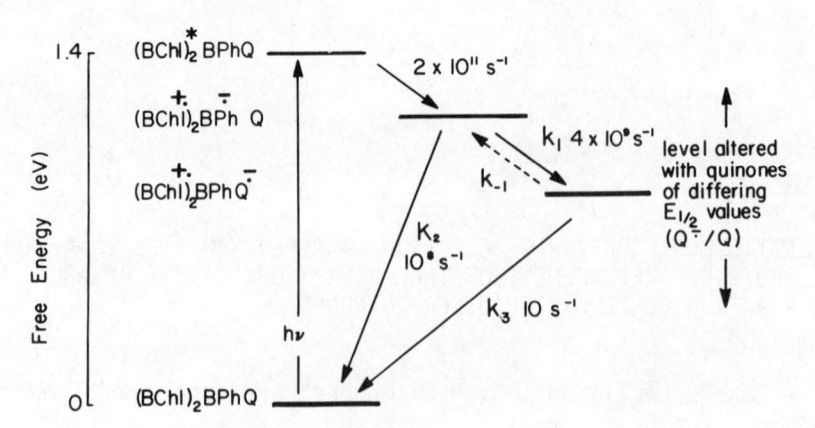

Figure 1. A schematic of the flash induced kinetics and the energy levels of the redox states of the components of reaction centers isolated from Rps. sphaeroides. The value for k_1 and k_3 are those seen in the native reaction centers, and in this paper we will present evidence that they change when ubiquinone is replaced by other quinones. The value for k_2 seems to remain constant under a variety of conditions, including the Q-extracted state (W.W. Parson, personal communication).

temperature independent down to 4°K. (See ref. 2 for detailed discussions). In native RC the 10^8 s^{-1} (k_2) decay of the $(BChl)_2^+$ BPh$^-$Q state back to the ground state (3) does not compete with the 4×10^9 s^{-1} (k_1) forward reaction in which BPh$^-$ reduces Q to form $(BChl)_2^+$ BPh Q$^-$ (4, 5, 6), so k_2 is not a significant pathway unless the forward reaction (k_1) is blocked by removal or chemical reduction of the Q. In the isolated native RC the $(BChl)_2^+$ BPh Q$^-$ state relaxes to the ground state apparently by direct exothermic electron transfer from Q$^-$ to $(BChl)_2^+$. The dashed line, indicating an alternative route to the ground state involving an initially endothermic step from $(BChl)_2^+$ BPh Q$^-$ to $(BChl)_2^+$ BPh·Q, is generally considered a remote possibility in unmodified RCs because the energy difference between these states is presumed too large and k_{-1} cannot compete with k_3, even at 300°K. In this paper we report the results of systematically altering the $(BChl)_2^+$ BPh Q$^-$ energy level with respect to the neighboring $(BChl)_2^+$ BPh$^-$Q and $(BChl)_2$ BPh Q states by replacing the natural ubiquinone with a series of anthraquinones of differing E_m (or $E_{1/2}$) values of the AQ$^-$/AQ couple. We have observed changes in the amount of Q reduced and the rate of $(BChl)_2^+$ reduction, providing the opportunity to better characterize the factors that govern the rates of electron transfer (k_1, k_{-1}, and k_3) in the photochemical reaction center.

Techniques for the reversible removal of ubiquinone from RCs were introduced some years ago (7, 8), when it was also shown that several different quinones, both physiological and non-physiological [namely tetramethylbenzoquinone (8), naphthoquinone (7), and anthraquinone (8)], could function in place of ubiquinone.

MATERIALS AND METHODS

RCs were isolated from Rhodopseudomonas sphaeroides (blue green mutant; R26) using the detergent lauryl dimethylamine-N-oxide (LDAO) (see 1) and purified by ion exchange column chromatography (Whatman DE 52) with final ammonium sulfate fractionation (precipitation), or by the recent modification of Wraight (9); no differences between the methods were encountered. The ubiquinone was removed by the method of Okamura et al. (8). Replacement quinones obtained commercially were purified by treatment with activated carbon, and recrystallized once or twice from ethanol. The quinones, in ethanol, were added to the RCs suspended in 10 mM Tris HCl (pH 8.0); ethanol concentration was always ∼0.5% v/v. Electron transfer kinetic measurements were made in a Johnson Foundation dual wavelength spectrophotometer with a rise time of ∼1 μs. The reactions were activated by a 15 ns Q-switched ruby laser; the flashes were intense enough to activate nearly all the RCs in the reaction cuvette. The electrochemical properties of the replacement quinones were measured in dimethylformamide (DMF) by cyclic voltametry using a Princeton Applied Research Instrument (Model 175).

RATIONALE

The experiments reported here were done by measuring the $(BChl)_2$ oxidation at the wavelength pair 605-540 nm. Since the response time of the spectrophotometer is ∼1 μs, reactions faster than this (k_1 and k_2) will not be recorded.

Figure 2 shows a schematic kinetic trace that could be encountered experimentally, chosen to show all the aspects of the measurements. The dashed part of the trace occurs before the instrumental response time. Within this time a flash generates $(BChl)_2^+ BPh^- Q$ in all RCs; the electron on the BPh^- then partitions between $(BChl)_2^+$ (k_2) and Q (k_1). If the electron goes directly back to $(BChl)_2^+$ the reaction will certainly not be recorded by our spectrophotometer, but if it goes on to the Q the resultant $(BChl)_2^+ BPh Q^-$ state will be more stable and will probably be seen by our spectrophotometer. Thus the yield of the $(BChl)_2^+$ formed stably in the $(BChl)_2^+ BPh Q^-$ state will depend on the rate of k_1 with respect to k_2. There are other major factors that a priori we may expect to influence the yield. For example, the yield will diminish measurably if a redox equilibrium between Q^-/Q and BPh^-/BPh is such that the $E_m(Q) - E_m(BPh)$ is a small (i.e., ∼60 mV) or a negative number. The yield will also depend on the proper positioning and affinity of the AQ in the RC, and therefore, under limiting conditions, on the concentration of the reactants.

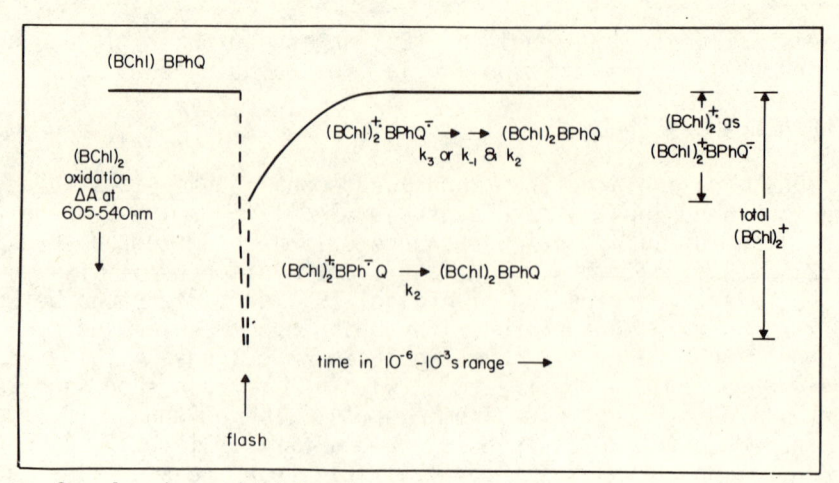

Figure 2. A schematic kinetic trace of flash induced $(BChl)_2$ oxidation and reduction to demonstrate the experimental rationale. The dashed lines represent events that occur before the instrumental reponse time. The solid line is what may be observed in an actual kinetic trace.

The rate constants k_{-1} and k_2 or k_3 will govern not only the lifetime of the $(BChl)_2^{\ddagger}$ BPh Q^{-} state but also the pathway of the return of the electron from Q^{-} to $(BChl)_2^{\ddagger}$. The electron will return to $(BChl)_2^{\ddagger}$ via the state $(BChl)_2^{\ddagger}$ BPh^{-}Q if $k_{-1} \gg k_3$, and will return directly to the ground state if $k_3 \gg k_{-1}$.

RESULTS AND DISCUSSION

Yield of $(BChl)_2^{\ddagger}$ and the Kinetics of Reduction: Figure 3 shows the laser-induced kinetics of $(BChl)_2$ oxidation and subsequent reduction in RCs reconstituted with the series of substituted anthraquinones (AQ) with differing E_m values. At the top of each column are two control traces showing the amount of $(BChl)_2^{\ddagger}$ generated in (a) ubiquinone-extracted reaction centers to which no additions have been made; and (b) after reconstitution with ubiquinone (UQ-6). Comparison of the two traces indicates that ~10% of the reaction center population did not have their ubiquinone removed during extraction. The other traces, starting from the dichloro AQ (top left column) to the diamino AQ (bottom right column), are arranged in order of decreasing E_m value. Two trends are revealed that are shown more clearly in Figures 4 and 5: (a) the rate of $(BChl)_2^{\ddagger}$ reduction becomes faster with lower E_m value of the AQ; (b) the amplitudes of flash-generated $(BChl)_2^{\ddagger}$ diminish from levels that are similar to the ubiquinone control to levels similar to the control in which no addition was made.

Binding Studies: Extensive association-dissociation studies for the RC with each AQ have clearly shown that the kinetic and

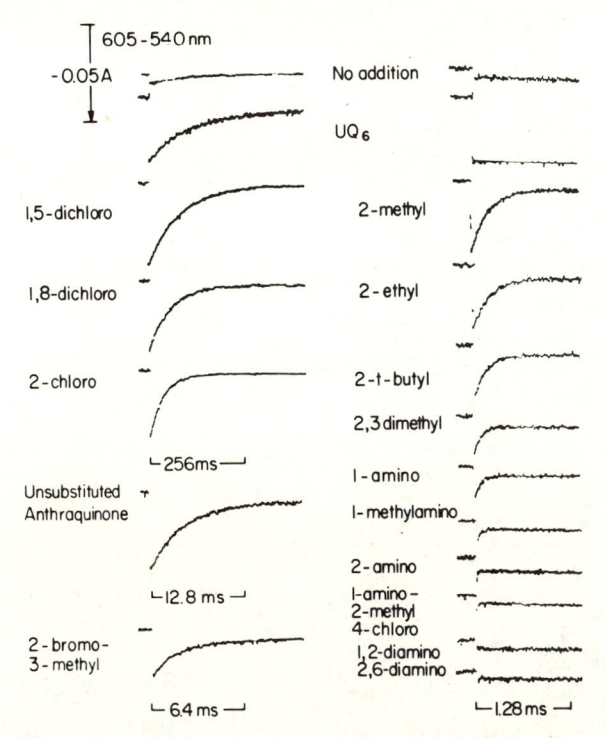

ELECTRON TRANSFER KINETICS IN PHOTOSYNTHESIS 565

Figure 3. Kinetic traces of laser flash induced absorbance changes representing (BChl)$_2$ oxidation-reduction. Quinone-extracted RC (4.6 μM) and replacement quinone (18.4 μM) in 10 mM Tris-HCl, 0.02% LDAO, pH 8. Each trace is the average of 2 flashes 3-5 minutes apart. T = 300°K.

yield trends are not related to binding phenomena. The K$_d$s for the high potential AQs that reconstituted (BChl)$_2$ oxidation were measured directly and were found to be in the 10^{-7} M range. The K$_d$s for the low potential AQs that failed to reconstitute (BChl)$_2$ oxidation were measured indirectly from their effect on the K$_d$ of AQ; the low potential AQs behaved like typical "competitive inhibitors" displaying affinities for the functional quinone site in the RC similar to those of their high potential counterparts. The experiments reported here were done at concentrations above the K$_d$s, and the kinetics obtained were independent of the AQ:RC ratio from 1:1 to 8:1. Thus it is reasonable to consider that the effects seen in Figure 3 arise in some way from the differing E$_m$ values of the AQs, and hence altered energy levels in the RC.

What Governs the Kinetics of (BChl)$_2^{+\cdot}$ Reduction?

As indicated in the Rationale, when the (BChl)$_2^{+\cdot}$ reduction rate is much slower than the 10^8 s^{-1} of k$_2$, the species (BChl)$_2^{+\cdot}$ BPh Q$^{-\cdot}$

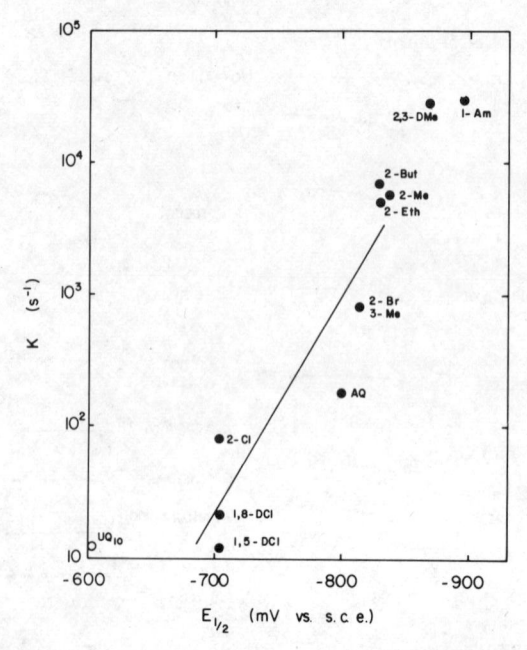

Figure 4. The measured first order rate constant of $(BChl)_2^{+\cdot}$ reduction as a function of the $E_{1/2}$ (E_m) value (measured in DMF) of the quinone/anionic semiquinone couple of the replacement quinone. The theoretical line is discussed in the text.

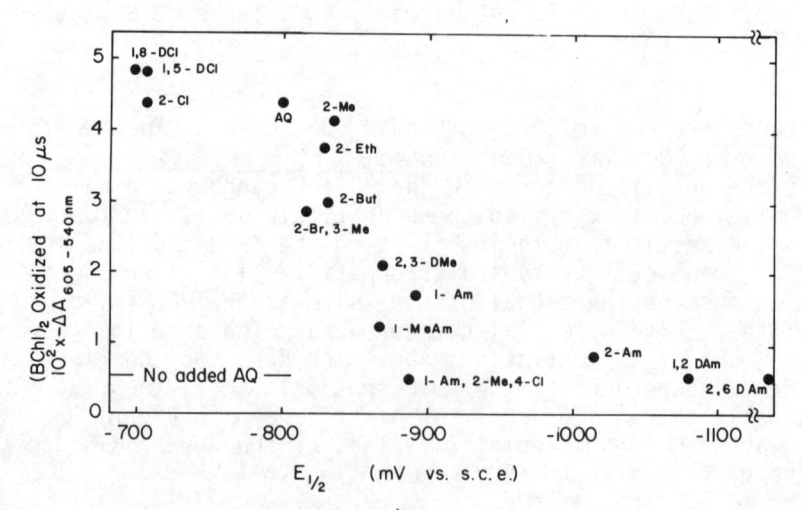

Figure 5. The amount of $(BChl)_2^{+\cdot}$ measured 10 μs after the actinic flash as a function of the $E_{1/2}$ (E_m) of the quinone/anionic semiquinone couple of the replacement quinone.

is formed. This species can decay back to the ground state either directly, via k_3, or via the endothermic reaction k_{-1} to $(BChl)_2^+$ BPh^-Q, followed by k_2. If the electron takes the former route, the reduction of $(BChl)_2^+$ will be a measure of k_3. If instead it takes the latter route, the reduction of $(BChl)_2^+$ can be shown to be $(k_{-1} \cdot k_2)/(k_2 + k_1)$.

In Figure 4 the $(BChl)_2^+$ reduction rate varies from 2×10^4 s^{-1} for the AQs with the lowest E_m values to 10 s^{-1} for the AQs with the highest E_m values. If these represent the endothermic route, these reactions should show temperature dependent rates with activation energies proportional to the "energy gap" between $(BChl)_2^+$ BPh^-Q^- and the transition state. Although much more work needs to be done, temperature dependencies of the reaction rates have been measured between 0 and 40°C on selected cases that cover the range of rates observed. Energies of activation of 280, 325, 335, 370 and 230 mV were obtained for 2,3-dimethyl AQ (E_m -868 mV), 2-t-butyl AQ (E_m -829 mV), 2-methyl AQ (E_m -836 mV) AQ, (E_m -800 mV) and 2-chloro AQ [E_m -703 mV (all referred to the Standard Calomel Electrode in DMF)] respectively. If we subtract the energy of activation (representing the energy difference between the $(BChl)_2^+$ BPh^-Q^- state and the transition state) from the E_m values (representing the $(BChl)_2^+$ BPh^-Q^- level with respect to the $(BChl)_2^+$ BPh^-Q level) we should obtain a number which is the sum of the E_m of the BPh^-/BPh couple (referred to the Standard Calomel Electrode in DMF) plus the energy gap between the $(BChl)_2^+$ BPh^-Q state and the transition state. This number is found to be 1148, 1154, 1171, 1170 and 933 mV for the AQ's listed above. With the exception of the case of 2-chloro AQ, these values are remarkably constant, suggesting that the different AQs do not significantly perturb the transition state or the BPh. We can account for the "anomalous" 2-chloro AQ data by suggesting that in this case the energy gap is sufficiently large (approaching that of UQ_{10} in Figure 4) that both the endergonic route and k_3 are significant. It is quite possible that in the limited temperature range studied we have not resolved the individual reactions, ending up with the average value of 230 mV for the heat of activation. While the k_3 activation energies are not known for all the AQs, it has been shown (8) that the unsubstituted AQ supports $(BChl)_2$ oxidation at helium temperatures with a $(BChl)_2^+$ reduction rate of ~ 5 s^{-1} which is almost certain to be k_3, obviously requiring little or no activation energy. We therefore expect that each AQ will have a characteristic temperature below which the endergonic route becomes slower than k_3. Assuming a k_3 activation energy of 0 mV for unsubstituted AQ; the temperature in this case can be calculated to be 240°K.

Thus, in summary the available evidence tends to support the idea that k_{-1} is feasible with the AQs at ambient temperatures while at lower temperatures k_3 ultimately becomes dominant.

What Governs the Yield of $(BChl)_2^{\ddot{+}}$?

There are two major explanations for the reduced yield of $(BChl)_2^{\ddot{+}}$ with decreasing E_m of the AQ. One is kinetic; that k_1 decreases with respect to k_2. The other is energetic; that the difference in energy levels between $(BChl)_2^{\ddot{+}}$ BPh·Q and $(BChl)_2^{\ddot{+}}$ BPh Q$\bar{\cdot}$ becomes small or negative. We can eliminate the latter because if the decrease in yield was caused solely by equal energy levels in rapid equilibrium, k_1 would not only have to be $\gg k_2$ but also equal to k_{-1} (i.e., $k_{eq} = 1$). If this were the case, the rate determining step for $(BChl)_2^{\ddot{+}}$ reduction would be k_2, and the entire reaction sequence would have been over too quickly to be seen in our spectrophotometer.

Let us now consider the first possibility, that the attenuation in the yield of $(BChl)_2^{\ddot{+}}$ as the E_m value of the AQ becomes lower is caused by a slowing of k_1 (and/or an increase of k_{-1}) so that the forward reaction to $(BChl)_2^{\ddot{+}}$ BPh Q$\bar{\cdot}$ becomes progressively less capable of competing with k_2. It is likely that k_2 is constant throughout the experiment because it varies < 2 fold under a variety of conditions, e.g., Q reduced or absent (W.W. Parson, personal communication). In the simplest case only k_1 will vary, and we may calculate k_1 from the yield (Figure 5) as shown in Figure 6. Taking 2,3-dimethyl AQ as an example, the $(BChl)_2^{\ddot{+}}$ yield is ~ 0.5 and k_1 will be the same as k_2 at 10^8 s^{-1}. The calculated values of k_1 for the different AQs approximate a ten-fold slowing per -60 mV change in E_m of the AQ.

The kinetic explanation for the decrease in yield is further supported by calculations of the back reaction, k_{-1}. The reduction of $(BChl)_2^{\ddot{+}}$ via the endothermic route is proportional to $k_2 \cdot [BPh\bar{\cdot}]$. Making a steady state assumption for $[BPh\bar{\cdot}]$, this expression becomes $([Q\bar{\cdot}] k_{-1} k_2)/(k_1 + k_2)$. Using data for 2-methyl AQ, this expression yields a value of 17 ms^{-1} for k_{-1}, and predicts the changes in rate for the different quinones shown in Figure 4 if k_{-1} remains constant over this range of AQ $E_{\frac{1}{2}}$.

CONCLUDING REMARKS

These preliminary experiments with reaction centers where the endogenous ubiquinone has been replaced with a series of anthraquinones have yielded some novel conclusions. One is that there is a change in the rate of electron flow from BPh$\bar{\cdot}$ to Q which is a function of the midpoint potential of the quinone. A second is that the endothermic reaction from Q$\bar{\cdot}$ to BPh, which has always been assumed to be irrelevant in native reaction centers, appears to be the major route for electron flow from Q$\bar{\cdot}$ to $(BChl)_2^{\ddot{+}}$ in RC possessing quinones of low E_m.

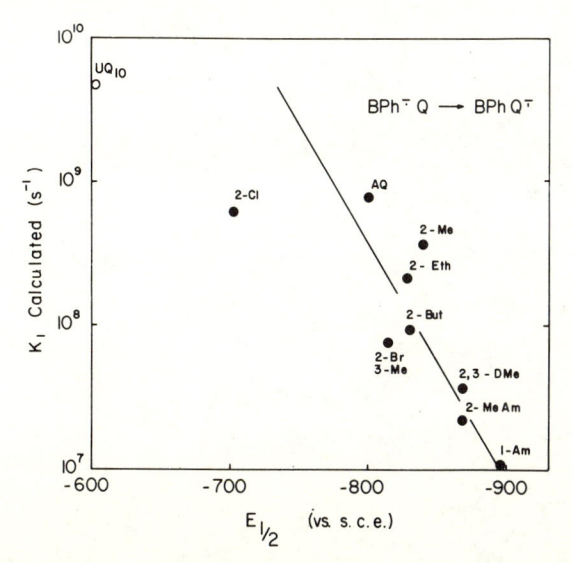

Figure 6. The rate constant (k_1) for the reaction $(BChl)_2^{\pm} BPh^{-}Q$ → $(BChl)_2^{\pm} BPh Q^{-}$ calculated from Figure 5 using the expression $k_1/(k_1 + k_2)$ = Yield of $(BChl)_2^{\pm}$ measured at 10 μs. The line is drawn for a change in rate of a decade per -60 mV change in $E_{\frac{1}{2}}$.

We are currently employing picosecond and nanosecond spectroscopy to directly measure k_1, and are measuring full temperature dependencies to get more information on k_{-1} and k_3. Furthermore we are planning to measure the E_m values of the AQs in vivo which should give us information on the functional E_m of $\overline{BPh^{-}}/BPh$.

Nevertheless, even our preliminary results indicate that replacing the endogenous quinones of the photochemical reaction center can provide a powerful experimental vehicle for testing theories of electron and nuclear tunneling in a biological system where the energetics of the system can be varied at will.

REFERENCES

1. Clayton, R.K. and Sistrom, W.R., in The Photosynthetic Bacteria, Plenum Press, New York (1978).
2. Blankenship, R.E. and Parson, W.W., in Photosynthesis in Relation to Model Systems (Barber, J., ed.) Elsevier, Ch. 3, (1979).
3. Parson, W.W., Clayton, R.K. and Cogdell, R.J., Biochim. Biophys. Acta 387:268 (1975).
4. Kaufman, K.J., Dutton, P.L., Netzel, T.L., Leigh, J.S. and Rentzepis, P.M., Science 188:1301 (1975).

5. Rockley, M.G., Windsor, M.W., Cogdell, R.J. and Parson, W.W., Proc. Natl. Acad. Sci. USA 72:2251 (1975).
6. Kaufman, K.J., Petty, K.M., Dutton, P.L., and Rentzepis, P.M., Biochem. Biophys. Res. Commun 70:839 (1976).
7. Cogdell, R.J., Brune, D.C. and Clayton, R.K., FEBS Letts. 45:344 (1974).
8. Okamura, M.Y., Isaacson, R.A. and Feher, G., Proc. Natl. Acad. Sci. USA 72:3491 (1975).
9. Wraight, C.A., Biochim. Biophys. Acta 548:309 (1979).

ACKNOWLEDGMENTS

This work was supported by grants DE-AC02-80 ER 10590 from the U.S. Department of Energy and PCM 79-09042 from the U.S. National Science Foundation.

DIRECT MEASUREMENT OF LIGHT INDUCED CURRENTS AND POTENTIALS

GENERATED BY BACTERIAL REACTION CENTERS

Nigel K. Packham, P. Leslie Dutton* and Paul Mueller

Eastern Pennsylvania Psychiatric Institute
Department of Molecular Biology
Philadelphia, PA 19129

*University of Pennsylvania
Department of Biochemistry and Biophysics
Philadelphia, PA 19104

INTRODUCTION

Reaction centers (RC) isolated from Rhodopseudomonas sphaeroides have been incorporated into planar phospholipid bilayers separating two aqueous phases by three methods: By fusion of RC containing vesicles with preformed planar bilayers (1); through apposition of phospholipid monolayers containing RCs (2); and through bilayer formation from RC-phospholpid-alkane solutions (3). In a previous report using the last method, we demonstrated by flash activation that the bilayers contain two populations of RCs with opposite orientation (3). Light excitation elicits net electrical currents and potentials only if one of these two populations is blocked either by the absence of electron donors or by chemical oxidation.

Here we report studies revealing which of the individual electron transfer steps associated with the RC contribute directly to the measured currents and potentials. We also report preliminary results on photo currents in asymmetrically oriented RC monolayers sandwiched between two transparent metal or semiconductor electrodes.

MATERIALS AND METHODS

Reaction centers containing a single ubiquinone, Q_I, were isolated as described in the accompanying manuscript. They were incorporated into a phospholipid mixture comprising phosphatidylethanolamine, -serine and -choline in the ratio of 3:1:3; for details see (3). Ubiquinone-10 was added to the membrane-forming solution to reconstitute the partial loss of bound Q_I incurred during the incorporation procedure, and to provide a secondary quinone pool. The bilayer membranes formed from this solution were illuminated through fiber optics with white light; electrical responses were recorded with a voltage or current lamp. Reaction center monolayers (no phospholipid) were spread as monolayers at an air-water interface, compressed to 40 dynes.cm^{-1} and deposited on a tin oxide or metal coated glass slide. A thin layer of platinum or aluminum was then deposited on top of the monolayer and electrical contact was made with the two conductors.

RESULTS AND DISCUSSION

1. <u>Planar Membranes Containing Reaction Centers</u>

   A. Contributions to the transmembrane electric current are made during electron transfer from (BChl)$_2$ to the primary Q but not from the primary to secondary Qs.

   Ferricyanide added to one side chemically oxidizes the (BChl)$_2$ of the RC population it can access; when the (BChl)$_2$ is oxidized, the RCs are functionally inactive. The oppositely oriented RCs are not affected by the ferricyanide and they are active; however, in the absence of an electron donor, they are restricted to a single turnover. Under these conditions, the current transient of Figure 1A is generated by the electron transfer between the (BChl)$_2$ and the added secondary quinones. With the cessation of the light pulse, the current in the opposite direction (reverse current) is most likely due to a back reaction involving the return of an electron from a secondary Q to the (BChl)$_2^{\ddagger}$, and its time course is the same as that recorded spectrophotometrically for (BChl)$_2^{\ddagger}$ reduction by secondary Qs in octane solubilized RCs (3).

   The additions of o-phenanthroline to RCs inhibit electron transfer from Q_I to secondary Qs (4, 5) but does not significantly alter the amplitude and integral of the photo current. The lack of signal attenuation indicates that electron transfer from (BChl)$_2$ to Q_I and not from Q_I to secondary Qs generates the membrane current. However, the inhibitor does decrease the reverse current decay halftime to approximately 60 ms. This halftime is the same as that of the back reaction involving electron transfer from $Q_I^{\bar{\tau}}$ to (BChl)$_2^{\ddagger}$ measured spectroscopically under similar conditions (6).

   B. Electron transfer between ferrocytochrome c and (BChl)$_2^{\ddagger}$ contributes to the transmembrane current.

   Figure 2 shows an experiment providing evidence that the electron transfer between ferrocytochrome c and (BChl)$_2^{\ddagger}$ contributes to

A. No additions

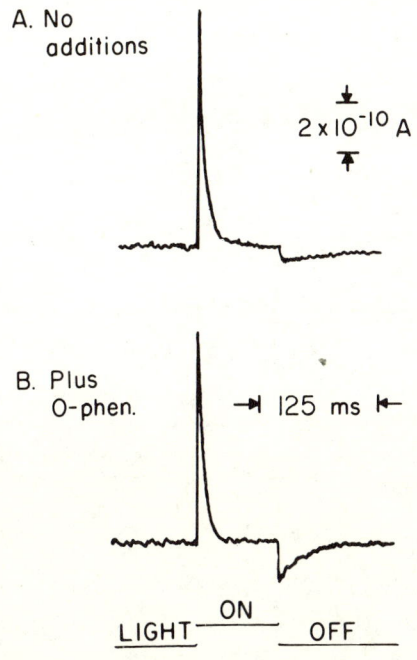

B. Plus O-phen.

→| 125 ms |←

2×10^{-10} A

LIGHT ——ON—— OFF

Figure 1. (A) Photoinduced single turnover current from a bilayer containing approximately 12.3 µM RC and 100 µM ubiquinone-10. The aqueous phase was distilled water pH 6. 2 mM ferricyanide was added to one side. (B) Same as (A), but with 2 mM o-phenanthroline added to the side containing the ferricyanide. Notice that the amplitude and integral of the initial current transient is identical in (A) and (B), whereas the reverse current at the end of illumination has a larger time constant in (A), presumably reflecting the slower rate of electron transfer from secondary Q to $(BChl)_2^+$. Membrane area was 10^{-2} cm^2.

the transmembrane current. RC membranes containing only Q_I (no added Qs) show, after addition of ferricyanide to one side of the membrane, a similar light response due to the electron transfer between $(BChl)_2$ and Q_I as membranes containing extra Qs plus o-phenanthroline. Addition of ferrocytochrome c to the side opposite the ferricyanide causes a 70-80% enhancement in the charging current integral suggesting that electron transfer between ferrocytochrome c and $(BChl)_2^+$ contributes to the measured charge separation.

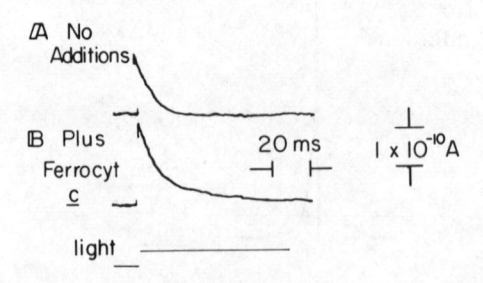

Figure 2. Effect of cytochrome c on the single turnover photocurrent. The membrane had the same RC concentration as in Figure 1 but with no added ubiquinone. Due to a partial loss of Q_I binding, the number of RCs contributing to the light-induced current is smaller than in Figure 1. Aqueous phase as in Figure 1A.
(A) Before,(B),after addition of ferricytochrome c to the side opposite the ferricyanide.

C. Summary of electrical data from planar membranes containing reaction centers.

The single turnover charging current integral of Figure 1 is approximately 8×10^{-10} A.sec.cm^{-2}, corresponding to the transfer of 5×10^9 electrons.cm^{-2}. The RC concentration in the membrane-forming solution was 12.3 µM, from which we can calculate the RC density in the membrane to be 7.4×10^9 cm^{-2}. This value matches the number of RCs which contribute to the charging current. Experimentally, the time constant of the current decay is determined by the light intensity. From the observed time constant of 8 ms (Figure 1) we can calculate an initial absorption of 125 photons. sec^{-1}.RC^{-1}. Assuming a quantum efficiency of 1, the initial current should be 6.5×10^{-8} A.cm^{-2}, a value close to that observed. Several other parameters can be calculated from the charging current integral. For example, a light flash that excited all the RCs within 150 ps, i.e., the halftime of electron transfer between (BChl)$_2$ and Q_I, would generate a current of 2A.cm^{-2}. A calculated electric potential of 2 mV has been confirmed in an experiment where the laser-flash induced electric potential was measured under current-clamp conditions. The magnitude of the light-induced electric potential is limited by the number of reaction centers present within the membrane. In the natural membrane, across which "single turnover" electric potentials of approximately 100 mV have been obtained using indirect measurements (7-10), the average RC density is 2×10^{11} cm^{-2} (7) which is approximately 50-fold greater than in the planar phospholipid membranes.

2. **Photo Currents and Potentials from RC Monolayers Sandwiched Between Two Electrodes**

The device shown in Figure 3 was made by deposition of an RC monolayer between two transparent electrodes as described in the METHOD section. The electrical resistance of this monolayer was 10^8 ohm.cm^2 and its capacity ~ 1 µF.cm^{-2}. Upon illumination the device gives transient currents of similar wave form to those observed in the bilayer, i.e., they have an early peak after onset of illumination and a smaller reverse current on cessation of illumination (Figure 4). There is no steady-state current. Whereas the rise of the current is faster than can be resolved with present instrumentation, the decay has a halftime of 60 msec which is commensurate with the back reaction of the electron from Q_I to the oxidized $(BChl)_2$. Calculated current densities from a 10 nsec flash were 1 A.cm^{-2} and the potential rise was faster than the 3 µsec resolution time of the voltage amplifier. The sign of the current indicates that the electrons move from the hydrophilic to the hydrophobic end of the molecule. Judging from the current integral, less than 1% of the RCs contribute to the current. This may reflect incomplete orientation and/or partial oxidation of the protein.

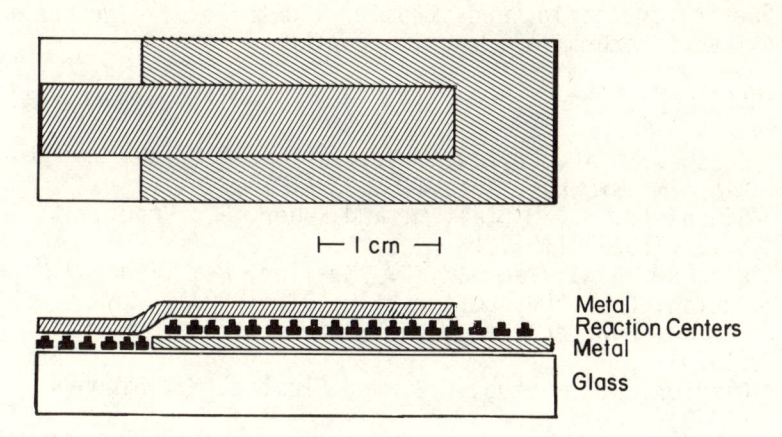

Figure 3. Schematic diagram of RC monolayer sandwiched between two metal electrodes.

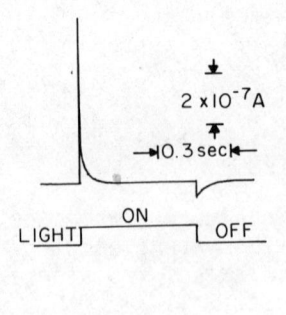

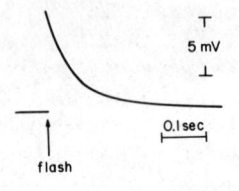

Figure 4. Photo current (upper trace) and photo potential (lower trace elicited by illumination of device. Potential generated by 10 μsec ruby laser pulse. The decay has the same lifetime as the back reaction.

CONCLUDING REMARKS

The results presented in this report support the idea derived from indirect measurements (8-10) that the (BChl)$_2$ is located within the membrane dielectric and that the electron transfer steps from (BChl)$_2$ to Q_I and the cytochrome c to (BChl)$_2^{+}$ cooperate to form an electric potential directed across the supporting membrane.

These techniques hold promise for future investigation into the kinetic, energetic, and structural details of electron transfer within the photochemical reaction center.

REFERENCES

1. Drachev, L.A., Frolov, V.N., Kavlen, A.D. and Skulachev, V.P., Biochim. Biophys. Acta 440:637 (1976).
2. Schonfeld, M., Montal, M. and Feher, G., Proc. Natl. Acad. Sci. 76:6351 (1979).
3. Packham, N.K., Packham, C., Mueller, P., Tiede, D.M. and Dutton, P.L.,FEBS Letters 110:101 (1980).
4. Parson, W.W. and Case, G.D., Biochim. Biophys. Acta 205:232 (1970).
5. Clayton, R.K., Szub, E.Z. and Fleming, M., Biophys. J. 12:64 (1972).
6. Clayton, R.K. and Yau, H.F., Biophys. J. 12:867 (1972).

7. Packham, N.K., Berriman, J.A. and Jackson, J.B., FEBS Letters 89:205 (1978).
8. Dutton, P.L., Biochim. Biophys. Acta 226:63 (1971).
9. Jackson, J.B. and Dutton, P.L., Biochim. Biophys. Acta 325:102 (1973).
10. Takamiya, K. and Dutton, P.L., FEBS Letters 80:279 (1977).

ACKNOWLEDGMENTS

This work was supported by grants DE-AC02-80 ER 10590 from the U.S. Department of Energy and GM 12202 from the U.S. National Institutes of Health.

PRIMARY AND ASSOCIATED REACTIONS IN PHOTOSYSTEM II

H.J. van Gorkom and A.P.G.M. Thielen

Department of Biophysics, Huygens Laboratory of the
State University, P.O. Box 9504, 2300 RA Leiden
The Netherlands

INTRODUCTION

   Photosystem II is the photosynthetic apparatus which oxidizes water and supplies photosystem I with an electron source which is less difficult to oxidize. In addition system II helps to create the proton concentration difference between the in- and outside of the photosynthetic membrane system, which provides the driving force for ATP synthesis. The proton translocation is due to the fact that water is deprotonated at the inside of the membrane system and the intermediate electron carriers between the two photosystems are protonated on the outside. The mechanism of water oxidation is poorly understood, but considerable progress may soon be expected, since it appears that the oxygen evolving enzyme has now been isolated [1]. The kinetics of the process have been studied extensively [2]. The liberation of one oxygen molecule requires four successive photoacts in the same system II reaction center. Of the corresponding oxidation states of the enzyme, designated S_0 to S_4, only S_0 and S_1 are stable in darkness, S_2 and S_3 have lifetimes in the order of a minute, and S_4 releases oxygen in a millisecond.

   Electron transport from system II to system I proceeds via plastohydroquinone, which carries two hydrogen atoms. This fact has a special significance in view of the recently discovered additional proton translocating cycle [3] which is now drastically changing the picture of the intersystem electron transport chain. The reduction of plastoquinone by system II takes two successive photoacts, the first of which produces a stable, bound plastosemiquinone. The semiquinone may disappear through reversal of system II electron transport if a higher S-state is present. The recombination with S_2 takes half a minute.

The primary reactions of photosystem II thus have to trap the excited state produced in the antenna chlorophylls by light absorption and transform it into a ~ 10^9 times more stable charge pair, consisting of an oxidant strong enough to oxidize water and a reductant which can reduce plastoquinone, situated on opposite sides of the membrane, with an observed overall quantum yield of about 80 %.

THE PRIMARY REACTANTS

The first reasonably stable products of system II photochemistry have been identified as the π-cation radical of a chlorophyll a, $P680^+$, and a plastosemiquinone anion, Q^- (for reviews, [4] and [5]). These molecules are tightly bound in the reaction center and their optical and electron spin resonance properties provide some information on their molecular environment. The reduction of Q is accompanied by a blue shift of absorption bands at 420, 545 and 685 nm, which we tentatively ascribed to a pheophytin a molecule, possibly involved as an intermediate electron carrier in the primary reaction like bacteriopheophytin in bacterial reaction centers. Klimov et al. have now accumulated strong evidence [6 and refs. cited there] that photosystem II, when Q is reduced or removed, does reduce a pheophytin a molecule. Its absorption bands have the peak wavelengths and halfwidths predicted from the bandshifts induced by reduction of Q. A close association between the two components is also strongly indicated by an ESR doublet signal which appears when both pheophytin and Q are reduced [6] and is similar in many respects to a signal in reduced bacterial reaction centers, ascribed to interaction of the spins on $BPhe^- UQ^-$ [7,8]. A close proximity of the pheophytin to P680, or at least chlorophyll, follows from the efficient quenching of chlorophyll fluorescence by Phe^- [9], which has considerable absorbance near 680 nm, extending well beyond 800 nm [10]. A similar quenching by $BPhe^-$ is observed in photosynthetic bacteria [11]. The presence of an intermediate between P680 and Q was also proposed on the basis of a delayed fluorescence emission from system II after illumination in the state Q^- [12] and a magnetic field effect on the fluorescence yield [13].

Thus it appears that the primary processes of excitation trapping and charge separation in photosystem II and those in bacteria like Rhodopseudomonas sphaeroides take place in very similar molecular structures. The primary reactions in bacteria from which the reaction centers can be isolated have become a favorite playground for picosecond spectroscopy [14] and fancy magnetic resonance techniques [15]. A detailed evaluation of the current model is given in [16].

EXCITATION TRAPPING AND CHARGE SEPARATION

By analogy to what has been shown for bacteria [11] the increase of the chlorophyll fluorescence yield upon reduction of Q might be ascribed to a delayed emission resulting from charge recombination: $P680^+$ Phe^- $Q^- \to P680^*$ Phe Q^- [9]. It should be noted, however, that the delay is probably small compared to the observed [17,18] 1.5 - 2 ns lifetime of the excited state. The reported 40 - 80 mV activation energy for this emission [19] awaits confirmation and may be taken as a maximum estimate [cf. 11,16]. The equilibrium constant K ($P680^+$ Phe^- $Q^-/P680^*$ Phe Q^-) is at most about 10. The absorption spectrum of P680 is similar to that of the antenna chlorophylls. If the excitation is thermally distributed over the various pigments it would spend less than 1 % of its lifetime on $P680^*$, and therefore less than 10 % of its lifetime in the state $P680^+$ Phe^-. The observed quantum yield of P680 Phe^- Q^- accumulation, 0.2 - 0.5 % [6], is consistent with the $P680^+$ reduction time of ~ 40 ns observed after dark adaptation [20,21]. In the absence of added reductants the same process should occur as well and charge recombination from the state Z^+ P680 Phe^- Q^- might be responsible for the 1 μs luminescence [12] as proposed earlier [22]. This would imply an equilibrium constant K (Z^+ P680 Phe^- Q^-/Z $P680^+$ Phe^- Q^-) of about 20 in the dark-adapted state. It may be less than 10 in the presence of a high S-state, as suggested by a minor 20 - 35 μs phase in $P680^+$ reduction [discussed in 22].

CHARGE RECOMBINATION

In order to explain the low maximum fluorescence yield and its correspondingly short lifetime, about one third of the in vitro values for chlorophyll a, a rapid decay of $P680^+$ Phe^- Q^- must be assumed. The quenching cannot occur in the antenna, because the observed 2.7 % fluorescence yield in the state P680 Phe Q [23], plus the doubtlessly concomitant 5.7 % triplet yield, leave little room for other losses in view of the high quantum yield of Q reduction (see below). If less than 10 % $P680^+$ Phe^- is present during the lifetime of the emission it should decay with a rate constant of more than 3.6×10^9 s^{-1}, assuming that in its absence the excited state would live 5 ns. The same process may be rate limiting for the deactivation of system II in general. In the presence of the inhibitor DCMU the state S_2 Q^- (α centers, see below) decays to S_1 Q with an activation energy of 600 mV and a lifetime of 3.6 s [24]. The frequency factor amounts to 6×10^9 s^{-1} and the activation energy is equal to the midpoint potential difference between Phe/Phe^- (-610 mV [25]) and Q/Q^- (-250 mV [26]) plus the 250 mV midpoint potential difference between S_2/S_1 and $P680^+/P680$ which follows from the ~ 200 μs recombination time of $P680^+$ Q^-. The number of delayed fluorescence quanta emitted per recombination of $S_2Q_\alpha^-$ is 0.08 times the emission yield in the state P680 Phe Q^- (van Gorkom, H.J. and de

Grooth, B.G., unpublished). Prompt and very short-lived delayed (singlet) emission should have a higher yield due to the phase memory of the electron spins in the charge pair, because it originates from a singlet excited state and deactivation of $(P680^+ Phe^-)^T$ via the triplet state of P680 is probably more rapid than the decay of $(P680^+ Phe^-)^S$ [cf. 16]. In addition the equilibrium constant $K ((P680^+ Phe^-)^S/P680^* Phe)$ may be larger in the absence of Q^-. Although a quantitative evaluation would require additional data it seems possible that the quenching by 'closed' system II centers and the recombination which limits the stability of the charge separation in system II in physiological circumstances are both largely due to deactivation of $P680^+ Phe^-$ via the triplet state of P680.

OVERALL PICTURE OF SYSTEM II ELECTRON TRANSPORT

The simplified scheme of system II electron transport given in Fig. 1 clearly illustrates that the reduction of Q is the main stabilizing step. It may also be noted that all electron transfers, except perhaps the oxidation of the component giving rise to ESR signal II v.f. in the presence of a low S-state [cf. 27 and 28], proceed to near equilibrium within the reaction time of the next transfer. Each component is an integral part of the reaction center structure and its actual operating potential may depend on the redox state of all others. The chemical identity of the components at the oxidizing side is still a matter of speculation. At least two redox components of intermediate potential, cytochrome b_{559} and a 400 mV component, are present and may in special conditions react with high quantum yield. Their function in physiological circumstances is not clear. Unless the 400 mV component is in the oxidized state [29,30,31] Q^- has a lifetime of nearly a millisecond. In the presence of Q^- no efficient charge separation can occur. Models invoking reasonably efficient 'double hits' or heterogeneity due to interchangeable conformational states have been proposed by various authors. We believe that much of the confusion is due to the fact that in higher plant chloroplasts a second type of photosystem II occurs.

PHOTOSYSTEM IIβ

Melis and Homann [32] postulated a heterogeneity of photosystem II on the basis of the apparently biphasic kinetics of the fluorescence rise due to Q reduction upon illumination in the presence of DCMU, which consist of a fast sigmoidal phase and a slow exponential one. The fast phase, ascribed to α centers in a matrix of antenna chlorophylls, titrates at -250 mV. The slow phase, ascribed to β centers occurring in separate units, titrates at about -50 mV [33,34]. The concomitant absorbance changes revealed no large differences between $Q\alpha$ and $Q\beta$ [35,36]. We have extended these studies and our results may be briefly summarized as follows.

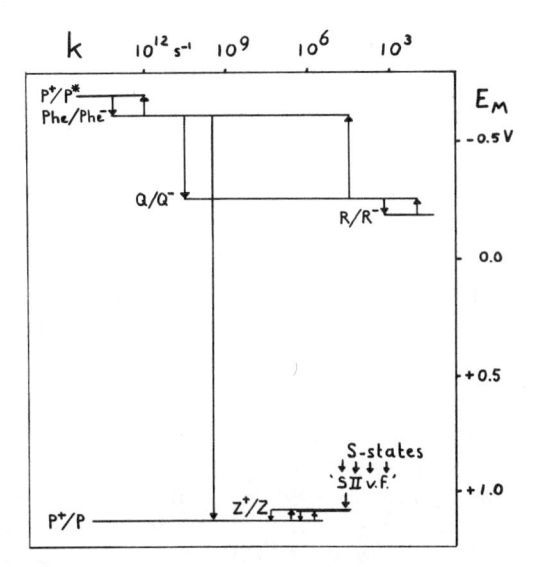

Fig. 1. Simplified scheme of redox reactions in photosystem IIα. The rate constants for Phe and Q reduction, and hence their reversal, are minimum values.

The possibility that β centers are slow because they do not efficiently stabilize the primary charge separation can now be excluded. We determined the quantum yields of Qα and Qβ photoreduction and P700 photooxidation in conditions where all three reactions took place simultaneously and almost irreversibly. As shown in the table the total quantum yield was identical in chloroplasts with very different relative amounts of α and β centers, obtained from wild type and two mutants of tobacco. Since the yield was even higher than

Table I. Concentrations of reaction centers in chloroplasts from tobacco (Nicotiana tabacum, cv John William's Broadleaf) wild type and mutants Su/su and Su/su Aurea, quantum yields of their primary reactions in 550 - 600 nm light.

	Su/su Aurea	Su/su	WT
P700/1000 chl	3.3	2.5	1.9
Qα /1000 chl	3.0	2.9	2.0
Qβ /1000 chl	15.4	5.0	1.3
Quantum yield P700	0.30	0.32	0.35
" Qα	0.12	0.38	0.48
" Qβ	0.59	0.28	0.16
" TOTAL	0.99	0.98	0.99

could be expected (suggesting a systematic error, possibly in the extinction coefficients) and illumination was at wavelengths between 550 and 600 nm, where absorption is approximately proportional to the number of chlorophylls, the antenna sizes for the three photosystems follow directly from these data. For the Su/su mutant, where the size difference between the antennae of system IIα and β is most pronounced, we determined partial absorption spectra of the three photosystems from their quantum yields in the red region (Fig. 2). The antenna of photosystem IIβ appears to consist mainly of a chlorophyll a form peaking at 683 nm. In contrast the presence of the chlorophyll a/b protein complex in the antenna of system IIα is obvious. This finding is consistent with the observations that a low chlorophyll b content is correlated with a low α/β ratio in the mutants and that Mg^{2+} depletion has no effect on the fluorescence of system IIβ.

Photosystem IIβ evolves oxygen and probably transports electrons to system I, but the intermediates may not be the same as in electron transport from system IIα to system I. No charge accumulation like that on R in system IIα seems to occur and except Qβ itself no plastoquinone may be involved at all. The concentration of cytochrome f is correlated with the abundance of system IIβ. The primary reaction, however, probably differs only in kinetic properties from that in photosystem IIα.

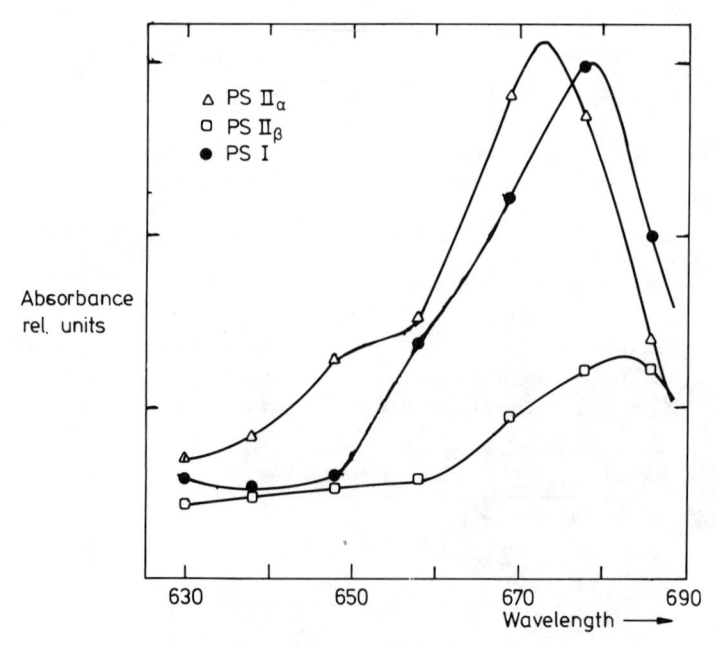

Fig. 2. Contribution by the three photosystems to the absorbance spectrum of chloroplasts from tobacco mutant Su/su, calculated from their quantum yields.

Acknowledgement

Financial support was given by the Netherlands Organization for the Advancement of Pure Research (ZWO), via the Netherlands Foundation for Chemical Research (SON).

REFERENCES

1. Spector, M. and Winget, G.D. (1980) Proc. Natl. Acad. Sci. USA 77, 957-959.
2. Joliot, P. and Kok, B. (1975) in: 'Bioenergetics of Photosynthesis' (Govindjee, ed.) pp. 387-412. Academic Press, New York.
3. Velthuys, B.R. (1978) Proc. Natl. Acad. Sci. USA 75, 6031-6034.
4. Van Gorkom, H.J. (1976) Thesis, University of Leiden.
5. Amesz, J. and Duysens, L.N.M. (1977) in: 'Primary Processes of Photosynthesis' (Barber, J., ed.) pp. 149-185. Elsevier, Amsterdam.
6. Klimov, V.V., Dolan, E. and Ke, B. (1980) FEBS Lett. 112, 97-100.
7. Prince, R.C., Tiede, D.M., Thornber, J.P. and Dutton, P.L. (1977) Biochim. Biophys. Acta 462, 467-490.
8. Okamura, M.Y., Isaacson, R.A. and Feher, G. (1979) Biochim. Biophys. Acta 546, 394-417.
9. Klimov, V.V., Klevanik, A.V., Shuvalov, V.A. and Krasnovsky, A.A. (1977) FEBS Lett. 82, 183-186.
10. Fujita, I., Davis, M.S. and Fajer, J. (1978) J. Am. Chem. Soc. 100, 6280-6282.
11. Van Grondelle, R., Holmes, N.G., Rademaker, H. and Duysens, L.N.M. (1978) Biochim. Biophys. Acta 503, 10-25.
12. Van Best, J.A. and Duysens, L.N.M. (1977) Biochim. Biophys. Acta 459, 187-206.
13. Rademaker, H., Hoff, A.J. and Duysens, L.N.M. (1979) Biochim. Biophys. Acta 546, 248-255.
14. Holten, D. and Windsor, M.W. (1978) Ann. Rev. Biophys. Bioeng. 7, 189-227.
15. Hoff, A.J. (1979) Phys. Rep. 54, 75-200.
16. Rademaker, H. and Hoff, A.J. (1980) Biophys. J., in press.
17. Sauer, K. and Brewington, G.T. (1978) in: 'Proc. 4th Int. Congr. Photosynth.' (Hall, D.O., Coombs, J. and Goodwin, T.W., eds.) pp. 409-421. Biochemical Society, London.
18. Searle, G.F.W., Tredwell, C.J., Barber, J. and Porter, G. (1979) Biochim. Biophys. Acta 545, 496-507.
19. Klimov, V.V., Allakhverdiev, S.I. and Pashchenko, V.Z. (1978) Dokl. Akad. Nauk. SSSR 242, 1204-1207.
20. Van Best, J.A. and Mathis, P. (1978) Biochim. Biophys. Acta 503, 178-188.
21. Sonneveld, A., Rademaker, H. and Duysens, L.N.M. (1979) Biochim. Biophys. Acta 548, 536-551.
22. Amesz, J. and van Gorkom, H.J. (1978) Ann. Rev. Plant Physiol. 29, 47-66.

23. Latimer, P., Bannister, T.T. and Rabinowitch, E.I. (1957) in: 'Research in Photosynthesis' (Gaffron, H., ed.) pp. 107-112, Interscience Publ., New York.
24. Bennoun, P. (1970) Biochim. Biophys. Acta 216, 357-363.
25. Klimov, V.V., Allakhverdiev, S.I., Demeter, S. and Krasnovsky, A.A. (1980) Dokl. Akad. Nauk. SSSR 249, 227-230.
26. Horton, P. and Croze, E. (1979) Biochim. Biophys. Acta 545, 188-201.
27. Blankenship, R.E., McGuire, A. and Sauer, K. (1977) Biochim. Biophys. Acta 459, 617-619.
28. Babcock, G.T., Blankenship, R.E. and Sauer, K. (1976) FEBS Lett. 61, 286-289.
29. Ikegami, I. and Katoh, S. (1973) Plant Cell Physiol. 14, 829-836.
30. Velthuys, B. and Kok, B. (1978) in: 'Proc. 4th. Int. Congr. Photosynth.' (Hall, D.O., Coombs, J. and Goodwin, T.W., eds.) pp. 397-407. Biochemical Society, London.
31. Bowes, J.M., Crofts, A.R. and Itoh, S. (1979) Biochim. Biophys. Acta 547, 320-335.
32. Melis, A. and Homann, P.H. (1976) Photochem. Photobiol. 23, 343-350.
33. Melis, A. and Homann, P.H. (1978) Arch. Biochem. Biophys. 190, 523-530.
34. Horton, P. and Croze, E. (1979) Biochim. Biophys. Acta 545, 188-201.
35. Melis, A. and Duysens, L.N.M. (1979) Photochem. Photobiol. 29, 373-382.
36. Melis, A. and Schreiber, U. (1979) Biochim. Biophys. Acta 547, 47-57.

BIOPHOTOLYSIS OF WATER FOR H_2 PRODUCTION USING IMMOBILISED AND SYNTHETIC CATALYSTS

David O. Hall, Paul E. Gisby and K. Krishna Rao

Plant Sciences Department, King's College
68 Half Moon Lane,
London SE24 9JF. England

INTRODUCTION

Biophotolysis of water for hydrogen production was first demonstrated by Krampitz[1] in 1972 when he observed hydrogen evolution from an illuminated mixture containing spinach chloroplasts, methyl viologen and Escherichia coli hydrogenase. A year later Benemann et al[2] reported that spinach chloroplasts mixed with ferredoxin and Clostridium kluyveri hydrogenase evolved H_2 in the light. In the biophotolytic reaction the electrons derived from water-splitting at chloroplast photosystem II are transported to photosystem I to reduce the ferredoxin (or methyl viologen), the reduced carrier in turn donates the electron to hydrogenase which converts the protons in the medium to molecular hydrogen.

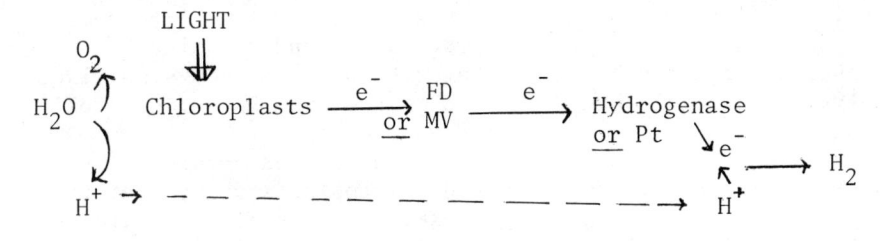

$$H_2O \xrightarrow{LIGHT} Chloroplasts \xrightarrow[\text{or MV}]{e^-} FD \xrightarrow{e^-} \text{Hydrogenase or Pt} \xrightarrow{e^-} H_2$$

The rate and duration of H_2 evolution obtained by Benemann et al. were very low and the authors cautioned that "the problems of ferredoxin autoxidation, hydrogenase inactivation and photosystem II instability must be resolved before photosynthetic hydrogen evolution can be considered for solar energy conversion". In the past few years we[3-5], and others[6], have done detailed investigations on the characteristics of this biophotolytic system and have improved the yield of hydrogen production from water more than hundred-fold. We will discuss in this paper the basic properties of the in vitro system, the problems encountered in continuous hydrogen production and some of the modifications adopted in order to partially overcome the problems.

METHODS

Chloroplasts were isolated according to Reeves and Hall[7]. Calcium alginate gel spheres were made by mixing chloroplasts with 1% sodium alginate in 50mM HEPES, pH 7.6, containing 1% BSA and then pouring the mixture as drops into 100mM $CaCl_2$[8]. In some experiments ferredoxins and/or hydrogenase were also incubated in the mixture. Better stability was achieved by covering a stainless steel grid with the sodium alginate-chloroplast mixture and dropping $CaCl_2$ on to this from a pipette[9]. Oxygen exchange measurements were carried out polarographically using a Clark-type oxygen electrode.

Ferredoxins and flavodoxins were isolated by standard techniques[10,11]. Cytochrome c_3 was prepared according to Bruschi et al[12]. Jeevanu particles were a gift of Dr. S. Ranganayaki. These particles were prepared according to Bahadur[13] by exposing to sunlight mixtures containing ammonium molybdate, diammonium hydrogen phosphate, formaldehyde, and solutions of mineral salts containing Na, K, Ca, Mn, Mg, V, etc. The molybdenum-iron-sulphur cluster $[Fe_6Mo_2S_8(SCH_2CH_2OH)_9]^{3-}$ dissolved in Tris buffer was synthesized and donated by Drs. G. Christou and C.D. Garner[14].

Hydrogenases were purified in our laboratory from C. pasteurianum[3], Desulfovibrio desulfuricans strains Norway[15] and 9974[4] (gift from Dr. J. Le Gall), Rhodospirillum rubrum[16], E. coli[17], Chromatium[18], Spirulina maxima[19] and Oscillatoria limnetica. Alkaligenes eutrophus hydrogenase was a gift from Dr. K. Schneider[20]. D. gigas hydrogenase bound to amino-spherosil was donated by Dr. C. Hatchikian. Platinum stabilised on polyvinyl alcohol (PVA-Pt) was obtained from Dr. M. Gratzel[21].

Hydrogen evolution was assayed as described before. The standard reaction mixture contained chloroplasts equivalent to 100 µg chlorophyll, 100 µmole glucose, 20 units glucose oxidase, 100 units catalase, 20 µℓ ethanol, 10 mg BSA and an appropriate electron mediator in a total volume of 2 ml made up with 50mM HEPES buffer, pH 7.5. The mixture in a 15 ml glass vial was capped with 'Suba seal' stoppers, flushed with N_2 to expel all O_2 and then incubated in a water bath at 25°C. Hydrogenase or platinum was then injected into the vial and illumination started with a set of 40W tungsten lamps giving a light intensity of 11000 lux at the base of the vials. The amount of hydrogen produced was measured at intervals by withdrawing samples from the gas phase and injecting into a gas chromatograph[4].

RESULTS AND DISCUSSION

By performing a series of experiments using chloroplasts isolated from various plant species, hydrogenases from different microorganisms and various natural electron carriers (Table 1) we were able to determine the best combination of catalysts to produce the maximum amount of hydrogen. The optimum rate of H_2 evolution of 50 µmoles per mg chlorophyll per hour continuous for 5 hours was observed with a system containing Chenopodium chloroplasts, Spirulina maxima ferredoxin and C. pasteurianum hydrogenase. Chloroplasts isolated from Chenopodium sp. were more stable than other plant chloroplasts[22] - the stability during storage was enhanced by the addition of BSA. Studies carried out in the presence and absence of oxygen (glucose + glucose oxidase) and peroxide (catalase + ethanol) traps showed that chloroplasts are sensitive to light inactivation and to oxygen and peroxide formed during the photolysis[4]. Glutaraldehyde fixation of chloroplasts improved the stability of chloroplasts during storage but did not prevent the photoinactivation[3]. Although immobilization of chloroplasts on calcium alginate gel films did not affect their oxygen exchange capacity, in the preliminary experiments the H_2 evolution rate from immobilised chloroplasts was found to be lower than that from free chloroplasts (Table 2). However, the fact that chloroplasts (and other catalysts) can be embedded on solid supports with retention of photosynthetic electron transport and H_2 evolution activity is promising for construction of heterogeneous biophotolytic systems.

Hydrogenase from C. pasteurianum is extremely active and can easily take up electrons from photosynthetically reduced ferredoxins and flavodoxins (Table 3). The enzyme is oxygen-labile but can be stabilised against oxygen inactivation by binding to ferredoxin covalently linked to AH-Sepharose[4]. Hydrogenases from Desulfovibrio gigas and D. desulfuricans can evolve H_2 using reduced cytochrome c_3 or methyl viologen as electron donor. These hydrogenases can be easily bound to aminospherosil or aminopropyl glass and the bound enzymes can be used repeatedly for H_2 evolution. Hydrogenases from cyanobacteria and photosynthetic bacteria, other than Thiocapsa, react poorly in the water photolytic system.

Our ultimate aim in this project is to replace one or all of the biological components by, preferably more stable, synthetic substitutes. We have reported that Jeevanu[8] and synthetic

Table 1. Relative rates of H_2 evolution from chloroplasts by various hydrogenases

Hydrogenase source	Electron mediator and concentration	µmole H_2 evolved
C. pasteurianum	Fd, 25 µM	100
	MV, 5µM	8.8
	Fld, 18µM	46.3
	Fd, 25µM	62.5*
D. desulfuricans Norway strain	Fd, 15µM	1.0
	MV, 25µM	52.5
	Jeewanu	0
	c_3, 0.65 nM	18.0
9974 strain	MV, 25µM	30.0
D. gigas, spherosil-bound	MV, 25µM	18.8
	c_3, 5µM	61.3
	c_3, 5µM	34.4*
A. eutrophus	Fd, 15µM	0
	MV, 50µM	19.7*
	MV, 50µM	30.0
	NAD, 600µM	1.3
Chromatium	MV, 50µM	12.5
E. coli	MV, 50µM	13.3
R. rubrum	Fd, 15µM	7.8

Fd, ferredoxin from Spirulina maxima or spinach; Fld, flavodoxin from Megasphaera elsdenii; MV, methyl viologen, c_3, cytochrome c_3 from D. gigas.
*No oxygen or peroxide trap. Hydrogenases were present in saturating amounts.

Fe_3MoS_4 dimers can substitute for ferredoxin in the chloroplast H_2 evolution reaction. We have also shown that Adam's catalyst (PtO_2) can replace hydrogenase in a biophotolytic system using chloroplasts or proflavine + EDTA for photon capture[23].
Recently Dr. Gratzel's group (Lausanne) has stabilised platinum on a variety of solids and these polymer-supported Pt catalysts were found to be better H^+ activators than PtO_2. Comparative rates of H_2 evolution using C. pasteurianum hydrogenase and PVA-Pt are given in Table 4.

Table 2. Oxygen exchange and hydrogen evolution rates of immobilised and non-immobilised chloroplasts (Ref. 9)

Alginate films were made by mixing chloroplasts containing 50 µg chlorophyll with 1% Na alginate + 50 mM HEPES pH 7.5 to a final vol. of 300 µl. A nylon grid was covered with 150µl of this mixture, and 3 ml of 0.1M $CaCl_2$ in 50mM HEPES were dropped on to the film. O_2 evolution was measured with ferricyanide as electron acceptor and O_2 uptake with methyl viologen[7]. Hydrogen evolution was assayed with C. pasteurianum hydrogenase and Spirulina ferredoxin. All assays at 25°C.

Activity	Free chloroplasts	Chloroplasts in calcium alginate
O_2 evolution: µmoles O_2/mg chl/h.	200	198
O_2 uptake: µmoles O_2/mg chl/h.	207	205
H_2 evolution: µmoles H_2/mg chl/h.	21	5

CONCLUSION

We have shown that a variety of natural and synthetic catalysts can be used together with illuminated chloroplasts to produce hydrogen from water. Chloroplasts can be immobilised on solid supports with retention of at least part of their hydrogen evolution activity. Future research is directed towards immobilization of the catalysts on solid supports and construction of heterogeneous catalytic reactors for hydrogen production with water as both the electron and proton donor.

SUMMARY

An aqueous mixture of chloroplasts plus hydrogenase and an electron transfer catalyst, on illumination liberates H_2; the source of electrons and protons for hydrogen is water. The rate and duration of H_2 production from such a system depends on the light- and oxygen-stability of the chloroplasts and other catalysts used. Both chloroplasts and hydrogenases can be stabilized to a certain extent by immobilization in gels or

Table 3. Efficiency of ferredoxins and flavodoxins as electron donors to hydrogenase in the chloroplast-hydrogenase system (Ref. 24)

The standard reaction mixture contained saturating amounts of C. pasteurianum hydrogenase and 22.5 µM ferredoxin or flavodoxin and spinach chloroplasts. nd, not determined.

Mediator	Em (MV)	µmoles H_2/mg chl/h.
Ferredoxins:		
Anacystis nidulans	nd	17.4
Chlorogloeopsis fritschii	-340	14.5
Nostoc Mac I	-350	16.0
Nostoc Mac II	-455	18.9
Spirulina maxima	-390	15.4
Chondrus crispus	nd	0.5
Gigartina stellata	nd	1.5
Porphyra umbilicalis	-380	20.5
Pisum sativum I	-425	13.2
Pisum sativum II	-410	15.3
Spinacia oleracea	-415	15.2
Flavodoxins:		
Nostoc Mac	nd	21.0
Anacystis nidulans	nd	27.2
Chondrus crispus	nd	1.3

solid supports and by addition of bovine serum albumin. Platinum as PtO_2 or as a dispersion in polyvinyl alcohol can replace hydrogenases. Synthetic clusters containing Mo, Fe and S and photochemically prepared 'Jeewanu' particles can replace ferredoxin in the hydrogen evolution system. Maximum rate of H_2 evolution so far observed is 50 µmoles H_2 per mg of chlorophyll per hour continuous for five hours. Our eventual aim is to construct a stable, synthetic and catalytic system (possibly mimicking the biological system) which will evolve H_2 photolytically from water.

ACKNOWLEDGEMENTS

This work was supported by grants from European Commission and U.K. Science Research Council. P.E.G. is supported by British Gas Corporation.

Table 4. Comparative activities of C. pasteurianum hydrogenase and PVA-Pt in the chloroplast H$_2$ evolution system. (Rao, K.K. and Hall, D.O. unpublished). The standard reaction mixture (see Methods) contained Chenopodium chloroplasts and e$^-$ mediators and hydrogenase or Pt as shown.

e$^-$ mediator	Proton activator	μmoles H$_2$ evolved/mg chl.			
		1h	2h	3h	4h
Spinach Fd, 25 nmole	C. pasteurianum hydrogenase, 100 units	27.4	54.9	79.1	96.5
Jeewanu (1980) 400 μg	"	7.9	15.0	21.7	26.9
[Mo-Fe-S] cluster, 210 nmole	"	20.9	31.6	34.0	34.3
Jeewanu, 400 μg	NONE	0	0		
NONE	C. pasteurianum hydrogenase	0.7	1.1	1.6	2.1
NONE	PVA-Pt, 96μg	1.1	1.3	1.6	2.1
Spinach Fd, 25 nmole	"	3.8	4.8	4.8	5.1
Jeewanu, 400μg	"	8.0	15.0	19.8	23.3
[Mo-Fe-S] cluster, 210 nmole	"	12.9	16.6	18.0	18.5
Methyl viologen, 100 nmole	"	3.2	6.4	9.1	11.5

REFERENCES

1. L.O. Krampitz. in "An Inquiry into Biological Energy Conversion (A. Hollaender et al., eds.) The University of Knoxville, U.S.A. (1972).
2. J.R. Benemann, J.A. Berensen, N.O. Kaplan, and M.D. Kamen. Proc. Natl. Acad. Sci. U.S.A. 70, 2317-2320 (1973).
3. K.K. Rao, L. Rosa and D.O. Hall. Biochem. Biophys. Res. Comm., 68. 21-27 (1976).
4. K.K. Rao, I.N. Gogotov and D.O. Hall. Biochimie, 60. 291-296 (1978).
5. K.K. Rao and D.O. Hall. in "Topics in Photosynthesis", (J. Barber, ed.) Vol. 3. pp.299-329. Elsevier, Amsterdam, (1979).
6. I. Fry, G. Papageorgiou, E. Tel-Or and L. Packer, Z. Naturforsch, 32c, 110-117 (1977).
7. S.G. Reeves and D.O. Hall, Biochim. Biophys. Acta, 314. 66-78 (1973).
8. K.K. Rao, M.W.W. Adams, P. Morris, D.O. Hall, S. Ranganayaki and K. Bahadur, in "Proc. Intl. Symp. on Biological Applications of Solar Energy", Madurai, (A. Gnanam et al. eds.) pp.201-204. The Macmillan Co. Madras (1980).
9. P.E. Gisby and D.O. Hall, Nature, submitted (1980).
10. D.O. Hall and K.K. Rao, in "Encyclopaedia of Plant Physiology Vol. 5. (A. Trebst and M. Avron, eds.) pp.206-216, Springer-Verlag, Berlin (1977).
11. G.N. Hutber, K.G. Hutson and L.J. Rogers. FEMS Microbiol Lett. 1, 193-196 (1977).
12. M. Bruschi, C.E. Hatchikian, L.A. Golovleva and J. Le Gall. J. Bact. 129, 30-38 (1977).
13. K. Bahadur, Zbl. Bakt. Abt. II. Bd. 130, S211-218 (1975).
14. M.W.W. Adams, K.K. Rao, D.O. Hall, G. Christou and C.D. Garner, Biochim. Biophys. Acta 589, 1-9 (1980).
15. W.V. Lalla-Maharajh, D.O. Hall, K.K. Rao and J. Le Gall. Biochem. J. submitted (1980).
16. M.W.W. Adams and D.O. Hall, Arch. Biochem. Biophys. 195, 288-299 (1979).
17. M.W.W. Adams and D.O. Hall. Biochem. J. 183, 11-22 (1979).
18. M.J. Llama, J.L. Serra, K.K. Rao and D.O. Hall. Biochem. Soc. Trans. 7. 1117-1118 (1979).
19. M.J. Llama, J.L. Serra, K.K. Rao and D.O. Hall. FEBS Lett. 98, 343-346 (1979).
20. K. Schneider and H.G. Schlegel. Biochim. Biophys. Acta. 452, 66-80 (1976).
21. J. Kiwi and M. Grätzel. J. Am. Chem. Soc. 101. 7214-7217 (1979)
22. K.K. Rao, P. Morris and D.O. Hall, in "Hydrogenases, their Catalytic activity, Structure and Function (H.G. Schlegel and K. Schneider, eds.) pp.439-452, E. Goltze, Gottingen, (1978).

23. M.W.W. Adams, K.K. Rao and D.O. Hall. Photobiochem. Photobiophys. 1, 33-41. (1979).
24. M.P. Fitzgerald, L.J. Rogers, K.K. Rao and D.O. Hall. Biochem. J. in press (1980).

SOLAR ENERGY BIOCONVERSION AT THE ECOSYSTEM LEVEL

Paul Duvigneaud

Laboratoire de Botanique Systématique, Ecologie
et Génétique des Plantes Supérieures
Université de Bruxelles, Belgique

DEFINITIONS

In nature, the importance of the bioconversion of Solar energy depends on the type of ecosystem in which this bioconversion is taking place. But the concept of ecosystem is a wide one, covering a large scale of ecological units differing in extent, structure and functioning. Adequate definitions are necessary.

The punctiform field unit is the biogeocenosis (Isukachev, 1944, 1954) which incorporates an homogeneous biocenosis to an homogeneous environment (climatope, hydrotope and edaphotope forming the biotope); it corresponds to the plant association based on the constancy of dominant species (sometimes replaced by homologous species, sensu Kuhnholz-Lordat, 1949) and on definite proportions of plant ecosociological groups bioindicating the conditions of their habitat (Duvigneaud, 1946). In the biogeocenosis, the plant association incorporates a trophic network of sedentary animals and microorganisms, forming in the soil the pedoflora and the pedofauna.

The three principal biogeocenoses in temperate regions are, following the Romans, and codified by Kuhnholz-Lordat (1945), the silva (natural or seminatural forest or tree plantation), the saltus (course of the flock) and the ager (field where are cultivated the food or industrial plants).

The so delimited biogeocenosis may have more or less tight ecological relations with neighbouring biogeocenoses, for example, transfers of water, materials and nutriments due to topography, wind, circulating animals (principally birds, fishes and mammals).

So are formed larger ecological units. The simplest example is the lake, ecosystem sensu stricto, whose tropho-dynamic functioning was codified in 1942 by Lindeman. A terrestrial example is the catena of soils and vegetation, much spread in tropical zones (Morison et al., 1948). An anthrophic example is the farm (agroecosystem), whose mosaic of differing crops and pastures is under planning and control of the farmer.

A still larger type of ecosystem is the ecological landscape with repetition and intermingling of well definite catenas and mosaics, with a regulation of the silva-saltus-ager equilibrium (equilibrium S.S.A.) which following the wisdom degree of men, may be ecologically harmonious or disrupted (desertus).

A very interesting type of ecological landscape is the ecochemical landscape (Perelman, 1964) where the different biogeocenosis are linked together by their common dependance on a special limiting edaphic factor which is the excess or deficiency of a toxic or indispensable chemical element; good examples are landscapes bound to marno-calcareous, dolomitic, serpentinic, gypseous, zinciferous or cupriferous, or calcium, or phosphor-lacking, bed rocks.

Numerous ecological landscapes result from a coherent human action: urban and industrial ecosystems. In what concerns the energy flow, human action in agroecosystems and urban and industrial ecosystems is principally an important addition of subsidiary energy obtained by combustion of fossil fuels.

The ecoregion is an ensemble of definite ecological landscapes where socio-economical factors surimpose their action to those of regional climates, bed-rocks and soils, vegetation, agriculture and industrial systems and human traditions. The functioning of the ecoregion leads to its ecodevelopment, the base of which is the solar energy conversion to biomasses of many types.

A hierarchy of ecoregions ends in the biosphere concept (Vernadskii, 1924) which is a very large ecosystem ("we have only one earth," Ward and Dubos, 1972), which must be divided in biogeosphere (the continents), biohydrosphere (the oceans) and aestuarisphere, this immense and so important ecotone (very high solar energy bioconversion) separating but joining ecologically the two previous ones.

The biosphere is divided into zonobiomes, like tundra, taiga, temperate caducefoliate forest, a.s.o. (Walter, 1977) parallel to the equator and separated by large megaecotones (tundra-taiga, forest-steppe, a.s.o.), their regularity is broken by the mountain orobiomes, and the large edaphobiomes, principally extended in arid zones with edaphic dominance of salt and gypsum.

An accurate cartography of terrestrial zonebiomes is progressing but in the oceans, the division in zonobiomes and their cartography are still to be made.

BIOGEOCENOSIS AND THE BIOCONVERSION OF SOLAR
ENERGY COMPLEXITY OF THE "BIOMASS" NOTION.

Biogeocenosis

Biogeocenosis is the basic ecosystemic unit whose study in the field may lead to some accurate data and results. Its photosynthetic efficiency depends, of course, on the climatope, hydrotope and edaphotope. But it depends also on its structure:

(i) in space (chorologic): canopy rugosity, stratification in photosynthetic and not photosynthetic layers, regulating the quantitative and qualitative distribution of PAR in the whole thickness of the aerial subsystem.

(ii) in time: length of the vegetation period evergreen or tropophyllous, which may depend on the constant replacement (or not) of the leaves of the dominant species (example equatorial forest) or on the succession of phenophases adapted to different seasons (example winter Brassica or Valerianella cultures dérobées', or winter weeds, in temperate regions).

Monsi (1959) and Walter (1964) have classified the biogeocenosis in six categories of productive structures whose functioning may be quite different:

1. Annual biogeocenoses
2. Bisannual biogeocenoses
3. Perennial geo- or hemicryptophytic herbaceous biogeocenoses
4. Perennial evergreen herbaceous communities
5. Caducifoliate forests or heathers
6. Evergreen forests or heathers

Starting from a nude or denuded soil, the vegetation varies in the course of years, going from a pioneer stadium to a definite climaxic phase, passing through a sere (série) of intermediary stages; a well known succession system is the old field — Hyparrhenia savanna — Pinus taeda intermediary forest — Oak-Hickory forest succession in U.S.A.

This progressive sere may be also regressive. Often, the sere is blocked by man at a chosen stage; this stage or the climax, may be replaced by a partially or totally artificialized biogeocenosis. So, the climax notion explodes in a great number of climax units, whose classification has been attempted by Clements (1916).

Also, following meteorological oscillations, the climax is submitted to more or less important fluctuations, which sometimes may be catastrophic: tornados, fires put in by storms or man, which destroy vegetation and initiate a new sere. The degree of resistance of the biogeocenosis to the destroying factor and the rate of reconstruction after destroying factor and the rate of reconstruction after destruction, make part of a new field ecosystemic ecology which may be called resilience.

Bioconversion of Solar Energy

At the level of the different photosynthetic layers of green plants (autotrophs, producers), the biogeocenose converts the solar energy and the atmospheric CO_2 into phytomass. The efficiency of this bioconversion is bound to the temperature, the water balance and the biogeochemical cycles of mineral nutriments, but also of toxic elements.

What is so important and so often neglected is that a more or less big part of the phytomass, living or dead, feeds a lot of consumers of quite different types (herbivores, carnivoures, parasites, detritivores or decomposers); added to what remains of the phytomass, their zoo- or microbiomass contribute to the formation of the total biomass of the system.

Biomass (B) is thus the abondance of organisms present in the biogeocenose at a definite moment; it may be expressed in dry weight or kcal, kJ or tep (energy) per surface unit; it contains principally organic matter, but also mineral matters which may be evaluated from the nitrogen and ashes analysis (mineralomasses; perennial) attached or standing, form the necromass (N) sensu Kestemont, 1970 (standing dead wood in forests, mulch in steppes, for instance).

At the soil level may accumulate a litter of dead organic matter (Ll), made of the annual litter fall (LFl); generally the litter is quickly or slowly consumed and respired by the soil decomposers, but part of it may be transformed into one or other forms of humus. In the litter fall is comprised the litter of roots and other organs dying in the soil.

For numerous authors, the biomass is the whole of all the organic matter of biologic origin, living or dead contained in the biogeocenosis. At least in ecological studies, it seems preferable to speak of biogeocenosis total organic matter (TOM). So,

TOM = B + N + L + H
TOM = B + MOM, where MOM represents the dead organic matter.

Productivity is the rate of biomass production. It takes place in a matter or energy flow initiated by solar energy and developed along trophic chains anastomosed in a trophic network. The unit of time, for ecologists, is generally 1 year; sometimes, they consider the period of vegetation only. The chosen unit of surface is sometimes 1 cm^2; it seems that 1ha is more realistic.

Bruto productivity PB is the total assimilation A of the photosynthetic system; part of it (RMF) is burned in the maintenance respiration of the system; the remaining assimilates (surplus productivity PS) are translocated to the non photosynthetic system, whose maintenance is assured by RMnF (Fig. 1).

The net assimilates productivity PNA, PNA = PS − RMnF, is used for the construction of new organs and the elaboration and stockage of reserves; it supplies the energy (RC) necessary for those syntheses.

The total respiration losses are

RA = RMF + RMnF + RC, RA being called the respiration of autotrophs

What remains of PB is the net primary productivity PN1:

PN = PB − RA

PN1 is thus the vegetal matter produced by the biogeocenosis in surplus of what has been lost by respiration.

In reality, phenomena are much more complicated and some of them are not always negligible: there are matter transfers from old organs to new ones and vice versa; organic substances are washed by rain from the leaves to the soil.

The fate of net primary productivity.

All along the energy flow, PN1 sustains losses by mortality (M1), consumption by herbivores and parasites (C2) and exportation (Ex1). After the end of the vegetation period subsists a certain phytomass quantity, which will also die if the biogeocenosis is annual or bisannual, or will be added to initial phytomass if the biogeocenosis is perennial. In this case, we may write

T1 = PN1 − (M1 + C2 + Ex1),

T1 being the increment (added living tissues).

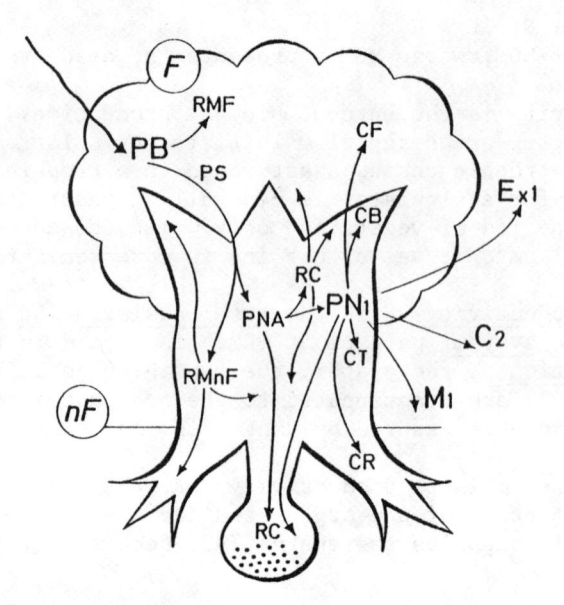

Fig. 1. Solar energy bioconversion in a perennial biogeocenosis. For explanations, see text.

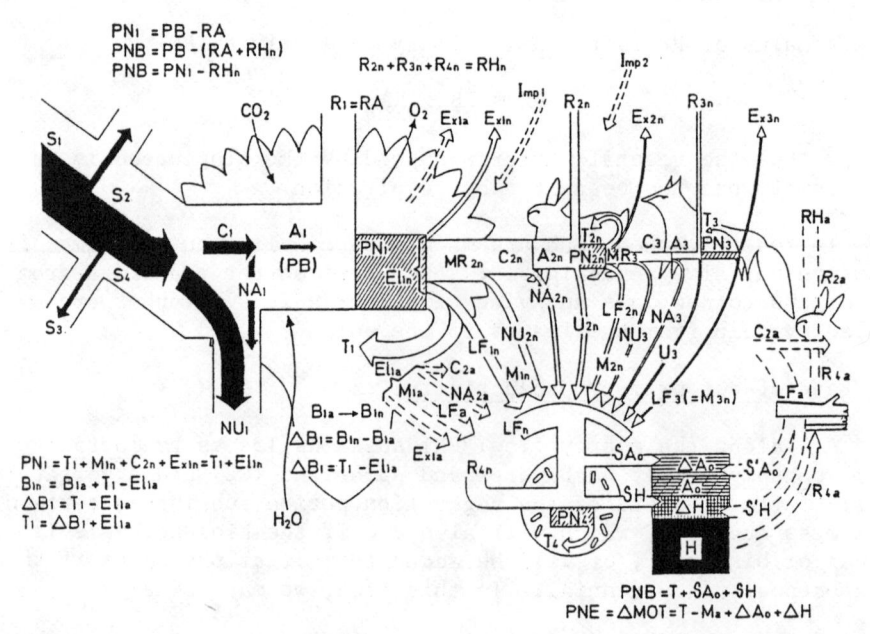

Fig. 2. Energy flows and budgets in a simplified biogeocenosis. For explanations, see text.

SOLAR ENERGY BIOCONVERSION AT THE ECOSYSTEM LEVEL

If the biogeocenosis is annual, all PN1 dies, and is delivered to decomposers. But annual biogeocenoses are often cultivated ones, and a large fraction of PN1 is collected and exported by man. One may call this fraction <u>yeald</u> (Y1), which may in turn be divided: for instance, grains and straw for cereals (Yg, Ys); what remains on the field may be called <u>field residues</u> (FR), delivered to decomposers. So

$$PN1 = Y1 + FR$$

Note that Y1 is only part of Ex1; for instance, grains may be exported by birds, tubercules by wild boars; it means that C2 due to herbivores coming from outside must be considered as exportation Ex1.

All the complexity of the biogeocenosis and organic matter production appears in the fate of M1 and C2; both of them feed complex trophic chains, ending in a partial <u>recycling</u> (rapid, slow or very slow) of the elements forming the initial biogeocenosis; the size of this recycling depends on the importance of exportation.

Figure 2 inspired by Odum, 1959, represents the principal events which characterize the energy flow initiated by solar energy and going through an idealized perennial biogeocenosis reduced to four trophic levels, each of them reduced to a single population: producer, herbivore, top carnivore, decomposer (a Bacteria supposed to act simultaneously as destructor, humifactor and remineralisator); all the parameters of <u>this</u> energy flow are sublined by an <u>n</u> index (for <u>new</u>), because, as we shall see later, the flow may be completed by <u>reimportation</u> or <u>autochtonous</u> dead material (<u>a</u> index for <u>ancient</u>, <u>old</u>), and sometimes by importation of <u>allochtonous</u> organic material. From the photosynthetically active radiations (PAR = light energy) striking the phytocenosis and designed here by S, only a small fraction C1 (C for "consumed") is retained by the photosynthetic subsystem and a still smaller fraction A1 (2-4%) is assimilated. A1 = PB. PB — RA = PN1.

We have seen what happens with PN1 (photosynthetic efficiency 1-2%) which divides in T, LF1, C2 and Ex1: herbivores are often <u>wasters</u>, and do not consume the whole of the removed material (MR2); the wasted part of it (NU2, for "not utilized"), join the litter fall to form M1, <u>not consumed mortality</u>.

From the feed consumed C2 ("consumed mortality") or ("ingestion"), only one fraction is assimilated (A2), the non assimilated fraction being rejected as <u>faeces</u> (NR2 = F2); from A2, a part is respired (R2) and a part is rejected as urine U2 (Nitrogen metabolism); those <u>rejecta</u> (F2 + U2) of excrements join the litter fall, and may be most important in pasture ecosystems dominated by cattle or sheep; what remains is PN2, the net secondary productivity, formed of new tissues and new individuals, which may emigrate (exportation Ex2). PN2 is distributed between T2 (increment of herbivorous level), MR3 (material removed by the carnivore) and mortality M2, joining the primary litter fall and rejects.

So, Pn2 = T2 + MR3 + LF2 + Ex2

Those phenomena repeat themselves at the carnivore level, which being a top carnivore is not consumed by any other predator.

All the dead organs, corpses and excrements produced during the chosen period constitute that may be synthetically called the total litter fall LF, LF is delivered to the decomposers; it contributes to decomposers productivity PN4 and increment T4, respiration R4, and also to the phenomena of humification δH (incorporation of humus to the soil for a more or less long time); a not much decomposed part may be added to the existing litter (δAo).

At the end, what remains of PN1 is the biocenosis net productivity PNB made of the biocenosis annual growth increment T, to which is added the augmentation of soil dead organic matter δAo + δH.

$$T = T1 + T2 + T3 + T4$$
$$PNB = T + \delta Ao + \delta H$$

Energy Balance in the Biogeocenosis. Reimportation.
Biomass Augmentation.

During the chosen period, part of the initial phytomass may disappear by exportation (Ex1a), consumption (C2a), mortality and fall of "big" litter (M2a) principally made of branches, roots and trunks.

Part of M2a is lost by decomposers respiration, but the remnants may be transformed in humus (δ'Ao and δ'H). It is practically impossible to distinguish between T2a and T2n, T3a and T3n, T4a and T4n; so, it is necessary to combine the two trophic chains.

Reimportation is the cause of two distinct phenomena: the energy flow and the energy balance (or organic matter flow or balance) which may be assessed at the producer's level. At the end of the period, the initial phytomass B1a has become

B1a — (M2a + C2a + Ex1a) or B1a — E11a (E1 for elimination).

To B1a has been added the increment T1.

So, the new phytomass is

$$B_{1n} = B_{1a} - E_{l1a} + T_1$$

The phytomass balance is:

$$\Delta B_1 = B_{1n} - B_{1a}$$
$$\Delta B_1 = T_1 - E_{l1a}$$
$$\text{or } \Delta B_1 = PN_1 - (E_{l1n} + E_{l1a})$$

This annual augmentation of phytomass is the difference between life and death; for Soviet ecologists (Mina, Rodin and Bazilevich), this dialectical interpenetration is the <u>true annual increment</u>.

At the entire biocenotic level, the energy balance is complicated by the mortality of old consumers and decomposers; the consequences are generally negligible, except for pasture biogeocenosis, organized by man for a large productivity of big mammals (cattle, sheep) or for cultivated ponds. We may write:

$$\Delta B = PN_1 - (RH_n + RH_a)$$

The <u>organomass balance</u> ΔTOM takes into account the fact that part of M_{1a} (and M_{2a}, a.s.o.) is not consumed and respired by heterotrophs but is <u>transformed</u> into $\delta'A_o$ and $\delta'H$ and added to pre-existing soil organic matter.

$$\Delta TOM = \Delta B + \Delta A_o + \Delta H$$

Avoiding subtilties, we may fuse A_o and H in H, and so we have

$$\Delta TOM = \Delta B + \Delta H$$

It is the total organomass living or dead added to the system. It approaches what Woodwell has called the <u>net productivity</u> of <u>the ecosystem</u> NPE (or PNE).

SYNTHESIS

The complexity of the facts and concepts bring us to synthesize them in one graphic (Fig. 3) representing two differing phenomena but linked together by the flow and the balance of energy in a perennial natural biogeocenosis; the graphic may be adapted to other types. One sees the difference between PB, PN1, PNE, ΔB and ΔTOM (=PNE). At climax, the biomass B is maximal but fluctuates following climate variations, ΔB is null. Let us insist on the fact that mortality, as well M_{1n} as M_{1a}, is divided in a fraction lost by respiration and a fraction conserved,

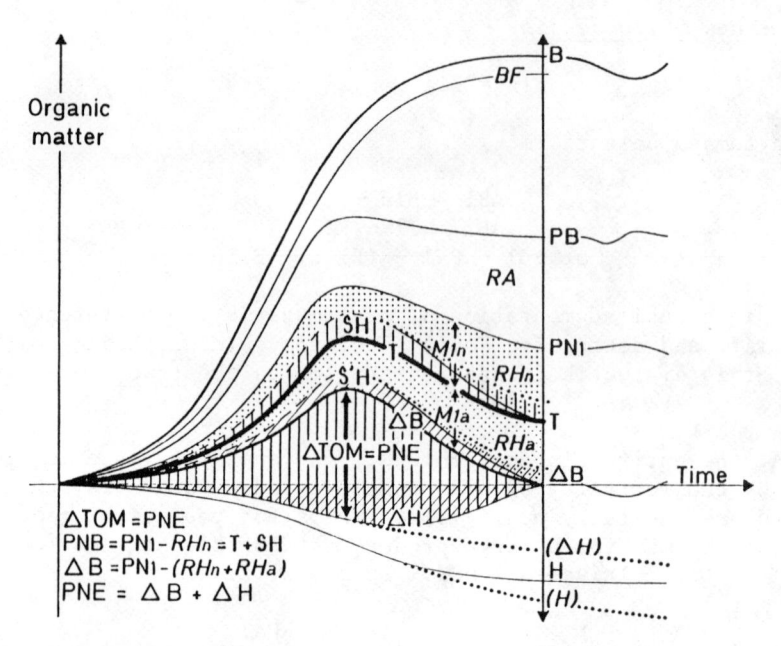

Fig. 3. Schema of ecosystem dynamics. Evolution of biomass and different types of productivity and energy budget, from initial stages to climax (after Duvigneaud, 1980).

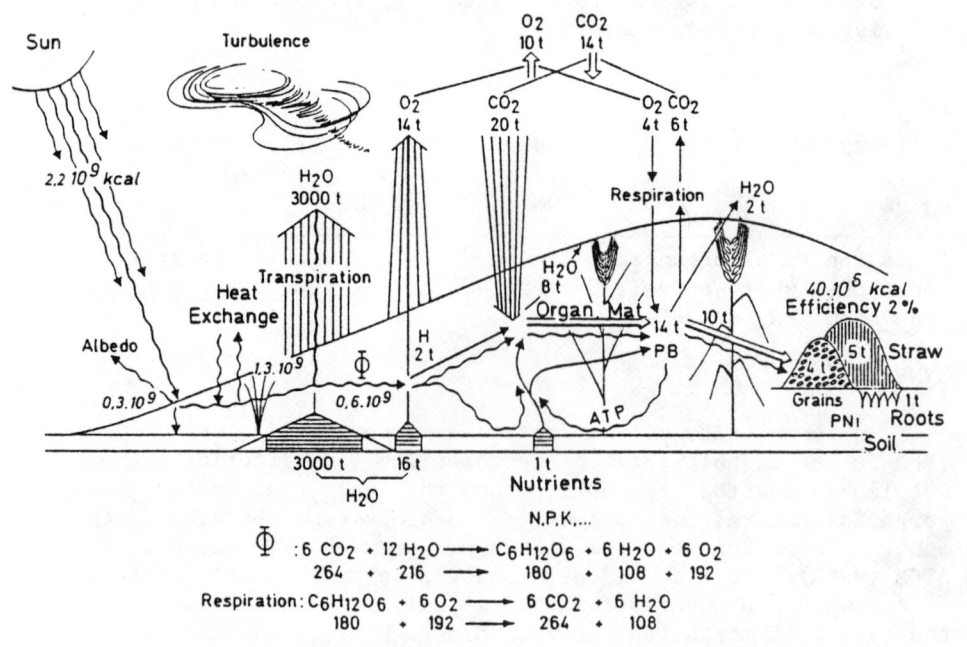

Fig. 4. Functioning of a temperate annual "wheat field" biogeocenosis (inspired by Nichiporovich, 1967).

but _transformed_ into humus. The graphic shows the importance of the dynamics of the biogeocenosis, the phenomena being not very different in the case of a whole succession, or in the case of the growth to adult stage of a single population of trees (plantation). The rate of biomass (phytomass) production, or organic matter production, is slow in the first _juvenile_ stages, rapid in young to adult stages (_full growth_) and again slow to nul in _adult_ to _senile_ stages (climax).

PN1, T, PNE, ΔB and ΔTOM pass by a maximum in the period joining youth to adultness in spite of the fact that leaf biomass (BF) remains constant up to the climax, maintaining nearly constant the level of PB.

So, in the field, because of the numerous actions of consumers and decomposers, the solar energy bioconversion at the biogeocenotic or ecosystemic level leads to a multiplicity of issues in the production of organic matter, living or dead, which on the theoretical as well as the practical planes may be inadequate or dangerous to be designated by the general term of _biomass_.

In the last decades in addition to the foresters' work, a huge quantity of quantitative research has been realized by ecologists, on the productivity of terrestrial ecosystems (biogeocenoses!) in natural or polluted conditions; two international programs, IBP and MAB have been organized on a large scale; SCOPE is trying to synthesize what is now known.

The accent has been principally put on phytomass (B1) and net primary productivity PN1; Russian ecologists have also studied their _true annual increment_, which is ΔB1.

Not much is known about the ecology of secondary productivity and excreta, but data may come from agriculturists.

ENERGY FROM BIOMASS

All the concepts, methods and data on the biologic productivity of biogeocenoses and ecosystems of all types are today very important because, besides the fact that they are scientific bases for a renovation of agriculture and forestry, they are closely related to the actual trend to utilise "biomass" as a source of energy and raw material for industry; in surplus of the traditional use of firewood and charcoal, one sees the transformation of biomass in solid, liquid and gaseous fluids as an _alternative_ to the growing penury of nonrenewable fossil fuels and to growing difficulties of the nuclear energy.

Recently, Chartier and Megaux (1980) have given a quasi-exhaustive enumeration of the energetic possibilities of the biomass. Smith

(1980) has resumed the conclusions of the Bio-Energy World Congress and Exportation (Alberta, Georgia, June 1980) attended by 1700 scientists, business men and policy-makers, "to discuss various means of squeezing usable energy of tress, crops, manure, seaweed, algae and urban waste."

Two well known facts appear: (i) the vagueness of the term "biomass" which reassemble living and dead materials of very different signification for the ecosystems; (ii) the accent put on the usable energy, and not on the usable biomass.

The problem is: which biomass? produced by which type of biogeocenosis or ecosystem? in which usable quantity (and quality) compatible with a sufficient fuel production, with the maintenance of ecosystem structure and functioning carbon, nitrogen and other biogeochemical cycles with economic rentability.

In all cases, the possibilities in "biomass" depend on the type of biogeocenosis and on the region where this biogeocenosis is, or may be, largely developed. Forest comes first, because the wood and also leaves mineralomass is very low (this is not true for bark!); many wastes may be exploited (small wood, of a diameter inferior to 7 cm (coppices), dead wood or dying trees leaf litter, saw dust, a.s.o.); most interesting are energy plantations of trees having a rapid growth and, in preference, fixing atmospheric nitrogen (Leucaena, Erythrina, Robinia)or of short revolution coppices. Then come strong objections from paper-makers, who apprehend a shortage of building and paper products, and a price rising of these. Highly productive herbeceous biogeocenoses (wild or cultivated) may also be utilized: in Brazil, 330,000 cars function with a mixture of water and ethanol from sugarcane or cassava, Nevertheless one fears that alcohol from grains or tubers should increase food prizes, principally in less developed countries.

The wasts of agro- and urbo-ecosystems are transformed in biogas (CH_4) and already 35 millions Chinese peasants use it. For our regions, Chartier insist on the great importance of straw.

From all this we may deduce the prior necessity of biogeocenoses and ecosystems studies. For forests, diverse authors used PN1 as the base of their calculations for estimating regional biomass energy possibilities, this is quite exaggerated.

What is really needed is T1, or Y, or PNB or PNE; for biogas, productivity of litter (LF) or excreta (NA2) must be known.

What is thus needed is the knowledge of the functioning of the productivity ecosystem and his resilience to exploitation, if one wants to act for the conservation of this renewable resource. What is needed is the knowledge of certain types of biogeocenoses being in the same time highly productive and very frugal in Nitrogen (or fixing if from air) and mineral nutriments, accepting a badly structured soil, easily harvestable; the approach may be driven from phytogeochemical studies. For instance, in tropical Africa, Protea leaves (dry weight) contain only 1% ashes (Duvigneaud et Denaeyer, 1963).

In Belgium, interesting biogeocenoses responding to the previous criteria are those made of "high fallow forbs" (Artemisia, Tanacetum, Solidago gigantae, Lactuca seriola, a.s.o.) which invade waste lands in towns and around villages; PN1 may be as high as 22t/ha/y, and Y1 may go up to 16t/ha/y; the dominating species may be rich in terpenic oils or latex, what puts them near the Calvin's petroliferous plantations.

A poorly understood ecological phenomenon is the compensation "water instead of mineral nutriments" in highly productive biogeocenoses developed on humid but poor soils; in Belgium, one may find examples in Acer — Fraximus — Prunus avium forests or coppices on very poor glen soils, or in Fagus forests on fresh, well drained, poor sandy soils, most characteristic is the very high productivity of Populus plantations on hydromorphic soils in the whole lowlands east of the river Meuse, and even inside the town of Brussels; like in equatorial forests, the nutriments may come from the rain, which gives a new aspect to atmospheric pollution; also the long distance influence of saharian dust has been recently assessed (Morales, 1974).

At this place of this paper would come detailed examples of typical biogeocenoses, principally some of those studied by us in Belgium. But there is no room, and only a few selected schemata could be given at the end of this chapter.

INTEGRATED ADAPTATIVE BIOMASS SYSTEMS

It is clear that in intertropical less developed countries energy from biomass (firewood, charcoal, alcohol, biogas) is an important way if not the only one, to cover energy needs.

In temperate, developed and crowded countries, biomass energy is subjected to discussions which generally give not much hope for the future. The possibilities in U.S.A. have been studied and ciphered in detail by Burwell (1978) who is moderately optimistic. In reality, the discussions are often sterile, because much more is known about the transformation of biomass in energy than about the bioconversion of solar energy (and CO_2!) in biomass (or dead organic matter of wastes) by the ecosystems.

Starting from the fact that biomass may be utilized as food, feed, fuel, paper and derivated products, textiles, furniture, construction materials, and may be transformed by thermic or biologic conversion into synthetic industrial products (also ethylene which is the base of plastics industry) Lipinsky (1978) has proposed the constitution of integrated adaptative biomass systems which may produce all kinds of desired products and could be modified following the evolution of needs and constrains of the population concerned.

A polyvalent system of polyvalent biogeocenoses would lead to thief modifications of territory management, but isn't it the way to recover the lost paradise?

For the tropics, Janzen (1973) proposes sustained yeald tropical agroecosystems (SYTA) = mixture of biogeocenoses, or new artificial ecosystems whose structure would take into account the resistance to climate, bad soils and pests and would realize during the whole year the needed bioconversions for food or money.

THE GLOBAL CARBON CYCLE

From a good estimation of the respective areas of the principal biogeocenoses characteristic of the different zonobiomes and megaecotones conveying the continents (FAO or vegetation maps) and from data on their Bl and PN1, it is possible to get approximate values of biogeospheric biomass and productivity (solar energy bioconversion). One may compare the values obtained by Bazilevich et al. (1971), Whittaker and Likens (1975), Duvigneaud (1979, 1980) (Table 1).

Bazilevich et al. has calculated (and mapped) the potential Bl and PN1 from an indirect method of climatologic indexes; Whittaker and Likens, and also Lieth, have considered the situation in 1950, when human action was not yet too strong, Bl and PN1 are means from personal measurements and literature data; Duvigneaud (1979) has considered the situation in 1976, Duvigneaud (1980) has combined zonobiomes cartography and climax data to get theoretical values of Bl and PN1 for a biosphere where human actions would cease (plesioclimax).

Table 1. Biomass and Productivity of the Biosphere (10^9t organic matter, dry weight)

	Bl 10^9t	PN1 10^9t/y	
Potential	2402	172	Bazilevich, 1971
Plesioclimax	2043	125	Duvigneaud, 1980
In 1950	1841	118	Whittaker et al., 1975
In 1976	1289	136	Duvigneaud, 1979

SOLAR ENERGY BIOCONVERSION AT THE ECOSYSTEM LEVEL 611

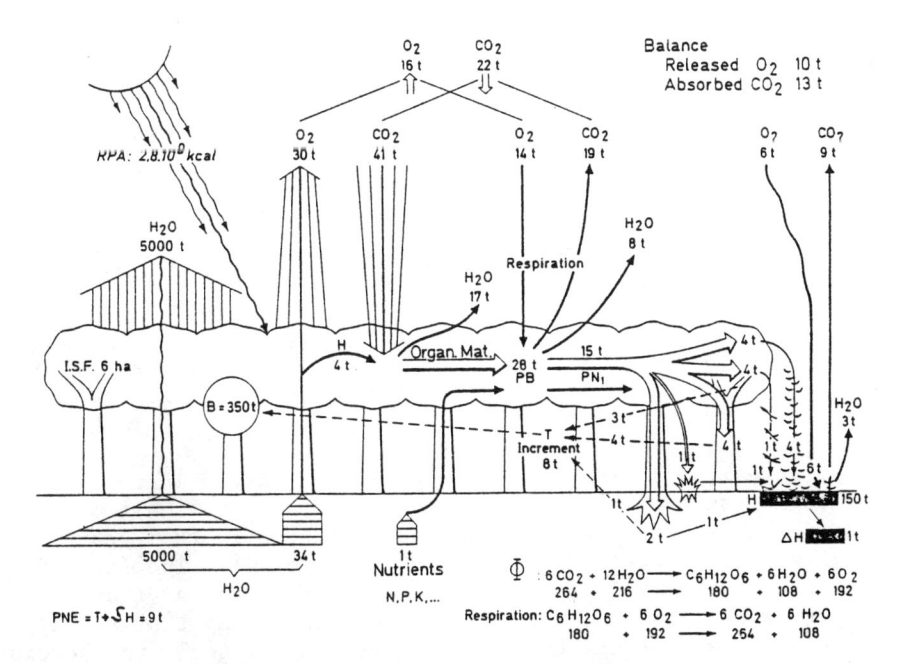

Fig. 5. Functioning of a temperate deciduous forest biogeocenosis in Belgium.

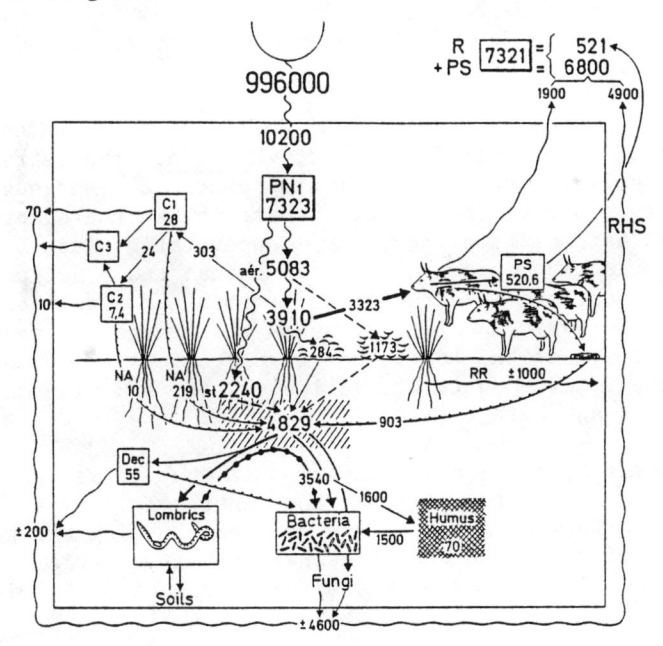

Fig. 6. Functioning of a temperate "permanent pasture" biogeocenosis. Data from Ricou, 1979.

Table 1 illustrates the well known fact that actually man diminishes the phytomass of the continents to increase their primary productivity.

The difficulties of mapping the actually existing situation of the biospheric biocenoses and estimating their solar energy bioconversion activities come from millenary human actions which are continuing year by year to commit their ravages: deforestation, savanisation, steppification, erosion, desertification; compensations are reforestation and agrarisation. The area of erodable soils dimishes, but what is progressing in all regions is the impoverishment of soils fine particles (clay and loam) and a consequent loss of fertility in the U.S.A. $\simeq 20$ t/ha/an are lost following more and more elaborate tillage techniques, the solution being an agriculture without tillage.

Every year, $\simeq 18 \cdot 10^9$ t CO_2 ($5 \cdot 10^9$ t C) are emitted in the atmosphere as a result of fossil fuels combustion. About half of it remains in the atmosphere, accumulating dangerously from year to year (glassroom effect and climate modifications?). The fate of the disappearing $2.5 \cdot 10^9$ t C is disputed: oceanologists say that oceans may only absorb $1 \cdot 10^9$ t C; terrestrian ecologists say that actually forests and savanna trees burning for shifting agriculture, heating and cooking in the tropics and humus oxydation in agroecosystems are adding biospheric CO_2 (may be $5 \cdot 10^9$ t C) to the fossil fuels CO_2 emission towards atmosphere (Woodwell et al., 1978); for yet ignored reasons, the oceans would be able to capture much more CO_2 as generally admitted.

A question whose answer is also controversial is: even if their surface is slightly diminished, why <u>do not</u> the terrestrial ecosystems <u>profit</u> of the excess of atmospheric CO_2, in augmenting their primary productivity? Some authors invoke the adaptation of plants to an optimal atmospheric CO_2 concentration (see Lemon, 1977), which nearly corresponds to the actual one (330 ppm); others* say that this optimum is situated at 5 times the actual concentration (1000-1500 ppm). Really, it seems that there are many ecological reasons, and that solar energy bioconversion depends first of all on the total amounts of available water and mineral nutrients (see Goudriaan and Altaj, 1979).

Nevertheless, the actual situation is what it is; and we share the opinion of Adams et al. (1977) that in the last decades most of the excess CO_2 from fossil fuel and net wood burning has gone into rapid circulation between the atmosphere and terrestrial biomass with a net lowering of the mean residence time of carbon in the land biomass.

---

*See Strain 1978, Wittwer, 1978.

SOLAR ENERGY BIOCONVERSION AT THE ECOSYSTEM LEVEL 613

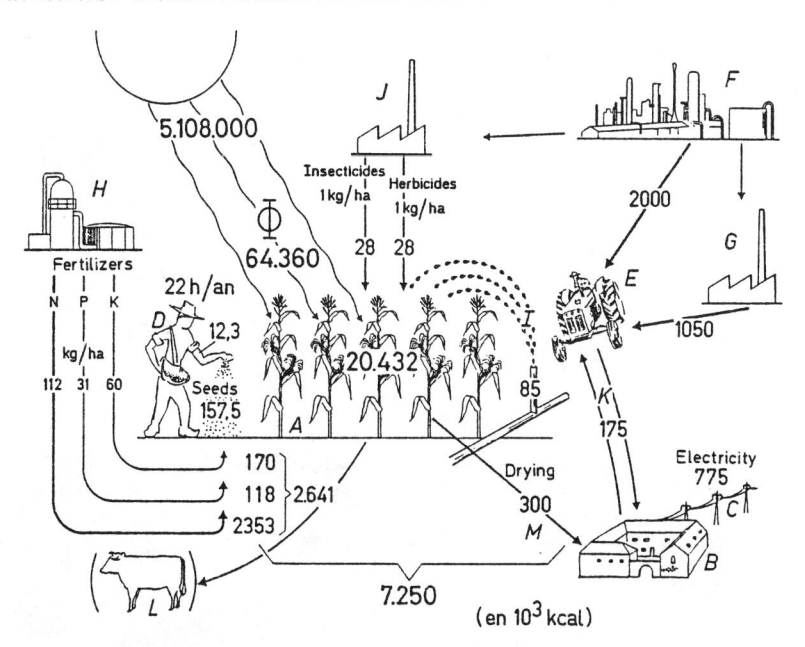

Fig. 7. Import of subsidiary energy in an American "corn field" biogeocenosis. Data from Pimentel 1973.

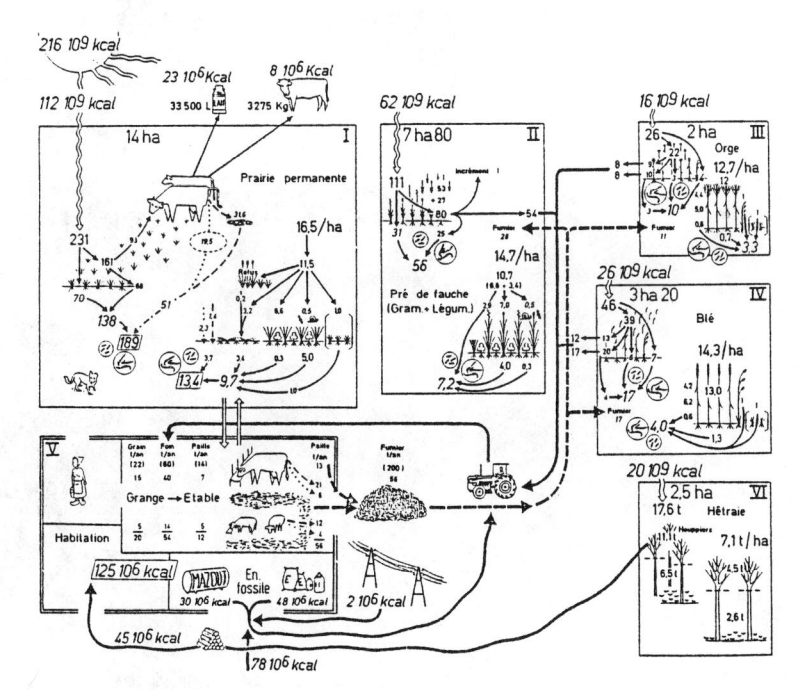

Fig. 8. Agroecosystem. A farm in the Belgian Ardennen (after Duvigneaud et al., 1978).

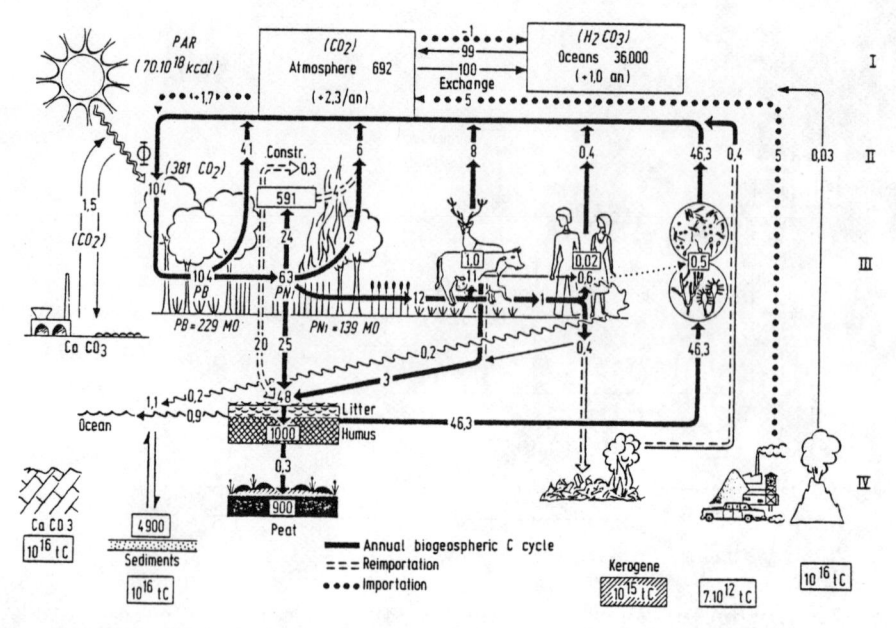

Fig. 9. The global Carbon Cycle on the continents (after Duvigneaud, 1980). For explanations, see text.

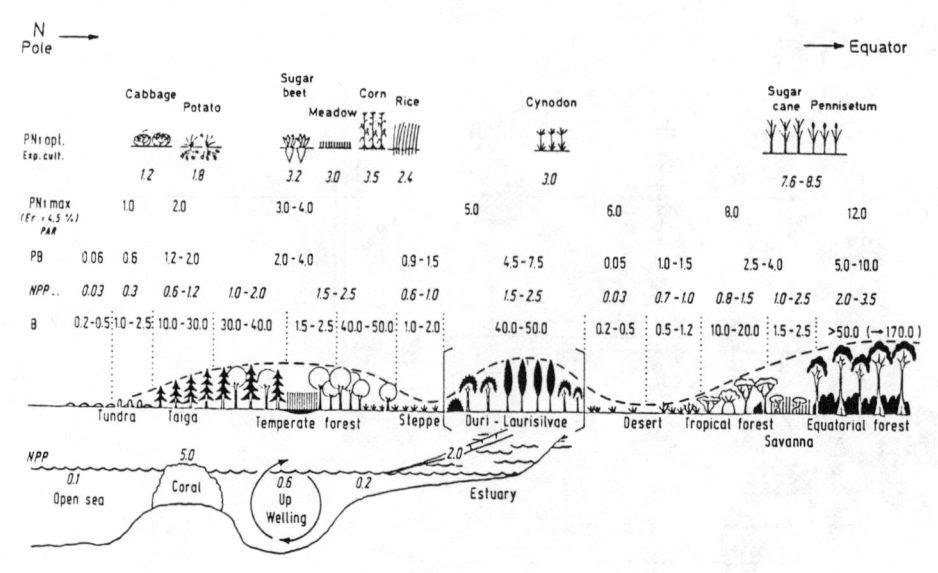

Fig. 10. North-to-South zonation of biomes, phytomasses and productivities (inspired from Bazilevich et al., 1971 and date of Nichiporovich, 1967).

SOLAR ENERGY BIOCONVERSION AT THE ECOSYSTEM LEVEL 615

The actual CO_2 accumulation in the atmosphere may be attributed to a momentary "laziness" of terrestrial ecosystems (Duvigneaud, 1980) easily explained by accelerated destructive human actions, caused by fast growing population and modern technology development; but if the existence of the biogeosphere is to continue, this anthropogenic laziness of terrestrial ecosystems must and will cease, not only because of men's instinct of conservation but principally because man will develop an intelligence of conservation (Peccei, 1976).

Fig. 9 is an attempt to represent the actual global carbon cycle, the mode of representation being inspired by Nichiporovich (1967).

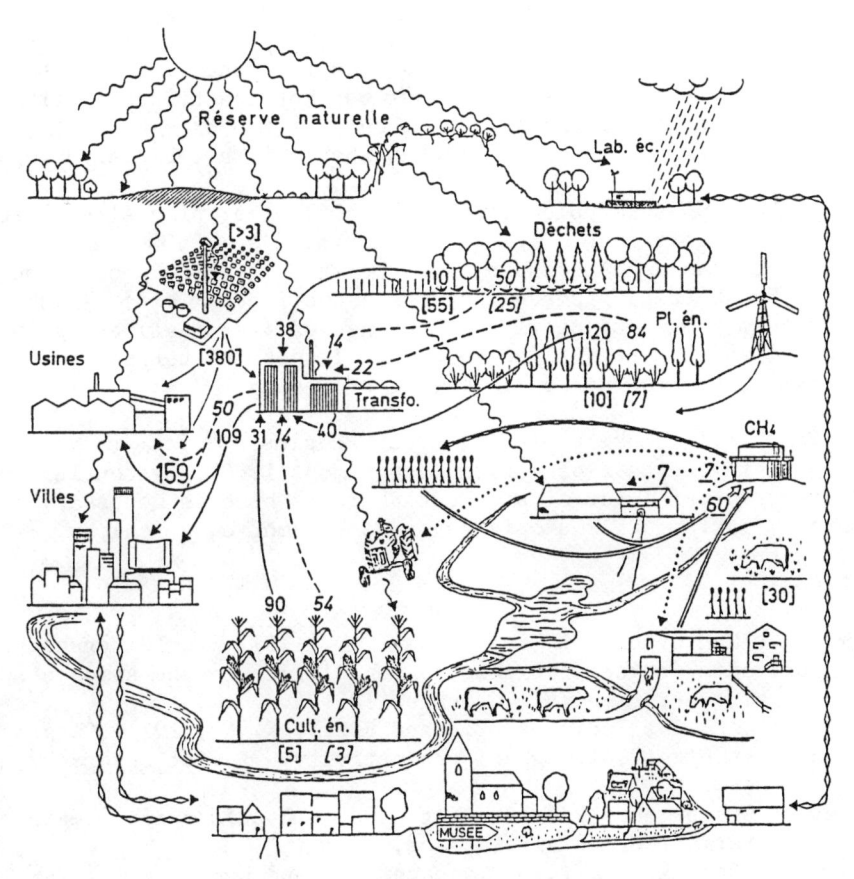

Fig. 11. The recovered paradise? Plan Alter for France. "Tout solaire" unit of 135,000 ha, providing 159,000 tep biomass. Full lines: solid fuels; dashed lines: liquid fuels; between brackets: surfaces. (Data from project Alter à long terme pour la France, 1978.)

THE NOOSPHERE

Figure 10 gives an optimistic note on the future of biospheric solar energy conversion to biomass. Following Nichiporovich (1967) and many others, genetics and agronomy could raise to 4.5% the field photosynthetical efficiency, which mean is actually 0.5-2%, for PAR. Some good results have already been obtained in research stations.

This better use of solar energy and atmospheric CO_2 coupled with better water budget and mineral cycling to produce many more quantities and qualities of biomass may lead to a new world, which should be the recovered paradise and should approach the Vernadskii's noosphere.

Yes, solar energy bioconversion utilized all together for the production of food, feed, technological materials, chemicals of all types, raw material for a renovated industry, energetic fuels, (remember Lipinsky and Janzen) and last but not least, beauty of natural and anthropic landscapes, should be the base of what they call an ecodevelopment based on what they now call biotechnology.

Two examples for biomass energy only: the "plan Alter," tout solaire pour la France (Fig. 11, illustrates a 130,000 Ha unit; 265 these units would be necessary for the whole territory of France), and the Sol Sverige program (Johansson and Steen, 1977); in those middle term programs, nearly all the needed (terminal) energy would come from solar energy bioconversion. I spoke of terrestrial ecosystems but oceans may be still more promising.

For "well thinking people," for "Panglosses" for whom everything goes well in the best of worlds (see Simon, 1980), my conclusions may appear utopia; but at a moment where it becomes necessary to innovate, part of this utopia may become tomorrow's reality.

BIBLIOGRAPHY

Adams, J., Mantovani, M., and Lundell, L., 1977, World Versus Fossil Fuel as a Source of Excess Carbon Dioxide in the Atmosphere. A preliminary Report. Science, 196:54.

Bazilevich, N. I., Rodin, L. Y., and Rozov, N. N., 1971, Geographical aspects of biological productivity. Soviet Geography, 293.

Bellevue, Groupe de, 1978, Project Alter: Esquisse d'un régime à long terme tout solaire, Syros, Paris.

Bolin, B., Degens, E. T., Duvigneaud, P., and Kempe, S., 1979, The global Biogeochemical Carbon Cycle in Bolin et al., The Global Carbon Cycle, Scope 13, Wiley, Chichester:1.

Burwell, C. C., 1978, Solar Biomass Energy. An overview of U.S. potential. Science, 199:1041.

Chartier, P. et Meriaux, S., 1980, L'énergie de la biomasse. La Recherche, 113:766.
Duvigneaud, P., 1946, La variabilité des associations végétales. Bull. Soc. Roy. Bot. Belg., 78:107.
Duvigneaud, P., 1953, Les savanes du Bas-Congo. Essai de Phytosociologie topographique. Lejeunia, Liège, Mém. 10.
Duvigneaud, P., 1975, Structure, Biomasses, Minéralomasses, Productivité et Captation du Plomb dans quelques associations rudérales. Bull. Soc. Roy. Bot. Belg., 108:93.
Duvigneaud, P. et Kestemont, P., 1977, Productivité biologique en Belgique. Duculot, Gembloux.
Duvigneaud, P., 1980, La synthèse écologique, Doin, Paris.
Glazovskaya, M. A., 1968, Geochemical Landscapes and Types of Geochemical Soil Sequences, 9th Int. Cong. Soil Sc. Trans., Adelaïde, 4:303.
Janzen, D. H., 1973, Tropical Agroecosystems. Science, 182:1212.
Johansson, T. B. and Steen, P., 1978, Solar Sweden, Ambio; 7:20.
Kuhnholtz-Lordat, G., 1945, La silva, le saltus et l'ager de garrigue. Ann. Ec. Nat. Agric., Montpellier, 26:1.
Kuhnholz-Lordat, G., 1949, La cartographie parcellaire, INRA, Paris.
Lieth, H. and Whittaker, R. H., 1975, Primary Productivity of the Biosphere, Ecol. Stud. 14, Springer, New York, Heidelberg.
Lipinsky, E. S., 1978, Fuels from Biomass. Science, 199:644.
Monsi, M., 1960, Dry-matter Reproduction in Plants. 1. Schemata of Dry-matter reproductions, Bot. Mag., Tokyo, 73:81.
Morison, C. G. T., Hoyle, A. C., and Hope-Simpson, J. F., 1948, Tropical soil-vegetation catenas and mosaics, J. Ecol., 36:1.
Nichiporovich, A. A., 1967, Photosynthesis of productive systems, Editor, Jerusalem.
Olson, J., Pfuderer, H., and Yip-Hoi, C., 1978, Changes in the Global Carbon Cycle and the Biosphere. Oak Ridge Nat. Lab.
Simon, J. L., 1980, Resources, Population, Environment: An over supply of False Bad News, Science, 208:1431.
Smith, R. F., 1980, Experts endose Biomass Energy, Science, 208:1018.
Sukachev, V. N., 1954, Quelques problèmes théoriques de la phytosociologie, Essais de Botanique, Acad. Sc. URSS, 310.
Walter, H., 1977, Vegetations zonen und Klima (3. Aufl.), Ulmer, Stuttgart.
Walter, H., 1964, Die Vegetation der Erde, in Oko-Physiologischer Betrachtung, Fischer, Jena (Bd. 1).
Woodwell, G. M. and Whittaker, R. H., 1968, Primary Production in Terrestrial Communities, Am. Zool., 8:19.
Woodwell, G. M., Whittaker, R. H., Reiners, W. A., Likens, G., Delwiche, C., and Botkin, D., 1978, The Biota and the World Carbon Budget, Science, 199:141.

UTILIZATION OF SOLAR RADIATION BY PHYTOPLANKTON

J.F. Talling

Freshwater Biological Association
Ambleside
Cumbria, England

INTRODUCTION

Of the solar radiation which reaches the surface of the earth, much the greater part encounters water-masses devoid of macroscopic plant cover. Instead, only very dilute suspensions of planktonic and mainly unicellular algae are significant as biological interceptors and transformers. Their performance as photosynthetic systems has been studied intensively and (geographically) extensively by marine and freshwater biologists during the last 30 years. Most of this work was inspired by the possibility of using measurements of photosynthesis to assess community productivity, thereby escaping certain formidable difficulties in obtaining such assessments from population increments. Although the results are obviously influenced by many environmental variables, the varying input and attenuation of solar radiation are all-important, and their relation to patterns of photosynthetic utilization is far from simple. The present contribution aims to illustrate this relation, rather than to review a numerology of productivity estimates. Most examples are taken from experience in freshwaters, but all the problems are common to marine areas.

PHOTOSYNTHESIS - DEPTH PROFILES : DIVERSITY AND UNITY

A convenient starting point is the depth-variation or vertical profile of photosynthetic activity, assessed per unit volume of water in terms of carbon uptake or oxygen evolution. The area below a profile then represents the depth-integral of photosynthesis, expressed per unit area (usually 1 m^2) of habitat. Such profiles can be of very varied shape in stratified waters where local accumulations of algae exist in discrete layers as a result of sinking, active migration, differential mortality, or growth in situ. Thus deep-seated maxima of algal biomass are apparently common in the northeast Pacific Ocean, and introduce some localized 'weighting' into depth-profiles of photosynthesis measured in these clear waters (e.g. Anderson, 1969). If such bias is eliminated by recalculating activity per unit of biomass-index (e.g. chlorophyll a: Fig. 3), or if the photosynthetic zone is more uniform from turbulent mixing, the photosynthesis - depth profile is of rather predictable shape. It can be divided into successive regions of light-inhibition, near-saturation, and near-exponential decline at greater depth. These may dominate profiles even in stratified populations (e.g. Tilzer et al. 1975, Fig. 1), although here population shifts may spuriously amplify the apparent 'surface inhibition'.

In long-maintained studies on two British lakes (L. Neagh, L. Leven) in which vertical mixing is prevalent, Bindloss (1974, 1976), Jewson (1976), and Jones (1977) have demonstrated seasonal series

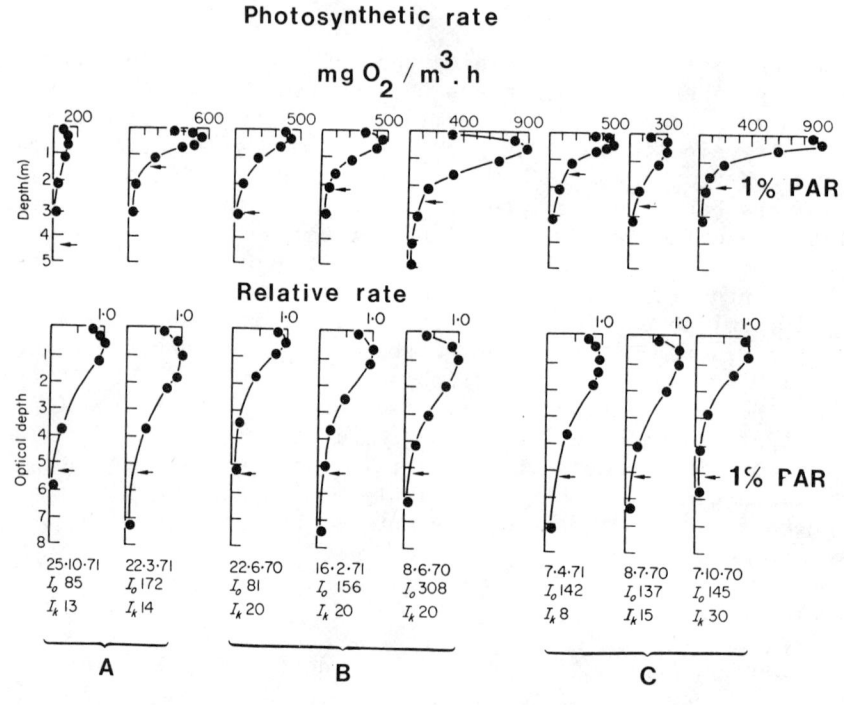

Fig. 1. Depth-profiles of photosynthesis by phytoplankton in Loch Leven, Scotland, shown (above) on absolute scales and (below) on relative scales. Downward shifts accompany increase in surface irradiance I_0 (A, B) or decrease in the saturation parameter I_k (C). From Bindloss (1976).

of profiles of regular shape. These undergo vertical displacement from increasing surface irradiance (I_o) or decreasing onset of light-saturation (I_k parameter) (see Fig. 1). The latter effect is often associated with lower temperature, although some sun-shade differentiation or 'adaptation' may also be involved, and the saturation (I_k) parameter may often reflect changes in a temperature-sensitive or age-sensitive photosynthetic capacity (Talling, 1957a; Yentsch and Lee, 1966; Lastein and Gargas, 1978).

As a very rough first approximation, neglecting spectral complexities, the depth-axis can be taken to represent an exponential decline of irradiance (PAR). Consequently, if rates of photosynthesis are assembled as a function of irradiance (e.g. Capblancq, 1972, Fig. 30) and the latter plotted on a logarithmic scale (e.g. Pyrina 1979, Fig. 5), the resulting curve reproduces the main features of the photosynthesis-depth profile. In particular, the basis of the near-exponential decrease of photosynthesis at greater depths is evident. It is in this region that the greatest efficiency of energy (or quanta) utilization can be expected.

Although the irradiance variables I_o (surface-incident flux) and I_k (saturation-onset) affect the shape of photosynthesis-depth profiles, much bigger variability in nature is caused by the vertical compression of profiles in waters of high light attenuation and by their horizontal expansion in waters of high volume-based activity. These two shifts are usually strongly correlated, because of the attenuation associated with unit concentration of algal biomass; examples are provided by wide-ranging surveys in East African lakes (Talling 1965, Fig. 5) and in Irish lakes (Jewson and Taylor 1978, Fig. 7). They do not, however, alter the intrinsic (scale-independent) shape of the profiles. This can be extracted by replotting profiles with activity relative to the maxima and depth on an optical scale (e.g. Talling 1965, Fig. 6; Bindloss 1976, Fig. 8) or more simply by a double logarithmic plot (Talling 1975, Fig. 10.3). The latter brings out a fundamental limitation of photosynthetic productivity per unit area, arising from the impossibility of combining very high activity per unit volume with very deep profiles. This limitation centres on the ratio between photosynthetic capacity per unit biomass (P_{max}) and the increment of attenuation coefficient per unit biomass (ε_s), which have opposing effects on the area enclosed by a depth-profile (i.e. areal productivity).

In recent years the second (specific attenuation) parameter, ε_s, has been evaluated from a number of field and laboratory studies. These include (Fig. 2) work on L. Leven in Scotland (Bindloss, 1976) L. George in Uganda (Ganf and Viner 1973, Ganf 1974), and L. McIlwaine in Zimbabwe (Robarts, 1979), in which ε_s is defined as the increment in the minimum vertical attenuation coefficient over the spectrum (ε_{min}) per unit increment of biomass concentration as chlorophyll a.

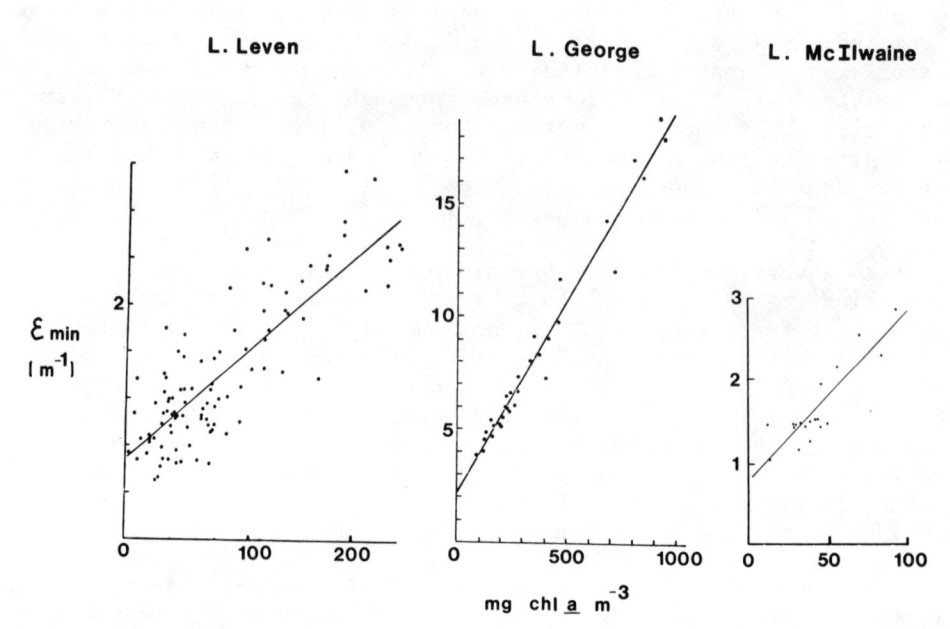

Fig. 2. Increase in the (spectral) minimum attenuation coefficient ε_{min} with increase of phytoplankton concentration (as chlorophyll a) in three lakes. From Bindloss (1976), Ganf and Viner (1973), Robarts (1979).

Values from these sites range from 0.008 to 0.021 m^{-1} (mg m^{-3})$^{-1}$ [i.e. m^2 mg^{-1}], broadly representing the range obtained elsewhere or by other methods (e.g. Harris, 1979, Fig. 26). In many analyses the vertical attenuation of PAR is represented, without spectral resolution, by an 'average' attenuation coefficient. In these cases the biomass-specific coefficient ε_s has a composite spectral basis and is liable to change with depth, in relation to the spectral modification of underwater radiation and the absorption spectra of algal cells in vivo (Atlas and Bannister, 1980).

UNDERWATER RADIATION : SOME EFFECTS OF SPECTRAL MODIFICATION

The spectral modification of underwater radiation has many consequences, of which one has been noted above. Another concerns the decrease with depth of irradiance (PAR), which is often expressed by a single 'average' attenuation coefficient but which - even in unstratified water - is actually steeper near the surface because of the eclipse of spectral regions characterised by high attenuation coefficients. At greater depths, the gradient will gradually

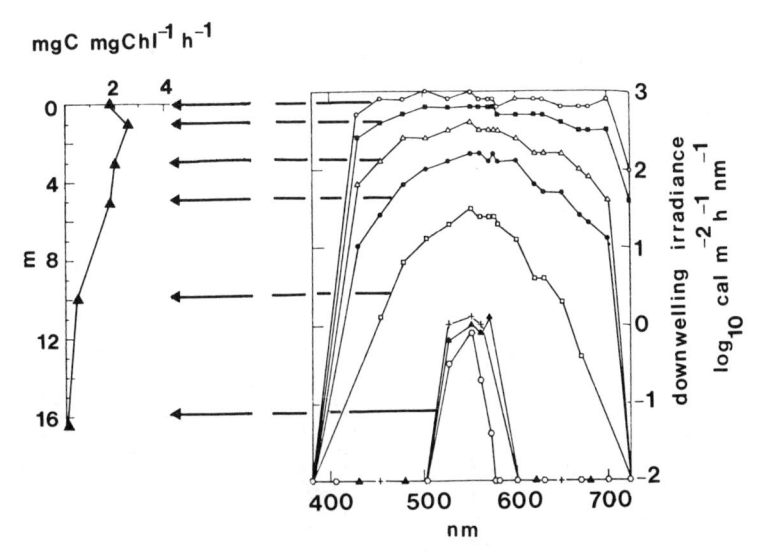

Fig. 3. Depth-profile of photosynthetic activity in L. Kinneret, with the spectral distribution of irradiance measured at corresponding depths. Adapted from Dubinsky and Berman (1976).

approach that defined by the (spectral) minimum attenuation coefficient, ε_{min}. This coefficient is therefore useful in defining a dimensionless scale of optical depth, e.g. as $\varepsilon_{min} \cdot z / \ln 2$, where z is depth in metres, to which diverse profiles of photosynthesis and irradiance (PAR) can be referred and standardised (Figs 1, 5). It has also been used extensively in the mathematical formulation of photosynthetic productivity per unit area (e.g. Talling 1957b, 1965; Bindloss, 1974, 1976; Jewson, 1975, 1976; Jones, 1977). In the latter aspect, an overall (PAR-related) coefficient of (1.20 to 1.33) times ε_{min} will often provide a close match to the behaviour observed.

The attenuation spectra of natural waters, measured <u>in situ</u>, usually show a single broad 'window' of low attenuance bounded by increases towards the lower and upper limits of the visible spectrum. In very clear waters, such as the offshore Pacific or Crater Lake, Oregon (Tyler and Smith, 1967; Smith and Tyler, 1967), the window is a property of H_2O itself, and lies in the blue region, but it tends to shift to progressively longer wavelengths as the overall attenuance increases. Stages in this red-shift are illustrated by Lake Kinneret the Sea of Galilee (Dubinsky and Berman, 1976; see Fig. 3) and Lake

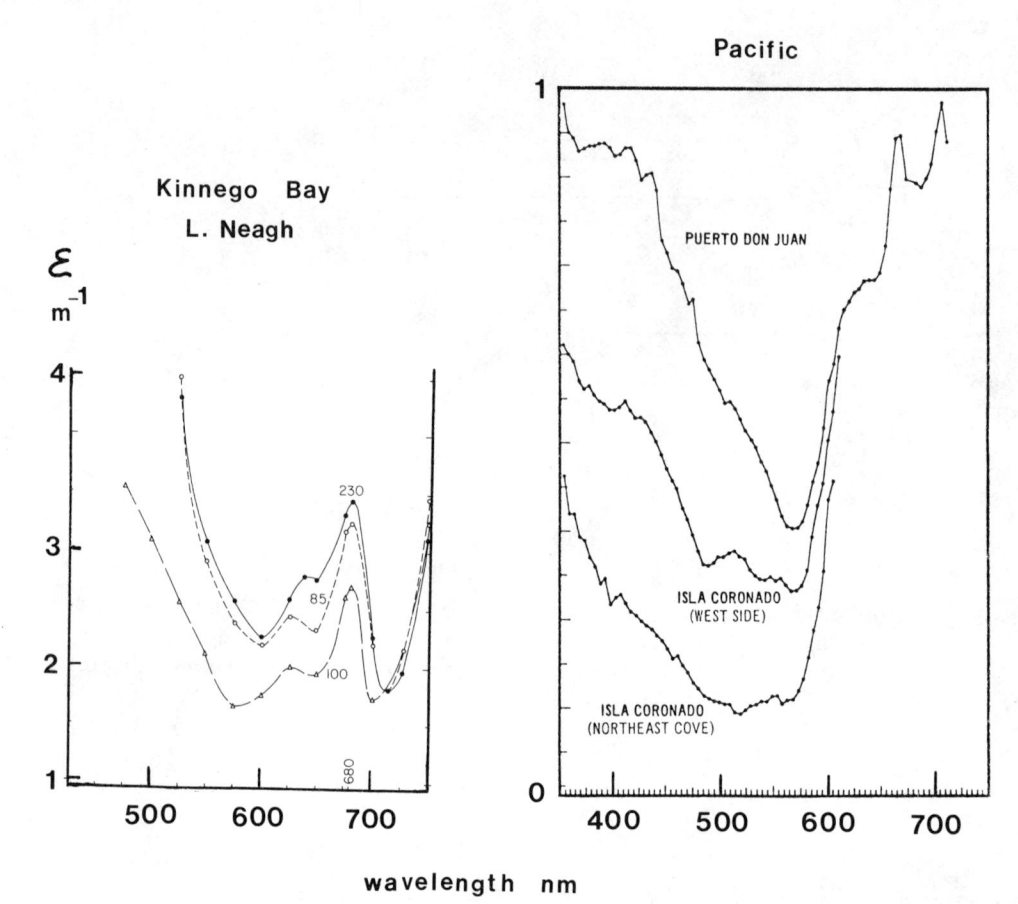

Fig. 4. Spectral variation in the vertical attenuation coefficient
ε : (left) in a productive bay of L. Neagh, Ireland, on
three dates with varying concentrations of chlorophyll a
shown in mg m^{-3}, (right) in three coastal Pacific waters off
California. From Jewson and Taylor (1978), Tyler and Smith
(1967).

George, Uganda (Ganf, 1974), in transitions from offshore to coastal sea water (Jerlov, 1976), and from unproductive to productive lakes within local areas (e.g. Talling, 1971; Jewson and Taylor, 1978; Kirk, 1979).

The red shift of minimum attenuance is often associated with more abundant plankton and photosynthetic activity, although extracellular pigments (e.g. 'yellow substance', related to humic materials : cf. Kirk, 1976) may be equally or more important than photosynthetic pigments in directing the shift. In particular, they may dominate over the blue absorption maximum of chlorophyll a. Its red absorption peak receives less competition from other pigments (excepting H_2O), and is often resolved (see Fig. 4) by suitable spectroradiometers in productive natural waters (e.g. Tyler and Smith, 1967; Jewson and Taylor, 1978; Kirk, 1979). Attenuance maxima related to the phycocyanin of blue-green algae have also been observed (Jewson and Taylor, 1979). Unlike chlorophyll a, such accessory pigments often have absorption maxima well-placed in the transmission window of natural waters; the fucoxanthin of diatoms, and peridinin of dinoflagellates, are other important examples.

RADIANT ENERGY : PARTITION AND EFFICIENCY OF UTILIZATION

Long ago, G.L. Clarke (1939) interpreted the framework of planktonic photosynthesis in terms of a competition for energy between photosynthetic and residual pigments in a water-column. In recent years, attempts to quantify this competition (or partition) have intensified. Most have used a specific attenuation coefficient referred to unit chlorophyll a, ε_s (also often denoted k_c), multiplied by the chlorophyll a concentration to obtain the component of attenuation coefficient referrable to algal cells, and the ratio of the latter to the total attenuation coefficient as a measure of the fractional absorption of PAR by algal cells. This treatment neglects the inherent difference in scattering media between attenuation and true absorption, and also the wavelength (and hence depth) - dependence of both algal and total attenuation coefficients. 'Sieve effects' and varying contents of accessory pigments also introduce real variations in ε_s unrepresented by any standard value. The results, therefore, can be judged as no more than useful first approximations.

Three examples relate to the overall light-interception within a water-column. Megard et al. (1979, Fig. 8) have illustrated the competitive rôle of background non-algal attenuation in a series of lakes, showing how increasing concentrations of biomass are required for a given relative photosynthetic performance in the more turbid waters. Lemoalle (1979, Fig. F12) found that the magnitude of photosynthetic productivity per unit area in Lake Chad was closely proportional to the percentage of total attenuation ascribable to phytoplankton. This relationship would not be expected in other waters (such as Loch Leven : Bindloss, 1974) where increased biomass density is associated with reduced photosynthetic capacity. Lastly, Bannister (1974a, b) has derived a comprehensive set of theoretical formulations for depth-integrated photosynthesis, although his original choice of a single representative value for the specific attenuation parameter (ε_s) was later qualified (Atlas and Bannister, 1980).

Estimates of energy-flux (or quantum-flux) partition can obviously be extended to assess the _efficiency_ of photosynthetic yield. The earlier calculations for phytoplankton were concerned only with overall efficiency as the dimensionless ratio (or percentage) between the total input and output energy fluxes per unit area. They showed (e.g. Winberg, 1948 - corrected for an arithmetic error) that short-term (hourly or daily) values were generally below 1%. Annual means would be much lower at the latitudes concerned, although this is not necessarily true for tropical regions (e.g. Talling, 1965; Ganf, 1975), from which many of the highest estimates of photosynthetic yield (c. 10 g C m^{-2} day^{-1}) are derived.

These efficiency estimates are clearly much lower than the maxima (about 20 - 25%) likely to be attainable by the algal photosynthetic apparatus under daylight (PAR) conditions. Sub-factors of efficiency, combined multiplicatively in the overall estimate, include (i) fraction of PAR in the total short-wave solar radiation, (ii) fractional surface transmission after loss of radiant energy by reflection and back-scattering, (iii) fractional interception of underwater radiation (PAR) by photosynthetic pigments, (iv) fractional area occupied by a photosynthesis-depth profile in relation to the area of a profile unlimited by light-saturation and -inhibition, (v) maximum efficiency set by quantum yield at low irradiance and spectral composition. Of these sub-factors, (i) and (ii) are relatively invariable and average about 0.46 ($\pm$ 0.05) and 0.90 ($\pm$ 0.05) respectively. Sub-factor (iv) depends chiefly upon the ratio between surface-penetrating irradiance I_o and the light-saturation parameter I_k (Talling, 1957b; Goldman, 1979) and values of 0.2 to 0.4 are probably typical for moderately bright daylight. If these values are considered together with the estimates of observed overall efficiency and of (v) cited earlier, it seems inescapable that low values for the interception sub-factor (iii) strongly influence the low overall efficiencies commonplace in nature. This is still true even if somewhat lower values for the quantum yield subfactor (v) are judged to be more realistic under natural conditions, such as the estimate by Bannister (1974a) of 0.06 mol C Einst^{-1}. Compare, however, this estimate with the values of 0.0019 to 0.0061 mol O_2 Einst(PAR)$^{-1}$ measured directly by Melack (1979) for overall photosynthetic yields per unit area by phytoplankton in productive East African lakes.

In the last decade a similar analysis of photosynthetic efficiency has been applied to observed yields per unit volume of strata within photosynthesis-depth profiles. If such a yield (m^{-3}) is taken in a direct ratio to the corresponding irradiance (m^{-2}), the resulting 'efficiency' (Platt, 1969; Tilzer et al., 1975) is a complex and dimensional (m^{-1}) quantity. However, if radiant power consumption per unit volume is used as the reference quantity, a dimensionless true efficiency is obtained. In a depth-element δz, power consumption is equal to the product of irradiance I_z, attenuation coefficient ε, and thickness δz, less a correction (usually small and neglected) for backscattering (Smith and Tyler, 1967).

Thus if the cellular component of attenuation can be estimated, by the product of biomass concentration and a chosen (rarely measured!) ε_s value, the corresponding cellular component of power consumption can be estimated. This, in turn, can be used - with reservations already mentioned - to calculate conversion efficiencies in strata throughout the profile.

This approach has been most effectively used when detailed spectroradiometric data on underwater radiation were available, as in the work by Dubinsky and Berman (1976) on L. Kinneret and by Morel (1978) on some sea areas off Central America and North Africa. These authors have compared their estimated efficiencies with those expected from highest quantum yields; in one case (Morel, 1978, Charcot cruise) the former were possibly 'uncomfortably high'. All these calculations are limited by the choice of a representative value for the specific attenuation parameter (ε_s). However, the important limitation of incomplete light interception by photosynthetic pigments is again demonstrated. This, and other efficiency sub-factors, are illustrated (Fig. 5) by a recalculation of data from two African lakes.

RELATIVE RADIATION INPUT AND GROWTH EFFICIENCY

A final comment concerns relative and often dimensionless measures of radiation input and growth of phytoplankton. Absolute measures of incident solar radiation (I_o) are very non-linearly related to the integral photosynthesis in a water column, so reducing the analytical value of direct quotients or normal overall efficiencies. However, a logarithmic transformation of I_o relative to a fraction of the I_k (saturation) parameter - $\ln I_o/0.5\, I_k$ - often restores proportionality; it can be also integrated with time and used to assess both the effective light-climate and photosynthetic efficiency (Talling, 1957b, 1965, 1971; Ganf, 1975; Jewson, 1975, 1976; Bindloss, 1976; Jones, 1977).

Likewise, growth yields of phytoplankton often relate fundamentally to exponential increase, and logarithmic transformations are commonly used in analyses of population dynamics. Thus, in most deep temperate waters, the winter minimum of radiation income controls a winter minimum of population density and hence of photosynthetic productivity. During spring the increased radiation income often induces a prolonged phase of exponential growth. In this the overall (absolute) conversion of solar energy is initially low, even though the relative yield as cell divisions (a logarithmic derivative) is typically high - and may constitute the major relative population change of the yearly cycle. Thus, an apparent lag between increased input and absolute biological response can mask an important relative conversion.

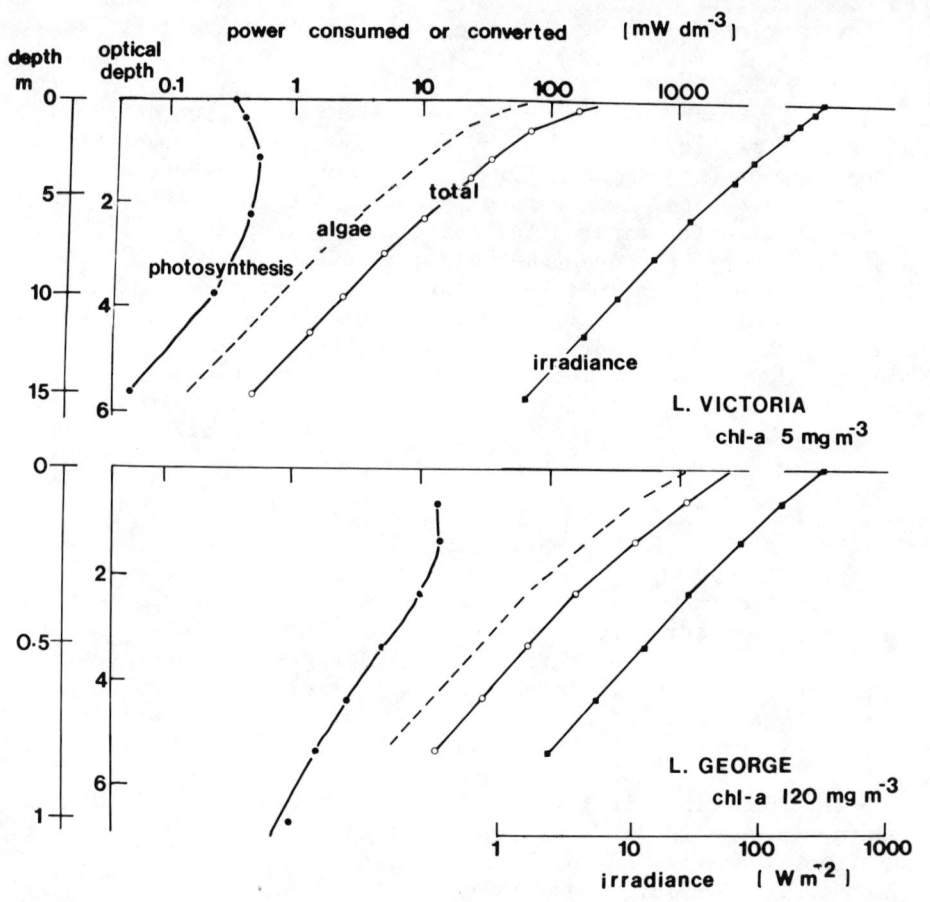

Fig. 5. The vertical attenuance of irradiance in two African lakes of differing productivity, and of power consumed or converted in unit volume (1 dm^3) ascribable to the total absorption, interception by algae if $\varepsilon_s = 0.016$ m^2 mg^{-1}, and photosynthesis based on the equivalence factor of 14.2 J (3.4 cal) [mg O$_2$]$^{-1}$. From data of Talling (1965, expts 13 and 30) and Ganf (1974).

REFERENCES

Anderson, G.C., 1969, Subsurface chlorophyll maximum in the northeast Pacific Ocean, Limnol. Oceanogr., 14:386-391.

Atlas, D., and Bannister, T.T., 1980, Dependence of mean spectral extinction coefficient of phytoplankton on depth, water colour, and species, Limnol. Oceanogr., 25:157-159.

Bannister, T.T., 1974a, Production equations in terms of chlorophyll concentration, quantum yield, and upper limit to production, Limnol. Oceanogr., 19:1-12.

Bannister, T.T., 1974b, A general theory of steady state phytoplankton growth in a nutrient saturated mixed layer, Limnol. Oceanogr., 19:13-30.

Bindloss, M.E., 1974, Primary productivity of phytoplankton in Loch Leven, Kinross, Proc. R. Soc. Edinb. B., 74:157-181.

Bindloss, M.E., 1976, The light-climate of Loch Leven, a shallow Scottish lake, in relation to primary production by phytoplankton, Freshwat. Biol., 6:501-518.

Capblancq, J., 1972, Phytoplancton et productivité primaire de quelques lacs d'altitude dans les Pyrénées, Annls Limnol., 8:231-321.

Clarke, G.L., 1939, The utilization of solar energy by aquatic organisms, in: "Problems of Lake Biology", F.R. Poulton, ed., Am. Ass. Adv. Sci. Publ. no. 10, pp. 27-38.

Dubinsky, Z., and Berman, T., 1976, Light utilization efficiencies of phytoplankton in Lake Kinneret (Sea of Galilee), Limnol. Oceanogr., 21:226-230.

Ganf, G.G., 1974, Incident solar irradiance and underwater light penetration as factors controlling the chlorophyll a content of a shallow equatorial lake (Lake George, Uganda), J. Ecol., 62:593-609.

Ganf, G.G., 1975, Photosynthetic production and irradiance - photosynthesis relationships of the phytoplankton from a shallow equatorial lake (Lake George, Uganda), Oecologia, 18:165-183.

Ganf, G.G., and Viner, A.B., 1973, Ecological stability in a shallow equatorial lake (Lake George, Uganda), Proc. R. Soc. B., 184:321-346.

Goldman, J.C., 1979, Outdoor algal mass cultures - 2. Photosynthetic yield limitations, Wat. Res. 13:119-136.

Harris, G.P., 1979, Photosynthesis, productivity and growth: the physiological ecology of phytoplankton, Ergebn. Limnol., 10:1-171.

Jerlov, N.G., 1976, "Marine optics", Elsevier, Amsterdam.

Jewson, D.H., 1975, The relation of incident radiation to diurnal rates of photosynthesis in Lough Neagh, Int. Rev. ges. Hydrobiol. Hydrogr. 60:759-767.

Jewson, D.H., 1976, The interaction of components controlling net phytoplankton photosynthesis in a well-mixed lake (Lough Neagh, Northern Ireland), Freshwat. Biol., 6:551-576.

Jewson, D.H., and Taylor, J.A., 1978, The influence of turbidity on net phytoplankton photosynthesis in some Irish lakes, Freshwat. Biol., 8:573-584.

Jones, R.I., 1977, Factors controlling phytoplankton production and succession in a highly eutrophic lake (Kinnego Bay, Lough Neagh). II. Phytoplankton production and its chief determinants, J. Ecol., 65:561-577.

Kirk, J.T.O., 1976, Yellow substance (gelbstoff) and its contribution to the attenuation of photosynthetically active radiation in some inland and coastal south-eastern Australian waters, Aust. J. mar. Freshwat. Res., 27:61-71.

Kirk, J.T.O., 1979, Spectral distribution of photosynthetically active radiation in some south-eastern Australian waters, Aust. J. mar. Freshwat. Res., 30:81-91.

Lastein, E., and Gargas, E., 1978, Relationship between phytoplankton photosynthesis and light, temperature and nutrients in shallow lakes, Verh. int. Verein. Limnol., 20:678-689.

Lemoalle, J., 1979, "Biomasse et production phytoplanctoniques du Lac Tchad (1968-1976). Relations avec les conditions du milieu", ORSTOM, Paris.

Megard, R.O., Combs, W.S., Smith, P.D., and Knoll, A.S. 1979, Attenuation of light and daily integral rates of photosynthesis attained by planktonic algae, Limnol. Oceanogr., 24: 1038-1050.

Melack, J.M., 1979, Photosynthetic rates in four tropical African fresh waters, Freshwat. Biol., 9:555-571.

Morel, A., 1978, Available, usable, and stored radiant energy in relation to marine photosynthesis, Deep-Sea Res., 25:673-688.

Platt, T., 1969, The concept of energy efficiency in primary production, Limnol. Oceanogr., 14:653-659.

Pyrina, I.L., 1979, Primary production of phytoplankton in the Volga, in: "The River Volga and its life", Mordukhai-Boltovskoi, Ph.D., ed., pp. 180-194, Junk, The Hague.

Robarts, R.D., 1979, Underwater light penetration, chlorophyll a and primary production in a tropical African lake (Lake McIlwaine, Rhodesia), Arch. Hydrobiol., 86:423-444.

Smith, R.C., and Tyler, J.E., 1967, Optical properties of clear natural water, J. opt. Soc. Am., 57:589-595.

Talling, J.F., 1957a, Photosynthetic characteristics of some freshwater plankton diatoms in relation to underwater radiation, New Phytol., 56:29-50.

Talling, J.F., 1957b, The phytoplankton population as a compound photosynthetic system, New Phytol., 56:133-149.

Talling, J.F., 1965, The photosynthetic activity of phytoplankton in East African lakes, Int. Rev. ges. Hydrobiol. Hydrogr., 50:1-32.

Talling, J.F., 1971, The underwater light climate as a controlling factor in the production ecology of freshwater phytoplankton, Mitt. int. Verein theor. angew. Limnol., 19:214-243.

Tilzer, M.M., Goldman, C.R., and de Amezaga, E., 1975, The efficiency of light energy utilization by lake phytoplankton, Verh. int. Verein. Limnol., 19:800-807.
Tyler, J.E., and Smith, R.C., 1967, Spectroradiometric characteristics of natural light under water, J. opt. Soc. Am., 57:595-601.
Winberg, G.G., 1948, The efficiency of utilisation of solar radiation by plankton, Priroda, 12:29-35. (Russ.)
Yentsch, C.S., and Lee, R.W., 1966, A study of photosynthetic light reactions, and a new interpretation of sun and shade phytoplankton, J. mar. Res., 24:319-337.

LIMITING FACTORS IN PHOTOSYNTHESIS- FROM THE CHLOROPLAST

TO THE PLANT CANOPY

        Jean-Louis Prioul

        Structure et Métabolisme des Plantes
        Bât. 490 - Université de Paris-Sud
        91405 Orsay Cedex France

INTRODUCTION

    The efficiency for solar energy conversion in the most efficient crop like sugar-cane approaches 2 % when expressed on a yearly basis (Varlet-Granchet et al 1978). This yield could seem rather low when compared to the efficiency of the photochemical apparatus which is higher than 30 % under optimal conditions. Despite their low efficiency, plants are still interesting energy converters because they have solved a major problem for solar energy conversion which is the storage of energy under an easily usable form for various purposes (Calvin 1980, this volume). The aim of this paper is to review the main causes of decrease in yield from the chloroplast to the canopy and to look at the possible improvements. This analysis is limited to plants grown under non-limiting conditions for water, temperature and mineral nutrition. The effect of water stress and mineral deficiency is very important in natural ecosystems and is discussed more specifically by Talling (1980, this volume) for fresh water population and by Divigneaud (1980, this volume) for terrestrial biocenosis.

    Four levels of organization will be considered : photochemical system, chloroplast, leaf and canopy .

PHOTOCHEMICAL SYSTEM

    A minimum of 8 quanta is needed to extract 2 electrons from the inner face of the thylakoid membrane to outer face and to form 2 $NADPH_2$ and 2 to 4 ATP (Hall 1980, this volume). With red actinic light (660 nm) the energetic yield is higher 30 %. The limiting factors of this reaction are light, chlorophyll content and organization and electron acceptor regeneration (Junge 1980 this volume).

CHLOROPLASTS AND CARBON CYCLES

C_3 and C_4 plants differ very clearly as far as energy requirement for CO_2 reduction is concerned. In C_3 plants, carbon dioxide is fixed in the chloroplast by ribulose-bis phosphate carboxylase (RUBPCase), the PGA formed are reduced using 2 $NADPH_2$ and 2 ATP and one more ATP is needed to regenerate the acceptor. The quantum requirement is at least 10 quanta/CO_2 but this high energetic yield (22 %) can only be obtained under low oxygen conditions (1-2 %) because the RuBPCase acts also as an oxygenase and is the first in photorespiratory cycle. During photorespiration part of the fixed CO_2 is evolved and then recycled but ATP and $NADPH_2$ are consumed creating a high energy loss. Hatch (1970) evaluated that, if photorespiration is 50 % of gross photosynthesis, at least 8 ATP and 4 $NADPH_2$ are needed for one CO_2. Experimental values, obtained by Viil and Parnik (1979), confirm and even exceed these figures under certain conditions.

In C_4 plants (like maize and many tropical native grasses) CO_2 is fixed in the mesophyll cell cytoplasm by phosphoenolpyruvate carboxylase and the C_4 acid formed is reduced into malate by $NADPH_2$. Then malate goes to bundle sheath chloroplasts where it is decarboxylated. The evolved CO_2 is re-fixed in the Calvin cycle. As bundle sheath chloroplasts possess an incomplete photochemical apparatus half of the PGA formed has to be reduced in the mesophyll. The functioning of that cycle needs 2 more ATP than the Calvin cycle because of phosphoenolpyruvate regeneration. The quantum requirement is about 14 quanta/CO_2 (energetic yield 18 %) but there is no oxygen effect due to the CO_2 concentrating mechanism occuring at the RuBPcase sites.

LEAF GAS EXCHANGE

Low Light Conditions

Under low light, net leaf photosynthetic rate is linearly related to absorbed light and the slope of that straight line is the apparent quantum yield (mol CO_2/mol absorbed quanta). Ehleringer and Björkman (1977) have shown by comparing six C_3 to five C_4 wild species (table 1) that the better efficiency of C_3 at low oxygen concentration disappears at ambient concentration and that there is no O_2 effect on C_4 apparent yield.

The magnitude of the counteracting effect of photorespiration on quantum yield is very dependent upon temperature and is much less important for C_3 grasses or cereals grown under their normal temperature conditions.

Table 1. Quantum requirement (absorbed mol quanta/mol CO_2) and yield for CO_2 uptake of intact leaves from C_3 and C_4 species under low light (leaf temperature 30°C, partial CO_2 pressure 325 µbar, light 400-700 nm). Recalculated from Ehleringer and Björkman 1977.

	2 % O_2		21 % O_2	
	quantum requirement	energetic yield	quantum requirement	energetic yield
C_3 species mean (standard deviation)	13,6 (0,15)	17 %	19,1 (0,49)	12 %
C_4 species mean (standard deviation)	18,6 (0,37)	12 %	18,8 (0,38)	12 %

The difference between the observed yield at leaf level and the theoritical chloroplast yield comes mainly from the absorption of light by non photosynthesing structure. Another loss in yield originates from the fact that part of the incident light (approximatively 20 % depending upon leaf surface properties) is reflected.

Light and CO_2 Response Curves

Under natural conditions plants do not work very often in low-light conditions and when irradiance increases the response is no longer linear and a levelling off is observed. This plateau is linearly dependent on CO_2 concentration. So as originally discussed by Blackman (1905) it can be said that CO_2 is the limiting factor under high light whereas light is the limiting factor under low light.

In CO_2 limiting conditions, CO_2 diffusion and carboxylation control the process. Using resistance analogy 3 resistances may be defined along CO_2 pathway from ambient air to chloroplast : r_a, boundary layer resistance, r_s stomatal resistance and r_m mesophyll resistance. Connecting together the resistance network with a model for Calvin cycle, Chartier et al 1970 derived a general equation which fits very well with the C_3 plant experimental curves. All the parameters may be determined experimentally and give a quantitative picture of the limiting factors at the leaf level (fig. 1).

When comparing light and CO_2 response curves from C_3 and C_4 plants some striking differences appear : there is no true light

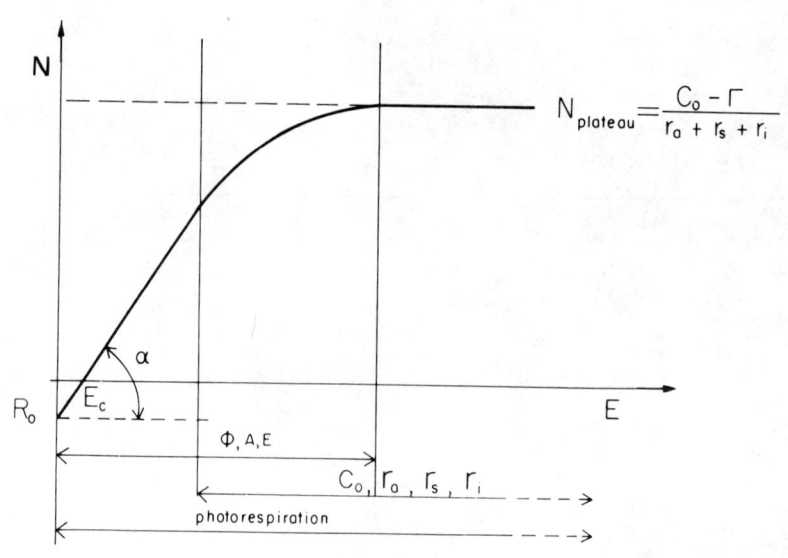

Fig. 1. Net photosynthesis (N) - light (E) response curve for a C3 plant. R_o : dark respiration, E_c : light compensation point, ϕ quantum yield, A : light absorption coefficient, C_o : ambient CO_2 concentration, Γ : CO_2 compensation point, r_a : boundary layer resistance, r_s : stomatal resistance, r_i : intracellular resistance (From Chartier and Bethenod 1977). The arrows in the lower part of the figure recall the zone of influence of each parameter.

saturation in C4 plants (fig. 2 left), and the CO_2 is saturating at ambient CO_2 concentration (330 v.p.m.). The total resistance to CO_2 (expressed by the reciprocal of the initial slope of N-CO_2 curve) is much lower in C4 than in C3 plants (fig. 2 right). To take into account C4 responses slight modifications of the model are necessary (Chartier and Bethenod 1977).

Comparison of experimental values of stomatal and intracellular resistances from various C3 and C4 plants (table 2) shows that the main barriers to CO_2 diffusion and fixation are intracellular in C3 plants whereas in C4 plants the stomata barrier is important. Whether the high intracellular resistance in C3 plants is mainly due to low carboxylation efficiency or to resistance to CO_2 transfer between cell walls and carboxylation sites is still under controversy (see Prioul and Chartier 1977 for further discussion on this point).

CANOPY LEVEL

At this level the yield is controlled by two types of factors : physical and biological. The efficiency of crop energy conversion (ϵ = chemical energy stored in biomass/Incoming solar energy) may

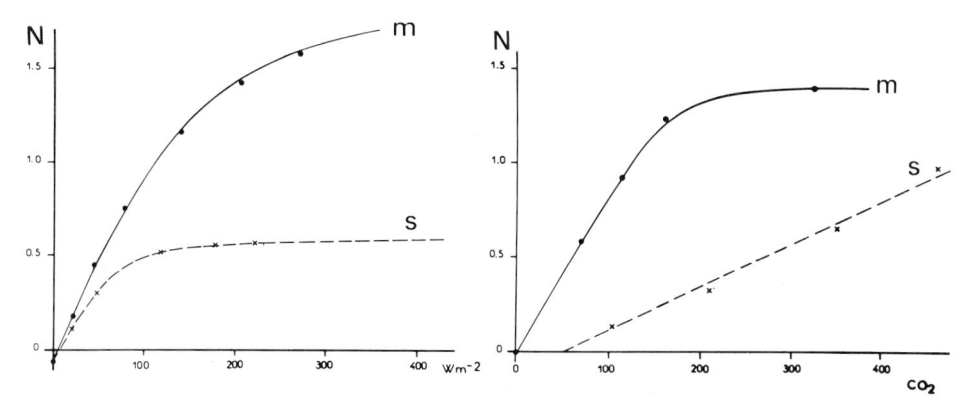

Fig. 2. Comparison of net photosynthesis light and CO_2 response curves in a C_3 plant (sugar-beet = s) and C_4 plant (Maize = m) redrawn from Chartier and Bethenod 1977.

Table 2. Minimal values recorded for stomatal (r_s) and intracellular (r_i) resistances to CO_2 transfer of some C_3 and C_4 species. Boundary layer resistance in these experimental conditions was lower than 90 $m^{-1}s$ (high ventilation).

	r_s ($m^{-1}.s$)	r_i ($m^{-1}.s$)	References
C_3 species			
Phaseolus vulgaris	110	720	Chartier et al 1970
Lolium multiflorum	250	670	Prioul 1971
Calapogonium mucunoïdes	108	300	Ludlow & Wilson 1972
Vigna luteola	470	370	Ludlow & Wilson 1972
Beta vulgaris	330	560	Chartier & Bethenod 1977
C_4 species			
Sorghum almum	125	136	Ludlow & Wilson 1972
Panicum maximum	73	63	Ludlow & Wilson 1972
Pennisetum purpureum	56	113	Ludlow & Wilson 1972
Zea mays	200	60	Chartier & Bethenod 1977

be conveniently partitioned in three components (Varlet-Grancher et al 1978) :

- ε_c climatological efficiency = photosynthetic active radiation, PAR/total solar radiation.
- ε_i interception efficiency = absorbed PAR/Incoming PAR
- ε_b biological efficiency = chemical energy stored in biomass/absorbed PAR.

ε being equal to $\varepsilon_c \times \varepsilon_i \times \varepsilon_b$.

Physical Factors

About 52 % of incident light is not usable for photosynthesis (ε_c = 0.48 Varlet-Grancher et al 1978) which causes a two-fold drop in energetic yield. The absorption of PAR radiation by the canopy depends upon interception properties and mainly leaf area, which, for convenience, is expressed per unit ground area and is called L.A.I. (leaf area index). The relationship between interception efficiency (ε_i) and LAI (fig. 3) shows that a leaf area 3 to 4 times that of the ground is necessary to get full interception (closed canopy). For annual crop a critical point for energy conversion efficiency is the time necessary to get complete interception i.e. the leaf area growth rate. For LAI < 3 the differences in ε_i between species are due to differences in canopy structure : leaf angle and orientation. The influence of leaf angle at low LAI is illustrated in fig. 4 and shows that the effect is less important than previously thought by first authors. The advantage of erectophile geometry disappears later in the season.

Physiological Factors

- Net photosynthesis. Some of the factors important at leaf level are still apparent at the canopy level. As reviewed by Gifford (1974) canopy light curves of C_3 and C_4 plants cultivated in their normal environment, temperate for C_3 and tropical for C_4, retain most of the characteristics observed on individual leaf : under low light and moderate temperature C_3 canopy is more efficient whereas C_4 canopy perform much better under high light and high temperature. This reflect the good adaptation of each metabolism to its usual growth condition. By difference to leaf response curves canopy response curves have a lower curvature due to the leaf penetration in the successive leaf layer of the stand.

- Respiration becomes a very important factor in canopy carbon balance because of the part played by non photosynthetising organs : roots, petioles, stems and storage tissues. For a long time it was assumed that respiration was proportional to dry matter leading to the erroneous concept of optimum LAI. In 1966-1970 Mc Cree and coworkers demonstrated that night respiration R may be partitioned

LIMITING FACTORS IN PHOTOSYNTHESIS   639

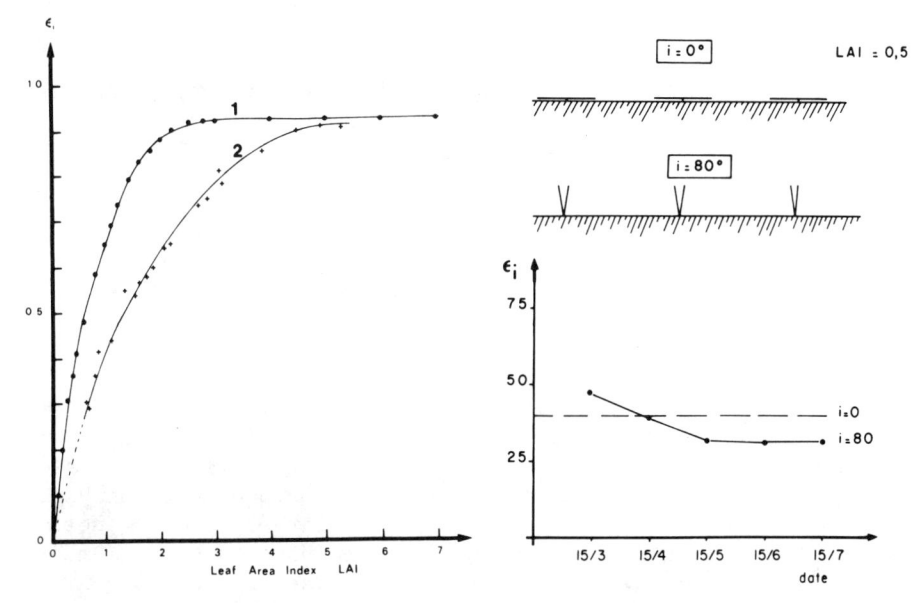

Fig. 3. Interception or trapping efficiency (ε_i) as a function of leaf area index (LAI) for 2 tropical crops : C_3 legume (cowpea) and C_4 grass (Sugar cane). From Varlet-Grancher et al 1978.

Fig. 4. Trapping efficiency as a function of date in the growing season (latitude 50° North) for horizontal (i = 0°) or erected leaves (i = 80°) at the same LAI (0.5). From Chartier

in two parts : one growth component being proportional to cumulated gross photosynthesis of the previous day (D) and one maintenance component proportional to dry weight (W expressed in CO_2 units for sake of homogeneity) :

$$R = kD + cW$$

k is the growth coefficient and c the maintenance coefficient. The following values have been observed for k and c :

k	c	plant	reference
0.25	0.015	White clover	Mc Cree 1970
0.19	0.012	Sub-clover nodulated	} Silsbury 1977
0.14	0.014	Sub-clover non-nodulated	

and mean that between 14 % to 25 % of the CO_2 fixed one day is lost the following night and the more than 1 % of dry weight CO_2 content is lost each day. The growth coefficient k expresses the cost of growth;

for example k which is higher in nitrogen fixing legumes than in nitrate assimilating plants quantify the cost of nitrogen reduction.

Figure 5 illustrates the importance of respiratory losses in controlling growth rate. Day-time and night gas exchanges were compared during growth for 2 canopies differing in initial planting density (420 and 4800 plants.m^{-2}). The initial difference in net photosynthesis disappears when leaf area, in low density crop, is sufficient to get full light interception (LAI < 3) but there is still a difference in respiration rate (triangles) which continue on a long period such as that the cumulated differences in respiration explain the better final growth rate and production of low density stands.

- <u>Dry matter distribution</u> between harvestable and non harvestable part is a factor of paramount importance in crop economy. This factor is actually beyond the scope of photosynthesis *stricto sensu* but is partly under the dependence of light through photomorphogenesis. However the dry matter partitioning is largely under genetic control, this property has been and is still used for improvement through plant breeding. Table 3 illustrates the critical importance of grain : straw ratio and so-called harvest index.

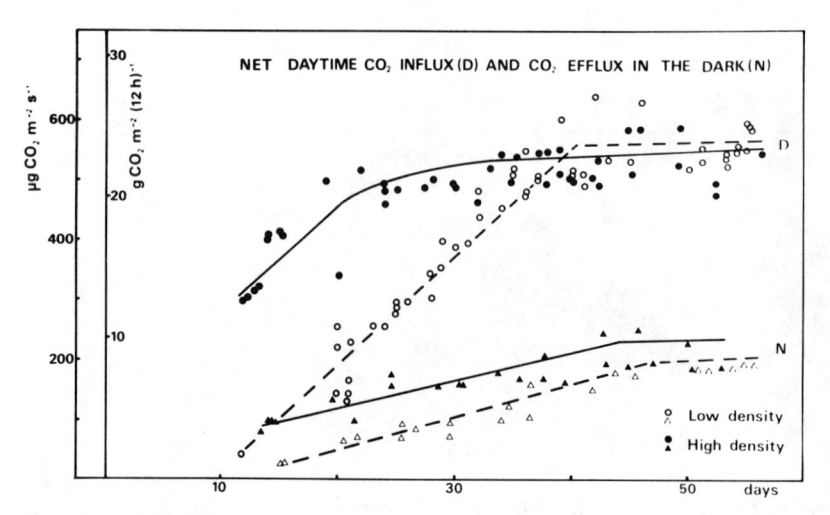

Fig. 5. Net day time CO_2 uptake (**D**) and dark day night efflux (**N**) during the growth of Subterraneum Clover canopies, with low planting density (o △) and high planting density (● ▲).

Table 3. Genotypic variation in total dry matter and harvestable dry matter in Rice (from Dossou-Yovo 1980).

Genotypes	Cigalon	G 30	G 34 A
Total dry matter g.plant^{-1} (roots included)	34.5	49.5	63
Grain dry matter g.plant^{-1}	10.9	14.5	12.6
Harvest Index (grain/straw)	0.57	0.57	0.33

CONCLUSION

The main limiting factors at each of the four levels of organization considered are summarized in table 4. Several features appear : a factor limiting at one level is not necessarily important at the superior one and may be offset by the emergence of new limitation.

Table 4. Summary of main limiting factors from chloroplast to plant canopy. * indicated possible improvements.

Level of organization	Maximum yield	Limiting factors
Photochemical system	30-33 % (low light)	-Light, chlorophyll organization -Regeneration of electron acceptors
Chloroplast	C_3 22 % (low light-1% O_2)	-CO_2 diffusion - carboxylation -Photorespiration -Light
	C_4 18 % (low light)	- Light
Leaf	C_3 17 - 12 % (1% O_2 21% O_2 low light)	(*?)-CO_2 diffusion-carboxylation (*?)- Photorespiration -Light : leaf absorption
	C_4 12 % (low light)	-Light (*) -Stomatal resistance
Canopy	1 - 2 % (year $^{-1}$)	(*)-Light interception at young stage:leaf area expansion (*)-Respiratory losses : when closed (*)-Dry matter distribution : harvest index

For example the role of chlorophyll and photochemistry disappears progressively from the chloroplast to the canopy whereas CO_2 diffusion, and respiration and, then, carbon distribution become predominant.

The large drop in maximum yield between leaf and canopy originates from the fact that time scale is changed and that one departs from linear conditions. The final observed yield is the resultant of the variable conditions of light and climate encountered by the crop during the growing season. In term of the energetic yield the period between germination and full light interception is very critical.

So perennial crops would possess a greater advantage. On this basis, Gifford 1974 suggested that the better high production of C_4 plants could be explained by the perennial character rather than by biochemical ability. In fact comparative measurements of annual C_3 and C_4 under tropical or temperate climate do indicate an advantage of C4 plant considered (table 5).

Table 5. Comparison of maximum efficiency of C3 and C_4 plant under temperate or tropical climate. ε_C : climatological efficiency = 0.48 ε_i = interception efficiency ε_B biological efficiency. From Varlet-Grancher et al 1978 and Varlet-Grancher 1980 personal communication.

Site	Species	Cycle length days	ε_I	ε_B	ε
West Indies (Guadeloupe) latitude 16°N	Sugar cane (C_4)	340	0.70	0.06	0.020
	Cow-Pea (C_3)	76	0.70	0.032	0.010
North France (Mons) 50°N	Maize (C_4)	202	0.41	0.057	0.011
France (Versailles) 49°N	Alfalfa (C_3 perennial)	80	0.51	0.049*	0.012*

* shoot only

REFERENCES

Blackman, F.F., 1905, Optima and limiting factors, Ann. Bot. 19 : 281.
Ehleringer and Björkman, O., 1977, Quantum yields for CO_2 uptake in C_3 and C_4 plants, Plant Physiol., 59 : 86.
Chartier, P., Chartier, M., and Catsky, J., 1970, Resistances for carbone dioxide diffusion and for carboxylation as factors in bean leaf photosynthesis, Photosynthetica 4 : 48.
Chartier, P., and Bethenod, O., 1977, La productivité à l'échelle de la feuille, in : "Les Processus de la Production Végétale Primaire" A. Moyse, ed., Gauthier-Villars, Publisher, Paris.
Dossou-Yovo, S., 1980, Variabilité génétique, productivité et réponse à l'ombrage chez le riz. Thesis n° 415 Université Paris-Sud ORSAY.
Gifford, R., 1974, A comparison of potential photosynthesis, productivity and yield of plant species with differing photosynthetic metabolism, Austr. J. Plant Physiol., 1 : 107.
Hatch, M.D., 1970, Chemical energy costs for CO_2 fixation by plants with differing photosynthetic pathways, p. 215 in "Prediction and Measurement of Photosynthetic Productivity", PUDOC, Publisher, Wageningen.
Ludlow, M.M., and Wilson, G.L., 1972, Photosynthesis of tropical pasture plants IV. Austr. J. Biol. Sci. 25 : 1133
Mc Cree, K.J., 1970, An equation for the rate of respiration of white Clover plants grown under controlled conditions, p 221, in "Prediction and Measurement of Photosynthetic Productivity", PUDOC, Wageningen.
Prioul, J.L., 1971, Réactions des feuilles de Lolium multiflorum à l'éclairement pendant la croissance et variation des résistances aux échanges gazeux photosynthétiques, Photosynthetica 5 : 364.
Prioul, J.L., and Chartier, P., 1977, Partitioning of transfer and carboxylation components of intracellular resistance to photosynthetic CO_2 fixation : a critical analysis of the methods used, Ann. Bot. 41 : 789.
Silsbury, J.H., 1977, Energy requirement for symbiotic nitrogen fixation, Nature 267 : 149.
Varlet-Granchet, C., Bonhomme, R. and Chartier, P., 1978, Efficiency of sugar cane and cowpea as solar energy converters, p 797, in "International Solar Energy Congress" vol. 1, ISES, (ed), Solar Energy Soc. India, Publisher, New Delhi.
Viil, J., and Parnik, T., 1979, Parameters of reductive pentose phosphate cycle and of the glycolate pathway under different concentrations of oxygen, Z. Pflanzenphysiol. 95 : 213.

BIOCONVERSION OF SOLAR ENERGY

Melvin Calvin

Laboratory of Chemical Biodynamics
University of California
Berkeley, California 94720*

Photosynthesis is examined as an annually renewable resource for material and energy. The direct photosynthetic production of hydrocarbons for fuel and materials from known plant sources is examined, as well as the production of fermentation alcohol from sugar cane. Experiments are underway to analyze the hydrocarbons from Euphorbias, and other hydrocarbon-containing plants, with a view toward determining their chemical components. In addition, several experimental plantings of Euphorbia lathyris have been completed. There are indications that we may expect a yield of at least 10 barrels of hydrocarbon material per acre, together with enough fermentable sugar to produce about 12 barrels of ethanol, in a 7-month growing period at a competitive cost.

Also, based upon what we know of the primary quantum conversion sequence in photosynthesis, a number of synthetic systems have been constructed which are designed to capture and store quanta in the form of stable chemical materials, some as energy sources and some as special and valuable products difficult to obtain in any other way.

---

*The work described was sponsored, in part, by the Office of Basic Energy Sciences, Division of Chemical Sciences, and the Biomass Energy Systems Division of the Department of Energy under Contract W-7405-eng-48.

1. INTRODUCTION

A review of the status of the green plants' use directly as energy capturing, converting and storing machines is appropriate as well as discussion of the use of the products of those green plants. Once it is learned how the green plant captures the quanta from the sun and converts them into long-lived chemical energy-containing and material products it should be possible to synthetically construct those parts of the plant that are the quantum-capturing and storing apparatus without the need for soil, water, agriculture or bioprocesses.

Remains of ancient photosynthesis form the basis for the current supplies of oil, gas and coal. These supplies have taken over 100 million years to generate and are currently being depleted at an ever faster rate. We should make efforts to develop methods of using the "current income" from the sun for materials and fuel, which can be done through the agency of the green plants which are on the surface of the earth today.

2. DEVELOPMENT OF PETROLEUM PLANTATIONS

The principal product of green plant solar collection is carbohydrate, that is, half-reduced carbon. There are some plants which store energy as soluble, directly fermentable carbohydrate. These are sugar cane, sugar beets and many of the food grains. That type of fermentable carbohydrate can be converted into a higher calorie-containing material by fermentation, whereby a solid is converted into a liquid without a substantial energy loss, from a material which is very useful biologically (sugar cane, corn, etc.) to one which has a broader use industrially (fermentation alcohol).

a. Alcohol for fuel

The Brazilians have recently adopted conversion of sugar cane to alcohol as a possible means of solving part of their energy problem, and since Brazil is the largest cane growing country in the world, they have a real opportunity.

In 1974, Brazil produced 19 million tons of raw sugar and 700 million liters of fermentation alcohol from the molasses left from the sugar. In 1975, the Brazilian government made the decision to go this route as a major source of fuel and chemicals. In 1978, the Brazilian production of fermentation alcohol from the sugar cane is 2.4 billion liters, and the projection is 4 billion liters in 1979.

In 1980 five companies in Brazil will produce 300,000 automobiles to run on alcohol alone and will convert some 100,000 standard cars to alcohol with the addition of a small heat exchanger converter.

In the United States "gasohol" (10% alcohol in gasoline) is now widely available in the Midwest and even in California a portion of the state-owned automobiles run on gasohol. In Nebraska, the production of gasohol from corn actually is an energy positive process, as the distillers' dried yeast which is a byproduct of fermentation supplements cattle feed, replacing the fraction of the corn crop which was diverted to alcohol. Cattle actually gain more weight per pound of feed from the distillers' dried yeast than they did from corn alone.

b. Hydrocarbon-producing plants

It should be possible to find green plants that carry the reduction of carbon all the way down to hydrocarbon. There are, in fact, whole families of plants that do that and the one that is commercial today is the Hevea rubber tree (a member of the Euphorbiaeceae family). This is not a useful species for the United States and we must find plants which will grow under our climatic conditions and produce similar hydrocarbon materials.

In the Euphorbiaeceae family there is a genus Euphorbia, almost every species of which produces the same kind of hydrocarbon in the form of latex (an emulsion of oil and water) but the molecular weight of the hydrocarbon in that emulsion is much lower than in Hevea and, therefore, is useful as an alternate energy and fuels source.

We have used two species of Euphorbia in experimental plantings in Southern California (and on our own ranch in Northern California). The first of these is an annual, Euphorbia lathyris (gopher plant) (Figure 1) which propagates from seed and comes to the harvesting stage in about 6 to 7 months with 20 inches of rainfall. The plant before harvest is about 4 feet high and it is cut, crushed and extracted for its hydrocarbon content. Another species, Euphorbia tirucalli, is a perennial and takes 2 to 3 years to come to harvest and become a tree (bush) which can be trimmed or tapped. As yet it is not known how long one of the E. tirucalli trees will be useful as a source of hydrocarbon-like materials. In Okinawa, a Japanese plastics company has begun the development of a plantation of E. tirucalli, and they have been very successful with their agronomic practices. The yields have been of the order of 15 barrels of oil equivalent/acre. They have now reached the state, after two years of development, of constructing a prototype extraction plant, and the first "green oil" has become available on a very small scale.

Figure 1. Euphorbia lathyris, Southern California

Both the Euphorbia species mentioned above contain in their latex a phorbol ester which is a toxic substance, being an irritant to the eyes and mucuous membranes of the nose. This irritant property can be removed with processing, and causes no problems with machine harvesting of these plants.

The Euphorbia lathyris plantation in Southern California has yielded about 10 barrels of oil/acre with unselected seed and no agronomic experience. The office of Arid Land Studies of the University of Arizona, in conjunction with the Diamond Shamrock Corporation, estimates for E. lathyris 34 barrels of hydrocarbon-equivalent/acre with small agronomic development, with 16 inches of water on semiarid desert soil. The estimated price of this "green oil" is $40/barrel, including the capitol costs for processing equipment. In addition, the cellulosic residues after the processing of the dried E. lathyris can be converted to ethanol, making this entire process energy positive.

The material extracted from the Euphorbia lathyris resembles crude oil and has been subjected to cracking by special catalysts. This was done by the Mobil Oil Corporation, with the following results: ethylene, 10%; propylene, 10%; toluene, 20%; xylenes, 15%; oelfins, 21%; coke, 5%; alkanes 10% (approximately) and fuel oil, 10%.

Euphorbias grow everywhere in the world and we learned from the literature that the French grew E. resinefera in Morocco in 1938, obtaining 3 tons of hydrocarbon/hectare with a total planting of 125,000 hectares. We also learned that in Ethiopia the Italians had plans for using various species of Euphorbia (specifically E. abyssinica) to produce a gasoline-like liquid called "vegetable gasoline". They actually planned a Euphorbia gasoline refinery at Agodat but it is not clear what actually happened to this endeavor. There are many species of Euphorbias which would be suitable for growth, extracting and processing of the oil-like materials they contain, particularly as a source for materials. All of these plants can grow in semiarid land, with a minimum of water, and do not require land which is suitable for food production.

Other plants also produce oils, such as Eucalyptus, which is very common in Australia, California, Spain and Portugal. It can be cut and regrown every two or three years and produces wood as well as oil. The oil from the Eucalyptus is a mixture of lower terpenes, whereas the oil from the E. lathyris is mostly higher terpenes.

We encountered a particular species of tree in Brazil (Copaifera) which, upon tapping, produces an oil which can be used as diesel fuel directly. It is possible to obtain this material every six months and it can be used directly in an automobile with no further processing. This tree would be useful in the tropics where it is found all the way from below São Paulo to north of Panama, but does not represent a useful source in the more temperate climates.

It has been determined that the development of hydrocarbon-producing plant plantations will be economically feasible, given the current (and projected costs of imported petroleum). In addition to the capital required for processing and extraction of plants, it will be necessary to develop a substantial acreage for cultivation of these plants (hundreds of millions of acres). We will be able to improve their yield by selection and genetic manipulation and will also be able to manipulate the plants to make more desirable compounds from the point of view of materials and fuel. I think it will be possible to choose the plants for large-scale development in the near future, and the choice will depend on growth rates and habits, hydrocarbon productivity and harvest adaptability as well as process development.

3. SYNTHETIC CHLOROPLASTS

The other aspect of this discussion is: How the green plant can take the quanta from the sun and convert it into useful material and how we can use this information from the natural system to create entirely synthetic systems for this purpose, using only the energy of the sun to produce the necessary chemicals and raw materials. The plant uses a membrane structure in the chloroplasts which presumably has the two quantum acts operating on opposite sides. The chloroplast membrane has two different kinds of surfaces, one with large and the other with small particles. The particles consist of pigments and proteins, but it is not known which protein is part of the chloroplast lamellar particles; the pigment, however, is chlorophyll.

The membrane uses two quanta in succession, one to raise the electron to the first level, after which it falls back; then another quantum raises the electron once more, and finally it falls into the reductase system and reduces the carbon dioxide to carbohydrate (Figure 2). These two successive quantum acts presumably take place on opposite sides of that membrane.

How is it possible to reproduce that function synthetically. We know a few of the catalysts involved: a quinone in the center, a cytochrome (an iron porphyrin) and another pigment. And on the hydrogen side of the membrane there is an iron-sulfur protein acceptor. On the left hand side of the membrane the electron comes from the water and the catalyst is some kind of manganese compound, whose structure is still unknown.

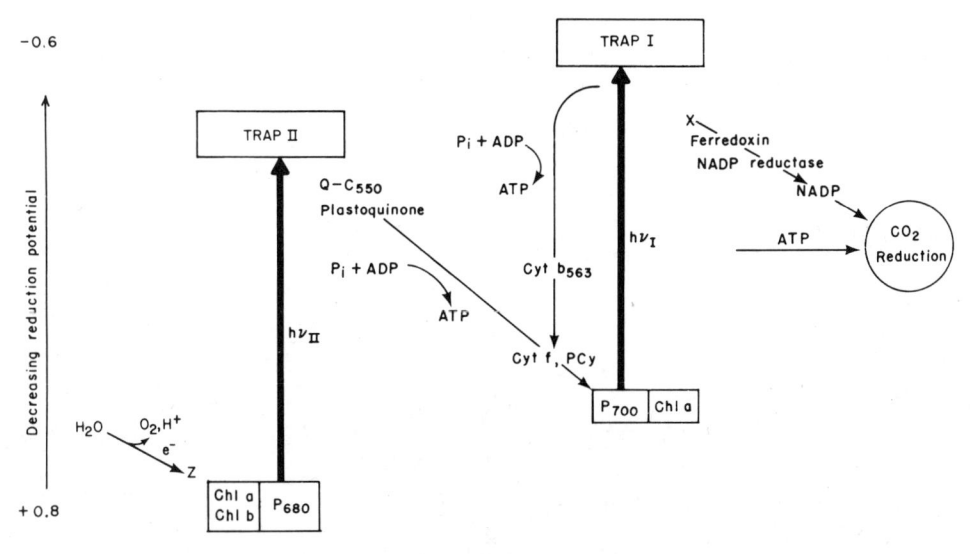

Figure 2. Photosynthetic electron transfer scheme.

a) Construction of micelles

Instead of trying to reconstruct a flat membrane which would allow performance of the two photoacts, accomplishment of these functions has been attempted in a somewhat different way. A diagram of the photosynthetic membrane with only the essential functional parts is shown in Fig. 3. Visible are the membrane and representative phospholipid molecules required to generate a bilipid membrane, a donor catalyst (D), the manganese (in the case of the natural materials) and the acceptor catalyst (A), the iron-sulfur protein (in the case of the natural materials), and the two pigments (P_I and P_{II}) on the surface of the membrane. However, the construction of a flat membrane which is only 200 to 300 Å thick would be a difficult task. Thus we consider an easier method of generating a thin membrane upon which the experiments could be performed. The membrane was rolled up into a small sphere, called a liposome or vesicle, which is a much easier process. These vesicles are made by shaking a phospholipid

in water, and the structure of a hollow sphere is formed
spontaneously, with water on the inside, and water on the outside,
of a phospholipid membrane in the form of a sphere. By constructing the vesicle properly the surfactant pigments can be put on
either side, the acceptor, on the inside, and the donor, on the
outside, thereby generating the entire system. Also, the two
surfaces of the vesicles could be separated in the form of two
separate micelles. The sensitizing pigments in nature are
chlorophyll molecules in different environments. Chlorophyll is
really a surfactant dyestuff, that is, a dyestuff with a hydrophilic head and a hydrophobic tail.

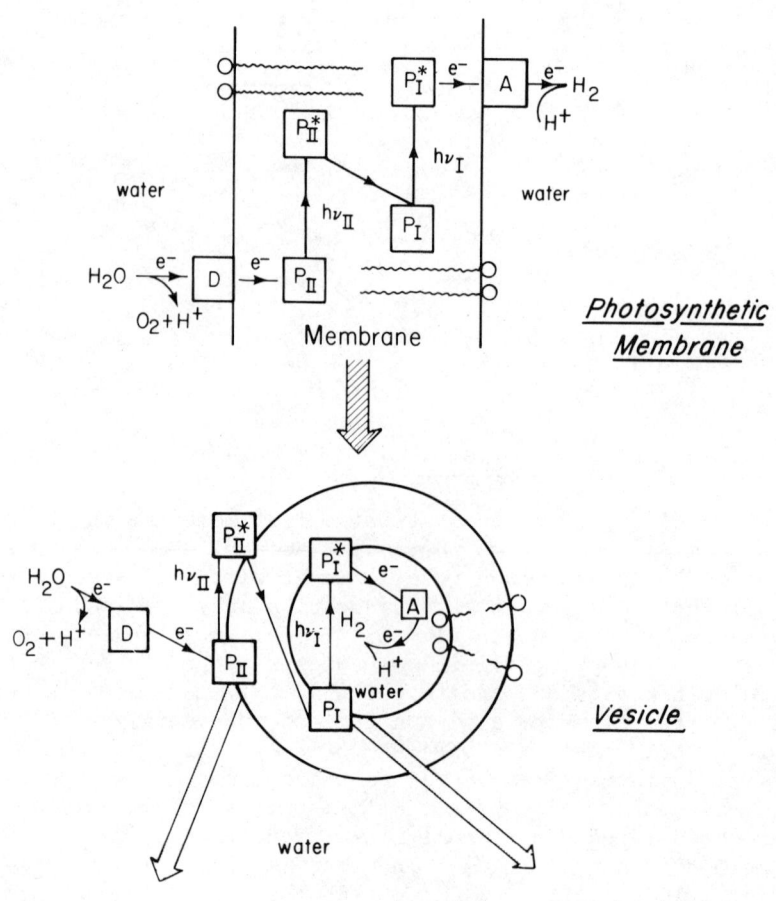

Figure 3. Photoelectron transfer scheme (vesicle).

We have constructed surfactant pigments which automatically find their way to the surface when such a vesicle-type membrane is made. They lie on the surface of a water-oil interface. In fact we have gone one step further. It is possible to separate the two surfaces into two separate micelles, so one reaction can be performed on one micelle and the other reaction on the other micelle. Recently it was learned that it is not necessary to break the functions down to that level in order to get the kind of information needed to eventually construct the synthetic vesicle which will make hydrogen on one side and oxygen on the other. If the two photoelectron transmembrane transfers could be demonstrated separately (one for the hydrogen side and one for the oxygen side), ways could be found to couple the two separate transferring acts similar to the way it is done in nature.

b) Photoelectron transfer

Photoelectron transfer occurs from a donor other than water; in the first instance, with a donor which is irreversibly oxidized when photoelectron transfer is performed so the back reaction is not a problem. These experiments were done to learn the factors which control the kinetics of such a system. A preliminary fluorescence quenching experiment was done with a porphyrin sensitizer-donor (n-alkyl-tetrapyridyl porphyrin), using anthraquinone sulfonate as acceptor. Clearly, when the dyestuff is put on a vesicle, the quinone is a very good quenching agent. When the porphyrin is placed in a homogeneous solution, the quinone is not a good quencher. This experiment shows that the membrane structure is an important component of the system for efficient photoelectron transfer. With ehtylenediaminetetriacetic acid (EDTA) as an irreversible donor and methyl viologen as an acceptor, and the surfactant dyestuff as a sensitizer, it is possible to demonstrate a sensitized photoelectron transfer. The dyestuff picks up the light and in some way sensitizes the oxidation of EDTA to ethylenediamine-triacetic acid; the reduced product is a free radical (of methyl viologen) which is colored. This reaction is uphill by 20 calories, so that the reaction requires the light to make it go. That reaction results in stored energy and will go even across a membrane which separates the reactants from each other. The results of this experiment indicate that the reaction will work with the donor system on the inside of the vesicle (membrane) and the acceptor on the outside, with the zinc present on the outside to block any leaked EDTA.

(c) Electron exchange across the membrane

Thus it has been demonstrated that transfer of photoelectrons across the membrane wall uphill by 20 calories is possible. A number of factors are involved in the membrane structure. There is a transfer of an electron from the surfactant dye to the surfactant acceptor and then that electron is transferred to the bulk dissolved acceptor, an isoelectronic transfer with no excitation. The whole reaction is in the surface of the membrane; and an exchange takes place across the membrane and between the surface of the membrane and the bulk acceptor in the water.

An isoelectronic exchange across 20 Å is apparently very easy, and the result is that there is an oxidized dye on the membrane adjacent to the bulk EDTA. This oxidized dye can extract an electron from the EDTA. The only component that could not be taken out of the membrane was the electron acceptor, the surfactant viologen. That had to be in the membrane in order for the entire reaction to take place rapidly. Neither a proton carrier nor an electron carrier across the membrane was needed, but the surfactant viologen acceptor system on either side of the membrane had to be present. This transmembrane electron transfer was fast. If it had required another quantum, a square law dependence on light intensity would exist. However, the results appear to be linear with light intensity. The reaction is therefore a fast non-photo isoelectronic exchange across the membrane.

Here is the point where reduction of the methyl viologen is possible. If platinum black is put in contact with reduced viologen, it generates hydrogen, with concomitant oxidation of the viologen. Thus, we have the hydrogen side of the cell. This is a very important result. It should be possible in the future to replace the viologen in the reaction with an iron-sulfur protein (as in nature) or it could be replaced with a dimeric rhodium compound to achieve hydrogen generation.

(d) Manganese requirement for oxygen production

What about the donor side of the reaction? Ultimately water must be used as the donor, and to do this oxygen must be generated. With electrons and protons removed from water, oxygen is left. However, it is necessary to have a catalyst which will generate oxygen (manganese compounds in the natural system). This catalyst is _the_ difficult substance to recognize and identify. What is the manganese compound in nature which performs this operation? Identification of the manganese compound has been sought for over 20 years, ever since the plant physiologists demonstrated that manganese was required for the oxygen generating capacity of the chloroplasts. Each time the manganese was extracted as a complex, it came out as Mn^{II}. There appears to be, as yet, no way to extract the manganese in its natural form.

So it was decided to find a spectroscopic method which would look at the functioning chloroplasts to see if the nature of the manganese compound can be determined without taking the chloroplasts apart. The only method that has appeared so far to be useful is that of extended X-ray absorption fine structure (EXAFS). There are only one or two manganese atoms per 500 chlorophyll molecules, so it is very difficult to find them. Whatever the spectroscopic method, it must be very sensitive and very specific for manganese. The EXAFS method has been used both on model substances to guide interpretation of the X-ray fluorescence which is very complex and, of course, to look at the functioning chloroplast.

When the same experiment was performed with the natural material from chloroplasts which generate oxygen, the frozen chloroplast material was inserted into the linear accelerator and the X-ray fluorescence spectra was taken. The information received was very much the same as in the model: in the natural system the manganese exists at least in pairs. There are at least two manganese atoms working together in that functioning chloroplast. This is the first unequivocal evidence for this idea and the observation suggests that we should make a compound which has two manganese atoms flexibly bound together so they could separate and allow water to enter their coordination sphere. After the electrons are removed from the atoms and the remaining protons ejected they could come together and oxygen generation could occur. The following set of reactions give what we think the manganese function actually is (Fig. 19). The important thing to recognize is that there are two manganese atoms cooperating in this sequence of events. When the dimeric manganese compound is synthesized, Mn^{III}, the stable form results; and, upon illumination, we get Mn^{II}. The only way to go from +3 Mn to +2 Mn is to generate oxygen. Thus oxygen has not yet been found and the two manganese atoms found have not been tied together. The porphyrin used now is a simple one, a tetrapyridylporphyrin, and on each one of the porphyrins is a pyridine ring. We will have two of these compounds hooked together by a chain of appropriate length, probably five carbon atoms, so that the manganese atoms can come together and separate as well.

In connection with this aspect of the work a dimer was synthesized, and we realized that a potent incipient oxygen atom with only six electrons was present in the oxidized monomer. By suitable chemical manipulation it would be possible to hand over an oxygen atom to a Lewis base and obtain ± Mn. It would then be possible to use any compound as an acceptor that has an unshared pair of electrons, or a pair of electrons capable of accepting an oxygen atom with six electrons. We used various acceptor systems and were able to perform essentially an oxygen-transfer reaction. The result of this effort has been the ability to generate "hot" oxygen atoms with light from water, which is potentially much more important than generating molecular oxygen from water. The byproduct of the reaction from "hot" oxygen atoms is water, which is essentially free economically (Figure 4).

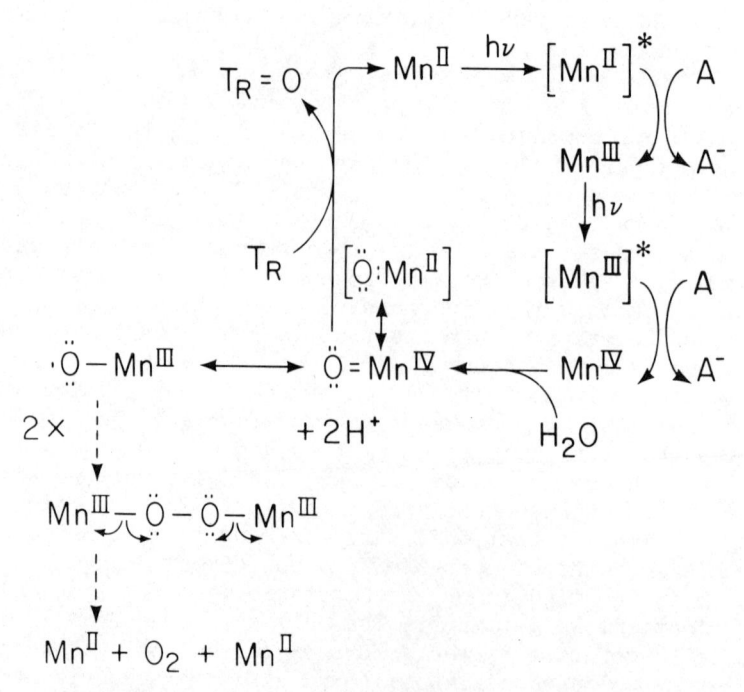

Figure 4. Proposed photochemical generation of oxygen sensitized by manganese porphyrins (general scheme).

There are other ways to make oxygen and hydrogen using different catalysts and acceptors, and experiments have been underway

BIOCONVERSION OF SOLAR ENERGY

in laboratories not only in the United States but in Japan, Great Britain and Switzerland as well as Germany.

It is now possible to put the hydrogen generating side and the oxygen generating side of the reaction together, with a carrier (quinone) between them. This could be done in an apparatus resembling a hollow fiber about 0.1 micron thick. The hollow fibers replace the spherical vesicles, one on each side (Figure 5). Out of one fiber will come hydrogen and out of the other fiber will come oxygen. Between the hollow fibers will be a membrane which is a barrier to ions but which will let electrons go through. Oxygen appears on one side, hydrogen on the other, and electricity in the middle.

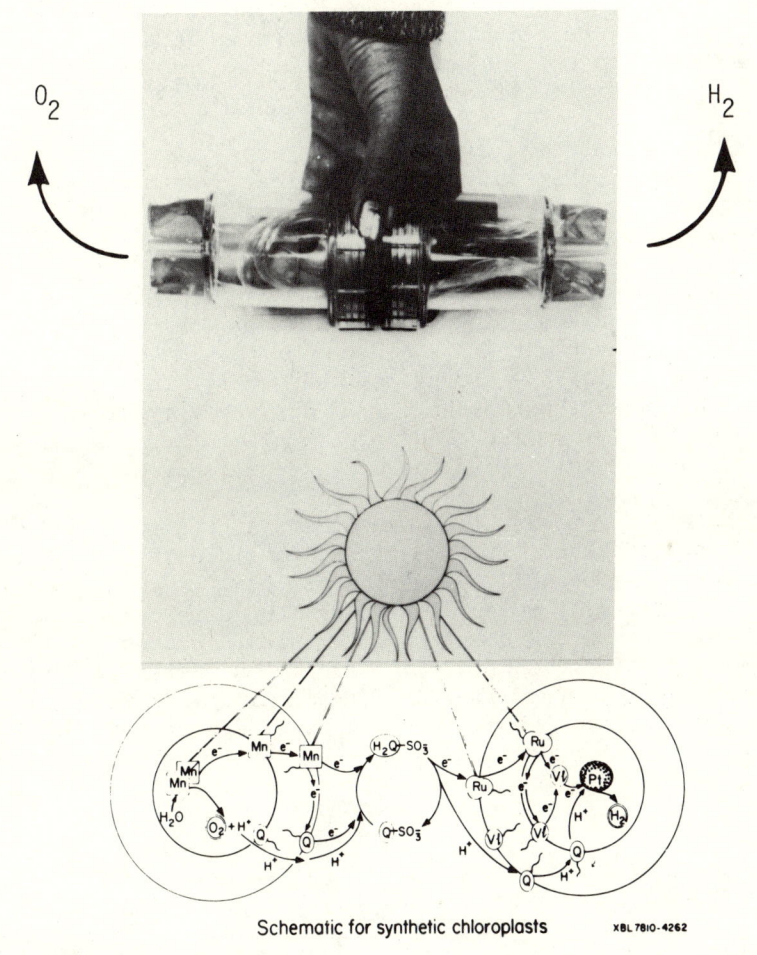

Figure 5. Schematic for synthetic chloroplasts.

CONCLUSIONS

We can look forward to the creation of an artificial system for photosynthesis based upon the synthetic chloroplast. This system will mimic the way in which the green plant takes (four) quanta to generate oxygen on one side and reduce power (either hydrogen or reduced carbon dioxide) on the other side of the synthetically constructed membrane. We have extended our experiments to demonstrate photoinduced electron transfer directly from the vesicle by measuring photoreduction of the quinone in solution.

It is possible to imagine a synthetic membrane system (possibly constructed of hollow fibers) in which the photoelectron transfer occurs across the phase boundary, producing the oxidant in one phase and the reductant in the other. This could be followed by actual physical phase separation together with separation of the potential reactants.

These reactants could then be separately stored and brought together in a suitably constructed physical system to recover the energy of the back reaction in some convenient form. The cycle would then be repeated.

These systems would accomplish both of the necessary requirements for useful solar energy devices, namely (i) the capture of the quantum and its conversion to some other energy form and (ii) the storage of that energy for indefinitely long periods, with the possibility of its recovery at will in some convenient form. The green plant, of course, has always performed these functions in the service of mankind.

While working toward the development of the purely synthetic devices just discussed, it is entirely possible that the green plant itself will be able to provide at least part of the energy needed in the form of materials through the development of "petroleum plantations" where specific plant species can be harvested to produce materials and fuel, thus releasing the use of fossilized photosynthetic fuel for applications where its own peculiar chemical characteristics are best fitted.

REFERENCES

1. M. Calvin, Petroleum plantations for fuel and materials 29, 533 (1979) and references cited therein.
2. M. Calvin, Hydrocabrons via Photosynthesis. Energy Res. 1, 299 (1977).
3. M. Calvin, Hydrocarbons from Plants: Analytical Methods and Die Naturwissen., in press.

4. E.K. Nemethy, J.W. Otvos and M. Calvin, Analysis of Extractables from One Euphorbia, J. Amer. Oil Chem. Soc., 56, 957 (1979).
5. E.K. Nemethy, J.W. Otvos and M. Calvin, Hydrocarbons from Euphorbia lathyris, to be published in Pure and Applied Chem.
6. J.D. Johnson and C.W. Hinman, Oils and Rubber from Arid Land Plants, Science, 208, 460 (1980).
7. S.G. Coffey and M. Halloran, Higher Plants as Possible Sources of Petroleum Substitutes. Search, 10, 423 (1980).
8. D.O. Hall, Solar Energy Use through Biology. Solar Energy, 22, 307 (1979).
9. J.R. Bolton and D.O. Hall, Photochemical Conversion and Storage of Solar Energy. Ann. Rev. Energy, 4, 353 (1980).
10. M. Calvin, Synthetic Chloroplasts. Energy Res. 3, 73 (1979).
11. N.N. Lichtin, Fixing Sunshine Abiotically. Chemtech (1980).
12. H. Gerischer and J.J. Katz (eds), "Light-Induced Charge Separation in Biology," Verlag-Chemie, New York (1979).
13. H. Kühn, Synthetic Molecular Organizates, J. Photochem. 10, 111 (1979).
14. I. Willner, W.E. Ford, J.W. Otvos and M. Calvin, Simulation of Photosynthesis as a Source for Energy. In "Bioelectrochemistry," F. Gutmann and H. Kuyzer (eds). Plenum, New York (1980).

LIST OF CONTRIBUTORS

ARNDT-JOVIN Donna J.	Max-Planck-Institüt für Biophysikalische Chemie, Göttingen, Germany
ASSENMACHER Ivan	Université de Montpellier, France
AVERBECK Dietrich	Institut Curie, Paris, France
BACHL Norbert	Institute for Applied Physiology, Weidling, Austria
BAUER H.	Institute for Applied Physiology, Weidling, Austria
BILEK P.	Institute for Applied Physiology, Weidling, Austria
BINAGHI Michel	Université Paris-Val-de-Marne, Créteil, France
BLUM A.	Illinois Institute of Technology, Chicago, U.S.A.
BOISSIN J.	Centre d'Etudes Biologiques des Animaux Sauvages, Chizé, Niort, France
BRENDZEL A.M.	Illinois Institute of Technology, Chicago, U.S.A.
BRETON Jacques	Centre d'Etudes Nucléaires de Saclay, Gif-sur-Yvette, France
BUDOWSKY Edward I.	Shemyakin Institute of Bioorganic, Chemistry, Moscow, U.S.S.R.
BULTS Gérard	The Weizmann Institute of Science, Rehovot, Israel
CAHEN David	The Weizmann Institute of Science, Rehovot, Israel
CALDAS Luis R.	Federal University of Rio de Janeiro, Rio de Janeiro, Brazil
CALVIN Melvin	University of California, Berkeley, U.S.A.
CAPLAN S. Roy	The Weizmann Institute of Science, Rehovot, Israel
CHABRE Marc	Centre d'Etudes Nucléaires de Grenoble, Grenoble, France
CHERRY Richard J.	E.T.H.-Zentrum, Zürich, Switzerland
CHUN Corliss	University of California, Berkeley, U.S.A.
COSCAS Gabriel	Université Paris-Val-de-Marne, Créteil, France

DALL'ACQUA Francesco	Padova University, Padova, Italy
DUTTON P. Leslie	University of Pennsylvania, Philadelphia, U.S.A.
DUVIGNEAUD P.	Université Libre de Bruxelles, Bruxelles, Belgique
EL SAYED Mostapha A.	University of California, Los Angeles, U.S.A.
ERLANGER, Bernard F.	Columbia University, New York, U.S.A.
FAVRE Alain	Université de Paris VI, Paris, France
FITZPATRICK Thomas B.	Massachusetts General Hospital, Boston, U.S.A.
FRAUENFELDER Hans	University of Illinois at Urbana-Champaign, Urbana, U.S.A.
SARTY Haim	The Weizmann Institute of Science, Rehovot, Israel
GEACINTOV Nicholas E.	New York University, New York, U.S.A.
GISBY Paul E.	King's College, London, England
GREITER Franz	Institute of Applied Physiology, Weidling, Austria
GROSSWEINER Leonard I.	Illinois Institute of Technology, Chicago, U.S.A.
GUNNER M.R.	University of Pennsylvania, Philadelphia, U.S.A.
GUTTMANN Giselher	Instituted for Applied Physiology, Weidling, Austria
HALL David O.	King's College, London, England
HEARST John E.	University of California, Berkeley, U.S.A.
HÖNIGSMANN Herbert	University of Innsbruck, Innsbruck, Austria
HYDE John E.	University of California, Berkeley, U.S.A.
ISAACS Stephen T.	University of California, Berkeley, U.S.A.
JÄCKLE Herbert	University of Texas, Austin, U.S.A.
JALLAGEAS, M.	Université de Montpellier, France
JOVIN Thomas	Max-Planck-Institut für Biophysikalische Chemie, Göttingen, Germany
KALTHOFF Klaus	University of Texas, Austin, U.S.A.
KRIPKE Margaret	Frederick Cancer Research Center, Frederick, Maryland, U.S.A.
KROYER G.	Institute for Applied Physiology, Weidling, Austria
LENCI Francesco	Biophysics Institute, C.N.R., Pisa, Italy
LONGWORTH James W.	Oak Ridge National Laboratory, Oak Ridge, Tennessee, U.S.A.
MADERTHANER Reiner	Institute for Applied Physiology, Weidling, Austria
MALKIN Shmuel	The Weizmann Institute of Science, Rehovot, Israel
McENTEE Kevin	Stanford University, Stanford, California, U.S.A.

CONTRIBUTORS

MENEZES Salatiel	Federal University of Rio de Janeiro, Rio de Janeiro, Brazil
MICHELSON, A.M.	Institut de Biologie Physico-Chimique, Paris, France
MOHR Hans	University of Freiburg, Freiburg, Germany
MOUSTACCHI Ethel	Institut Curie, Paris, France
MUELLER Paul	Eastern Pennsylvania Psychiatric Institute, Philadelphia, U.S.A.
PACKHAM Nigel K.	Eastern Pennsylvania Psychiatric Institute, Philadelphia, U.S.A.
PAILLOTIN Guy	Centre d'Etudes Nucléaires de Saclay, Gif-sur-Yvette, France
PARRISH John A.	Massachusetts General Hospital, Boston, U.S.A.
PRIOUL	Université de Paris-Sud, Orsay, France
PRINCE Roger C.	University of Pennsylvania, Philadelphia, U.S.A.
PROKOP L.	Institute for Applied Physiology, Weidling, Austria
PROKOPENKO Jury	A.N. Sysin Institute of General and Communal Hygiene, Moscow, U.S.S.R.
PRUE Daphne Vince	Glasshouse Crops Research Institute, Littlehampton, England
QUAIL Peter H.	University of Wisconsin, Madison, U.S.A.
QUENTEL Gabriel	Université de Paris-Val-de-Marne, Créteil, France
RAO Krishna, K.	King's College, London, England
RAPOPORT Henry	University of California, Berkeley, U.S.A.
RAULIN François	Université Paris-Val-de-Marne, Créteil, France
RIEDERER P.	Institute for Applied Physiology, Weidling, Austria
RUPP W. Dean	Yale University School of Medicine, New Haven, Connecticut, U.S.A.
SCHWEIGER Hans-Georg	Max-Planck Institut für Zellbiologie, Ladenburg, Germany
SISSON Thomas R.C.	Perth Amboy General Hospital, Perth Amboy, New-Jersey, U.S.A.
SMITH Harry	Leicester University, Leicester, England
SMITH Kendric C.	Stanford University, Stanford, California, U.S.A
SPIKES John	University of Utah, Salt Lake City, U.S.A.
STEINER Ingrid	Institute for Applied Physiology, Weidling Austria
SUTHERLAND John C.	Brookhaven National Laboratory, Upton, New Jersey, U.S.A.
SZAFARCZYK A.	Université de Montpellier, France
TAKEBE Hiraku	Kyoto University, Kyoto, Japan
TALLING John F.	Freshwater Biological Association, Ambleside, England

THIELEN A.P.G.M.	The State University, Leiden, The Netherlands
TYRREL Rex M.	Federal University of Rio de Janeiro Rio de Janeiro, Brazil
VAN GORKOM, H.J.	The State University, Leiden, The Netherlands
VERMEGLIO André	Centre d'Etudes Nucléaires de Saclay, Gif-sur-Yvette, France
WASHÜTTL Josef	Institute for Applied Physiology, Weidling, Austria
WASSERMANN N.H.	Columbia University, New York, U.S.A.
WOLFF Klaus	University of Innsbruck, Innsbruck, Austria

INDEX

Acetabularia, 439 et seq.
Acetylcholine, 86-87
Acetylcholinesterase, 87
Acne, 322
Acridine, 109-110
  orange, 116, 349, 354
Acriflavine, 354
Actinomycin D, 520
Adaptability, 475
Adrenocortical system, 451-452
Alcohol dehydrogenase
  inactivation, 355
Alkalingenes eutrophus
  hydrogenase, 588, 590
Amino acids
  quantum yield
    for photolysis, 70
    for photionisation, 70
Angelicin, 267-269, 271, 296
  derivatives, 269 et seq., 301
  mutagenicity, 274
  photoaddition to DNA, 269,
    271, 273
  photoreactions, 273
Anthocianin, 501, 516 et seq.
Anthraquinone, 563, 653
Ataxia telangectasia, 222, 232
ATP, 134-135
ATPase, 260-261
  Ca^{++}, 48
Azobenzene, 83, 85, 89
  cis, 83-84
  trans, 83-84

Bacteriochlorophyll, 557 et seq.,
    572, 574
  oxydation, 563 et seq.

Bacteriochlorophyll (continued)
  reduction, 564, 567
Bacteriopheophytin, 557, 561, 580
Bacteriorhodopsin, 1-2, 5, 8, 39,
    44, 81, 399, 406, 424,
    427, 429, 431
Bacterioruberin, 427
Bathorhodopsin, 401 et seq.
Benzodipyrones, 268
Bilirubin, 339 et seq.
  unconjugated, 339-340
Biocenosis, 604 et seq., 633
Bioconversion of solar energy,
    597 et seq., 645 et seq.
Biogeocenosis, 597 et seq.
Bioid, 130
Bioluminescence, 133 et seq.
Biomass, 599 et seq., 625, 627
Biphotonic processes, 72, 76
Bisulfite, 95
Blue green Algae, 425, 625

Calvin cycle, 634-635
cAMP, 359
Carbon cycle, 614
Carboxypeptidase A, 69
Carotenoids, 89, 117, 431, 527-
    528
Catalase, 143, 145
Cataract, 315
cGMP, 400, 408
Chamaenerion angustifolium, 507
Chenopodium album, 506-507, 589
Chemical evolution, 123 et seq.
Chemiluminescence, 134-135, 143
Chemophotochemotherapy, 316
Chirality, 128-129

Chlamidomonas reinhardtii, 425 et seq.
Chloramphenicol, 441
Chlorophyll, 516-517, 543
  a, 524, 540, 550
  b, 524, 540
  fluorescence yield, 553, 581
Chloroplasts, 25-26, 554, 633
  fluorescence yield, 553, 556
  synthetic, 650
Chlorpromazine, 110
Cholesterol, 473
Chromosomal aberrations, 316, 353
Chromosomes, 54-55, 57, 93
Chymotrypsin, 83, 86, 89
Circadian rhythm, 341, 437 et seq. 451 et seq.
Circaea lutetiana, 507
Circannual rhythm, 451
Circular dichroism, 13
  magnetic, 13-14
Clostridium kluyveri, 587
Clostridium pasteurianum
  hydrogenase, 588 et seq.
Coacervates, 124
Concanavalin A, 59-61
3 CPs (3-carbethoxypsoralen), 296 et seq.
Creatine kinase, 134
Cross links, 94 et seq., 267 et seq., 296 et seq.
Crustacyanin, 115
Cryptochrome, 517
Cryptomonas, 428, 430, 431
Cutaneous mastocytosis, 314
Cycloheximide, 302, 442-443, 448, 520
Cytochrome, 650
  b_5, 488
  C_2, 561
  C_3, 588
  f, 595
  ferro C, 572
Cytochrome oxydase, 46

Deletion, 228
den V gene
  T4, 214
  cloning, 214

Desulfovibrio gigas
  hydrogenase, 588-590
Desulfovibrio desulfuricans
  hydrogenase, 588-590
Dichroism
  transient, 43 et seq.
Diffusion
  lateral, 46
  rotational, 43
Difurobenzene, 268
Dinitrochlorobenzene, 260
Dinoflagellates, 625
Diphteria toxin, 221-222
DNA
  interstrands cross links, 267, 274, 296-297, 301
  lesions, 158
  protein cross links, 95, 158
  recombination, 212
  repair, 205
  strand breaks, 158, 353
DNA polymerase
  I, 200
  III, 190, 192, 201
Dose effects relationship, 217
Dose response relationship, 218-219
Dosimetry
  solar, 165

Ecosystem, 597
Eczema, 322
EDTA (as electron donor), 653, 654
EGF (epidermal growth factor), 58
Electron
  hydrated, 69 et seq.
  yield, 73 et seq.
Electron microscopy, 16
  freeze fracture, 46
Electrophorus electricus, 87-88
Energy transfer, 539, 541
Environmental factors, 375
Eosin, 109-110
  iodoacetamido, 48
  maleimide, 46-47
Error free repair, 164, 221-222
Error prone repair, 160-161, 189 et seq., 221-222, 297, 302
  regulation, 193, 199
  gene, 197

INDEX

Erythema, 254, 310
  action spectrum, 254
  delayed, 254, 326 et seq.
Erythrocyte membranes, 46, 47
Erythropoietic protoporphyria, 255
Ethyl methane sulfonate (EMS), 161
Etioplast, 516, 522
Euglena gracilis, 425 et seq.
Eumelanins, 258
Euphorbia
  abyssinica, 649
  lathyris, 645 et seq.
  resinefera, 649
  tirucalli, 647
EXAFS (extended X ray absorption fine structure), 15, 655
Excision-repair system, 356
Excision strains, 162, 165, 170
Excitons, 539-540
  excitons annihilation, 550 et seq.
Extra retinal photoreception, 457
Extraversion, 475

FAD, 134
Far U.V. mutagenesis, 164
Ferredoxin, 587 et seq.
Filamentation, 197
Flash photolysis, 33 et seq., 44, 111
  laser, 70
  of aromatic residues in enzymes, 71
Flavins, 110, 431
Flavodoxins, 588, 592
Flow
  analysis, 51 et seq.
  sortin, 51 et seq.
Fluorescein sensitizers, 116
Fluorescence
  delayed, 15
  lifetimes, 13, 557
  spectroscopy, 14
  time resolved, 1
  yield
    of chlorophylls, 553
    of chloroplasts, 553, 556
FMN, 134

Fucoxanthin, 625
Furocoumarins (*see* Psoralens)

Genetic engineering, 205
Glutathione reductase, 521
Glyoxal, 127
Growth delay, 156, 165, 419, 420
  in *E.coli*, 414
  mechanism, 418
  targets for, 416
Growth factor
  nerve, 58
  epidermal, 58
Guanine
  8 aza-, 219 et seq.
  6 thio-, 219 et seq.

Halobacterium halobium, 25 et seq., 81, 425 et seq.
Halophilic bacteria, 89
Hemes, 33 et seq.
Hemoglobin,
  carbonmonoxy, 1-2, 8
  deoxy, 7
Heliotherapy, 322, 329
Herpes
  simplex, 350, 362
  genitalis, 350
H_2O_2, 110, 137 et seq.
Human growth hormone (HGH), 341
  level of, 344
17-Hydroxycorticoide, 473
Hyperbilirubinemia, 339 et seq.
Hypsorhodopsin, 404

Immediate pigment darkening, 259, 260
Immunologic alterations, 236
Inactivation
  of cells, 155
  enzymes, 68
  quantum yield, 68-69
Infrared spectroscopy, 35
Inhibitors, 109 et seq.
Insulin, 58
Intellectual ability, 475
Iodoacetamidoeosin, 48
Iodoxuridin, 350
Isorhodopsin, 402

Jaundice, 339

Karyotypes, 54, 57-58
Keratinocytes, 255 et seq., 325
  melanosomes, 259

Lactase deficiency, 341
Langerhans cells, 254 et seq.
Laser
  in ophtalmology, 367
  pulsed system, 2-3
lex A
  gene, 189 et seq., 207-208, 211
  mutant, 194
  product, 190,
  protein, 189 et seq., 207-208, 213
  repressor, 197
lex C 113 mutation, 192
LHCP (light harvesting Chl a/b protein), 524, 526
Lichen planus, 315
Liposomes, 112-113
Luciferase, 134 et seq.
Luciferin, 134 et seq.
Lumichrome, 353
Luminol, 135 et seq.
Lymphocytes, 236
Lysosomes, 257, 259
Lysosyme, 37, 70, 275

Malonaldehyde, 127
  photosynthesis, 126
Maxicell, 209 et seq.
MED (minimal erythemogenic dose), 254, 326, 327
Melanin pigmentation, 258
Melanocytes, 255 et seq.
Melanogenesis, 258-259
Melanoma, 156, 240, 465, 466
Melanosome, 258 et seq.
Melatonin, 457-458
Mercurialis perennis, 507
Methylene blue, 350 et seq.
Methyl methane sulfonate (MMS), 161
Methylviologen, 28, 127, 587, 653-654
Minicells, 208-209

Mitochondria, 257
Monophotonic processes, 72
Monophotonic photoionisation, 72 et seq.
  quantum yield, 76
5MOP (5-methoxypsoralen), 304, 317
8MOP (8-methoxypsoralen), 114, 275-304
Myoglobin, 33, 36-37
Mutagenesis, 155, 197
  far U.V., 164
  radiation induced, 217, 223
Mutations
  frequency, 218 et seq.
  induction, 299, 301
  "petite", 297, 300
  photodynamically induced, 353
  somatic, 217
Mycosis fungoides, 314

NADH, 134-135
NADPH, 134-135
Naphtoquinone, 563
Near U.V.
  mutagenic action, 161
  damages repair, 157
Neutral red, 350
N-formylkynurenine, 115
NGF (nerve growth factor), 58
Nontronite, 126
Norfurazon, 527

Odland bodies, 257
O-methylhydroxylamine, 94-95
O-phenanthroline, 572
Oscillatoria limnetica
  hydrogenase, 588
Ouabain, 221
Oxygen
  molecular, 349
  singlet, 111 et seq., 275, 302, 354, 356
  quenchers of, 111, 116
  quenching, 117

PADPC, 84-85
  cis, 84, 86
  structure, 84
  trans, 84, 86
  U.V. spectra, 85

INDEX

Parapsoriasis, 322
Peridinin, 625
Peroxidase, 134 et seq.
Phage
  λ, 160
    cloning on, 212
  MS_2, 95, 97
  PM_2, 95
  Sd, 94-95
  T_4, 160, 214
Phenol red, 350
Phenothiazine, 110, 113
Phenylalanine ammonia lyase, 520 et seq.
Pheomelanin, 258
Pholas dactylus, 136
  luciferase, 137
  luciferin, 137 et seq.
Phormidium uncinatum, 425 et seq.
Phosphorescence depolarization, 43
Photoacoustic spectroscopy, 13-14, 21 et seq.
Photoallergy, 378
Photocalorimetry, 26
Photocarcinogenesis, 238
Photocarcinogenic action, 114
Photochemotherapy, 114, 267
  oral psoralen, 309, 321
Photochromic compounds, 83, 86
Photochromicity, 502
Photodynamic dyes
  classification, 351
Photodynamic reactions, 109-110
  damages in proteins, 353
  sites for, 352
Photodynamic therapy, 349
Photoelectrolysis of water, 128
Photoinactivation of enzymes, 109
Photoionization, 72, 76
  biphotonic, 76
Photokinesis, 423
Photomorphogenesis, 515 et seq.
Photomovements, 421 et seq.
Photooxidation, 110
Photophobic response, 423 et seq.
Photophosphorylation, 130
Photopigments, 424, 431
Photoreactivation, 158, 170 et seq., 173 et seq., 191

Photoreactivating enzyme, 182, 184
Photoreactivating light, 175
Photoreactivating treatment of DNA, 177
Photoregulation, 81
  of growth, 413
  in membranes, 86
  of neuroendocrine rhythms, 451, 458
  of reproductive cycles, 455
Photosensitization, 109, 112
Photosensitized formation of Py <> Py, 178
Photosensitizers in biological systems, 110
Photosynthesis
  bacterial, 561
  circadian rhythm of, 437
  of malonaldehyde, 126
  model systems of, 127
  planktonic, 625
  primary processes in, 549
Photosynthetic antenna, 542, 543
Photosynthetic bacteria, 557
Photosynthetic pigments, 539
Photosystem
  I, 556, 579
  II, 579, 582, 587
  IIα, 583-584
  IIβ, 582, 584
Phototaxis, 423-424
  action spectrum, 428
  positive, 428
Phototherapy, 321
  U.V., 322, 329
  of jaundice, 339
Phototoxicity, 378
Phototransduction, 399
Phototropism, 423
Phycocyanin, 625
Phycoerithrin, 428, 431
Phycomyces, 428, 431
Phytochrome, 81-82, 89
  destruction, 497
  location, 485
  and gene expression, 515
  role of, 501
Phytohemagglutinin, 344
Phytoplankton, 619 et seq.
Pigmentation, 310

Pineal, 457
Pityriasis
  lichenoide, 315
  rosea, 322
Plasmohydroquinone, 579
Plastoquinone, 580
Plastosemiquinone, 580
Pneumoconiosis, 377
Pol A strains, 160
Pollution, 375, 503
Porphyrins, 110
  sensitizers, 116
Post replication repair, 192, 222
Prebiotic photochemistry, 123 et seq.
Proflavin, 350
Prophage induction, 197-198
Protein
  DNA cross links, 95
  inhibition of synthesis, 177
  photochemistry, 67
  polynucleotides cross links, 93, 96
  protein cross links, 115
  ribosomal S$_1$, 94
  RNA cross links, 177
Protochlorophyll(ide), 523 et seq.
Protein pump system
  of bacteriorhodopsin, 5
  of purple membrane, 81
Protoporphyrin, 255
Psoralens, 110 et seq., 239, 256-257, 267 et seq., 279 et seq., 295 et seq., 309 et seq.
  binding parameters to DNA, 288
  carcinogenic activity, 297
  derivatives, 280 et seq., 295 et seq., 309 et seq.
  extinction coefficients, 288
  mechanism for reaction with DNA, 286
  mutagenic activity, 297
  non covalent binding to DNA, 286
  photoaddition to DNA, 289
    kinetic parameters, 291
    quantum yields, 290
  photobiology, 295
  photobreakdown, 292
    kinetic parameters, 291

Psoralens (continued)
  photobreakdown (continued)
    quantum yields, 290
  phototherapy, 309 et seq.
  solubilities of, 288
  synthetic path of derivatives, 280
Psoriasis, 114, 268 et seq., 309 et seq., 323, 329
  photochemotherapy, 309
  phototherapy, 324
Psychophysiological parameters, 472
Psychosomatic parameters, 465
Pulse radiolysis, 111
Puromycin, 442, 520
Purple membrane, 25, 30-31, 44-45, 81, 425
PUVA
  carcinogenesis, 315
  dermatose responsive to, 314
  side effects, 315
  therapy, 309 et seq., 321
Pyranocoumarins, 268
Pyrimidine dimers, 158, 173, 257
  excision repair of, 222
  photoenzymatic splitting of, 182
  in RNA, 177-182
    photosensitized formation of, 178
Pyrimidine hydrates, 182

Radiationless decay, 23
Reaction center, 557, 561, 571
rec A
  gene, 160, 167, 189 et seq., 205 et seq.
    cloning of, 205
    product, 190, 193
  mRNA, 207
  mutants, 194
  operator, 208
  promoter, 208
  protein, 189 et seq., 206 et seq.
    protease activity of, 206
    proteolytic cleavage of, 194
  strains, 167, 209
rec B, 192
rec C, 192
rec F, 192

rec L, 192
Reflectance spectra, 24-25
rel A⁻ strains, 419
Repair
  competent strains, 160-161, 164, 168
  deficiency, 229
  deficient
    cells, 222
    strains, 167
  DNA, 205
  error prone, 160-161, 189, 193
  post replication, 192
  proficient strains, 167
  Rec dependent, 358-359
  of U.V. damages, 157
Repressor of λ phage, 189, 195, 197
Resonance Raman, 2, 4, 5
  time resolved, 1, 2
Retina, 399
Retinal, 401, 404, 406-407
  11-cis, 5, 401, 404
  13-cis, 6
  conformation, 403
Retinoblastome, 217
R/FR ratio (ζ), 490 et seq., 505 et seq.
Rhodopseudomonas
  sphaeroides, 558, 561 et seq., 571, 580
  viridis, 558
Rhodopsin, 6-7, 81, 399 et seq.
  structure, 405
  photoexcitation, 407
Rhodospirillum rubrum, 29, 588, 590
Rhythms
  circadian, 341, 437 et seq., 451 et seq.
  circannual, 451
  neuroendocrin, 451 et seq.
Ribonuclease A, 69
Ribosomal protein S_1, 94
Ribosomal subunit
  30S, 94, 101-102
  50S, 101
Ribosome, 99
  70S, 97, 101, 104, 439
  80S, 439 et seq.

Ribulosebiphosphate carboxylase, 521
Rifampicin, 197, 441
RNA
  inhibition of synthesis, 184
  protein cross links, 96
  quantum yield, 96
  16S, 94, 96, 101
    cross links to tRNA, 96
  23S, 102
Rose bengal, 110

Saccharomyces cerevisiae, 296, 351
Sendai virus, 97
Sensitized photoreactions
  type I, 109-110, 112, 115
  type II, 111, 115
Serum albumin
  bovin, 275
  human, 275
Sinapsis alba, 88, 507 et seq., 522
Singlet-singlet annihilation, 552, 554
Singlet-triplet annihilation, 554
Sister chromatid exchange, 316
Skin, 253
  cancer, 229 et seq., 235 et seq.
  color, 232
  diseases, 321
  typing, 310
Skotomorphogenesis, 515-516
Smittia spec, 173 et seq.
SOS
  functions, 189 et seq.
  responses, 189 et seq.
Spillover, 541-542
Spirulina maxima, 588-589
spoT mutant, 419
SSB (single stranded DNA binding protein), 192
Staphylococcus
  albus, 360
  aureus, 493, 495
Stentor coerulens, 426 et seq.
Stentorin, 427 et seq.
Stratum corneum, 253-254, 259, 326, 333
Suberythemal dose of U.V., 379, 380

Suberythemal dose of U.V. (continued)
 prophylactif effects, 378
Subtilisin carlsberg, 68-70
Sun protection factor (SPF), 386-388
Sunburn cells, 256-257
Sunscreens, 385-386
Superoxide
 dismutase, 135
 radical, 135
 radical anion $O_2^{\cdot-}$, 110, 137, 141-143
Suppressing mutation, 161, 168
Suppressor
 cells, 237, 238
 T lymphocytes, 236, 239
 U.V. induced cells, 236
Synchrotron radiation, 11-15

Tar, 330, 333
 photosensitization, 331
 vehicles, 333
Tetramethylbenzoquinone, 563
Teucrium scorodonia, 507
Thiopyronin, 116-117, 353, 356
5-Thiouridine, 156, 414-420
Thymine glycol, 158
tif-1, 191
 mutation, 189, 192, 194, 202
 protein, 200-201
Time resolved
 emission spectra, 15
 fluorescence, 1, 8
 Fourier transform I.R. spectroscopy, 35
 resonance Raman, 1, 5, 6
Toluidine blue, 116-117
Transformation
 of cells, 218
 frequency, 220-222
 induction, 221
 malignant, 218
 neoplastic, 218, 235-236, 239
Translocation, 218
Transposons, 209, 212
Trichloblast, 519
Trinitrochlorobenzene, 260
Trinitrophenyl, 237
Triplet probes, 43

tRNA
 photochemistry of, 414
 as target for growth delay, 416
Trypsin, 86
 photosensitized inactivation, 355
Tunneling
 molecular, 34-35
 quantum, 35
Two photons absorption, 550
Tyrosinase, 258-259

Ubiquinone
 10, 561-562, 575
 QI, 571
UDS (unscheduled DNA synthesis), 229, 258
umu C
 gene, 189
  product, 201
 mutation, 191, 197
Uracil
 5 bromo, 98
 4 thio, 98
U.V. carcinogenesis, 222
U.V. damages
 lethal consequences of, 158
 repair of, 157-158
U.V. induced tumors, 235-236, 239
uvr
 gene, 210, 213
  cloning of, 208
  products, 213
uvr A
 gene, 160, 212, 213
  products, 191, 210
 protein, 210-211
 strains, 162, 167
uvr A rec A strains, 167
uvr B
 gene, 212-213
  products, 191, 210
 promoter, 212-213
 protein, 210-211
uvr C
 gene, 212
  products, 191, 212

Vitamin D, 257, 465
Vitiligo, 114, 309

W-mutagenesis, 189 et seq.
W-reactivation, 189 et seq.

Xeroderma pigmentosum, 217, 222, 229 et seq.
  excision deficient, 231

X-rays, 13
  cristallography, 16
  microscopy, 15, 17
  small angle scattering, 15, 17
  spectroscopy, 16